The Palgrave Handbook of Just Green Transitions in the Western Balkans and Beyond

Erblin Berisha · John Moodie · Ledio Allkja ·
Marija Jeftić
Editors

The Palgrave Handbook of Just Green Transitions in the Western Balkans and Beyond

palgrave
macmillan

Editors
Erblin Berisha
Interuniversity Department of Regional and Urban Studies and Planning (DIST)
Politecnico di Torino
Torino, Italy

Ledio Allkja
Co-Plan Institute for Habitat Development and Polis University
Tirana, Albania

John Moodie
Nordregio Research Institute for Regional Development and Planning
Stockholm, Sweden

Marija Jeftić
Faculty of Geography
University of Belgrade
Belgrade, Serbia

ISBN 978-3-031-95074-2 ISBN 978-3-031-95075-9 (eBook)
https://doi.org/10.1007/978-3-031-95075-9

Cover illustration: Park Studios. AL, 2022

This Palgrave Macmillan imprint is published by the registered company Springer Nature Switzerland AG
The registered company address is: Gewerbestrasse 11, 6330 Cham, Switzerland

Foreword by Zorica Nedovic-Budic

A singular focus on economic and industrial progress without environmental protection has made the Earth less inhabitable, unhealthy, and unattractive. The 20th and 21st centuries have witnessed the destruction of our climate and environment but also seen positive initiatives aimed at repairing the damage that has been done. These initiatives have taken the form of climate and environmental policy actions and laws aimed at preserving and protecting the Earth's valuable natural resources, including air, water, soil, flora, and fauna. The transition to a greener and more climate-friendly planet has begun and is ongoing.

We still have a long way to go to rectify past mistakes and address damaging practices. This volume aims to assist us. The Handbook focuses on the relatively recent label of 'just green transitions' (JGTs), connecting it to the existing overarching goals of social, economic, and environmental sustainability and resilience. 'Green' signals the need to return to a more pristine, livable, and clean Earth, as well as adopt non-polluting practices. 'Just' refers to the collective responsibility to respect the equity and rights of all inhabitants to live in healthy and safe places.

The editors and contributors provide a systematic overview of Just Green Transitions within the European context, with a particular focus on the Western Balkans. They present the concept and dimensions of just green transitions, including the policy sphere and themes, as well as their related socio-economic and governance challenges and opportunities. They also provide examples of just green transition policies and good practices. From the introduction to the conclusions, the authors inform us of the challenges that occur at all governance levels and policymaking and planning aspects. The world's regions and their constituent countries are diverse with heterogeneous institutions, histories, ways of life, cultures, legal frameworks, political regimes, levels of development, climate, and natural and built landscapes. JGT policies, practices, and technologies cannot be easily transferred from one context to another and attempts to do so could do more damage than good or lead to unexpected and undesired outcomes. Each chapter of this volume reminds us that while we can learn from others, we must ensure that policy solutions are tailored and adjusted to different local circumstances.

Hence, transitions are multiple, and they often occur simultaneously. Transitions should be considered as a process that requires its participants to take it as a journey and trust it. While countries embark on different transition pathways, it is important to take stock, evaluate, and compare new and innovative JGT approaches, experiences, and outcomes. This Handbook makes an important contribution to raising our knowledge and understanding of JGT challenges and opportunities along the way. There is no doubt that the Western Balkans lag behind other parts of Europe in terms of environmental quality, green practices, and targets. Also, most West Balkan countries are still experiencing ongoing socio-political and economic transitions and rebuilding after years of communist rule and, in some cases, wars and destruction. However, we can take some positives from this complex situation. The state of flux of institutions offers some flexibility for building new organisations and capacities. Learning from neighbours in Europe and having a common general framework is crucial for the just green transition of the Western Balkans countries.

There is an abundance of JGT policy framing documents—green agendas, deals, plans and strategies—but what is more important is their implementation. A significant portion of this volume is devoted to dissecting and evaluating implementation across different JGT policy themes (e.g. waste management, coal mining, climate change/coastal risk prevention, heat, spatial planning, energy), Western Balkan regions (Adriatic-Ionian, Danube valley, central European, southern and central Balkans, mountainous and arid, etc.) and other European countries. There are also several informative comparative cases, with a welcome attempt to identify common JGT ground and principles that can be assessed in relation to different contexts and potentially applied broadly.

While finding their own solutions and building their own capacities and institutional/governance structures is necessary for advancing just green transition goals in the Western Balkans, its countries need implementation support from the European Union and its more developed countries. Such support should be just and driven by climate and environmental needs and circumstances in each European country or region and not the politics of accession. The 'stick and carrot' method has its advantages but also drawbacks. The environment knows no boundaries, and the authors alert us to that fact by exploring cross-border issues and practices.

There are targets that just green transitions across Europe and the Western Balkans are aspiring to meet, but there is no endgame. The positive impacts of climate improvements for advancing quality of life and environment are felt and will be felt across the Western Balkans, Europe, and the globe. This volume highlights the importance of these transitions, takes stock of their progress, identifies their challenges and opportunities, and offers valuable lessons for future JGT endeavours, from individual to global level. Most importantly, the open discussion and information make it clear that we are on the right path towards the common good.

Zorica Nedović-Budić, Ph.D.
Professor Emerita, University of Illinois @ Urbana-Champaign
Champaign, IL, USA

Visiting Full Professor
University College Dublin
Dublin, Ireland

Foreword by Pantoleon D. Skayannis

The Western Balkans, belonging entirely to the former Yugoslavia, except for Albania, experienced a traumatic transition to a market economy with significant upheavals, civil unrest, drain of human resources and significant internal and international challenges. In a world still shaken by dangerous antagonisms, where issues of revision not only of borders and international arrangements and the very existence of national entities are constantly being raised, but also of other now apparent facts such as climate change that has led to the climate crisis, most of the countries of the Western Balkans are trying to find a new pace through rapprochement with the European Union.

Here, a very peculiar contrast arises: On the one hand, the countries of the Western Balkans have moved in a relatively rapid manner to the capitalist market economy, which, however, has the problem of raising severe environmental issues since what we would call capitalist modernisation is not self-evidently in line with the eco-environmental conditions pursued today. On the other hand, the previous situation in these countries, due to low development, had the 'privilege' of maintaining environmental conditions at a relatively good level, except in highly industrialised areas where no attention was paid to eco-environmental conditions, such as the notorious Timisoara in Romania.

Let us take the energy sector as an example. The accession of countries to the European Union, driven by new policies, has led to a recent restructuring of energy source distribution, with a shift towards the use of green solutions, especially in response to the climate crisis. These developments create a picture of significant reshuffles.

Today, the question that arises is whether the developmental modernisation that is being carried out can maintain a good level of eco-environmental status but also make leaps so that the development can lead to a society and economy that become as environmentally neutral as possible in order to keep pace with the global indicators and agreements of new environmental aspirations worldwide as they are expressed in the Paris Agreement and the Conference of the Parties (COP) that follow (more or less successful—efficient, productive, promising).

In this context, these countries are trying to coordinate with European aspirations and goals, not only because these goals are logical and internationally innovative, as Europe is one of the most advanced regions in the world in actions to protect the natural environment, but also because these countries themselves are seeking to become members of the European Union. Some of them are already at a more advanced stage than others.

Therefore, the more they coordinate with European policies, the faster they believe, and rightly so, that they can achieve their accession to the European Union.

However, the issue is not simply a technical matter of achieving goals and indicators based on the European acquis. The issue also takes on other dimensions that are deeply social. In other words, it is not simply a question of a green transition. The European Union itself, through decisions and policies, raises the issue of a 'just' green transition, and the word 'just' is a keyword and very characteristic of the policies and the very ideology of the European Union, which has a particular sensitivity, perhaps not as much as we would always like. Nevertheless, it does have some sensitivity to social issues. Because, of course, the green transition overturns many facts in both production and consumption. Of course, in terms of way of life, these upheavals are not the same for the entire population of a country or a region and all activities and all regions. For example, the abolition of lignite mines in regions that are economically based on coal mining creates a considerable loss of jobs and an economic recession in these regions, in contrast to other regions that do not undergo these changes. Therefore, we have an additional level of vulnerability in various regions and population groups, not from the climate crisis itself but from the measures to address it!

Therefore, the changes that underlie this transition to a green, circular, etc. economy are self-evident, and they have different levels of resilience to face (or overcome?). The resilience of these areas immediately leads us to reflect on the sustainability of the interventions, that is, whether these interventions lead to sustainable development from an eco-environmental, social, economic, and institutional point of view. These interventions are carried out within complex contexts that are decisive for their development. The complexity of contexts, which are national, regional, international, or even global, often involves great uncertainties. We have recently seen the case of the last COP in Baku, which many consider a failure, and the withdrawal of the United States of America from the Paris Agreement. These are very characteristic of the uncertainty that leads to the risk of the very project of halting climate change, mitigating its effects, and adapting, especially socially sensitive adaptation.

Given the circumstances, the future world cannot be entirely visible, and a foresight exercise could lead to multiple scenarios, many of which may not be that pleasant. The European Union and the countries that accompany it, as the Western Balkans seem to want to be, are making a difficult struggle to follow a path that leads to a better perspective, both technically and socially. The broader effort also supports these countries with various financial and institutional tools to the extent possible through the self-commitments that the EU itself creates and has due to the socio-economic environment in which it operates, namely modern European capitalism. However, it has nothing to do with the unregulated American neo-capitalist model already taking the first steps towards creating risks for the global climate balance. In addition, an additional drawback is the inability to implement the policies that are decided, mainly depending on each state.

In this sense, the book at hand is an important contribution to the discussion on the just green transition in the Western Balkans.

This is not only because it raises the issue of the just green transition but also because of the space it refers to. Namely, the Western Balkans. It is a complex undertaking in a challenging region with very different political, social, and intergovernmental traditions from Western Europe, which is the initiator and proponent of these policies. Can custom-made policies for a set of countries with shared characteristics apply to another set of countries with different characteristics? In other words, can one claim that these policies have the element of being generic and not context specific? Of course, this remains to be seen as through the path towards the European Union, many particularities will tend to lose their sharp points, and a homogenisation of the countries will be achieved little by little (without, of course, losing their identity).

I think this is one of the significant challenges that this book poses. By examining various dimensions of the green transition, it gives a comprehensive picture of this complex issue.

Pantoleon D. Skayannis, Ph.D.
Professor Emeritus
Department of Planning
and Regional Development
University of Thessaly
Thessaly, Greece

Editor's Preface

> *Justice and equity must form an integral part of the transition towards a low carbon world*
>
> (UNRISD, 2018)

The transition towards a greener, more sustainable future is one of the defining challenges of our time. As societies across Europe and beyond work to decarbonize their economies, the concept of a Just Green Transitions is emerging as a crucial policy and academic conceptual framework. While the imperative to shift towards carbon-neutral technologies is clear, the ways in which this transition unfolds—its governance, socio-economic implications, and territorial impacts—remain complex and context specific. This handbook, Just Green Transitions in the Western Balkans and Beyond, seeks to unpack these complexities by offering a multi-dimensional analysis of just green transitions from theoretical, policy, and practical perspectives. The genesis of this handbook lies in the collaborative research conducted under the Horizon Europe GreenFORCE Project (2022–2025). As part of this initiative, scholars and practitioners from across Europe have come together to explore the many facets of just green transitions, with a particular focus on the Western Balkans. This region, characterised by its dynamic yet challenging socio-political and economic landscape, provides a compelling case for examining how just green transition policies can be tailored to support sustainable development while addressing socio-spatial disparities. The main motivation for this volume is the recognition that while just green transition policies are discussed at the European level—especially within the framework of the EU's Green Deal and the Just Transition Mechanism—their implementation varies significantly across regions and localities. This handbook aims to bridge the gap between policy discourse and real-world practice by presenting both conceptual explorations and case studies that highlight how different territories are navigating their transitions.

Just Green Transitions (JGTs) address the shift to sustainable, low-carbon economies while ensuring fairness and inclusivity. They are central to global policies, such as the Paris Agreement and the European Green Deal, but their interpretation and application remain diverse and complex. JGTs are inherently multidimensional, encompassing social, economic, spatial, environmental, technical, and governance dimensions. These elements interact dynamically, shaping the opportunities and challenges associated with transitioning to sustainable economies and societies. The social dimension emphasises justice and equity, ensuring that the burdens and benefits of green transitions are distributed fairly. This involves addressing vulnerabilities, empowering marginalised communities, and fostering participatory decision-making. Economically, JGTs provide opportunities for regional economic restructuring and diversification, including circular economies and green innovation, but they also entail substantial costs, particularly for high-polluting industries shifting to carbon-neutral practices. Spatial disparities influence the effectiveness of transition strategies, as urban and rural areas experience differing challenges. While some regions may benefit from new economic pathways, others risk economic stagnation or decline. The environmental dimension focuses on achieving long-term sustainability through carbon neutrality, biodiversity preservation, and resilience-building. Meanwhile, governance plays a pivotal role in fostering multi-level collaboration, policy coherence, and citizen engagement. Without adaptive governance structures, JGTs risk becoming fragmented and ineffective. Historically, just transition policies evolved from labour movements advocating for worker protections during industrial shifts. Over time, the framework has expanded to encompass broader socio-economic and environmental considerations. The European Union's policy landscape, particularly the Just Transition Mechanism, reflects this evolution, aiming to mitigate disparities between regions and ensure that no one is left behind in the transition to sustainability. This handbook is structured into five key parts:

- Part I: Theorizing Just Green Transitions—Establishes the conceptual underpinnings of JGTs and their policy relevance.
- Part II: Exploring Key Just Green Transition Policy Themes—Examines critical policy areas such as renewable energy, circular economy, and sustainable transport.
- Part III: Just Green Transitions in the Western Balkans—Investigates regional challenges and opportunities, providing case studies from across the Balkans.
- Part IV: Just Green Transition Practices in Europe—Showcases examples of policy implementation from European countries, illustrating best practices and governance strategies.
- Part V: Future Directions for Just Green Transitions—Looks ahead to emerging trends and policy recommendations that can guide future transition efforts.

This handbook is the result of a collective effort involving 55 scholars and practitioners from diverse disciplinary backgrounds. We are deeply grateful to all contributors for their valuable insights and dedication. We also extend our appreciation to our institutional partners, including Co-PLAN, Institute for Habitat Development (Albania), CEA (North Macedonia), UB-GEF (Serbia), Nordregio (Sweden), and Politecnico di Torino (Italy), for their support in fostering research excellence on just green transitions. By bringing together a diverse set of voices—including leading scholars, policy experts, and practitioners—we hope to provide a robust knowledge base that informs both academic research and policy design. We hope that this volume serves as a valuable resource for policymakers, academics, students, and practitioners engaged in the ongoing dialogue on just green transitions. By exploring the intersections of sustainability, justice, and territorial governance, we aim to contribute to a more inclusive and resilient green transition across Europe and beyond.

Torino, Italy	Erblin Berisha
Stockholm, Sweden	John Moodie
Tirana, Albania	Ledio Allkja
Belgrade, Serbia	Marija Jeftić

Reference

UNRISD. (2018). *Mapping just transition(s) to a low-carbon world*. United Nations Research Institute for Social Development. Report—JTRC-2018.pdf.

Acknowledgements

The preparation of this Handbook would not have been possible without the joint effort of all the contributors involved. A project of this magnitude, involving experts from diverse disciplines and backgrounds, requires extensive coordination, commitment, and interdisciplinary cooperation. The value of such a collaborative effort extends beyond the publication itself, as it fosters knowledge exchange, strengthens international partnerships, and enhances the understanding of complex territorial and environmental challenges from multiple perspectives.

Bringing together professionals from various fields—ranging from urban and regional planning to environmental sciences, economics, and political studies—has allowed for a truly integrated approach to addressing sustainability challenges. The diversity of expertise ensures that different perspectives are considered, leading to richer discussions, more robust findings, and comprehensive policy recommendations. The process of coordinating such a wide range of contributors has been a challenging but highly rewarding experience, requiring careful planning, open communication, and mutual respect for each other's work. The success of this endeavour is a testament to the power of collaboration across disciplines and borders.

Professionals from different countries have been instrumental in ensuring a comprehensive and well-rounded analysis, making this Handbook a significant resource for academics and practitioners. The project has provided a platform for comparative analysis and knowledge transfer by involving researchers and experts from various European regions, enriching the findings with diverse local experiences and methodologies. This exchange of ideas and best practices has not only strengthened the quality of the research but also fostered long-term collaborations that will continue beyond the completion of this publication. In this light, we would like to thank our Horizon Europe Green-Force[1] project partners: Co-PLAN, Institute for Habitat Development in Albania; University of Belgrade—Faculty of Geography (UB-GEF) in Serbia; Center for Economic Analyses (CEA) in North Macedonia; Nordregio, pan-Nordic research organization, based in Sweden; Politecnico di Torino, Italy; and the European Union for funding this initiative.

Moreover, a big thanks should go to the authors who have been involved: Aliaj Besnik, Bejko (Gjika) Anila, Böhme Kai, Brunetta Grazia, Buza Dea, Caldarice Ombretta, Cedergren Elin, Cotella Giancarlo, Damian Nicoleta, Dhrami Kejt, Djordjević Aleksandar, Finka Maros, Garvanlieva Andonova Vesna, Giacometti Alberto, Gjiknuri Ogerta, Gjoka Rodio, Gotovska Elena, Grigorescu Ines, Husár Milan, Imami Fiona, Jasso Matje, Katcharava Ketevan, Kraja Temali Irma, Krupalija Mustafa, Kurta Amela, Kurtovic Ilma, Maričić Tamara, Micu Dana, Mitric Bianca, Mohabat Doost Danial, Ndreka Elvis, Nikolov Ana, Nikolov Marjan, Novikau Aliaksandr, Oruc Nermin, Pantić Marijana, Popović Vladimir, Poro Enkela, Protić Branko, Puntillo Erika, Roznoviechi Irena, SeCerov Velimir, Shaker Yahya, Shaneh Aida, Tapia Carlos, Toptsidou Maria, Toska Merita, Valkenburg Govert, Vukovic Azra, Živanovic Zora, erban Paul-Răzvan.

Finally, our gratitude goes to Palgrave Macmillan Publishing for accepting to host the result of our work among their renowned publications, as well as to Rachael Ballard and Naveen Dass for dedicating their time to reviewing our work and developing a foreword for the book. We deeply appreciate their kind and inspiring words and hope that readers will share a similarly positive opinion.

Erblin Berisha
John Moodie
Ledio Allkja
Marija Jeftić

[1] This handbook is funded by the research project European HORIZON—WIDERA-2021-ACCESS-02-01 | Foster Research Excellence for Green Transition in the Western Balkans—GreenFORCE Twinning of the Western Balkans, grant number: 101059411.

Contents

Editors and Contributors

About the Editors

Erblin Berisha (Ph.D.—Urban and Regional Development) is an assistant professor at the Interuniversity Department of Regional and Urban Studies and Planning (DIST) of Politecnico di Torino. His research focuses on the evolution of Spatial Planning Systems and Territorial Governance in the Western Balkan region in the light of international influence. He is also interested in understanding the impact of Just Green Transitions in the Western Balkans.

John Moodie (Ph.D.—Political Science) is a senior research fellow at Nordregio Research Institute, specialising in EU governance and policymaking in the fields of regional development, green transitions, innovation, and marine policy. His most recent research has examined the concept of just green transition and its key dimensions. He also formed part of the research team that worked closely with Swedish national, regional, and local actors in developing and preparing EU Territorial Just Transition Plans for eligible Swedish Regions.

Ledio Allkja (Ph.D.—Spatial Planning) is a senior spatial planning expert and researcher at Co-PLAN, Institute of Habitat Development. Ledio is also a lecturer on spatial planning systems and EU policies at POLIS University in Tirana. His main areas of research focus on spatial planning systems, territorial governance, and Europeanization. Over the years, he has led and been part of several projects in different areas, such as spatial planning, social housing, land value capture, and, more recently, with a special focus on disaster risk reduction and resilience.

Marija Jeftić (Ph.D.—spatial planning) is a full professor at the Department for spatial planning at the Faculty of Geography, University of Belgrade. Her scientific and professional specialisations are oriented towards spatial and regional development, principles and methods of regionalisation, regional policy, and urban and rural renewal. Recently, she has been dealing with the implementation of the green agenda in spatial planning policies and green transition effects on spatial policies in the Western Balkans.

Contributors

Besnik Aliaj Co-PLAN, Institute for Habitat Development, Tirana, Albania; Faculty of Planning, POLIS University, Tirana, Albania

Ledio Allkja Co-Plan Institute for Habitat Development and Polis University, Tirana, Albania

Anila Bejko Co-PLAN, Institute for Habitat Development, Tirana, Albania

Erblin Berisha Interuniversity Department of Regional and Urban Studies and Planning (DIST), Politecnico di Torino, Turin, Italy

Kai Böhme Spatial Foresight, Heisdorf, Luxembourg

Grazia Brunetta Interuniversity Department of Regional and Urban Studies and Planning (DIST), Responsible Risk Resilience Centre (R3C), Politecnico Di Torino, Turin, Italy

Dea Buza Agricultural University of Tirana, Tirana, Albania

Ombretta Caldarice Interuniversity Department of Regional and Urban Studies and Planning (DIST), Responsible Risk Resilience Centre (R3C), Politecnico Di Torino, Turin, Italy

Elin Cedergen Nordregio Research Institute for Regional Development and Planning, Stockholm, Sweden

Giancarlo Cotella Interuniversity Department of Regional and Urban Studies and Planning (DIST), Politecnico di Torino, Turin, Italy

Nicoleta Damian Environmental Geography and GIS, Institute of Geography, Romanian Academy, Bucharest, Romania

Aleksandar Djordjević Faculty of Geography, University of Belgrade, Belgrade, Serbia

Danial Mohabat Doost Interuniversity Department of Regional and Urban Studies and Planning (DIST), Responsible Risk Resilience Centre (R3C), Politecnico Di Torino, Turin, Italy

Paul-Răzvan Serban Human Geography and Regional Development Department, Institute of Geography, Romanian Academy, Bucharest, Romania

Maros Finka Faculty of Management, SPECTRA Centre of Excellence of the EU, Comenius University, Bratislava, Slovakia

Vesna Garvanlieva Andonova CEA—Center for Economic Analyses, Skopje, North Macedonia

Alberto Giacometti Nordregio Research Institute for Regional Development and Planning, Stockholm, Sweden;
School of Business and Law, University of Stavanger, Stavanger, Norway

Ogerta Gjiknuri POLIS University, Tirana, Albania

Rodion Gjoka Co-Plan, Institute for Habitat Development, Tirana, Albania

Elena Gotovska CEA—Center for Economic Analyses, Skopje, North Macedonia

Ines Grigorescu Environmental Geography and GIS, Institute of Geography, Romanian Academy, Bucharest, Romania

Milan Husár Spectra Centre of Excellence of the EU, Slovak University of Technology in Bratislava, Bratislava, Slovakia

Fiona Imami Co-PLAN Institute of Habitat Development, Polis University, Tirana, Albania

Matej Jaššo Spectra Centre of Excellence of the EU, Slovak University of Technology in Bratislava, Bratislava, Slovakia

Marija Jeftić Faculty of Geography, University of Belgrade, Belgrade, Serbia

Ketevan Katcharava Interuniversity Department of Regional and Urban Studies and Planning (DIST), Politecnico Di Torino, Turin, Italy

Mustafa Krupalija Department of Political Science and International Relations, International University of Sarajevo, Ilidža, Bosnia and Herzegovina

Amela Kurta Center for Development Evaluation and Social Science Research, Sarajevo, Bosnia-Herzegovina

Ilma Kurtovic Center for Development Evaluation and Social Science Research, Sarajevo, Bosnia-Herzegovina

Tamara Maričić Institute of Architecture and Urban & Spatial Planning of Serbia, Belgrade, Serbia

Dana Micu Department of Climatology, National Meteorological Administration (Meteo Romania), Bucharest, Romania

Bianca Mitrică Human Geography and Regional Development Department, Institute of Geography, Romanian Academy, Bucharest, Romania

John Moodie Nordregio Research Institute for Regional Development and Planning, Stockholm, Sweden

Elvis Ndreka Co-PLAN Institute of Habitat Development, Polis University, Tirana, Albania

Ana Nikolov ICBC Institute, AEBR Balkans, Belgrade, Serbia

Marjan Nikolov CEA—Center for Economic Analyses, Skopje, North Macedonia

Aliaksandr Novikau Department of Political Science and International Relations, International University of Sarajevo, Ilidža, Bosnia and Herzegovina

Nermin Oruc Center for Development Evaluation and Social Science Research, Sarajevo, Bosnia-Herzegovina

Marijana Pantić Institute of Architecture and Urban & Spatial Planning of Serbia, Belgrade, Serbia

Vladimir Popović Faculty of Geography, University of Belgrade, Belgrade, Serbia

Branko Protić Faculty of Geography, University of Belgrade, Belgrade, Serbia

Erika Puntillo Interuniversity Department of Regional and Urban Studies and Planning (DIST), Politecnico di Torino, Turin, Italy

Irena Roznovieţchi Human Geography and Regional Development Department, Institute of Geography, Romanian Academy, Bucharest, Romania

Velimir Šećerov Faculty of Geography, University of Belgrade, Belgrade, Serbia

Yahya Shaker Interuniversity Department of Regional and Urban Studies and Planning, Politecnico Di Torino, Turin, Italy

Aida Shaneh Interuniversity Department of Territorial Sciences, Planning and Policies, Politecnico Di Torino, Torino, Italy

Carlos Tapia Nordregio Research Institute for Regional Development and Planning, Stockholm, Sweden

Irma Kraja Temali Planning and Territorial Development Department, Municipality of Shkodra, Shkodër, Albania

Maria Toptsidou Spatial Foresight, Heisdorf, Luxembourg

Merita Toska POLIS University, Tirana, Albania

Govert Valkenburg Department of Interdisciplinary Studies of Culture, Norwegian University of Science and Technology NTNU, Trondheim, Norway

Azra Vukovic GreenHOME, Podgorica, Montenegro

Zora Živanović Faculty of Geography, University of Belgrade, Belgrade, Serbia

List of Abbreviations

ACT IP	Accelerated Coal Transition Investment Plan
AF	Alternative Fuel
BCR	Benefits-to-Costs Ratio
BiH	Bosnia and Herzegovina
CB	Cross-Border
CBA	Cost-Benefit Analysis
CBC	Cross-Border Cooperation
CEE	Central-East European
COP	Conference of the Parties
COVID	Coronavirus disease
CPS	Cross-border Public Service
CRO	Croatia
CSOs	Civil Society Organizations
CTCN	Climate Technology Centre and Network
DSM	Demand-Side Management
DDR	Disaster Risk Reduction
EaP	Eastern Partnership
EBRD	European Bank for Reconstruction and Development
EC	European Commission
EE	Energy Efficiency
EGD	European Green Deal
EIB	European Investment Bank
ENPV	Economic Net Present Value
ERDF	European Regional Development Fund
EPA	Environmental Protection Agency
EPS	Electric Power Company of Serbia
ERO	Energy Regulatory Office
ES-CBA	Environmental and Social Cost-Benefit Analysis
ESCOs	Energy Service Companies
ESFP+	European Social Fund Plus
ESPON	European Spatial Planning Observatory Network
EU	European Union

EUGD	European Union's Green Deal
EUSDR	European Union Strategy for the Danube Region
EUSAIR	EU Strategy for the Adriatic and Ionian Region
FBiH	Federation of Bosnia and Herzegovina
FDI	Foreign Direct Investment
GAWB	Green Agenda for the Western Balkans
GHGs	Greenhouse Gas Emissions
GI	Green Infrastructure
GIZ	German Organization for International Cooperation
GVA	Gross Value Added
HLPF	High-Level Political Forum
HVAC	Heating, Ventilation and Air Conditioning
HVI	Heat Vulnerability Index
IAM	Integrated Assessment Model
ICFTU	International Confederation of Free Trade Unions
IEA	International Energy Agency
IED	Industrial Emissions Directive
IFI	International Financial Institutions
ILO	International Labour Organization
INECP	Integrated National Energy and Climate Plan
INTERREG	European Territorial Cooperation
IOA	Input-Output Analysis
IPA	Instrument for Pre-Accession Assistance
IPARD	EU Pre-Accession Assistance for Rural Development
IRR	Internal Rate of Return
ITUC	International Trade Union Confederation
JGT	Just Green Transition
JGTs	Just Green Transitions
JTR	Just Transition Roadmap
JTF	Just Transition Fund
JTM	Just Transition Mechanism
KEEA	Kosovo Energy Efficiency Agency
KEDS	Kosovo Energy Distribution System Operators
KEK	Kosovo Energy Corporation
KESIP	Kosovo Energy Strategy Implementation Program
KFEE	Kosovo Fund for Energy Efficiency
LWMP	Local Waste Management Plan
MESP	Ministry of Environment and Spatial Planning
MLP	Multi-Level Perspective
MW	Megawatt
NALAS	Network of Associations of Local Authorities of South-East Europe
NBS	Natural Based Solution
NDC	Nationally Determined Contribution
NECP	National Energy and Climate Plan
NEEP	National Energy Efficiency Plan
NERP	National Emission Reduction Plan
NEPA	Nature and Environment Protection Agency
NGEU	NextGenerationEU
NGO	Non-Governmental Organisation

NOx	Nitrogen Oxide
NPV	Net Present Value
NRRP	National Recovery and Resilience Plan
NUTS	Nomenclature of Territorial Units for Statistics
NWMP	National Waste Management Plan
NWMS	National Waste Management Strategy
OECD	Organization for Economic Co-operation and Development
PM	Particulate Matter
PNIEC	National Integrated Plan for Energy and Climate
PSLF	Public Sector Loan Facility
PUCs	Public Utility Companies
PVs	Photovoltaics
RCC	Regional Cooperation Council
RCI	Rotterdam Climate Initiative
RES	Renewable Energy Sources
RDF	Refused Derived Fuel
RNM	Republic of North Macedonia
RS	Republic of Serbia
RSD	Serbian dinar, currency of the Republic of Serbia
SDG	Sustainable Development Goals
SDR	Social Discount Rate
SFR	Federal Republic of Yugoslavia
SHPPs	Small Hydropower Plants
SFF	Solid Recovered Fuel
SMEs	Small and Medium Enterprises
SO_2	Sulphur Dioxide
SPRS	Spatial Plan of the Republic of Serbia
STEW	Sustainable Transition Epistemic Work
SWPR	Southwest Planning Region
TG	Territorial Governance
TJTPs	Territorial Just Transition Plans
TPP	Thermal Power Plant
UK	United Kingdom
UN	United Nation
UNDESA	United Nations Department of Economic and Social Affairs
UNDP	United Nations Development Programme
UNEP	United Nations Environmental Program
UNESCO	United Nations Educational, Scientific and Cultural Organization
UNFCCC	United Nation Framework Convention on Climate Change
VAT	Value Added Tax
WBs	Western Balkans
WBIF	Western Balkans Investment Framework
WCL	Christian World Confederation of Labour
WBGA	Western Balkans Green Agenda
WM	Waste Management
W2E	Waste to Energy

List of Figures

List of Tables

PART I

Setting the Context of Just Green Transitions

CHAPTER 1

Just Green Transitions - An Introduction

Erblin Berisha, John Moodie, Ledio Allkja, and Marija Jeftić

1.1 Introduction

Since the ratification of the Paris Agreement in 2015, the notion of Just Green Transitions (JGTs) has become an increasingly salient topic in contemporary research and academic discourse (Castona, 2024; Heffron and Heffron, 2021; Shaker & Berisha, 2025; Wang & Lo, 2021). JGTs are also a prominent feature in international climate and environment policy frameworks, including

E. Berisha (✉)
Interuniversity Department of Regional and Urban Studies and Planning (DIST), Politecnico Di Torino, Turin, Italy
e-mail: erblin.berisha@polito.it

J. Moodie
Nordregio Research Institute for Regional Development and Planning, Stockholm, Sweden
e-mail: john.moodie@nordregio.org

L. Allkja
Co-Plan Institute for Habitat Development, Tirana, Albania

Polis University, Tirana, Albania

L. Allkja
e-mail: ledio_allkja@universitetipolis.edu.al

M. Jeftić
Faculty of Geography, University of Belgrade, Belgrade, Serbia
e-mail: marija.jeftic@gef.bg.ac.rs

E. Berisha et al. (eds.), *The Palgrave Handbook of Just Green Transitions in the Western Balkans and Beyond*,
https://doi.org/10.1007/978-3-031-95075-9_1

the European Union's Green Deal (EUGD: European Commission, 2019), the United Nations 2030 Agenda and Sustainable Development Goals (United Nations, 2015, 2023), and the Just Transition Guidelines of the International Labour Organisation (International Labour Organisation, 2015). The EUGD is an ambitious policy framework designed to support the aim to make the European Union (EU) climate-neutral by 2050 by stimulating climate and environmental transitions within a range of policy areas, such as the circular economy, building renovation, biodiversity, farming, and innovation (European Commission, 2019).

The European Union's Just Transition Mechanism, introduced in 2021 under the EUGD, is directly committed to identifying and supporting Europe's most affected regions with the transition's technical, social, and territorial-specific impacts of a low-carbon economy (European Commission 2020a). Financial support is also made available to all member states through the European Union (EU) Just Transition Fund, focusing on those regions highly dependent on fossil fuels and carbon-intensive industries, with citizens most vulnerable to the socio-economic impacts of transition (Ibid). While the EU is seen as leading the way in promoting and supporting JGTs, the EUGD faces enormous implementation challenges due to the scope of facilitating multiple and simultaneous transitions across different countries and regions (Shaker & Berisha, 2024).

To become a climate leader, the EU is keen to export its new climate and environmental policy initiatives to neighbouring countries and accession and pre-accession states (European Commission 2022). To facilitate this, the EU provides policy and financial support for developing and implementing the EUGD in EU candidate and pre-accession countries, including those in the Western Balkans (WBs). Besides the mentioned JTM and JTF, the EU has also introduced several other JGT-oriented political measures and economic support packages for accession, candidate, and third countries, including the 2022 Enlargement Package of the European Neighbourhood Policy and Enlargement Negotiations (DG NEAR), the NextGenerationEU (NGEU), the JTM, the Fit-for-55 package, the REPower EU plan, the EU Reform Agenda, and the New EU Strategy for the Potential Enlargement of the European Union (European Commission, 2024). The EU also initiated the idea of creating the Green Agenda for the Western Balkans (GAWB) in 2020 (Berisha, 2025a, 2025b). The GAWB closely aligns with the framework outlined in the EUGD (European Commission, 2020b; Knez et al., 2022). The EU supports the implementation of the GAWB and national climate plans primarily through the Berlin Process and the new EU Growth Plan for the WB Region, providing funding via the EU Instrument for Pre-Accession Assistance and the Western Balkan Investment Framework (European Commission, 2023; Regional Cooperation Council, 2021).

The shift to carbon neutrality fostered by the EUGD and the GAWB could profoundly impact societies, economies, and governance processes across the

WBs and Europe (Ignjatović et al., 2024). As many European states are experiencing, the socio-economic challenges and benefits of JGTs are not necessarily homogenous, with transitions generating both positive and negative impacts across a diversity of regions and societal groups (Kortetmäki et al., 2025). How JGTs are designed and implemented will have an important effect on regional development across European countries, particularly in regions that historically suffer from slow or stagnant socio-economic growth, such as the WB Region (Rodríguez-Pose & Bartalucci, 2024).

This handbook has been prepared under the auspices of the GreenFORCE Horizon project, aiming to support the WB towards the transition from a carbon-based to a more free-carbon society. It offers a panorama of case studies and initiatives activated in the WB and beyond, showing the heterogeneity of JGT approaches. The handbook represents a first attempt to systematise the concept of JGTs and offers an entry point to understand where the WB stands on this topic. Besides this brief introduction, Sect. 1.2 presents a detailed discussion of the JGT concept, investigating its complexity and why it is a concept 'hard to grasp' (Shaker & Berisha, 2025). After that, Sect. 1.3 presents the state of the art of JGTs in the WB, pointing out its main issues. Section 1.4, instead, presents the handbook focus, aims, and objectives, followed by Sect. 1.5, which delves into the handbook contributions' content—here, each chapter is briefly presented, outlining the logic back to the handbook.

1.2 Just Green Transitions: A Concept Hard to Grasp

The term 'Just Green Transitions' is largely a policy and academic construct rarely used within a societal context or by citizens (Shaker & Persico, 2024). Since the 1950s, the notions of 'Green Transition(s)' and 'Just Transition(s)' have been used regularly and interchangeably in policy and academic discourse, along with other iterations, such as 'Green Transformation(s)' and 'Just Transformation(s)' (Shaker & Berisha, 2024). In recent years, however, it has been increasingly commonplace to find compound references combining both these notions, including 'Green and Just Transition', 'Green Just Transition', and 'Just Green Transition' (Cedergren et al., 2022).

Shaker (2025, in this volume) argues that the JGTs is a meta concept that embodies a rich tapestry of interconnected and concurrent concepts. He continues that within the JGTs terminology, there is no agreed consensus on what constitutes 'green' or 'just', and no clear blueprint for developing and implementing JGTs (Ibid). Furthermore, Shaker and Berisha point out that the use of the words transition(s) and transformation(s) in both singular and plural form creates confusion as to whether JGTs represent a singular phase or goal or several parallel and simultaneous phases and goals (Shaker & Berisha, 2024). Here, it might be helpful to distinguish between 'transition' as a goal, i.e. the shift to carbon-neutral societies and economies, and 'transitions' as a process, i.e. the multiple ongoing activities across multiple governance levels and policy spheres in helping to meet the transition goal.

The term JGT embodies a range of concepts and serves as an umbrella term that includes different policy themes and sectors, including climate, environment, energy, water, air, soil, and food. While there is no official agreed definition, throughout the Handbook, the term 'just green transition' is used to refer to strategies, policies, and measures that ensure no one is left behind in the transition to low-carbon and environmentally sustainable economies and societies (Tavares, 2022). While definitions of JGTs are extremely broad, Moodie et al. (2021) note that there are several key dimensions of JGTs that strategies, policies, and measures must try to address, highlighting the complexity and multidimensional nature of shifting to carbon-neutral economies and societies (also see Chapter 2 in this volume). The WBs are in the very early stages of developing and implementing JGTs (Imami et al., 2023), so policymakers can learn from existing policy challenges and good practices from across Europe and the world, in addition to academic contributions defining key JGTs dimensions and characteristics.

1.3 Just Green Transitions in the Western Balkans

The WB is lagging behind its European neighbours regarding implementing JGTs and meeting international climate and environmental goals and targets (Berisha, 2024). Energy systems in the region remain outdated, with coal-powered thermal energy plants accounting for between 60 and 95% of the region's electricity production (Bechev, 2023). Road transport as the dominant transportation mode in the WB countries significantly contributes to greenhouse gas emissions and air pollution. Jeftić et al. are stating that the total share of transport emissions in WB countries is dominated by CO_2 from road transport, which is most evident in large urban areas that also suffer from associated high pollution (Jeftić et al., 2025). In regions with lower population density and underdeveloped public transport, energy consumption per capita in urban transport is significantly higher. The role of public transport in reducing urban energy consumption is crucial, as cities with well-integrated systems show significantly lower per capita energy use (Vračarević, 2021). The slow speed of climate and energy transitions in the region runs counter to the recent policy commitments of national-level decision-makers (Regional Cooperation Council, 2021). All WB countries pledged to cut national greenhouse gas emissions in line with the 2015 Paris Agreement. In the Joint Statement on Clean Energy Transition adopted in February 2019 in Podgorica, ministers of energy and of environment confirmed their will to align as swiftly as possible with the EU's energy, climate, and environmental policies. At the Berlin Process Summit in Poznań in July 2019, WB leaders expressed their willingness and readiness to meaningfully promote and contribute to the implementation of the EUGD and support EU efforts to fight climate change, protect the environment, and unlock the economic potential of the green, low-carbon, and circular economy in the region (Ibid). In 2020, the leaders of the Six WB countries also signed the Sofia Declaration, which stated, 'we commit to work towards the 2050 target of a carbon-neutral continent together with the EU through mainstreaming a strict climate policy and reforming energy and transport sectors'. (Regional Cooperation Council, 2020). The Sofia Declaration laid out a number of climate and environment related actions, including a commitment of each WB country to align with EU Climate Law and EU Emissions Trading Scheme, set forward thinking 2030 energy and climate targets, prepare and implement climate mitigation strategies, participate in the European Climate Pact, and transpose the EUGD within all climate and environmental policies and interrelated policy sectors (Ibid).

In meeting these commitments, the 2020 Sofia Declaration outlined plans to develop a GAWB, as mentioned above (Regional Cooperation Council, 2020). Introduced in 2021, the GAWB is a growth strategy for the WB designed to support climate and energy transitions and the shift to sustainable carbon-neutral economies and societies across the WBs (Regional Cooperation Council, 2021). The GAWB and the corresponding Green Agenda for the Western Balkans Action Plan is expected to support the EU accession processes (Berisha, 2025a) by attempting to build policy links and cohesion between WB countries under key green agenda policy sub-themes closely aligned with the EUGD and other EU climate targets, including: (1) Decarbonisation (Climate, Energy, and Transport); (2) Circular Economy; (3) Depollution (Air, Water, and Soil); (4) Sustainable Food Systems and Rural Areas; and (5) Biodiversity (Protection and Restoration of Ecosystems) (Regional Cooperation Council, 2021). The joint framework seeks to enable coordinated action for WB countries to adopt EU standards while adapting to local contexts. The efforts are also motivated by access to funding and advancing the pre-accession dialogue and the EU accession process. In addition to the GAWB, each country has also developed EU-mandated national-level energy and climate plans (Regional Cooperation Council 2021).

As previously mentioned, JGTs will be crucial in shaping political, territorial, socio-economic, and environmental developments across the WB (Berisha, 2025a and b; Shaker, 2025, in this volume). On a political and territorial level, the extent to which WB align national and regional climate and environmental strategies with the goals and targets of the EU Green Deal will significantly impact on their chances of EU accession while also laying the foundations for enhanced macro-regional and cross-border cooperation and integration between Western Balkan countries (Tziarras, 2019). On a socio-economic and environmental level, JGTs present an opportunity to modernise energy infrastructures, improve energy security, promote economic growth through green foreign direction investment, and bring climate and environmental gains, including less pollution and cleaner air, particularly in larger urban areas, such as Skopje and Belgrade, where there are record high levels of pollution (Berisha, 2025a).

Despite closer alignment with EU climate and environment goals, the extent to which WB countries can maximise the opportunities and benefits presented by JGTs is open to question as the WB Region suffers from an implementation deficit in relation to the GAWB and national climate plans (Bechev, 2023). Operationalising JGTs in the WBs is a major challenge due to existing social, political, economic, and territorial specificities, as well as limited institutional capacities across the region (Bieri, 2015; Berisha & Cotella, 2021; Ignjatovic et al., 2024). Many WB countries are still heavily dependent on coal and thermal power plants for energy production (Berisha, 2025a). The energy infrastructure is dominated by large state-owned companies with undue political influence, so key sectors lack political will and commitment to change the status quo (Bechev, 2023). At a societal level, citizens are both unaware and somewhat resistant to JGTs; the term 'transition' has become loaded with negative connotations, as the WBs have been in a loop of post-socialism transition since the early 1990s, with the slow transition towards liberal democracies, free economies, and EU integration, creating disillusionment and 'transition fatigue' among citizens (Toto et al., 2023).

Overcoming these challenges will be pivotal if the WBs is going to develop and implement effective JGTs that meet regional needs and are accepted and supported by citizens. An increase in national-level political will and ambition, enhanced capacity building, and continued EU-level support and finance is considered essential for JGTs in the WB Region (Bechev, 2023). As the WBs are still in their relative infancy regarding JGTs, they can learn a lot from past and ongoing JGT policy experiences from across other WB and European countries. Shared knowledge and expertise, including best JGT policy practices, can contribute towards helping WB countries to achieve the climate goals and targets set out in the GAWBs and their national climate and environment plans.

1.4 Handbook Focus, Aims, and Objectives

This handbook explores the current JGT research and policy landscape in the WBs and across Europe. Even though the concept of JGTs is rapidly emerging within academic and policy circles, there is no standard definition of JGTs and no way to design and implement them effectively to support WB and European regions and societies. The handbook aims to enhance knowledge and understanding of JGTs by identifying and analysing the latest conceptual and policy developments. In meeting this aim, the handbook has three key objectives:

- Examine the concept of JGTs and its key dimensions
- Explore the core policy thematic areas at the heart of the JGTs and the socio-economic and governance challenges and opportunities within each policy theme

- Provide emerging good practice examples of just green transition policies developed and implemented across the Western Balkans and other European regions

The handbook contributes to the policy and academic debate by exploring the multidimensional aspects of JGTs related to circularity, sustainable development, spatial planning, territorial governance, regional economy, human geography, and spatial and social justice. There is no one-size-fits-all approach to JGTs, and policies must be carefully tailored to meet the needs of different regional contexts and specificities. The findings and results outlined in the Handbook can be used by JGT policymakers and practitioners (e.g. public authorities, industries, businesses, universities, NGOs, and civil society groups) across the WBs and Europe to support the development and delivery of targeted JGTs that address regional challenges and meet the needs of citizens.

1.5 Structure of the Handbook

This final section of the introductory chapter provides an overview of the handbook's structure and presents the individual contributions. The handbook is structured into four main parts:

- Part I: Setting the Context of Just Green Transitions
- Part II Conceptualising Just Green Transitions.
- Part III: Exploring Just Green Transition Policy Themes.
- Part IV: Just Green Transitions in the Western Balkans.
- Part V: Just Green Transition Practices from around Europe and beyond.

1.5.1 Part I—Setting the Context of Just Green Transitions

Setting the context of JGTs is an intricate endeavour. Countries and transnational institutions are currently developing various initiatives attempting to accompany their contexts towards a more sustainable economy and society. Efforts are made at various levels as JTGs seem to be a conceptual challenge as well as a political and technical issue. Part I of this handbook is organised to provide the reader with a minimum of contextualisation from a conceptual and more regional perspective. More in detail, in chapter 1, Berisha et al. (2025a) identify the conceptual gap and the reason why it focuses on the Western Balkans. Moreover, it describes the handbook's focus, aims, and main objectives, offering a better contextualisation. Finally, this chapter details the handbook structure and illustrates how the various chapters contribute to consolidating the JGTs' experiences at various levels and contexts.

In Chapter 2, Moodie et al. (2025a) discuss the multidimensional framework of JGTs that is integral to achieving climate neutrality while ensuring fairness and inclusivity. This framework encompasses social, economic, spatial, environmental, and governance dimensions, emphasising the need for a holistic approach to transitioning to sustainable, low-carbon economies. More in detail, the chapter discusses (i) the social aspect of prioritising justice and equity, addressing vulnerabilities and empowering marginalised groups through distributional, recognitional, and procedural justice; (ii) the economic aspects of JGTs providing opportunities for regional restructuring, circular economies, and innovation, despite the substantial costs associated with transitioning; (iii) the spatial aspects advocating for tailored policies to mitigate territorial disparities and integrate socio-cultural identities into transition strategies; (iv) the environmental dimension focuses on achieving long-term sustainability through carbon neutrality, biodiversity preservation, and resilience-building and (v) governance dimension underlying its pivotal role in fostering multi-level collaboration, citizen engagement, and policy coherence. Finally, Chapter 3, authored by Moodie et al. (2025b), focuses on the JGTs' policy landscape, providing a context for understanding the policy orientations and implementations at various levels, including national with a WB focus, EU and international levels. The chapter discusses the various attitudes and presents a preliminary landscape of policy frameworks and implementation.

1.5.2 PART II—Conceptualising Just Green Transitions

The concept of JGTs is gaining increasing attention in academic circles in response to recent policy developments supporting the shift to climate-neutral economies and societies. Part II of the Handbook critically reviews the notion of JGTs through conceptually focused chapters that develop our understanding of JGTs and its key elements. Chapter 4 by Shaker (2025) discusses the role of the meta-governance models in effectively developing and implementing JGTs' strategies across scales, actors, and sectors within EU Member States and Candidate Countries. The chapter advocates for a meta-conceptual approach to JGTs, which recognises the multiplicity, simultaneity, diversity, and complexity of transitions, emphasising the important role of spatial planning and territorial governance in achieving both institutional integration and carbon neutrality. The chapter finds that the so-called Just Green Transitions, as envisioned in the EUGD, offer a viable framework for the WB integration into the EU, but only if it is accompanied by a tailored meta-governance approach that addresses the macro-regional challenges and opportunities included in the GAWBs. Chapter 5 by Tapia et al. (2025) examines the evolution of the JGTs concept from a historical perspective, connecting its major conceptual and operational milestones to the development of the international climate policy agenda, from its origins to its current interpretation in practice, policy, and research. The chapter shows how the JGTs paradigm has not evolved in the same way as other strategic frameworks for sustainable development. Most of these were first developed on an abstract theoretical and conceptual level, to be then applied on a more practical level, whereas the JGTs was developed as a practical policy to address workers' needs before developing into an overarching policy framework and academic concept. Historical evolution of the JGTs policies, highlighting how the focus of JGTs has changed over time, moving from a predominantly preventative and mitigative perspective to a more adaptive and transformational policy. Different JGTs' approaches are not mutually exclusive but complementary, as JGTs policies may require a combination of mitigative, adaptive and/or transformative measures. This contribution leads nicely into Chapter 6 by Jaššo and Husár (2025), which discusses the role of participatory planning culture in promoting a just and sustainable transition. The authors argue that traditional spatial planning methods, which rely on regulations and legislation, are no longer sufficient in addressing the specific and unique features of JGTs, which require adaptable strategies and tools that engage local communities and citizens. They conclude that JGTs require the application of collaborative planning tools that promote citizen engagement and empower communities to shape JGTs through direct participation and increased education, information, and technology. In the final chapter of this section (chapter 7), Valkenburg (2025) explores the role of knowledge and expertise in JGTs policymaking and planning processes. The chapter highlights that JGTs pose complex challenges that need to be faced with an interdisciplinary approach, integrating a broad range of knowledge types. The chapter concludes that considerations of epistemic justice need to be at the heart of JGTs policy and governance processes to ensure that all relevant expertise and voices are heard and no one institution or actor comes to dominate JGTs proceedings.

1.5.3 PART III—Exploring key Just Green Transitions Policy Themes

The notion of JGTs has slowly permeated the policy discourse at all levels of governance (international, national, regional, and local levels). In policy documents, JGTs often acts as a broad umbrella term, which covers a range of key policy thematic areas, such as climate resilience and adaptation, clean and renewable energy, circular economy, environmentally friendly transport, sustainable agriculture and food production, depollution of air, water, and soil etc. This section of the handbook broadens a significant role of public policies in promoting fairness, both in the distribution of resources and in overcoming structural barriers. Based on in-depth analyses, chapters offer recommendations for shaping policies that integrate economic stability and environmental sustainability, drawing on experiences from the region and beyond.

In Chapter 8, Buza and Allkja (2025) discuss landscape design as a tool to foster spatial transformations considering just green transition in the WB. The aim is to analyse current policy frameworks and policy practices in the WB by taking Albania as a main case study in order to understand the potential that landscape planning and design can offer for JGT. It offers a mapping overview of how and if stakeholders are involved in the conceptualisation, development, and implementation of policies regarding the multiple transitions in the region, pointing out the limits of a top-down approach. Chapter 9 by Novikau and Krupalija (2025), instead, provides a detailed analysis of the current energy landscape in Bosnia and Herzegovina, discussing the reliance on coal and the significant environmental and health issues this reliance causes. They also examine the potential benefits of transitioning to renewable energy, such as improved air quality, job creation in new sectors, and alignment with EU environmental standards. The authors emphasise the importance of a just transition that includes retraining programs for workers, social safety nets, and community engagement to mitigate the negative impacts of this shift. They conclude that Bosnia and Herzegovina's green energy transition demands a multi-faceted approach which should be supported by collaboration between government agencies, international bodies, private investors, and local communities, etc. and future research as a base to explore best practices for decommissioning coal plants, fostering alternative industries, and developing sustainable, community-supported renewable energy projects. Following chapter 10 by Nikolov (2025) has an overview character, providing a thorough analysis of the legal frameworks and practical aspects of cross-border cooperation in the sensible border regions in the Western Balkans and also offering valuable insights into the potential for cross-border public services (CPS) supporting the Just Green Transition. The author concludes that cross-border cooperation is essential for supporting a Just Green Transition in the Western Balkans, particularly through cross-border public services that facilitate the implementation of the Green Agenda's goals. Furthermore, the author emphasises the importance of leveraging existing cross-border structures, programs, and bilateral agreements, while suggesting that the Regional Cooperation Council (RCC) should take on the role of regional CPS (cross-border public services) provider and coordinator of the Action Plan for the Western Balkans Green Agenda. A more detailed analysis of regional CPS bodies and the normative frameworks underpinning them is necessary to fully capitalise on the potential of existing regional institutions to support this transition. Chapter 11 by Maričić and Pantić (2025) provides an in-depth analysis of the socio-economic and environmental challenges associated with the green transition in coal-mining regions with a focus on Serbia. Coal-mining communities in economically less developed regions, including the Western Balkans, face significant challenges due to severe pollution, outdated technologies, and weak enforcement of environmental standards. In Serbia, where energy consumption heavily depends on coal, regions like Kolubara and Kostolac are economically significant (their municipalities are among the top-ranked in the country in terms of income) but environmentally vulnerable. The chapter offers practical strategies for transforming coal-mining regions, drawing on successful examples from Germany and Poland. Incorporating insights from stakeholder interviews adds depth to the analysis and highlights real-world challenges and opportunities. The authors conclude that while JGT can enhance environmental quality and overall life standards, it poses significant risks of economic deprivation in post-coal-mining regions. To prevent an unjust transition, strong commitments from both national and local governments are essential. The primary obstacle to implementing JGT in coal-mining communities lies in the conflicting interests of various social actors, such as trade unions, politicians, environmentalists, and mining companies. Lessons from other post-mining transitions highlight the importance of early preparation, financial resource allocation, and inclusive strategies to mitigate decarbonization risks. The labour market will inevitably be reshaped, requiring a coordinated approach, transparent decision-making, and broad stakeholder engagement. In Serbia, despite growing awareness of decarbonization, it is not yet integrated into policy systems or strategic thinking. Examples from Poland and Germany show that success in achieving JGT goals depends on clear, long-term restructuring policies. A place-based, bottom-up approach, involving local stakeholders in policy design and implementation, as seen in the Ruhr region and the UK, is recommended to ensure a smoother transition. In the final chapter of this section 12 by Brunetta et al. (2025) an innovative perspective is introduced by focusing on spatial justice, addressing geographical dimensions often overlooked in policy discussions. The authors provide a thorough critical standpoint of the EU's Just Transition Strategy, highlighting significant gaps in its approach to justice, offering clear recommendations for integrating spatial justice into the Green Transition and providing practical strategies for policymakers. They conclude that the Green Just Transition (GJT) must balance economic growth, social equity, and environmental protection while addressing systemic injustices that underlie environmental challenges. Current EU approaches overemphasise distributive justice and overlook procedural and spatial dimensions, which limit their effectiveness. A spatial framework is recommended to integrate local knowledge, ensure equitable resource distribution, and support disadvantaged areas through targeted infrastructure, job creation, and skills training. Challenges include political resistance, funding disparities, and rigid governance structures. The authors emphasise the need for holistic, place-based strategies and call for further research on the outcomes of the EU's Just Transition Mechanism (JTM) in addressing local-level inequities.

1.5.4 Part IV—Just Green Transitions in the Western Balkans

Since the beginning of the nineties, the Western Balkan countries have transitioned from a centralised to a market economy, which has led to significant institutional, social, and economic developments (Berisha et al., 2018, 2024; Tsenkova & Nedović-Budić, 2006). Within this context, JGTs have emerged as an important new policy theme for developing collaborative partnerships, enhancing socio-economic growth, and building regional resilience across Balkan countries. These ideas are manifested in the Green Agenda for the Western Balkans, which presents several challenges and opportunities that just green transition presents for the region. This part of the handbook critically examines different experiences of JGTs processes across the WB and their spatial, governance, economic, social, and environmental (direct or side) effects. The chapters offer different perspectives regarding processes of the overall JGTs in the WB as well as case studies from the different countries such as Albania, Serbia, Kosovo*,[1] Montenegro, North Macedonia, and Bosnia and Herzegovina. Below, there is a short description of what to expect from each of the chapters.

Chapter 13, written by Garvanlieva et al. (2025), presents the current conditions of the Just Green Transition in North Macedonia and, more specifically, in the mine-based SouthWest Economic Region. The socio-economic dimension of green transition is explored in this case focusing on the decarbonisation of the TTP Oslomej. The study uses different costbenefit analyses to offer a basis for policymakers on the impact that JGTs policies may have in the region. The recommendations are of high value for the entire region with a strong energy production base through fossils. Chapter 14, presented by Imami and Aliaj (2025), delves into the risks of purposive transition in Albania. This chapter is based on transition theories and tries to compare theoretical frameworks with the multiple transitions occurring in Albania. The concept of transition has been thoroughly analysed, offering a basis for a better understanding of the processes of change that have occurred in Albania since the early 1990s. This chapter not only offers a new perspective of understanding JGTs through transition theories in Albania but also serves as a stepping stone for further and deeper studies in the WB countries that face similar processes of change. Chapter 15 has been prepared by Živanovic et al. (2025) and focuses on the case study of Serbians' pathway to JGTs. Following a detailed analysis of the international and national policy frameworks on green transition, the chapter delves deeper into the policy analysis of Serbia. The chapter offers interesting perspectives on the limits of (top-down) policy transfer and their impacts in specific regions. Additionally, the chapter analyses the gap between policy objectives and results in reality. This gap is quite evident for most WB countries. Thus, the perspective opened by the discussion in this chapter serves as a tool for policymakers to reduce the gap between formal policy adoption and implementation. Chapter 16 by Oruc et al (2025) assesses the budgetary and distributional implications of expected job destruction of coal mines closures. To address the issue, the chapter estimates on the labour market impact of mine closure using a panel dataset of the BiH Labour Force surveys 2016–2021 and BiH occupational forecasting model by assessing the transition to (un)employment of redundant workers from coal mines, taking into account their education level, occupation and age. Then, the estimated job and wage losses were incorporated into the BiH tax-benefit microsimulation model to assess the impact of such losses on government revenues and poverty and inequality levels among households in BiH, by type of household and region. Chapter 17 focuses on the waste management sector as one of the key dimensions of just green transition. Toska et. al. (2025), analyse

[1] This designation is without prejudice to positions on status, and is in line with UNSCR 1244 and the ICJ Opinion on the Kosovo declaration of independence.

in detail the approach of the Montenegrin government and policies in terms of waste management related to green transition. The study focuses on the central region of Podgorica with the aim of understanding the potentials that Refused Derived Fuel and Solid Recovered Fuels can have as a source for energy production. Following a detailed analysis of current waste management practices in the region, the chapter makes a good case for applying principles of circular economy to the sector. This chapter makes a strong argument for moving towards the application of circular economy principles in the waste management sector and gives distinct recommendations for the advancement of green transition in the region. Chapter 18 is presented by Ndreka (2025) and focuses on energy transition in Kosovo. The energy sector is heavily based on fossil resources not only being directly linked to the green transition of the sector but also to deep environmental impacts in terms of air pollution and depletion of resources. The chapter makes a deep analysis of the policy and institutional framework of the energy sector in Kosovo. It focuses on renewable energy and increasing energy efficiency. For each the authors make a detailed analysis of the policy landscape, institutional framework as well as the impact that these policies have. Additionally, it delves into issues of waste management and the concept of waste to energy. This concept offers a great opportunity for rethinking the management system of these sectors and directing them towards a new path. It has a strong link with Chapter 15, which offers a similar perspective in terms of the potential of using waste as a source for energy production in the region. Chapter 19 is prepared by Temali (2025) and focuses on nature-based solutions and green technologies to mitigate climate risk in coastal areas. The article offers a different perspective in terms of green transition actions. Following a theoretical and international policy overview, the chapter delves into different international (European) case studies to create a framework of the successful application of nature-based solutions as a key tool in the hands of planners and architects to transform coastal areas towards resilient and adaptive space under the continuous and inherent threat of climate change impacts. The chapter afterwards focuses on the Albanian case studies comparing *vis-à-vis* international ones to understand the pitfalls and the potentials of green technology applications in the country. The chapter puts a strong emphasis on planning for the future and the role of site-specific projects that interrelate climate risks with green technologies. Berisha, with chapter 20, closes off this part with a roundup discussion on the state of the art of policy frameworks in the WB with regard to green transition and taking up of the Green Agenda for the Western Balkan. This chapter serves as a conclusion to this part as it somewhat makes a recollection of all individual contributions to make a coherent and integrated analysis of the status of policy development and policy implementation in WB countries. It shows how, although countries have taken similar steps in terms of policy development, they are in different stages of preparation. Nevertheless, one key component which unites this analysis is the difficulty that WB countries have to turn green transition into a priority for policy development.

1.5.5 Part V—Just Green Transitions Practices from Around Europe and Beyond

The European Union has made just green transitions a central policy objective as part of the EU Green Deal. The implementation of the Green Deal is supported through financial mechanisms, including the EU Just Transition Mechanism and Funds, which are designed to support the implementation of EU Just Transition Plans and other Green Deal policies related to the key environmental themes outlined above (e.g. climate resilience and adaptation, clean energy, circular economy, renewable energies, environmentally friendly transport, sustainable agriculture and food production, depollution of air, water, and soil, and protect biodiversity and ecosystems). This section of the handbook provides fresh contributions that analyse how green transition processes have been developed and implemented across EU Member States. These chapters illuminate good practice JGTs policy examples and evaluate the direct and indirect socio-economic and governance consequences of transition processes.

Berisha et al. (2024), in Chapter 21, examine the Just Transition frameworks of Italy, Poland, Sweden, and the Netherlands within the context of the European Green Deal. The comparative analysis highlights how each country's unique socio-economic and environmental priorities shape its approach to the JGTs approach, underscoring the importance of flexibility and inclusivity. The selected case studies offer insights into the EU's collective path to a sustainable future, stressing the need for tailored solutions that address local challenges while achieving EU-wide climate goals. In Chapter 22, Shaneh et al. (2025) discuss instead the territorial governance dynamics at play in JGTs planning processes. The place-based policymaking concept of territorial governance is employed to investigate the planning procedures implemented during the preparation of EU Territorial Just Transition Plans (TJTPs) in three Swedish regions—Gotland, Norrbotten, and Västerbotten. Finka (2025) in chapter 23, tries to mediate tacit knowledge on JGTs from Central European Countries' perspective by introducing a realistic view on past and current transition concepts as a base for the critical perception of the Green Deal as a political framework from the position of planning practice for multiple transitions in Central Europe, in order to outline possible conceptual and practical solutions suitable for Western Balkan as inherent part of integrated Europe, demonstrated on an example of Vision and Strategy of Development Slovakia 2030. It highlights the potential of just green transition as a running challenging process of learning by doing, bringing together multiple stakeholders with different interests, many innovative solutions and mistakes, as the process of mutual learning in which the Western Balkan countries can bring their own tacit knowledge as the Central Europe including Slovakia did. Mitrică et al. (2025), in Chapter 24, instead focus more on the Romanian sector of the Danube Valley challenges. This study aims to develop a multi-criteria vulnerability assessment by identifying the levels of socio-economic vulnerability by computing a Heat Vulnerability Index (HVI), which considers a set of indicators grouped into three key components of vulnerability (Exposure, Sensitivity, and Adaptive Capacity). The current research is aimed to be useful for academics by improving the general understanding of heat-related vulnerability in an area under the influence of one of Europe's most important rivers, but also for local decision-makers and planners to further contribute to the local climate change mitigation and adaptation strategies. Different from the previous contribution, Toptsidou and Böhme (2025), in Chapter 25, explore two distinct pathways towards green transition by 2050, taking the example of the Adriatic Ionian Region. Two territorial scenarios are explored, one focusing on the green transition being part of the quality of life, where health and social cohesion are prioritised. In contrast, the other focuses on a bolder idea of 'giving back to nature' where depopulated and environmentally vulnerable regions are restored as natural habitats. In the diverse landscape of the Adriatic Ionian Region, which faces socio-economic and geopolitical challenges, both scenarios highlight the need for tailor-made, place-sensitive approaches, where policy coordination, innovative solutions, community engagement, and rethinking of current actions need to be considered. The scenarios aim to offer food for thought to develop sound policy choices about the future. Katcharava (Chapter 26) offers a fascinating comparative contribution to the future trajectory of JGTs within Europe's external dimension, explicitly focusing on the Eastern Partnership (EaP) countries. The chapter examines the successful practices of WB, which, despite not being an EU member, has pioneered its own Green Agenda. The chapter further explores the policy and governance frameworks needed to build multi-level partnerships in EaP, capable of leading the complexities of the green transition. As the guiding tool, it proposes the specifically tailored JGTs' roadmap to align and deal with regional specificities within the EaP.

Finally, in chapter 27, the editors sum up the main lessons learned and identify a number of future research perspective pointing out the importance of taking into consideration the promotion of tailored region-specific strategies, participatory policymaking, and investment in education and workforce reskilling to ensure fair and sustainable transitions.

References

Bechev, D. (2023). The green transition and the Western Balkans, Carnegie Europe Article.

Berisha, E. (2025a). Moving towards an inclusive green agenda in the Western Balkans. Foundation for European Progressive Studies (FEPS). https://doi.org/10.13140/RG.2.2.12092.76162

Berisha, E. (2025b). Green EU integration—The role of the green agenda in accelerating Just Green Transitions in the Western Balkans. In E. Berisha, J. Moodie, L. Allkja, & M. Jeftić (Eds.), *The Palgrave Handbook of just green transitions in the Western Balkans and Beyond*. Palgrave Macmillan.

Berisha, E., Čolić Marković, N., Cotella, G., & Nedović-Budić, Z. (2024). Potential future enlargement: Spatial planning in the Western Balkans. *Spatial Planning Systems in Europe*, 216–245. https://doi.org/10.4337/9781839106255.00022

Berisha, E., Colic, N., Cotella, G., & Nedovic-Budić, Z. (2018). Mind the gap: Spatial planning systems in the Western Balkan region. *Transactions of the Association of European Schools of Planning, 2*, 47–62. https://doi.org/10.24306/TrAESOP.2018.01.004

Berisha, E., & Cotella, G. (2021). Territorial development and governance in the Western Balkans. In E. Berisha, G. Cotella, & A. Solly (Eds.), *Governing territorial development in the Western Balkans. Advances in spatial science*. Springer, Cham.

Bieri, M. (2015). *The western Balkans Between Europe and Russia* (CSS Analyses in Security Policy, N0. 170) Centre for Security Studies (CSS), ETH Zurich.

Brunetta G., Caldarice, O., & Mohabat Doost, D. (2025). Between light and shadow: Exploring spatial and multidimensional justice in the EU's green transition. In E. Berisha, J. Moodie, L. Allkja, & M. Jeftić (Eds.), *The Palgrave handbook of just green transitions in the Western Balkans and beyond*. Palgrave Macmillan.

Buza, D., & Allkja, L. (2025). Landscape design as a tool to foster spatial transformations in light of just green transition in the Western Balkan. In E. Berisha, J. Moodie, L. Allkja, & M. Jeftić (Eds.), *The Palgrave handbook of just green transitions in the Western Balkans and beyond*. Palgrave Macmillan.

Castano Garcia, A. (2024). Transitions for zero carbon futures: From just to generous. *Futures, 162*, Article 103430. https://doi.org/10.1016/j.futures.2024.103430

Cedergren, E., Tapia, C., Sánchez Gassen, N., & Lundgren, A. (2022). *Just green transition–Key concepts and implications in the Nordic Region*. Nordregio.

European Commission. (2019). Communication from the commission—The European Green deal, COM, 2019, 640 Final.

European Commission. (2020a), The Just Transition Mechanism: Making Sure No One Is Left Behind, Just Transition Mechanism Fact Sheet, January 2020.

European Commission. (2020b). Guidelines for the Implementation of the Green Agenda for the Western Balkans. SWD (2020) 223 final.

European Commission. (2022). 2022 Communication on EU Enlargement Policy. COM (2022) 528 final. 13 October 2022.

European Commission. (2023). New Growth Plan for the Western Balkans. COM (2023) 691 final.

European Commission. (2024). On Pre-Enlargement Reforms and Policy Reviews. Brussels, COM (2024) 146 final.

Finka, M. (2025). Synergy effects of just green transition—From overall challenges to innovative responds in the case of a Slovakia 2030. In E. Berisha, J. Moodie, L. Allkja, & M. Jeftić (Eds.), *The Palgrave handbook of just green transitions in the Western Balkans and beyond*. Palgrave Macmillan.

Garvanlieva Andonova V., Gotovska, E., & Nikolov, M. (2025). Just green transitions in North Macedonia: A place-based case study from Southwest planning region. In E. Berisha, J. Moodie, L. Allkja, &M. Jeftić (Eds.), *The Palgrave handbook of just green transitions in the Western Balkans and beyond*. Palgrave Macmillan.

Heffron, R. J., & Heffron, R. J. (2021). *What is the "just transition"?. Achieving a just transition to a low-carbon economy* (pp. 9–19).

Ignjatović, J., Filipović, S., & Radovanović, M. (2024). Challenges of the green transition for the recovery of the Western Balkans. *Energy, Sustainability and Society, 14*, 2.

Imami, F, & Aliaj, B. (2025). The risks of purposive transitions in Albania—What to ex-pect by the Just Green Transitions. In E. Berisha, J. Moodie, L. Allkja, & M. Jeftić (Eds.), *The Palgrave handbook of just green transitions in the Western Balkans and beyond*. Palgrave Macmillan.

Imami, F., Bejko Gjika, A., Giacometti, A., & Tapia, C. (2023). *Just green transition in the Western Balkans: Overcoming transition fatigue*. https://greenforcetwinning.net/wp-content/uploads/2024/02/POLICY-BRIEF-OVERCOMING-TRANSITION-FATIGUE.pdf

International Labour Organisation. (2015). *Guidelines for a just transition towards environmentally sustainable economies and societies for all* (p. 2015). ILO Publications.

Jaššo, M., & Husár, M. (2025). Participatory planning culture and its role and potential for fostering just green transition, In E. Berisha, J. Moodie, L. Allkja, & M. Jeftić (Eds.), *The Palgrave handbook of just green transitions in the Western Balkans and beyond*. Palgrave Macmillan.

Jeftić, M., Vračarević, B., Šećerov, V., Đorđević, A., Protić, B., Živanović, Z., Nikolić, T., & Popović, V. (2025, forthcoming): Public transport satisfaction and the path to a green transition in city of Kragujevac (Serbia), Europa XXI.

Knez, S., Štrbac, S., & Podbregar, I. (2022). Climate change in the Western Balkans and EU green deal: Status, mitigation and challenges. *Energy, Sustainability, and Society, 12*, 1.

Kortetmäki, T., Timmermann, C., & Tribaldos, T. (2025). Just transition boundaries: Clarifying the meaning of just transition. *Environmental Innovation and Societal Transitions, 55*, Article 100957. https://doi.org/10.1016/j.eist.2024.100957

Maričić T,and Pantić, M. (2025). Just green transitions of coal-mining cities. In E. Berisha, J. Moodie, L. Allkja, & M. Jeftić (Eds.), *The Palgrave handbook of just green transitions in the Western Balkans and beyond*. Palgrave Macmillan.

Mitrică, B., Roznovieţchi, I., Grigorescu, I., Micu, D., & Damian, N. (2025). The vulnerability of the Romanian Danube Valley to heat-related phenomena. A socio-economic index-based approach. In E. Berisha, J. Moodie, L. Allkja, & M. Jeftić (Eds), *The Palgrave handbook of just green transitions in the Western Balkans and beyond*. Palgrave Macmillan.

Moodie J., Berisha, E., Allkja, L., & Jeftic, M. (2025a). The multidimensionality of just green transitions. In E. Berisha, J. Moodie, L. Allkja, & M. Jeftić (Eds.), *The Palgrave handbook of just green transitions in the Western Balkans and beyond*. Palgrave Macmillan.

Moodie J., Berisha, E., Allkja, L., & Jeftic, M. (2025b). The just green transitions policy landscape. In E. Berisha, J. Moodie, L. Allkja, & M. Jeftić (Eds.), *The Palgrave handbook of just green transitions in the Western Balkans and beyond*. Palgrave Macmillan.

Moodie, J., Tapia, C., Löfving, L., Gassen, N. S., & Cedergren, E. (2021). Towards a territorially just climate transition—Assessing the swedish EU territorial just transition plan development process. *Sustainability, 13*(13), 7505.

Ndreka, E., & Imami, F. (2025). Implementation of the energy efficiency plan and waste management in Kosovo. In E. Berisha, J. Moodie, L. Allkja, & M. Jeftić (Eds.), *The Palgrave handbook of just green transitions in the Western Balkans and beyond*. Palgrave Macmillan.

Nikolov, A. (2025). Cross-border dimension of Just Green Transition in the Western Balkans—Case study of the Republic of Serbia. In E. Berisha, J. Moodie, L. Allkja, &

M. Jeftić (Eds.), *The Palgrave handbook of just green transitions in the Western Balkans and beyond*. Palgrave Macmillan.

Novikau, A., & Krupalija, M. (2025). Green energy transition in the Western Balkans: Challenges and opportunities for Bosnia and Herzegovina. In E. Berisha, J. Moodie, L. Allkja, & M. Jeftić (Eds.), *The Palgrave handbook of just green transitions in the Western Balkans and beyond*. Palgrave Macmillan.

Oruc, N., Imami, F., Bejko, A., Kurta A., & Kurtovic, I. (2025). Local labour market and fiscal impacts of decarbonisation in Bosnia-Herzegovina: The case of Zenica and Banovici. In E. Berisha, J. Moodie, L. Allkja, & M. Jeftić (Eds.), *The Palgrave handbook of Just green transitions in the Western Balkans and beyond*. Palgrave Macmillan.

Regional Cooperation Council. (2020), Sofia declaration on the green agenda for the Western Balkans, Bosnia and Herzegovina.

Regional Cooperation Council. (2021). Action plan for the implementation of the Sofia declaration on the green agenda for the western Balkans 2021–2030, Bosnia and Herzegovina 2021.

Rodríguez-Pose, A., & Bartalucci, F. (2024). The green transition and its potential territorial discontents. *Cambridge Journal of Regions, Economy and Society, 17*(2), 339–358.

Shaker, Y. (2025). The unknown knows of the just green transitions: A meta-governance approach for a possible enlargement of the European Union. In E. Berisha, J. Moodie, L. Allkja, & M. Jeftić (Eds.), *The Palgrave handbook of just green transitions in the Western Balkans and beyond*. Palgrave Macmillan.

Shaker, Y., & Berisha, E. (2025). Just Green Transitions: Between terminological inexactitudes, conceptual fragmentation, and the exigency for a theoretical framework. European Spatial Research and Policy, 32(1), 121–138. https://doi.org/10.18778/1231-1952.32.1.05

Shaker, Y., & Berisha, E. (2024). Transizioni verdi giuste: Tra imprecisioni semantiche e necessità di nuovi modelli di governance. Società Italiana degli Urbanisti: PLANUM, 193–203. https://hdl.handle.net/11583/2989395

Shaker, Y., & Persico, S. (2024). Mapping the just green transitions in the socio-political virtual space. *European Journal of Spatial Development, 21*(1), 61–82. https://doi.org/10.5281/zenodo.13817780

Shaneh, A., Moodie, J., Tapia, C., Giacometti, A., & Cedergen, E. (2025). Understanding territorial governance dynamics in the Swedish EU territorial just transition planning processes—Lessons for the Western Balkans. In E. Berisha, J. Moodie, L. Allkja, & M. Jeftić (Eds.), *The Palgrave handbook of just green transitions in the Western Balkans and beyond*. Palgrave Macmillan.

Tapia, C., Giacometti, G., Moodie, J. R., Cedergen, E., & Shaneh, A. (2025). From mitigation to transformation: An evolutionary perspective on the just green transition. In E. Berisha, J. Moodie, L. Allkja, & M. Jeftić (Eds.), *The Palgrave handbook of just green transitions in the Western Balkans and beyond*. Palgrave Macmillan.

Tavares, M. (2022). *A just green transition: Concepts and practice so far* (Future of the World, UN-DESA Policy Brief, No. 141).

Temali Kraja, I. (2025). Climate change adaptation green technologies for coastal risk prevention—Case studies summary from Coastal countries. In E. Berisha, J. Moodie, L. Allkja, & M. Jeftić (Eds.), *The Palgrave handbook of just green transitions in the Western Balkans and beyond*. Palgrave Macmillan.

Toptsidou, M., & Böhme, K. (2025). Green transition in the Adriatic-Ionian region: A scenario. In E. Berisha, J. Moodie, L. Allkja, & M. Jeftić (Eds.), *The Palgrave Handbook of just green transitions in the Western Balkans and beyond*. Palgrave Macmillan.

Toska, M., Gjoka, R., Gjiknuri, O., & Vuković, A. (2025). The Montenegrin waste management sector transformation and its impacts towards just green transitions. In E. Berisha, J. Moodie, L. Allkja, & M. Jeftić (Eds.), *The Palgrave handbook of just green transitions in the Western Balkans and beyond*. Palgrave Macmillan.

Toto, R., Imami, F., Poro, E., et al. (2023). Just Green Transition Conceptualisation in the Western Balkans. *GreenFORCE Horizon Project, Deliverable, 4*, 1.

Tsenkova, S., & Nedović-Budić, Z. (2006). *The Urban mosaic of post-socialist Europe*. Physica-Verlag. https://doi.org/10.1007/3-7908-1727-9

Tziarras, Z. (Ed.). (2019). *The new geopolitics of the Eastern mediterranean: Trilateral partnerships and regional security*. PRIO Cyprus Centre.

United Nations (2015). Transforming our World: The 2030 agenda for sustainable development, A/RES/70/1, 21st October 2015.

United Nations. (2023). A globally just transition perspectives from the committee for development policy, UN committee for development policy: Policy note, December 2023.

Valkenburg, G. (2025), Epistemic work and epistemic justice in green transitions. In E. Berisha, J. Moodie, L. Allkja, & M. Jeftić (Eds.), *The Palgrave handbook of Just green transitions in the Western Balkans and beyond*. Palgrave Macmillan.

Vračarević, B. (2021). Urban transport in world metropolises: A comparative analysis and key features of energy consumption. *Spatium., 45*, 46–58.

Wang, X., & Lo, K. (2021). Just transition: A conceptual review. *Energy Research and Social Science, 82*, Article 102291.

Živanović, Z., Popović, V., Protić B., Djordjević, A., Šećerov, V., & Jeftić, M. (2025). Signs by the Roadside on the road to just Green Transi-tions in Serbiass-Border dimension of just green transitions in the Western Balkans. In E. Berisha, J. Moodie, L. Allkja, & M. Jeftić (Eds.), *The Palgrave handbook of just green transitions in the Western Balkans and beyond*. Palgrave Macmillan.

CHAPTER 2

The Multidimensionality of Just Green Transitions

John Moodie, Erblin Berisha, Marija Jeftić, and Ledio Allkja

2.1 Introduction

The concept of Just Green Transitions (JGTs) encapsulates the interplay between achieving climate neutrality and fostering equitable socio-economic outcomes (Cedegren et al., 2022). Rooted in the labour movements of the 1970s, JGTs have now evolved into a comprehensive policy framework

J. Moodie (✉)
Nordregio Research Institute for Regional Development and Planning, Stockholm, Sweden
e-mail: john.moodie@nordregio.org

E. Berisha
Interuniversity Department of Regional and Urban Studies and Planning (DIST), Politecnico Di Torino, Turin, Italy
e-mail: erblin.berisha@polito.it

M. Jeftić
Faculty of Geography, University of Belgrade, Belgrade, Serbia
e-mail: marija.jeftic@gef.bg.ac.rs

L. Allkja
Co-Plan Institute for Habitat Development, Tirana, Albania

Polis University, Tirana, Albania

L. Allkja
e-mail: ledio_allkja@universitetipolis.edu.al

E. Berisha et al. (eds.), *The Palgrave Handbook of Just Green Transitions in the Western Balkans and Beyond*,
https://doi.org/10.1007/978-3-031-95075-9_2

addressing the complexities of transitioning to sustainable economies (European Commission, 2019). Tapia et al. (2025, in this volume) note that academic research on the concept of JGTs has failed to keep pace with policy developments. While there is no agreed consensus on what constitutes JGTs in academic and policy circles, and no universal blueprint for developing and implementing JGTs, the policy and academic discourse do reveal several key dimensions of JGTs that strategies, policies, and measures must try to address, highlighting the complex, multidimensional nature of shifting to carbon neutral economies and societies (Shaker 2025, in this volume). Indeed, JGTs embody key overlapping and interlinked dimensions, including environmental protection, economic restructuring, social equity, spatial planning, technological development, and governance reforms (Moodie et al., 2021).

The aim of this chapter is to provide an overview of the core dimensions of JGTs. Where possible, the discussion relates these dimensions to current JGT policy developments and the WB context. The chapter examines the social dimension of JGTs, emphasising justice frameworks such as distributional, recognitional, and procedural justice, and how these principles are designed to mitigate against socio-economic disparities. The economic imperatives of JGTs are presented, analysing the costs and opportunities inherent in transitioning to low-carbon economies. The spatial and environmental dimensions are also addressed, highlighting the territorial disparities that arise from transitions. The core elements of effective JGT governance structures are evaluated, outlining the importance of multi-level, multi-actor governance structures that promote collaboration between key stakeholders. The final section of the chapter synthesizes insights from the preceding discussions, emphasising cross-cutting temporal, technical, communicative, and resilience dimensions of JGTs. The multidimensional JGT framework underscores the need for adaptive, context-sensitive JGT approaches that prioritise long-term sustainability, while remaining flexible to evolving challenges.

2.2 Just Dimension

2.2.1 Social Component

The 'Just' element of JGTs focuses on ensuring that transition policies promote fairness and social equality (Tavares, 2022). This is crucial as new regional economic growth pathways stimulated by JGTs have the potential for generating different forms of societal inequality and exploitation (Eadson & van Veelen, 2023). Growing unemployment, limited access to skills and training, and the uneven distribution of available funding can, in some cases, lead to the exclusion and displacement of certain societal groups (Mackinnon et al., 2019). Sovacool et al. (2021), studying the justice and equity implications of low-carbon transitions in France, Germany, Great Britain, and Norway, point out that political responses to climate change may spur increasing economic inequalities, unemployment, human rights abuses and other types of externalised impacts on different societal groups. In this context, some individuals and societal groups are more vulnerable than others (Carley et al., 2018). Consequently, the notion of 'social justice' has become central to the debate surrounding JGTs, so no one region or person gets left behind (Wang & Lo, 2021).

The term'just' in just transitions relates to the concept of social justice developed by philosophers and thinkers, including Rawls, Locke, Rousseau, and Kant (Moodie et al., 2021). Social justice is a normative concept which implies that society and the state should strive towards achieving social equality for all citizens (Heffron & Heffron, 2021). There is no accepted definition of social justice in relation to JGT policies (Shaker & Berisha, 2024, 2025) but several different tenets of the broader social justice concept have been applied in academic discussions on JGTs (Cedergren et al., 2022). Distributional justice is viewed as a guiding principle of JGTs which advocates that resources and the benefits and burdens of policies are allocated fairly across society and to the advantage of the most vulnerable groups (Ibid). In their study of energy justice, McCauley and Heffron (2018) examine recognitional justice and procedural justice; the former explores which groups or sections of society are recognised and neglected in energy policymaking, and the latter emphasises the need for both informal and formal societal engagement in energy policymaking. They also examine cosmopolitan justice, which argues that principles of social justice should apply to all human beings regardless of the country and region they are located (Ibid). Valkenburg (2025, in this volume) highlights that epistemic justice is an often-overlooked element in JGTs, outlining the importance of including different types of knowledge in JGT policymaking processes to help promote both testimonial justice, i.e. justice with respect to different voices being heard, and hermeneutic justice, i.e. justice in respect to citizens being enabled to develop knowledge.

The 'just' elements are clearly reflected in the 2015 just transition guidelines developed by the International Labour Organization (ILO: International Labour Organisation, 2015). These outline the need for social dialogue and consultations between policymakers and affected groups; proactive labour market policies based on social protection and worker rights, including financial support and retraining opportunities for the unemployed; early-stage research into the potential socio-economic impacts of climate policies; and the development of local economic diversification plans (Ibid). The guidelines also make a distinction between social investment policies (e.g. active labour market policies, training and reskilling policies to increase workers' employability in a greener economy) and social protection policies (e.g. unemployment and minimum income benefits), which means both are needed in a just transition framework (Ibid). The 'just' element is also prominent in the EUs Green Deal and Just Transition Mechanism and Funds. The social elements are highlighted in the EU's commitment to worker protection rights and reskilling opportunities. There is also a strong emphasis on the social justice element of the transition in relation to open and inclusive social dialogue with citizens to promote the trust and acceptance of policies (European Commission, 2019).

While social justice is recognised as an essential element of transitions, the different dimensions of social justice are rarely explored in any great depth in JGT policymaking processes. Cedergren et al., (2022, p. 5) note that 'which criteria to apply when assessing and addressing justice concerns—in other words, what is "just" and for whom—depends on the philosophical and political positions taken and has implications for how policies are formulated and adopted'. In response to this challenge, Bennett and colleagues (2019) developed a criteria to consider when looking at justice in relation to concepts such as transformations and just transitions, elaborating on the different philosophical positions on equity, which in turn influence the construction of 'justice'. These different positions on preferred notions of equity include: (a) utility (actions that produce increases in aggregate goods or improve overall human well-being); (b) equality (all parties have equal opportunities); (c) proportionality (potential losses and gains should be proportional); (d) needs (preferential treatment should be given to the most vulnerable); (e) merit (different parties ought to receive what they deserve based on their effort or input, e.g. reaping the benefits of their actions or change processes, also known as 'just desserts'); and finally (f) rights (the treatment of different groups should ensure that a minimum threshold of basic human rights are respected) (Bennett et al., 2019). There is heterogeneous understanding of notions of equity and social justice across political and cultural spheres, so they need to be carefully considered in relation to different spatial contexts (Teitje & Cresap, 2018). Nordberg (2020, p. 48) argues 'geography exercises influence over social circumstances, and such circumstances are ignored if the spatial aspect goes unnoticed regarding the concept of justice'. Indeed, the social challenges presented by JGTs cannot be separated from the spatial dimension of JGTs, as the socio-economic impacts of transitions will be influenced and exacerbated by regional contexts and specificities, particularly existing territorial assets and weaknesses (Rodríguez-Pose & Bartalucci, 2024).

2.2.2 Democratic Component

Citizen acceptance and support for JGTs is largely dependent on inclusive, open, and transparent policymaking processes that proactively engage citizens across the policy cycle, through policy development, implementation, and evaluation (Balz et al., 2023). So far, there has been little meaningful citizen engagement in JGT processes, with policymaking dominated by EU, national, and sectoral actors (Georgieva & Ferry, 2023). Enhanced citizen involvement is vital for meeting the procedural, relational, and epistemic dimensions of social justice in JGT processes. Policymakers can implement more inclusive and transparent JGT processes using citizen engagement tools and methods that proactively involve citizens in the different phases of the JGT policymaking cycle. Citizen engagement needs to move beyond one-way information sharing towards interactive participatory methods which provide citizens with an opportunity to voice their views and influence the direction of JGT policies that meet local needs (Georgieva & Ferry, 2024). Enhanced participation is particularly important for those societal groups and citizens that face the worst socio-economic impacts of JGTs.

From a societal perspective, trade unions, civil society groups, NGOs, and community groups have an important role to play in representing citizen voices and perspectives in JGT policy processes (Eadson & Van Veelen, 2023). NGOs and societal groups have a particularly key role in protecting the most vulnerable groups in society from the negative socio-economic impacts of transitions (Buijs et al., 2024). The NGO Forum on the Green Agenda for Western Balkans (NGO Forum) has been created to provide a platform for NGOs across the WB to share knowledge and experience on JGT (Regional Cooperation Council, 2021). The NGO Forum primarily includes NGOs with knowledge and experience of working with JGT issues and will play a role in drafting JGT policies and recommendations, monitoring and evaluating the implementation of JGTs, raising citizen awareness, and engaging vulnerable groups in the JGT processes, such as Roma and other minorities (Ibid). Youth groups, including the Regional Youth Cooperation Office and the RCCs Youth Lab, have also been identified as important for young people to express their views and concerns regarding JGT (Ibid). There has been limited citizen engagement with JGT across the WB and policies and citizen consultation processes need to be framed in a language which raises citizen awareness, proactive engagement, and support (Frey, 2024; Ignjatović, 2024).

2.3 Economic Dimension

JGTs present both economic costs and opportunities (Kemfert et al, 2020). It is extremely expensive for large-scale industries and businesses to convert to carbon neutral internal processes and practices. Economic assistance is essential for regions and sectors to undergo JGTs, with high levels of investment needed in low-carbon technologies (Green & Gambhir, 2020). This is why national governments in EU member states, such as Sweden, devoted most of their EU Just Transition Funds to covering the costs of the development and implementation of carbon neutral technologies in the country's highest polluting industrial sectors (Moodie et al, 2021). The Swedish EU Just Transition Plans are driven by national level and sectorial perspective that EU funding should be prioritised to cover the economic costs of technological change, rather than alleviate the socio-economic impacts of the transition (Ibid). Consequently, trade unions and civil society groups have voiced concern that the Swedish just transition is too focused on the economic and technical dimensions of climate policies, while neglecting the potential regional social impacts of transition (Ibid).

While JGTs are costly, there is an economic imperative at the heart of the EU Green Deal that views decarbonisation as an opportunity for economic restructuring and growth across European regions (European Commission, 2019). Easdon and Van Veelen (2023) point out that new JGT economic pathway development policies need to move beyond traditional high-growth and production-oriented paradigms with a single sector focus. They argue that a multi-sector economic pathway development strategy, based on a degrowth perspective and economic models that support human and ecological well-being, might produce more favourable social and environmental impacts (Ibid). They continue that such an economic development approach might also create stronger synergies between industries and businesses which is essential for developing circular economy models central to delivering effective JGTs (Ibid). In the context of JGTs, regional economic development and business diversification processes are policy induced (Toto et al., 2023). Policy tools are, therefore, required that facilitate long-term regional resilience by promoting the development of new business innovation ecosystems and regional economic pathway creation (Grillitsch & Sotarauta, 2020). JGTs should be built around regional development strategies and EU smart specialisation strategies that prioritise regional and local economic diversification possibilities (Woien Meijer & Giacometti, 2024). In this regard, finding stronger connections between JGTs and EU Smart specialisation Strategies is essential to identify targeted links between transitions and economic and business diversification processes (Ibid).

The non-EU countries of the Western Balkan region are signatory to most international conventions, strategies, programmes, directives, and agendas that bind the field of JGTs, sustainable development, biodiversity, and conservation. However, investments in green transition and achieving EU standards in this respect for some of them are not priorities (Serbia for example), leading to ecological problems and demonstrating how unsustainable exploitation of natural resources and unbalanced tourism development that threatens biodiversity can jeopardise long-term economic potential (Protić et al., 2024). This is particularly evident in ski resorts and infrastructure, which contribute to the degradation of landscapes and biodiversity, thereby diminishing the region's long-term tourist appeal (Ibid). Vukoičić et al. are stating that sustainability of tourism development in protected areas relies largely on the ability of destination management to harmonise the activities of visitors, local communities, entrepreneurs, and other tourism actors with the primary aim of nature and landscape protection (Vukoičić, 2022).

2.4 Green Dimension

The 'green' element of JGTs emphasises the need for sustainable and resilient development that can help societies to mitigate and adapt to climate change, protect biodiversity, and stop environmental degradation (Cedergren et al, 2022). In the JGT policy sphere, the green dimension is most visible in the long-term climate and environmental targets set by international organisations and national governments, such as the EU's goal to be net zero by 2050 (European Commission, 2019). At the heart of the 'green transition' is the goal of developing a 'green economy' and 'green growth', which can help promote economic growth and social equality while relieving pressure on the environment and reducing ecological damage (Bogovic & Grdic, 2020). The World Bank (2014) defines green growth as the development of an economic system that is 'efficient in its use of natural resources, clean in that it minimises pollution and environmental impacts and resilient in that it accounts for natural hazards'. Green growth promotes carbon neutrality and the shift from linear to circular business models based on the efficient use of natural resources and sustainable patterns of production and consumption (Cedergren et al., 2022). The notion of green growth is central to the EU Green Deal, which seeks to strike a balance between economic growth and reaching climate neutrality by reducing waste and pollution, moving to circular and resource-efficient economies, and preventing the loss of biodiversity (European Commission, 2019).

Green growth is not only a theoretical concept but a practical model of development, which requires the integration of spatial planning, environmental policies, and sustainable economic activities. Serbia, for example, through the Spatial Plan and strategies for the preservation of natural reserves (Gatarić et al., 2022), is already laying the foundation for this kind of development, but further implementation and systemic support for green growth policies is necessary.

2.5 Spatial Dimension

The impacts of climate transformations will not be neutral from a territorial perspective (Rodríguez-Pose & Bartalucci, 2024). The costs and benefits of JGTs may not be shared equally, which could contribute to growing territorial disparities and inequalities between regions (Toto et al., 2023). Furthermore, urban and rural areas possess different territorial assets and weaknesses that can either strengthen or impede their capacities and ability to adapt to the challenges and opportunities presented by JGTs (Coenen et al., 2012). On a spatial level, the closure of a major industry in one region presents opportunities for other regions to strengthen and diversify their own economic base; however, it can also lead to negative spatial displacement effects with socio-economic and environmental burdens being transferred to other areas (Chlebna et al., 2022). JGT plans in one region, therefore, need to carefully consider the potential impacts and knock-on effects on neighbouring regions and municipalities, particularly those areas with an economy, manufacturing, or businesses that are closely connected to and rely on the materials provided by large-scale industries that face closure or restructuring (Moodie et al., 2021).

The sociology of territory and place is an important element of the spatial dimension of JGTs (Binz et al., 2020). JGTs have the potential to negatively impact on citizen's place-based socio-cultural identities, particularly in regions where communities have built up around large-scale fossil fuel industries (Eadson & van Veelen, 2023). Garvey et al., (2022, p. 2) note that local territorial identities will affect how citizens anticipate and experience industrial change, so there needs to be an: 'acknowledgement of divergent identities, cultural histories, and power dynamics, and how these may interact with proposed changes...Space is particularly important to recognition justice in terms of recognising specific place identities and how these may shape the acceptability of transition measures'. Space, place, and territory play a central role in framing socio-cultural identities, which will impact citizens' support and acceptance of JGT policies. Understanding these spatial identities and how they might affect citizens' responses to JGTs is essential for policymakers to frame and communicate their strategies effectively to local communities (Binz et al., 2020). It is, therefore, important that policymaking concepts that embrace territory and local knowledge guide JGT policymaking processes, including place-based policymaking, territorial governance, and active subsidiarity (Moodie et al., 2021).

In the Western Balkan countries, the socio-economic challenges of population decline, migration, unemployment, and urban shrinkage have become pressing issues, particularly in small and medium-sized towns. These trends reflect broader regional dynamics, shaped by post-socialist transition, economic restructuring, and uneven territorial development. Urban shrinkage is often accompanied by environmental degradation, yet paradoxically, the decline in industrial activity may temporarily reduce certain pollution sources. However, such outcomes do not align with long-term goals of green transition supporting sustainable economic development, population concentration, and agglomeration of economic activities, parallel with environmental care (Živanović et al., 2025). Addressing urban shrinkage and advancing the green transition require a strong territorial perspective (Berisha, 2025a, 2025b). The spatial disparities between shrinking small towns and growing metropolitan centres, such as Belgrade, Skopje, Tirana, etc., illustrate the need for integrated regional planning. Territorial cohesion policies should aim to balance development by fostering urban-rural synergies, decentralising governance, and promoting sustainable economic activities in peripheral areas (Jeftić et al., 2021). These measures would ensure that environmental care, economic growth, and social equity are achieved across all regions, preventing further polarisation and fostering sustainable territorial development.

2.6 Governance Dimension

The development and implementation of JGTs requires a holistic, multi-actor policymaking approach built on collaborative multi-level governance and planning processes (Avelino & Wittmayer, 2016). Public and private actors across different governance levels (EU, national, regional, and local) have a central role in determining the direction of JGT pathways and targets (Smith et al., 2005). As Geels (2002) explains, transitions occur as the outcome of linkages between developments at multiple levels. Consequently, proactive multi-level institutional and stakeholder cooperation, based on open and inclusive dialogue, is imperative in responding to the socio-economic challenges and opportunities presented by JGTs (Nikas et al., 2020; Sandmann et al., 2024). The importance of collaborative multi-level governance and planning in JGT relates to four main elements: (1) ensuring that JGT policies are well integrated and coherent across multiple levels of governance (transnational, national, regional and local levels); (2) empowering sub-national level institutions and actors in policymaking processes to ensure that JGT policies are carefully tailored to meet regional specificities and the needs of citizens in local communities; (3) developing platforms and tools for building connections, creating synergies, and sharing knowledge and ideas between key policymakers, sectors, businesses, researchers, NGOs, and societal groups; (4) raising public awareness and support for policies by harnessing the potential of digital tools for facilitating active citizen engagement in JGT policymaking processes (Cedegren et al., 2022; Moodie et al., 2021).

At the transnational level, the major challenge is to ensure that climate and JGT policies are in alignment with international climate goals, particularly the objectives of the EUGD (Ahlström & Sjåfjell, 2023; Shawoo et al., 2023). This is particularly important in a WB context where alignment with EUGD policies has become a central element of the EU accession process. An open and continuous multi-level dialogue with relevant European Commission services is important to ensure that EU climate and accession priorities are effectively translated into workable policies in WB countries with practical implementation and monitoring tools (Berisha et al., 2024). The GAWB reflects a first step in attempts to build transnational and cross-border JGT strategies, programmes, plans, collaborations, and partnerships (Regional Cooperation Council, 2021). Here, the Regional Cooperation Council (RCC) can play a role in building macro-regional partnerships across the WB. Existing intergovernmental institutions, such as the Regional Working Group on Environment and South-East Europe Biodiversity Task Force, can provide important platforms for creating a permanent dialogue between WB national governments on climate and JGT issues (Ibid). Western Balkans countries are highly fragmented, so existing cross-border institutions and structures provide important forums for overcoming any political, geographical, natural, economic, and socio-cultural obstacles to collaboration (Berisha & Cotella, 2021).

At the national level, recent research on the development of Swedish EU Territorial Just Transition Plans highlights that the role of sub-national level actors was constrained in a process dominated by the national governments and sectors (Moodie et al., 2021). Snell (2018) and Eadson and Van Veelen (2023) correctly observe that national governments and sectors inevitably have a large role to play as key drivers in JGT processes, given the size and scope of transitions and the primary focus on the biggest polluting industries. However, more can be done to promote place-based and bottom-up governance approaches to JGTs to ensure that policies meet the needs of citizens at regional and local levels (Shaneh et al., 2025, in this volume). Rainnie (2021) argues that there is an 'inclusionary bias' at the sub-national level, which means that participation in policy processes is restricted to stakeholders within existing regional and local networks. Opening up JGT policymaking beyond the 'usual suspects' at the national and regional levels is important to ensure that all different voices, perspectives, and knowledge are represented and heard (Valkenburg, 2025, in this volume).

Centralised governance systems across the WBs make enhancing the role of sub-national level actors challenging (Berisha et al., 2021). The WB has over 520 local governmental units (municipalities) that are important in addressing environmental and social issues. The Network of Associations of Local Authorities of South-East Europe (NALAS) provides an existing cooperation framework where municipalities can share JGT knowledge and best practices (Regional Cooperation Council, 2021). A strategic approach is required to outline how regions and municipalities can be better integrated into multi-level governance JGT policymaking processes to enable vertical coordination and ensure that EU and national-level JGT priorities recognise local needs and challenges (Radovanovic & Carapina, 2024). Furthermore, regional and municipal authorities need to work closely with key sectors to ensure that sectoral changes address wider local socio-economic issues. Sub-national authorities, therefore, need substantial support to strengthen their administrative capacity to facilitate participative processes and environmental assessments in strategic JGT planning, programmes, and projects (Ignjatović et al., 2024).

JGTs cannot be developed and implemented without effective cross-sector collaboration at regional and local levels (Guerrero & Hansen, 2021; Wälitalo et al., 2020). Transforming the internal processes of resource-intensive sectors in the WBs is crucial for meeting climate targets, including the construction, textiles, plastics, and electronics industries (Bechev, 2023). However, transformations in one industry can have significant knock-on effects on other sectors and businesses (Moodie et al., 2021). To mitigate against the knock-on effects of transitions, sectoral organisations, and groups (e.g. the Energy Community Secretariat, Transport Community Permanent Secretariat, the Standing Working Group on Regional Rural Development, and International Union for Conservation of Nature) can play an important role in helping sectors develop roadmaps for industrial transition processes and building networks between sectoral actors (Regional Cooperation Council 2021). Furthermore, the Balkans Chamber Investment Forum can help facilitate links between industries and businesses that can work together to promote industrial symbiosis and develop circular economy ecosystems (Angelis-Dimakis et al., 2022; Ignjatović et al., 2024).

2.7 Technical Dimension

The technical component of JGTs relates to the complex and costly technological systems and process changes required to facilitate the shift towards carbon neutrality (Bergek et al., 2008; Hekkert et al., 2007). Geels (2002) argues that transitions are an outcome of recurrent innovation and technology substitution processes, particularly the internal processes of high-polluting sectors and businesses. The high importance of the technical component of JGTs is reflected in the EU's commitment to investing large sums of money in carbon neutral and environmentally friendly production technologies, transport, renewable energies, and digitalisation (European Commission, 2019). An emphasis on restructuring the internal technological processes of the largest greenhouse gas emitting industries has given large sectors (e.g. mining, steel, cement, and mineral industries) a prominent position in JGT policymaking. Narrowing the focus of JGT plans on specific sectors is understandable, as transforming the internal processes of large-scale industries is vital for reducing overall emissions; however, one consequence of a limited sectoral approach is that the technical component becomes the predominant and overriding focus of JGT policies (Moodie et al., 2021). This narrow focus can lead to a situation where other key JGT dimensions are neglected, particularly addressing the potential socio-economic and spatial impacts of JGTs across regions and on different societal groups. Taking a broader territorial approach to JGT planning can lay firmer foundations for producing more balanced plans that carefully consider all JGT dimensions, which do not only prioritise sectors and the technical elements of shifting to carbon neutral societies and economies (Shaneh et al., 2025, in this volume).

2.8 Transversal Dimensions

2.8.1 Communicative Component

There is a lack of awareness, information, and knowledge on JGTs at the societal level in the WBs and across the EU (Shaker & Persico, 2024). The shift to climate neutrality presents both socio-economic challenges and opportunities for citizens, which require preparation, adaptability, and resilience at the societal level. Citizen awareness, understanding, and inclusion in JGT processes is essential for building citizen support and acceptance for JGT objectives, actions, and impacts; however, the semantic and terminological ambiguity surrounding JGTs risks creating communicative barriers that could lead to a further breakdown in trust between policymakers and citizens (Sanderson et al., 2024). Policymakers cannot automatically assume that all citizens know and understand what JGTs are and involve, so a communicative breakdown can be prevented by, firstly, ensuring that the concept of JGT is defined simply in a way that is understandable to citizens. Secondly, JGTs need to be communicated to citizens in a manner that the concept is easily recognised and perceived across all levels of society and adaptable to different territorial contexts (Panarello & Gatto, 2023).

Communicating JGTs effectively is particularly important in a WB setting where the term 'transition' has become loaded with negative connotations (Toto et al., 2023). Toto et al. (2023) argue that the WBs has been a loop of post-socialism transition since the early 1990s, with the slow transition towards liberal democracies, free economies, and EU integration, creating disillusionment and 'transition fatigue' amongst citizens across the WBs. They note that 'fatigue from a stagnating process of EU integration, conflicting narratives, dominant power relations, corrupted institutions, and huge territorial disparities have produced an agony of "never-ending transition", which suggests that resistance to change may be notably present for the green transition as well'. (Toto et al., 2023, p. 6) JGTs can be the catalyst for significant changes at regional and local levels, which can be a source of uncertainty and fear that can induce citizen opposition and conflict. To alleviate this fear, JGTs need to be conducted through a social dialogue that allows for debate, contestation, and consensus building (Abram, 2017).

Social media can play a significant role in raising public awareness of JGTs and be used as a tool for connecting citizens and policymakers (European Commission, 2017; Falkenberg et al, 2022). Social media has become a key tool for disseminating political information making an online space where policymakers can communicate information and get immediate feedback which can promote trust and increase transparency (Valdez, 2018). Shaker and Persico (2024) have shown that the term JGT is barely used in policy and social media discourse, with citizens predominantly using terms such as 'climate change', 'global warming', 'climate neutrality', 'decarbonisation', and 'sustainability'. It is vital that policymaker's start to use these terms and a more precise and simple language that citizens commonly understand. This can help promote citizen understanding of JGT policy discourse and lay the foundations for increased citizen involvement in policymaking (Eckert & Kovalevska, 2021; Fink et al., 2021).

2.8.2 Temporal Component

JGT policies are constructed by policymakers to occur within a specific time frame (Levin-Keitel et al., 2018). JGTs are, therefore, inherently temporal possessing both long-term and short-term time specific aims and objectives developed in line with politically established climate and environmental targets and policies (Berkhout et al., 2004). There are longer-term time sensitive targets, such as the EU's target to become carbon neutral by 2050, and shorter-term time sensitive policies, including the EU Just Transition Mechanism, that supports EU regions in the implementation of their EU just transition plans until the end of the current EU Cohesion Policy programme in 2027 (European Commission, 2019). However, Toto et al. (2023) argue that it is counterproductive to set JGT targets around specific time frames, as new equilibriums and changing contexts force modifications to plans to meet societal needs and demands. Consequently, JGTs need not be time specific, but considered as a continuous process that requires amendment and adaptability according to new circumstances as they arise (Ibid).

2.8.3 Resilience Component

JGTs are closely linked to the 'regional resilience' of local communities (McCandless & Onbargi, 2023). Resilience thinking has become widely used in relation to climate and environmental transformations, including an assessment of the impact of JGTs on communities (Neuhuber, 2025) Regional resilience refers to the ability of an entity or system to recover from and position elastically following a disturbance or disruption (Martin, 2012; Martin & Sunley, 2015). More specifically, it refers to a set of regional territorial capitals (e.g. economic, social, environmental, and institutional assets) that enable a region to respond to a policy shock and maintain system stability and durability while adapting to structural changes and moving to new development pathways (Giacometti & Teras, 2019). The shift to carbon neutrality does not constitute a conventional spontaneous policy shock, as JGTs and the need for climate action are well known and an established theme within EU, national, and regional level policymaking; however, the socio-economic impacts of these transformations and the technological changes and innovations required to implement them require high levels of regional resilience and adaptive capacity, with new JGT rules putting pressure on policymakers, industries, businesses, workers, and citizens alike (Rodríguez-Pose & Bartalucci, 2024). Territorial JGT planning interventions, made through EU, national, and local and regional plans, are affected by systemic factors (management and market) and shaped by political institutions (Jeftić et al., 2019.). JGTs plans built on a territorial governance logic that maximise the territorial asset base can serve to enhance regional resilience and adaptive capacity (Bianchi & Richiedei, 2023). This applies to the resilience and adaptive capacity of ecological systems, socio-economic structures, governance processes, and the responsiveness of citizens (Loorbach et al., 2017).

2.9 Conclusion

Despite their widespread adoption in policy and academic discourse, JGTs face definitional ambiguities, often compounded by varied interpretations of 'justice' and 'green' in different contexts. The JGT multidimensional framework presented in this chapter highlights the complexities of transitioning to sustainable, low-carbon economies while ensuring fairness and inclusivity. This chapter underscores that achieving the ambitious goals of JGTs requires addressing the intersecting social, economic, spatial, environmental, and governance dimensions. Each of these facets plays a crucial role in navigating the challenges and opportunities of the green transition, particularly in socio-economically and politically diverse regions like the Western Balkans.

Central to JGTs is the principle of justice, which seeks to mitigate the adverse impacts of transitions on vulnerable groups. Distributional, recognitional, and procedural justice frameworks provide valuable lenses for ensuring equity. However, translating these theoretical principles into actionable policies requires contextual sensitivity, particularly in regions facing socio-economic vulnerabilities. The social dimension of JGTs prioritises fairness and justice to ensure no one is left behind during the transition. By integrating principles of distributional, recognitional, and procedural justice, JGT policies can effectively address socio-economic disparities, minimise vulnerabilities, and empower marginalised groups. However, the lack of depth in exploring these justice dimensions in policymaking processes remains a critical challenge. Policymakers must adopt more inclusive strategies that incorporate diverse knowledge systems, foster equitable resource distribution, and enhance citizen engagement to build trust and legitimacy.

Economic restructuring lies at the heart of JGTs, presenting both opportunities and challenges. While investments in clean technologies and circular economies promise long-term benefits, they also necessitate significant upfront costs. Economically, JGTs represent both a significant investment and an opportunity for transformative growth. The costs of transitioning to low-carbon technologies and processes are substantial, but they are accompanied by the potential for regional economic restructuring, innovation, and the creation of circular economies. Balancing these economic imperatives with social equity remains paramount. The emphasis on localised economic diversification and regional resilience—supported by targeted policies and funding—can help mitigate socio-economic impacts and foster sustainable development.

The spatial and environmental dimensions of JGTs emphasise the need for place-based policies that account for territorial disparities. Urban and rural areas possess unique strengths and weaknesses, necessitating tailored approaches to spatial planning. Recognising the socio-cultural and identity-related implications of transitions is critical for policy acceptance at the local level. Concurrently, the environmental aspect of JGTs, encapsulated in goals like carbon neutrality and biodiversity preservation, demands that transitions align with broader ecological objectives to ensure long-term sustainability.

Governance plays a pivotal role in facilitating JGTs. Collaborative multi-level governance structures—spanning transnational, national, regional, and local levels—are essential for coherent and inclusive policymaking. Effective governance frameworks empower sub-national actors, integrate cross-sector collaborations, and promote citizen participation. In the Western Balkans, fragmented governance systems necessitate robust platforms for dialogue, cross-border cooperation, and alignment with international climate goals. Overcoming fragmented governance systems requires robust platforms for dialogue, fostering synergies between stakeholders. Equally important is citizen engagement, which builds trust and legitimacy. Transparent, inclusive processes, supported by effective communication strategies, ensure that JGT policies resonate with local communities. Finally, the technical and temporal dimensions of JGTs underscore the importance of adaptability and innovation. Transitioning to climate-neutral technologies requires significant investments and sectoral transformations. However, focusing solely on technical solutions risks neglecting the broader socio-economic and spatial implications of transitions. Balancing technical innovation with socio-economic considerations remains a critical challenge. Policymakers must adopt multi-sector strategies that prioritise human and ecological well-being, fostering resilience and inclusivity. Moreover, while time-bound targets, such as the EU's 2050 carbon neutrality goal, provide direction, flexibility is vital to accommodate evolving policy developments, socio-economic, contexts and citizen needs.

The findings of this chapter illustrate that JGTs are not merely technical or environmental endeavours but inherently socio-political processes. Achieving just and sustainable transitions necessitates an integrated approach that harmonises all dimensions, prioritises equity, and responds to local contexts. For regions like the Western Balkans, JGTs offer a pathway to resilience and sustainability, but their success depends on adaptive governance, inclusive participation, and sustained commitment to justice and fairness at every stage of the transition.

References

Abraham, J. (2017). Just Transitions for the Miners: Labor environmentalism in the Ruhr and Appalachian Coalfields. *New Political Science, 39*(2), 218–240.

Ahlström, H., & Sjåfjell, B. (2023). Why policy coherence in the European Union matters for global sustainability. *Environmental Policy and Governance, 33*(3), 272–287.

Angelis-Dimakis, A., Arampatzis, G., Alexopoulos, A., Vyrkou, A., Pantazopoulos, A., & Angelis, V. (2022). Industrial Symbiosis in the Balkan-Mediterranean region: The case of solid waste. *Environments, 10*(1), 1.

Avelino, F., & Wittmayer, J. M. (2016). Shifting power relations in sustainability transitions: A multi-actor perspective. *Journal of Environmental Policy & Planning, 18*(5), 628–649.

Balz, V., Verbeek, T., Qu, L., Baugh, R., Dąbrowski, M., Jungsberg, L., Ferry, M., Georgieva, N., Serdült, U., Trifonova, M., Amin, S., & Clouet, Y. (2023). DUST theoretical and conceptual framework, Deliverable 1.1. Horizon Europe DUST project.

Bechev, D. (2023). The green transition and the western balkans. Carnegie Europe Article.

Bennett, N. J., Blythe, J., Cisneros-Montemayor, A. M., Singh, G. G., & Sumaila, U. R. (2019). Just Transformations to Sustainability. *Sustainability, 11*(14), 3881.

Bergek, A., Jacobsson, S., Carlsson, B., Lindmark, S., & Rickne, A. (2008). Analyzing the functional dynamics of technological innovation systems: A scheme of analysis. *Research Policy, 37*(3), 407–442.

Berisha, E., & Cotella, G. (2021). Territorial development and governance in the Western Balkans. In E. Berisha, G. Cotella, A. Solly (Eds.), *Governing territorial development in the Western Balkans. Advances in spatial science*. Springer, Cham.

Berisha, E., Cotella, G., & Solly, A. (2021). Introduction: The Western Balkans between continuity and change. *Governing territorial development in the Western Balkans: Challenges and prospects of regional cooperation* (pp. 1–19). Springer International Publishing.

Berisha, E., Čolić Marković, N., Cotella, G., & Nedović-Budić, Z. (2024). Potential future enlargement: Spatial planning in the Western Balkans. In *Spatial Planning Systems in Europe* (pp. 216–245). https://doi.org/10.4337/9781839106255.00022

Berisha, E. (2025a). Moving towards an inclusive green Agenda in the Western Balkans. Foundation for European Progressive Studies (FEPS). https://doi.org/10.13140/RG.2.2.12092.76162

Berisha, E. (2025b). Green EU integration—The role of the green agenda in accelerating Just Green Transitions in the Western Balkans. In E. Berisha, J. Moodie, L. Allkja, & M. Jeftic (Eds.), *The Palgrave handbook of just green transitions in the Western Balkans and beyond*. Palgrave Macmillan.

Berkhout, F., Smith, A., & Stirling, A. (2004). Socio-technological Regimes and Transition Contexts. In B. Elzen, F. Geels, & K. Green (Eds.), *System innovation and the transition to sustainability* (pp. 48–75). Edward Elgar Publishing.

Bianchi, S., & Richiedei, A. (2023). Territorial governance for sustainable development: A multi-level governance analysis in the Italian context. *Sustainability, 15*(3), 2526.

Binz, C., Coenen, L., Murphy, J. T., & Truffer, B. (2020). Geographies of transition—From topical concerns to theoretical engagement: A comment on the transitions research agenda. *Environmental Innovation and Societal Transitions, 34*, 1–3.

Bogovic, N., & Grdic, Z. S. (2020). Transitioning to a green economy—Possible effects on the croatian economy. *Sustainability, 12*(22), 9342.

Buijs, A. E., de Koning, S., Mattijssen, T. J., Smeding, I. W., Smits, M. J., & Steins, N. A. (2024). Civil society for sustainable change: Strategies of NGOs and active citizens to contribute to sustainability transitions. *Journal of Environmental Planning and Management, 67*(12), 2863–2884.

Carley, S., Evans, T. P., Graff, M., & Konisky, D. M. (2018). A framework for evaluating geographic disparities in energy transition vulnerability. *Nature Energy, 3*(8), 621–627.

Cedergren, E., Tapia, C., Sánchez Gassen, N., & Lundgren, A. (2022). *Just green transition–Key concepts and implications in the Nordic Region*. Nordregio.

Chlebna, C., Martin, H., & Mattes, J. (2022). Grasping transformative regional development—Exploring intersections between industrial paths and sustainability transitions. *Environment and Planning A: Economy and Space*, 0308518X2211373

Coenen, L., Benneworth, P., & Truffer, B. (2012). Toward a spatial perspective on sustainability transitions. *Research Policy, 41*(6), 968–979.

Eadson, W., & van Veelen, B. (2023). Green and just regional path development. *Regional Studies, Regional Science, 10*(1), 218–233.

Eckert, E., & Kovalevska, O. (2021). Sustainability in the European Union: Analyzing the discourse of the European green deal. *Journal of Risk and Financial Management., 14*(2), 80.

European Commission, Directorate-General for Employment, Social Affairs and Inclusion. (2017). *Quality of public administration: A toolbox for practitioners* (Abridged Version). Publications Office.

European Commission. (2019). Communication from the commission—The European Green deal, COM, 2019, 640 Final.

Falkenberg, M., Galeazzi, A., Torricelli, M. D., Marco, N., Larosa, F., Sas, M., Mekacher, A., Pearce, W., Zollo, F., Quattrociocchi, W., & Baronchelli, A. (2022). Growing polarization around climate change on social media. *Nature Clinical Practice Endocrinology & Metabolism, 12*, 1114–1121.

Fink, S., Ruffing, E., Burst, T., & Chinnow S.K. (2021). Less complex language, more participation: How consultation documents shape participatory patterns. *Interest Groups & Advocacy, 10*(3), 199–220.

Frey, B. (2024). The energy transition in the Western Balkans: The Status Quo, Major Challenges. *Policy.*

Garvey, A., Norman, J. B., Büchs, M., & Barrett, J. (2022). A "spatially just" transition? A critical review of regional equity in decarbonisation pathways. *Energy Research & Social Science, 88*, Article 102630.

Gatarić, D., Đerčan, B., Živković, M. B., Ostojić, M., Manojlović, S., Sibinović, M., Lukić, T., Jeftić, M., Lutovac, M., & Lutovac, M. (2022). Can depopulation stop deforestation? The impact of demographic movement on forest cover changes in the settlements of the South Banat District (Serbia). *Frontiers in Environmental Science, 10*, Article 897201. https://doi.org/10.3389/fenvs.2022.897201

Geels, F. W. (2002). Technological transitions as evolutionary reconfiguration processes: A multi-level perspective and a case-study. *Research Policy, 31*, 1257–1274.

Georgieva, N. & Ferry, M. (2023), Identification and assessment of participatory processes in sustainability transition measures in case study regions, Deliverable 3.1. Horizon Europe DUST Project.

Georgieva, N., & Ferry, M. (2024), Factors impacting the scale and quality of participation of least engaged communities in deliberative governance of transition policies, Deliverable 3.1. Horizon Europe DUST Project.

Giacometti, A., & Teräs, J. (2019). *Regional economic and social resilience: An exploratory in-depth study in the Nordic countries.* Nordregio.

Giacometti, A., Wøien Meijer, M., & Salonen, H. (2024). *Who drives green innovation in the Nordic Region?: A change agency and systems perspective.* Nordregio.

Green, F., & Gambhir, A. (2020). Transitional assistance policies for just, equitable and smooth low-carbon transitions: Who, what and how? *Climate Policy, 20*(8), 902–921.

Grillitsch, M., & Sotarauta, M. (2020). Trinity of change agency, regional development paths and opportunity spaces. *Progress in Human Geography, 44*(4), 704–723.

Guerrero, J. E., & Hansen, E. (2021). Company-level cross-sector collaborations in transition to the bioeconomy: A multi-case study. *Forest Policy and Economics, 123*, Article 102355.

Heffron, R. J., & Heffron, R. J. (2021). What is the "just transition"?. *Achieving a just transition to a low-carbon economy* (pp. 9–19).

Hekkert, M. P., Suurs, R. A. A., Negro, S. O., Kuhlmann, S., & Smits, R. E. H. M. (2007). Functions of innovation systems: A new approach for analysing technological change. *Technological Forecasting and Social Change, 74*(4), 413–432.

Ignjatović, J., Filipović, S., & Radovanović, M. (2024). Challenges of the green transition for the recovery of the Western Balkans. *Energy, Sustainability and Society, 14*(1), 2.

International Labour Organisation. (2015). *Guidelines for a just transition towards environmentally sustainable economies and societies for all* (p. 2015). ILO Publications.

Jeftić, M. R., Šećerov, V., & Tošić, D. (2021). Development of the urban systems in Serbia—Towards new spatial plan of the Republic of Serbia. In *Collection of Papers of the 5th Serbian congress of geographers "Innovative approach and perspectives of the applied geography", Novi Sad* (pp. 194–202). University of Novi Sad—Faculty of Sciences, Department of Geography, Tourism and Hotel Management.

Jeftić, M., Tošić, D., & Nikolić, T. (2019). Resilient functional urban regions: Spatial planning in the light of climate change. In W. Leal Filho et al. (Eds.), *International perspectives on climate change, climate change management* (pp. 113–126). International Publishing Switzerland. https://doi.org/10.1007/978-3-030-03383-5_8

Levin-Keitel, M., Mölders, T., Othengrafen, F., & Ibendorf, J. (2018). Sustainability transitions and the spatial interface: Developing conceptual perspectives. *Sustainability., 10*(6), 1880.

Kemfert, C., Schäfer, D., & Semmler, W. (2020). Great green transition and finance. *Intereconomics, 55*, 181–186.

Loorbach, D., Frantzeskaki, N., & Avelino, F. (2017). Sustainability transitions research: Transforming science and practice for societal change. *Annual Review of Environment and Resources, 42*(1), 599–626.

Martin, R. (2012). Regional economic resilience, hysteresis and recessionary shocks. *Journal of Economic Geography, 12*(1), 1–32.

Martin, R., & Sunley, P. (2015). On the notion of regional economic resilience: Conceptualization and explanation. *Journal of Economic Geography, 15*, 1–42.

McCandless, E., & Onbargi, A. F. (2023). Just transitions and resilience in contexts of conflict and fragility: The need for a transformative approach. *Current Opinion in Environmental Sustainability, 65*, Article 101360.

Mackinnon, D., Dawley, S., Pike, A., & Cumbers, A. (2019). Rethinking path creation: A geographical political economy approach. *Economic Geography, 95*(2), 113–135.

McCauley, D., & Heffron, R. (2018). Just transition: Integrating climate, energy and environmental justice. *Energy Policy, 119*, 1–7.

Moodie, J., Tapia, C., Löfving, L., Gassen, N. S., & Cedergren, E. (2021). Towards a territorially just climate transition—Assessing the swedish EU territorial just transition plan development process. *Sustainability, 13*(13), 7505.

Nikas, A., Neofytou, H., Karamaneas, A., Koasidis, K., & Psarras, J. (2020). Sustainable and socially just transition to a post-lignite era in Greece: A multi-level perspective. *Energy Sources, Part B: Economics, Planning, and Policy, 15*(10–12), 513–544.

Nordberg, K. (2020). Spatial justice and local capability in rural areas. *Journal of Rural Studies, 78*, 47–58.

Neuhuber, T. (2025). One and the same or worlds apart? Linking transformative regional resilience and just transitions through welfare state policies. *Sustainability, 17*(2), 637.

Panarello, D., & Gatto, A. (2023). Decarbonising Europe: EU citizens' perception of renewable energy transition amidst the European green deal. *Energy Policy, 172*, Article 113272.

Protić, B., Lukić, B., Popović, V., Ristić, D., Šećerov, V., Jeftić, M. R., Đordević, D. S., & Vračarević, B. (2024). Balancing act: Assessing the impacts of winter tourism on natural heritage in Kopaonik national park and charting a sustainable path forward. *Sustainability, 16*(4), 1509. https://doi.org/10.3390/su16041509

Radovanovic, N., & Stevanovic Carapina, H. (2024). *Green transition and smart specialisation in the western balkans*. Tech. Rep., Joint Research Centre.

Rainnie, A. (2021). Regional development and agency: Unfinished business. *Local Economy, 36*(1), 42–55.

Regional Cooperation Council. (2021). Action plan for the implementation of the Sofia declaration on the green agenda for the Western Balkans 2021–2030. Bosnia and Herzegovina.

Rodríguez-Pose, A., & Bartalucci, F. (2024). The green transition and its potential territorial discontents. *Cambridge Journal of Regions, Economy and Society, 17*(2), 339–358.

Sanderson, R., Franklin, R., MacKinnon, D., & Matthews, J. (2024). Left out and invisible?: Exploring social media representation of 'left behind places.' *GeoJournal, 89*, 37.

Sandmann, L., Bülbül, E., Castano-Rosa, R., Hanke, F., Großmann, G., & R., K., Jiglau, G., Laakso, S, Nuorivaara, E, & Vornicu, A. (2024). The European green deal and its translation into action: Multilevel governance perspectives on just transition. *Energy Research & Social Science, 115*, Article 103659.

Shaker, Y., & Berisha, E. (2024). Transizioni Verdi Giuste': tra imprecisioni semantiche e necessità di un inquadramento teorico. In Formato E., Vingelli F. (Eds), Strumenti per il governo del valore dei suoli, per un progetto equo e non-estrattivo, Atti della XXV Conferenza Nazionale SIU "Transizioni, giustizia spaziale e progetto di territorio", Cagliari, 15–16 giugno 2023, vol. 09, pp.193–203. Planum Publisher e Società Italiana degli Urbanisti, Roma-Milano. https://hdl.handle.net/11583/2989395

Shaker, Y., & Berisha, E. (2025). Just Green Transitions: Between terminological inexactitudes, conceptual fragmentation, and the exigency for a theoretical framework. European Spatial Research and Policy, 32(1), 121–138. https://doi.org/10.18778/1231-1952.32.1.05

Shaker, Y., & Persico, S. (2024). Mapping the just green transitions in the socio-political virtual space. *European Journal of Spatial Development, 21*(1), 61–82. https://doi.org/10.5281/zenodo.13817780

Shaneh, A., Moodie, J., Tapia, C., Giacometti, A., & Cedergen, E. (2025). Understanding territorial governance dynamics in the Swedish EU territorial just transition planning processes—Lessons for the Western Balkans. In E. Berisha, J. Moodie, L. Allkja, & M. Jeftić (Eds.), *The Palgrave handbook of just green transitions in the Western Balkans and beyond*. Palgrave Macmillan.

Shawoo, Z., Maltais, A., Dzebo, A., & Pickering, J. (2023). Political drivers of policy coherence for sustainable development: An analytical framework. *Environmental Policy and Governance, 33*(4), 339–350.

Smith, A., Stirling, A., & Berkhout, F. (2005). The governance of sustainable sociotechnical transitions. *Research Policy, 34*(10), 1491–1510.
Snell, D. (2018). 'Just transition'? Conceptual challenges meet stark reality in a 'transitioning' coal region in Australia. *Globalizations, 15*(4), 550–564.
Sovacool, B. K., Turnheim, B., Hook, A., Brock, A., & Martiskainen, M. (2021). Dispossessed by decarbonisation: Reducing vulnerability, injustice, and inequality in the lived experience of low-carbon pathways.
Tapia, C., Giacometti, G., Moodie, J. R., Cedergen, E., & Shaneh, A. (2025). From mitigation to transformation: An evolutionary perspective on the just green transition. In E. Berisha, J. Moodie, L. Allkja, & M. Jeftić (Eds.), *The Palgrave handbook of just green transitions in the Western Balkans and beyond*. Palgrave Macmillan.
Tavares, M. (2022). *A just green transition: Concepts and practice so far* (Future of the World, UN-DESA Policy Brief, No141).
Teitje, L., & Cresap, S. (2018). Social justice simplified: How to teach a contested concept. *Journal for the Advancement of Educational Research International, 12*, 80–96.
Toto, R., Imami, F., Poro, E., et al. (2023). Just green transition conceptualisation in the Western Balkans. *GreenFORCE Horizon Project, Deliverable, 4*, 1.
Valdez, J. L. V. (2018). *Study on the role of social media and the internet in democratic development*. European Commission for Democracy Through Law (Venice Commission). Strasbourg, 21 November 2018, Cdl-la (2018)001 or. Angl. Council of Europe.
Valkenburg, G. (2025), Epistemic work and epistemic justice in green transitions. In E. Berisha, J. Moodie, L. Allkja, & M. Jeftić (Eds.), *The Palgrave handbook of just green transitions in the Western Balkans and beyond*. Palgrave Macmillan.
Vukoičić, D., Petrović, D., Gatarić, D., Božović, S., Ristić, D., & Jeftić, M. (2022). Assessing the tourist potential of cultural–historical spatial units of Serbia using comparative application of AHP and mathematical method. *Open Geosciences, 14*(1), 1170–1189. https://doi.org/10.1515/geo-2022-0411
Wälitalo, L., Robèrt, K. H., & Broman, G. (2020). An overarching model for cross-sector strategic transitions towards sustainability in municipalities and regions. *Sustainability, 12*(17), 7046.
Wang, X., & Lo, K. (2021). Just transition: A conceptual review. *Energy Research & Social Science, 82*, Article 102291.
World Bank. (2014). *YFR Macedonia green growth country assessment*. The World Bank Group Report.
Živanović, Z., Popović, V., Protić B., Djordjević, A., Šećerov, V., & Jeftić M. (2025). Signs by the roadside on the road to just green Transi-tions in Serbiass-border dimension of just green transitions in the Western Balkans. In E. Berisha, J. Moodie, L. Allkja, & M. Jeftić (Eds.), *The Palgrave handbook of just green transitions in the Western Balkans and beyond*. Palgrave Macmillan.

CHAPTER 3

The Just Green Transitions Policy Landscape

John Moodie, Ledio Allkja, Marija Jeftić, and Erblin Berisha

3.1 Introduction

The core dimensions of JGTs, as outlined in Chapter 2, are wide-ranging, which means the concept is understood, interpreted, and defined heterogeneously by different institutional actors, sectors, and other key stakeholders, including international organisations, industries, unions, environmental lobbyists, and civil society groups. Consequently, the methods for achieving JGTs are sometimes unclear and dependent on how different actors perceive transitions in terms of scale, context, and time (Shaker & Berisha, 2025). The

J. Moodie
Nordregio Research Institute for Regional Development and Planning, Stockholm, Sweden
e-mail: john.moodie@nordregio.org

L. Allkja (✉)
Co-Plan Institute for Habitat Development, Tirana, Albania
e-mail: ledio_allkja@universitetipolis.edu.al

Polis University, Tirana, Albania

M. Jeftić
Faculty of Geography, University of Belgrade, Belgrade, Serbia
e-mail: marija.jeftic@gef.bg.ac.rs

E. Berisha
Interuniversity Department of Regional and Urban Studies and Planning (DIST), Politecnico Di Torino, Turin, Italy
e-mail: erblin.berisha@polito.it

E. Berisha et al. (eds.), *The Palgrave Handbook of Just Green Transitions in the Western Balkans and Beyond*,
https://doi.org/10.1007/978-3-031-95075-9_3

high level of conceptual differentiation is reflected in the evolution of the JGT term at the international, national, regional, and local policy levels, where the overall focus on different JGT dimensions has changed and developed over time. The chapter provides an overview of the evolution of JGT policies at the supranational level and also within a Western Balkan context.

3.2 Just Green Transition Policy Evolution at the Supranational Level

The origins of the term 'just transition' go back to the United States during the 1970s and the American Oil, Chemical, and Atomic Workers Union leader, Tony Mazzocchi (Moodie et al., 2021). During the 'ban-the-bomb' peace movement, Mazzocchi lobbied for labour unions and the national government to support workers whose jobs might be threatened by disarmament, receiving support comparable to that provided by the GI Bill of Rights, which helped millions of World War II veterans' gain peacetime education and employment (Tapia et al., 2025, in this volume). Later, in the 1980s, Mazzocchi worked actively to bring labour unions and environmental movements together, arguing that a 'super fund' for workers was required to provide financial, educational, and training support for blue-collar workers at risk of unemployment in industries threatened by new environmental legislation (Roessler, 2016). In the 1990s, the 'super fund' was retroactively named the 'just transition fund' and officially endorsed by North American labour organisations. These early JGT developments focused solely on the social dimension of JGTs and the need to address the potentially negative impacts of environmental regulations and industrial closures on workers and citizens, including immediate job losses and long-term unemployment (Thomas, 2021). Mazzocchi deserves credit for considering social impacts during a period where climate policy discussions largely focused on the environmental and economic dimensions of JGTs (Tapia et al., 2025, in this volume). The 1972 United Nations Conference on the Human Environment in Stockholm introduced climate governance and environmental issues onto the global agenda, leading to the development of the United Nations Environmental Program (UNEP). During the 1980s, climate and environmental 'sustainability' became a popular term and goal in policy circles due to the 1983 Brundtland Report produced by the UN Commission on Environment and Development (Tavares, 2022). The 1992 United Nations Conference on Environment and Development (also known as the Earth Summit) led to the establishment of the UN Framework Convention on Climate Change (UNFCCC) to combat human interference with the climate and stabilise greenhouse gas emissions. The first UNFCCC Conference of the Parties (COP) was held in Berlin in 1995, which started climate negotiations that led to the creation of the 1997 Kyoto Protocol (COP III Kyoto Summit) that established legally binding obligations for developed countries to reduce greenhouse gas emissions. The term 'just transition' was not central to these discussions. The environmental dimension of climate transitions dominated these international climate policy debates focusing on promoting long-term environmental sustainability through targeted reductions in greenhouse gases (Tavares, 2022). The carbon reduction targets outlined in the Kyoto Protocol were a highly contentious topic for international trade unions as they were concerned about the societal impacts that these climate targets would have for primary and secondary jobs. Trade unions were faced with the difficult dilemma of addressing the eminent environmental threat to society while protecting the interests of their industrial affiliates and workers (Thomas, 2021).The merger between the International Confederation of Free Trade Unions (ICFTU) and the Christian World Confederation of Labour (WCL) into the International Trade Union Confederation (ITUC) in 2006 represented an attempt to strengthen the capacity of international trade unions to influence climate policies (Ibid). With this merger, the ITUC gained member status within the UNFCCC, and the concept of just transition returned to the forefront of the global climate debate. As Johansson (2023) notes, working in collaboration with the International Labour Organisation (ILO), the ITUC, the term 'just transition' was used to illustrate the ITUC's support for strengthened climate action was compatible with its core mission to secure social justice for workers and affected communities. Once again, the social dimension of transitions became central to the debate around climate and environmental policies. This was evident at COP 13 in Copenhagen where the notion of just transition was included in the draft text for negotiation (Johansson, 2023). Although the just transition was not included in the final text from the Copenhagen summit, it was included in the Cancun Agreement at COP16, which noted that 'addressing climate change requires a paradigm shift towards building a low-carbon society [...] while ensuring a just transition of the workforce that creates decent work and quality jobs.' The COP also recognised 'the importance of [...] promoting a just transition of the workforce [and] the creation of decent work and quality jobs.' The commitment of the ITUC to representing the social dimension in debates within the UNFCCC culminated in the term's inclusion in the 2015 Paris Agreement (COP 21 Paris Summit). Johansson (2023) notes that during discussions over the Paris Agreement an alliance between the ITUC and other civil society organisations advocated for 'a common article introducing a rights-based approach to climate action with reference to ensuring the respect, protection, promotion, and fulfilment of human rights, including indigenous rights, gender rights, intergenerational equity, food sovereignty, and a just transition of the workforce that creates decent work and quality jobs.' In the final agreement, the term just transition was separated in the text and given its recognition in the phrase '[t]aking into account the imperatives of a just transition of the workforce and the creation of decent work and quality jobs in accordance with nationally defined development priorities'. The ITUC has used the just transition concept to strike a balance between meeting climate and environmental objectives while protecting and equipping workers most at risk from climate interventions. This is confirmed by Rosemberg (2017), who notes that the just transition 'can be understood as the conceptual framework in which the labour movement captures the complexities of the transition towards a low-carbon and climate-resilient economy, highlighting public policy needs and aiming to maximise benefits and minimise hardships for workers and their communities in this transformation'. In 2015, the ILO published its own guidelines for a just transition where it defined just transition as 'greening the economy in a way that is as fair and inclusive as possible to everyone concerned, creating decent working opportunities and leaving no one behind.' (International Labour Organisation, 2015) The guidelines outline the need for high levels of investment in low-carbon technologies, a social dialogue and consultations between policymakers and affected groups; proactive labour market policies based on social protection and worker rights, including financial support and retraining opportunities for the unemployed; early-stage research into the potential socio-economic impacts of climate policies; and the development of local economic diversification plans (Ibid). The guidelines also make a distinction between social investment policies (e.g. active labour market policies and training and re-skilling policies to increase workers' employability in a greener economy) and social protection policies (e.g. unemployment and minimum income benefits), meaning that both are required within a just transition framework (Moodie et al., 2021). The primary focus of the guidelines is, therefore, on addressing the social dimension of transitions and the need to protect workers' jobs and provide retraining opportunities. However, the environmental, economic, and technical dimensions of JGTs are recognised in the guidelines through the need for greening the economy through high levels of investment in carbon-neutral technology. The democratic, governance, and communicative dimensions come through the guidelines as they argue that transitions require a collaborative effort based on inclusive and transparent dialogue with a range of actors and citizens. After the publication of the ILO guidelines, the multi-dimensional nature of JGTs became increasingly evident in other key international policy documents, including the UN 2030 Agenda and the corresponding 17 Sustainable Development Goals published in 2015 (United Nations, 2015). The SDGs advocate for transformative climate and environmental actions, highlighting connections between sustainable development's environmental, social, and economic dimensions. The COP24 event entitled 'Just Transition for All' held in Katowice in 2018 resulted in the multi-dimensional JGT approach outlined in 'Solidarity and Just Transition Silesia Declaration' (Tavares, 2022). The Declaration emphasised the social and environmental dimensions, stating that transitions should focus on creating work opportunities and quality jobs, while addressing environmental goals. The technical and economic dimensions are evident in references to JGTs enhancing economic growth and the need to ensure climate-resilient infrastructure. The governance and democratic dimensions of JGTs are present in the Declaration, which notes 'the importance of a participatory and representative process of social dialogue involving all social partners.' This reference showcases that transitions are a joint collaborative venture but falls short of suggesting direct citizen engagement in the process. The spatial dimension is also recognised in the Declaration, which acknowledges 'the specific needs and special circumstances of developing countries, especially those that are particularly vulnerable to the adverse effects of climate change.' Finally, the resilience dimension is referenced in the Declaration, which states that there is a need to enhance resilience in countries most vulnerable to the impacts of JGTs. The various agreements, guidelines, and declarations referenced above are strong on just transition rhetoric, but difficult to put into practice as they do not come with an implementation plan and funding. In response to these implementation obstacles, at the UN Climate Action Summit held in New York 2019, 46 countries committed to developing practical just transition strategies (Tavares, 2022). Similarly, the 2020 European Union Green Deal (EUGD) became the most comprehensive effort to deliver JGTs providing both implementation plan mechanism and substantial funding (European Union 2019). The EUGD follows the EU's long history of supporting regions in industrial transition. The original European Coal and Steel Community provided technical and economic assistance for different regions undergoing industrial reconstruction. More recently, the 2017 Initiative for Coal Regions in Transition focused on providing guidance for European coal regions in transition. As part of this initiative, the EU Just Transition Platform was designed to facilitate an exchange of best transition practices, strategies, and projects that could guide the transition process in declining coal regions. Building on these initiatives, the Just Transition Mechanism (JTM) was introduced as part of the EUGD to ensure that 'no person or place is left behind' because of climate transitions (European Commission, 2021). The JTM is split into three main pillars. The Just Transition Fund (Pillar 1) provides financial support to regions and sectors dependent on fossil fuels. The InvestEU Just Transition Scheme (Pillar 2), delivers budgetary support for private sector investments that support transitions. Finally, the Public Loan Facility (Pillar 3), supports public sector transition investments. Financial support is available to all member states but focuses on the most carbon-intensive regions and citizens most vulnerable to transitions (European Commission, 2021). The JTM will protect them by:

- supporting the transition to low-carbon and climate-resilient activities;
- creating new jobs in the green economy;
- offering re-skilling opportunities;
- investing in public and sustainable transport;
- providing technical assistance;
- investing in renewable energy sources;
- improving digital connectivity;
- providing affordable loans to local public authorities;
- improving energy infrastructure, district heating, and transportation networks (European Commission, 2021).

EU member state access to transition funding, implemented within the remit of EU Cohesion Policy, is predicated on the development of Territorial Just Transition Plans (TJTPs). TJTPs outline regional just transition processes up until 2030, identifying (1) which EU member state regions should require transition support; (2) the main climate-related challenges these regions face; (3) targeted technical, socio-economic and environmental transition actions; and 4) governance mechanisms needed to help implement these actions. EU member states have now developed and ratified their TJTPs which gives them access to dedicated financing under the pillars of the JTM. (European Commission, 2021). The aims and objectives of the JTM and TJTPs indicate an attempt to balance the technical, economic, environmental, and social dimensions of JGTs. The technical, economic, and environmental dimensions are reflected in the EU's commitment to investing in carbon-neutral and environmentally friendly production technologies, transport, renewable energies, and digitalisation, whereas the social dimensions are highlighted in the worker protection rights and re-skilling opportunities offered (Moodie et al., 2021). The spatial dimension of the just transition is embedded within the JTM as the Commission highlight the important role of regions and cities in guiding just transition processes: 'Citizens, depending on their social and geographic circumstances, will be affected in different ways. Not all Member States, regions and cities start the transition from the same point or have the same capacity to respond. These challenges require a strong policy response at all levels.'(European Commission, 2019). The territorial focus of the TJTPs is in keeping with the Commission's recent emphasis on territorial governance and place-based policymaking, such as the development of regional smart specialisation strategies, in addition to applying the concept of 'active subsidiarity' advocating a central role of regions in EU policymaking (Moodie et al., 2021).

JGT processes require effective institutional and stakeholder collaboration across multiple levels of governance. The governance, democratic and communication dimensions are highlighted in documents relevant to the EUGD, JTM and JTF (European Commission, 2019, 2021). The European Commission notes that TJTPs should be developed and implemented through open and inclusive processes based on local and regional knowledge and expertise to ensure that climate transition policies meet the needs of citizens and leave no one behind (European Commission, 2021). In a key passage within a communication document on the EU Green Deal, the Commission further outlines the core democratic and communication dimensions of the just transition: 'This transition must be just and inclusive. It must put people first and pay attention to the regions, industries and workers who will face the greatest challenges. Since it will bring substantial change, active public participation and confidence in the transition is paramount if policies are to work and be accepted. A new pact is needed to bring together citizens in all their diversity, with national, regional, local authorities, civil society and industry working closely with the EU's institutions and consultative bodies (European Commission, 2019) In this excerpt, there is a strong emphasis on the democratic and communication dimensions of the transition, particularly the need for open and inclusive social dialogue with citizens to promote trust, acceptance and support for TJGT policies. While the Commission strongly advocates that climate transition processes should be open and inclusive of all stakeholders and citizens, particularly underrepresented minority groups, in practice, there has been no significant citizen participation in the preparation of the TJTPs across the EU's transition regions (Georgieva & Ferry, 2023, 2024). There is, therefore, a need for further analysis on how best to include citizens directly in transition processes from a governance, policy, and communication perspective (Balz et al., 2023). Furthermore, TJTPs have been criticised by trade unions and civil society groups who have voiced concern that plans are too focused on the economic and technical dimensions of transitions while neglecting the potential social dimensions. This is largely because TJTP processes have been dominated by national governments and sectoral actors, with a constrained role for sub-national level institutions and other key stakeholders, including NGOs and societal groups (Moodie et al., 2021). This highlights the importance of ensuring that all stakeholders have equal access to the process, so different voices, ideas and perspectives are recognised in the final plans.

3.3 Just Green Transition Policies in the Western Balkans

In their process for EU accession, the WB countries have started to take measures in the framework of Just Green Transition (Berisha, 2025a, 2025b). These can be seen across the four main pillars of JGT: governance, technological, socio-economic, and ecological. The different measures can be categorised in aspects such as administrative and governance changes, policy development to support green transition, and direct investment in infrastructure and technology. This section of the book's introductory chapter will discuss some of the main drivers supporting these measures, including international and EU frameworks, national policy frameworks, and regional or local ones.

The Paris Agreement has been signed by 5 of the WB countries and offers a wider framework for intervention in adaptation and mitigation of climate change. Meanwhile, Kosovo*,[1] although not a direct signatory of the agreement due to its status and representation within the United Nations Framework, through bottom-up initiatives, is showing its commitment to decarbonisation and climate action, including the voluntary preparation of the NDC (Dietachmair et al., 2024). These measures have been turned into national legislation and strategies, leading to a direct impact in orienting governance and policies. While the focus is tackling climate change, they contribute to achieving also JGT.

[1] This designation is without prejudice to positions on status, and is in line with UNSCR 1244 and the ICJ Opinion on the Kosovo declaration of independence.

The *UN Sustainable Development Goals* and the *Sendai Framework* 'On Disaster Risk Reduction' (DRR) are other important global actions supporting green transition. Both frameworks have been ratified at the national level, leading to legal, governance, and policy changes in the WB countries. The Sendai Framework has brought changes in the conception of civil protection sectors, going from a practice of managing 'emergencies/disasters' to managing and reducing risks. From a sectorial approach that deals with disaster events and manages the consequences, it has changed towards an integrated and comprehensive approach that reduces and manages risk from disasters. These have brought changes and established new legal frameworks, institutions, financing mechanisms, and instruments that can be used for policy development (Allkja et al., 2022). For example, now countries at national, regional, and local levels must develop Risk Assessment Strategies as well as Strategies for Disaster Risks Reduction. It is important to understand that some of the main 'natural' disasters are exacerbated by climate change, therefore there is a direct link between DRR policies and adaptation to climate change, thus contributing to JGT. The EU is one of the main drivers of JGT. Besides policies at the EU level, a tailored policy document has been prepared to orient the roadmap for green transition for the WB, respectively '*Guidelines for the Implementation of the Green Agenda for the Western Balkans*'. The guideline offers policy and technical solutions for WB countries to follow in their efforts of green transition and supports governance and financing guidance. In addition, an Economic and Investment Plan has been adopted, which will support the WB countries to implement the Green Agenda. The EU has committed 1.25 billion Euros to support implementation through four flagship pillars: the Environment and Climate, Clean Energy, Sustainable Transport and Private Sector Development. Besides the above, the EU, through the Western Balkan Investment Framework (WBIF), provides additional assistance to support the green transition with direct investments in strategic projects. These funding opportunities tackle project preparation (design), capacity building, and direct investment in project implementation. The Instrument for Pre-Accession Assistance (IPA) is another way the EU impacts WB in Just Green Transition. In this aspect, the most notable are territorial and cross-border cooperation programs (CBC). On the one hand, CBC programs are either bilateral or trilateral, and territorial cooperation includes large macro-regional areas such as the Adriatic–Ionian, Danube, Mediterranean, and Balkan Med. All these programs have priority axes or objectives related not only to climate change adaptation and environmental preservation (i.e. green transition) but also towards social inclusion and job creation, primarily focusing on new skills in youth or deprived regions (i.e. just dimension of transition) (Pinnavaia & Berisha, 2021; Solly & Berisha, 2021). As part of IPA programs, countries, regions, or local authorities can benefit through technical assistance and improvement in governance, technological development, ecological projects, as well as social ones, thus cross-cutting among the main pillars of JGT. Table 3.1 offers an overview of the main IPA programs and their main priority axis, which are interlinked with JGT.

The above table shows overall coherence in territorial cooperation programs with the WB Green Agenda and EU targets on Green Transition. Although, in many cases, it is not explicitly stated that the program contributes to green transition, this can be associated with priorities on environment, energy, youth skills, support of SMEs and risk management. Only a few of the programmes have an explicit priority on green transition such as for example Greece–North Macedonia, Serbia–Croatia, Serbia–Hungary, Greece–Albania, Albania–Italy–Montenegro, and Danube. Another important aspect covered in most territorial cooperation programs is increasing labour skills, especially improving youth skills to adapt to new economic needs. This shows that now, not only the ecological component of JGT is considered but also the socio-economical one to balance the impacts of transition. While these programs do not directly apply to JGT, they influence and drive local policies towards these aims through financial conditionality. Additionally, the financing of projects that serve as best practices and highlight these priorities directly influences not only local authorities (i.e. institutions) but also communities, citizens, and SMEs to understand and contribute to JGT.

3.4 National Policy Frameworks

There is no direct institution in the WB countries which deals explicitly with Green Transition. However, there are different institutions which deal with different components of JGT. For example in Albania there are different Agencies which deal with various dimensions such as the National Territorial Planning Agency (shapes spatial planning policies); National Agency of Civil Protection (deals with Disaster Risk Reduction (DRR) policies and climate change adaptation); National Environment Agency (deals with environmental policies and climate change adaptation); Energy Efficiency Agency (support energy transition); General Directorate of Road Transport (support decarbonisation of transport); Albanian Investment and Development Agency (support SMEs and attract FDI); National Agency on Employment and Skills (support new employment opportunities and increase of skills in vulnerable communities); Municipalities (cross-sectorial policies on JGT). Although, as previously mentioned, there is no explicit policy or strategic document which deals with JGT and supports the implementation of the Green Agenda. In an analogous manner, it can be said that also in Kosovo, several institutions deal with various aspects of JGT spanning from spatial planning to energy transition, waste management and employment (Dietachmair et al., 2024). In Serbia, also, several institutions play a key role in various aspects of the JGT, even though no single institution covers all its dimensions. As the central state body for environmental protection policy, the Ministry of Environmental Protection leads Serbia's efforts in the fight against climate

Table 3.1 IPA territorial cooperation programs in the Western Balkan

IPA program	*Priority Axis linked to JGT*	*WB countries Involved*
Danube	A more competitive and smarter Danube Region A greener, low-carbon Danube Region A more social Danube Region	BiH; Serbia
Adriatic Ionian	Blue Growth, Connecting the Region, Environmental Quality, Sustainable Tourism	Albania; BiH; North Macedonia; Montenegro; Serbia
Mediterranean	Strengthening an innovative sustainable economy, Protecting, restoring, and valorising the natural Environment and heritage, Promoting green living areas, and Enhancing sustainable tourism	BiH; Albania;
Balkan Med	Entrepreneurship and Innovation, Environment	Albania; North Macedonia
Kosovo–Albania	Protecting the environment, promoting climate change adaptation and mitigation, and risk prevention and management	Kosovo; Albania
Albania–Montenegro	Protecting the environment, promoting climate change adaptation and mitigation, risk prevention and management; Promoting employment, labour mobility, and social and cultural inclusion across the border	Albania; Montenegro
Albania–Montenegro–Italy	A smarter South Adriatic programme area, by promoting innovative and smart economic transformation A greener South Adriatic programme area, by promoting clean and fair energy transition, green and blue investment, the circular economy, climate change adaptation, and risk management A more connected South Adriatic programme area by enhancing mobility and regional connectivity A more social South Adriatic programme area	Albania; Montenegro; Italy

(continued)

Table 3.1 (continued)

IPA program	*Priority Axis linked to JGT*	*WB countries Involved*
Albania–Greece	A greener and more resilient Greece-Albania cross-border region A more accessible Greece-Albania cross-border region An economically and socially sustainable Greece-Albania cross-border region	Albania; Greece
Albania–North Macedonia	Enhancing competitiveness, business, trade, and investment (needs and opportunity-driven) Protecting the environment, promoting climate change adaptation and mitigation, risk prevention and management (needs driven)	Albania; North Macedonia
Serbia–Hungary	A greener, low-carbon transitioning towards a net zero carbon economy and resilient Europe by promoting clean and fair energy transition, green and blue investment, the circular economy, climate change mitigation and adaptation, risk prevention and management, and sustainable urban mobility A more social and inclusive Europe implementing the European Pillar of Social Rights	Serbia; Hungary
Serbia–Croatia	Cooperating for a smarter programme area Cooperating for a greener and climate change resilient programme area Cooperating for healthier and more inclusive programme area Cooperating for more sustainable and socially innovative tourism and culture	Serbia; Croatia
Serbia–BiH	Encouraging tourism and cultural and natural heritage Investing in youth, education, and skills	Serbia; BiH

(continued)

Table 3.1 (continued)

IPA program	*Priority Axis linked to JGT*	*WB countries Involved*
BiH–Montenegro	Environment protection, climate change adaptation and mitigation, risk prevention and management Encouraging tourism, cultural and natural heritage	BiH; Montenegro
Serbia–Montenegro	Employment, labour mobility, and social and cultural inclusion across borders Encouraging tourism and cultural and natural heritage	Serbia; Montenegro
Serbia–North Macedonia	Promoting employment, labour mobility, and social and cultural inclusion across the border Encouraging tourism and cultural and natural heritage	Serbia; North Macedonia
Kosovo–North Macedonia	Competitiveness, business environment and SME development, trade, and investment Environmental protection, climate change adaptation and mitigation, risk prevention and management	Kosovo; North Macedonia
Kosovo–Montenegro	Promoting employment, labour mobility and social and cultural inclusion across the border; Protecting the environment, promoting climate change adaptation and mitigation, and risk prevention and management; Encouraging tourism and cultural and natural heritage;	Kosovo; Montenegro
Greece–North Macedonia	Transition to a low-carbon Economy Support and upgrade of Health & Social Services	Greece; North Macedonia
North Macedonia–Bulgaria	Increase the competitiveness of the local economy and improve the business environment Development of an attractive, all-season tourism product by means of smart solutions that ensure universal access and participation	North Macedonia; Bulgaria

(continued)

Table 3.1 (continued)

IPA program	*Priority Axis linked to JGT*	*WB countries Involved*
Serbia–Romania	Environmental protection and risk management Social and economic development	Serbia; Romania
Serbia–Bulgaria	Sustainable Tourism Youths Environment	Serbia; Bulgaria

Source Authors elaboration

change. The ministry covers key sectors such as decarbonisation and the transition to clean energy. It is responsible for implementing national strategies related to environmental protection and sustainable resource management. Several other ministries also contribute to different aspects of JGT within their sectoral competences (Ministry of Environmental Protection—climate change efforts, decarbonisation and environmental protection issues; Ministry of Mining and Energy—promotion of renewable energy sources; Ministry of Agriculture—Forestry and Water Management ensures sustainable agricultural practices and manages water resources; Ministry of Construction, Transport and Infrastructure—development of green infrastructure and decarbonisation of transport, etc.). In addition to ministries, various national agencies, public companies, and institutes also play a significant role in the public domain. The Environmental Protection Agency of Serbia monitors the quality of the environment and implements policies to mitigate climate change. The Institute for Nature Protection of Serbia is dedicated to preserving Serbia's biodiversity and natural habitats. The Institute leads efforts in ecosystem restoration and monitoring the effects of environmental degradation. The Energy Agency of the Republic of Serbia regulates the energy sector, promoting energy efficiency and sustainability. Finally, the public company 'Serbia Forests-Srbija Šume' manages forest resources and contributes to the protection of biodiversity. At the same time, UNDP Serbia supports broader sustainability initiatives with a focus on climate change resistance and green energy. In the private sector, the Serbian Chamber of Commerce plays a key role in the promotion of sustainable development. It focuses on areas such as the circular economy and rural development, supporting businesses in transitioning to greener practices. The Serbian Chamber of Commerce is also working on improving the production system and encouraging the use of sustainable resources. The German Organization for International Cooperation (GIZ) plays a key role in supporting the energy transition in Serbia. Focused on decarbonisation, GIZ cooperates with local and national actors in the implementation of sustainable energy solutions. The organisation also supports international cooperation in the Western Balkans, offering expertise in energy efficiency and adaptation to climate change. Thus, as it can be seen from a governance and institutional perspective, the lack of an overarching institution and policy framework means that the JGT is primarily driven by institutional and sectorial coordination and the willingness of the different stakeholders to achieve this. Meanwhile, while there is a lack of coordination at the national level, municipalities have a far greater opportunity in achieving JGT as compared with the national level, and institutional coordination (at least in theory) is easier to achieve. Considering the above-mentioned situation, it is interesting to discuss the contribution of the different sectors to JGT. Understanding the diverse types of policies that contribute to JGT offers a good opportunity to also understand gaps in policies which may hinder achieving JGT. This analysis, combined with the contribution from the different chapters with case studies from WB in this book as well as comparative research with EU regions, could lead to recommendations on further policy and institutional improvement.

3.4.1 Spatial Planning

As an overarching and cross-sectorial policy, it plays a significant role in achieving climate goals (Davoudi, 2020) and promoting green transition. At national, regional, and local levels, policies of spatial planning can play an integral role. When looking at the WB, most national spatial plans were prepared before the emergence of the JGT concept in the policy milieu. Therefore, there is little direct and explicit reference to it. In Albania, the General National Territorial Plan was prepared in 2016, and it has policies referring to climate change adaptation, promotion of renewable energy and energy efficiency, for instance. In the Republic of Serbia, the latest Spatial Plan of the Republic of Serbia 2020–2035 (SPRS),[2] which is still in the drafting phase (it is expected to be adopted in the following period), touches on numerous issues of environmental protection and sustainable development but does not specifically focus on 'just green transition' (JGT) in the sense in which this term is known internationally, especially when talking about aligning decarbonisation with social justice. While the plan contains elements related to sustainable use of resources, protection of natural resources and environmental impact assessment, greater emphasis is placed on the broad goals of infrastructural and regional development, including energy efficiency, transport infrastructure and environmental protection. However, it lacks a detailed framework to ensure that the green transition is implemented in a socially just manner, especially for vulnerable sectors and regions that could be affected by the transition from fossil fuels to sustainable energy sources. This represents a key challenge for the development of certain regions in Serbia, especially those whose economy relies on the exploitation of coal and other fossil fuels, urban areas with high emissions from traffic and heating or rural areas with weak energy infrastructure and limited access to renewable energy sources. However, the lack of

[2] More detailed in: Jeftić et al. (2021).

a clear regional policy in the Republic of Serbia, which results from a slow transition from dominant centralism and unitarism in the socio-political and economic system, has led to the marginalisation of the role of some regions in national development. This directly affects their ability to adapt to transformative processes, such as the green transition. The neglect of the principle of functionalism is particularly visible in the existing NUTS 3 nomenclature, which does not sufficiently take into account the network of settlements and the functional organisation of urban centres in Serbia (Jeftić et al., 2019). Administratively defined regions often ignore their functional characteristics. The introduction of a functional approach, based on the specific economic, ecological and social needs of the region, is crucial for a successful transition and sustainable regional development. Also, as the key carriers of regional development, urban centres should play a central role in shaping sustainable and balanced spatial development. Without a functionalist approach to regional planning, the potential of regions in Serbia to significantly contribute to the green transition remains underutilised. This problem becomes even more pronounced in the context of the green transition, where it is crucial to direct resources and development measures according to the real needs and possibilities of the region. Statistical units are currently used in the allocation of EU structural funds, which often leads to insufficient use of those funds. Some of the urban centres in rural areas in Serbia have already been recognised as key priorities for the allocation of these funds, but an additional improvement of the functional approach is necessary to enable the even inclusion of smaller regional centres (Jeftić et al., 2021.). Only integrating a functional approach can ensure sustainable regional development and successful implementation of the green transition. Separate initiatives in Serbia, often in partnership with international organisations such as UNDP, directly address a JGT, with an emphasis on mitigating the socio-economic consequences of decarbonisation. These initiatives are usually complementary to national development strategies but are not explicitly part of the Spatial Plan. In Kosovo, the National Spatial Plan 2010–2020 too has policies referring to climate change adaptation and energy transition. It remains to be seen whether the new spatial plan will have dedicated policies in terms of JGT. Similarly, when looking at the local level, policies try to deal with different sectorial issues in terms of energy, transport, waste management, but due to the time of preparation and the emergence of the concept of JGT, it is difficult to find policies and attempts which deal with this aspect. An interesting study prepared by UN-Habitat (Caka, 2021) in Kosovo highlight some of the challenges in dealing with climate change which are true also for JGT. For example, 84% of municipal staff in Kosovo are undertrained and have no experience in dealing with climate change issue. Thus, it can be said that this may hold true also for JGT considering that it is a new concept which needs to integrate knowledge from different disciplines, as for example staff in municipalities have medium to low level of knowledge and expertise in green cities, resilient cities and low-carbon cities. An adequate system of protection, education and sustainable management of natural resources (especially forests, water, land, etc.) is the basis for the long-term stability of the area, which directly contributes to the employment of the local population and investments in environmental protection (Gatarić et al., 2022.). This approach supports the concept of green growth and can be integrated into regional development and spatial planning strategies. According to the declaration of the Serbian Chamber of Commerce, the greatest demand for personnel is expected to be in the sectors of renewable energy sources, circular economy and sustainable business. In order to respond to these demands, it is necessary to develop new educational programs and profiles in Serbia, with a focus on 'green' professions. This process includes further development of dual education, improvement of professional practice and master's training, as well as other forms of acquiring practical knowledge and skills, in accordance with the needs of Industry 4.0 and the challenges of 'green' and digital transition (Živanović et al., 2025.). These findings, although with differences, have similar implications for local authorities in the other WB countries. Traditional planning systems and approaches have mostly focused on urban areas and urban regulatory frameworks (Berisha et al., 2018, 2024). Thus, the transition towards spatial planning and towards a wider comprehensive and integrated approach, which is necessary for JGT, is yet in process and faces challenges of capacity building and lack of expertise.

3.4.2 Disaster Risk Reduction

It is a new concept which has penetrated also in the WB countries. The Sendai Framework has led to changes in terms of revising legal and institutional frameworks. In Albania, for example, there is a new Agency of Civil Protection and a new Strategy on Disaster Risk Reduction. One of the innovative aspects of this strategy is the introduction of the Nature Based Solution (NBS) in the national policy framework which contributes towards JGT. Currently, municipalities in Albania have started to prepare their local DRR Strategies. From a content perspective it can be seen that there are attempts to promote NBS as well as climate change adaptation policies and priority projects. Similarly, in Kosovo, the legislation has changed, and a DRR Strategy has been prepared, leading to the introduction of new concepts and policies which contribute to JGT. Serbia's institutional framework for disaster risk reduction includes multiple levels—from legislative frameworks and state institutions to local initiatives and international cooperation, which contributes to the importance of capacities in the country. Serbia adopted the Integrated National Climate and Energy Plan (INECP) and National Strategy for Disaster Risk Reduction until 2030 which sets goals for strengthening disaster resilience and integrating risk reduction measures into all sectors of society and the economy. Law on Climate Change includes measures to introduce green technologies and promote sustainable development, which is key to mitigating the consequences of climate disasters. At the local level, local risk reduction plans are developed by municipalities based on risk assessment, which include early warning systems, risk mapping, evacuation routes and training for the population. For example,municipalities with large watercourses implement measures such as the construction of defensive leaves, restoration of natural floodplains and local early warning systems. Nature Based Solutions (NBS) are also part of the local governments‘ implementation procedures that include green space and watercourse restoration to mitigate the effects of floods and droughts, improving community resilience.

3.4.3 *Climate Change Adaptation*

It has introduced new legislation and policies in the WB. In Albania, the law 155/2020 On Climate Change was approved in 2020, while in Kosovo, law 08/L250 On Climate Change was endorsed in January 2024. These legislations provide a framework for the reduction of GHG emissions in line with the Paris Agreement as well as new opportunities for policy development. Both countries have also prepared action plans for Climate Change Adaptation which prioritise (in Albania) among other the encouragement of economic transition towards a low-carbon economy, increase of capacities as well as integration of climate change in sectorial policies. Meanwhile, in Kosovo, besides the promotion of a low-carbon economy, there is also a dedicated objective on mobility planning and decarbonisation of the transport sector and a more tailored approach to the energy sector in terms of renewable energy use, energy efficiency and district heating. Similarly, at the local level there are efforts from municipalities to prepare Local Strategies of Climate Change Adaptation. In Albania, these efforts have been supported so far by UNDP, while in Kosovo, municipalities are being supported by UN-Habitat to prepare the strategies which contribute directly to JGT. In Serbia is adopted the Law on Climate Change prepared by the Ministry of Environment, the Climate Change Act (2021), National Climate Change Adaptation Plan (NAP), The Law on Environmental Protection (which basically covers all pillars of GT), Air Protection Law, The Law on Waters, The Law on Soil Protection (Depollution), The Law on Nature Protection, National strategy for sustainable use of natural resources and goods and Biodiversity Strategy of the Republic of Serbia (Biodiversity), Law on Agriculture and Rural Development and The Law on planning and construction (Sustainable Food Systems and Rural Areas). National strategy for sustainable development, Law on Nature Protection, National action plans for the sector (agriculture, forestry, and energy), etc.

The current overview of the policies that are concerning the issue of JGT in Serbia are mainly on a national level but there are examples of policies on the regional and local one. Significant regional policies and practices are related to the administrative areas of the City of Belgrade and the autonomous region of Vojvodina (Sustainable Energy and Climate Action Plan for the City of Belgrade—SECAP, Sustainable Urban Mobility Plan for the City of Belgrade, Green City Action Plan for the City of Belgrade, AP Vojvodina Environmental Protection Program for the Period 2016–2025 etc.) while local activities and policies are mostly concentrated in larger and better developed municipalities and cities (Novi Sad, Niš, Kragujevac, Kraljevo, Subotica etc.). In general terms, Serbia has strategies and plans for the green transition, but they are not sufficiently coordinated and mutually integrated. In this sense, better integration of different policies and plans at the national and local levels is needed. The European Commission's reports on Serbia's progress often emphasise the need to speed up the implementation of policies in the field of environmental protection and climate change, pointing to challenges in effective implementation. North Macedonia has the advanced policy and strategic framework, as well as a well-defined governance structure, and is evaluated by the EC (EC Country Progress Reports) as a country with some level of preparation for the environment and climate change. However, the country needs to accelerate the implementation of the Green Agenda and the economic and investment plan, including the just transition process. Moreover, the country needs to improve its governance and institutional capacities for policy implementation. Montenegro's climate governance is managed through national ministries and specialised agencies working collaboratively to address environmental and climate challenges. The Ministry of Ecology, Spatial Planning, and Urbanism is central, overseeing climate policies and international cooperation through its Directorate for International Cooperation, EU Integration, and Climate Change. Additional pivotal ministries include the Ministry of Economic Development and Tourism and the Ministry of Capital Investments. The Statistical Office of Montenegro (MONSTAT) plays a crucial role in data collection and reporting, while the State Statistical Office and Environment Protection Agency are responsible for preparing emission inventories. Furthermore, the Working Group for Mitigation and Adaptation to Climate Change within the National Council for Sustainable Development provides strategic guidance and coordination. Law on Protection from Adverse Impacts of Climate Change with relevant by-laws already drafted and adopted, sets the framework for climate action in Montenegro. Also, Law on Amendments to the Law of Adverse Impacts of Climate Change will further refine and strengthen the legal framework and it will be soon adopted. Strategic documents such as the National Climate Change Strategy 2030 and the National Sustainable Development Strategy 2030 demonstrate Montenegro's alignment with global sustainability goals as outlined in the UN Agenda 2030. The revised Nationally Determined Contribution (NDC) is particularly notable, setting an ambitious target of at least a 35% reduction in national GHG emissions (excluding land use, land-use change, and forestry—LULUCF) by 2030 compared to 1990. Furthermore, the planned reduction in biodegradable municipal waste disposable to landfills highlights Montenegro's commitment to sustainable waste management practices.

3.4.4 *Transport and Mobility*

It is one of the main sources of greenhouse gas emissions, as well as nitrogen oxides (NOx), in the WB region. Combined with the energy sector, transport is responsible for two-thirds of total emissions. Within that sector, the dominant contribution is carbon dioxide (CO2) originating from road transport, which accounts for more than 90% of total emissions from transport. Sustainable and Smart Mobility Strategy for the Western Balkans through its three key components is the foundation for the JGT in the future. Smart mobility implies the integration of digital technologies and automation into transport systems with the aim of increasing the efficiency and sustainability of passenger and cargo mobility. For the successful implementation of this strategy, economies in the region must transform their transport systems through the introduction of multimodal ticketing systems, the 'mobility as a service' model, paperless transport and the application of software, 5G networks and artificial intelligence to optimise transport activities. The political landscape of urban mobility in WB is gradually shifting towards sustainability, driven by national strategies, local initiatives, and international support. There is a need to strengthen institutional capacities and better coordination between different levels of administration and sectors. The lack of an integrated approach hinders the effective implementation of policies. Strengthening the capacity of institutions involved in the implementation of the green transition is crucial, including the training of personnel and the provision of adequate resources. For example, institutional and legislative framework for transport and urban mobility in Serbia includes legislation (Law on transportation in road traffic, Law on road traffic safety, Law on Railways, Law on Planning and Construction, Law on Energy Efficiency), national strategies (National Traffic Development Strategy (2015–2030), National Climate and Energy Plan (NECP), Strategy of sustainable urban mobility, National railway development strategy), plans (for example Sustainable Urban Mobility Plan for the City of Belgrade, Novi Sad, Niš, Užice, Subotica, Green City Action Plan for the City of Belgrade, the General Urban Plans of the larger cities such as Belgrade, Novi Sad and Niš include measures that support a just green transition) and institutions (Ministry of Construction, Transport and Infrastructure, Traffic Safety Agency, Energy Efficiency Agency, Local governments) that regulate various aspects of transport, with an increasing focus on sustainability, energy efficiency and pollution reduction.

3.4.5 *Energy Transition*

Most of the WB countries are still significantly dependent on coal as a primary source of energy. Much of the electricity is produced in coal-fired thermal power plants, which contribute to significant CO_2 emissions and air pollution. This dependence slows down the process of transitioning to cleaner energy sources. On the other side, WB region has significant potential for the use of renewable energy sources. The region is rich in natural resources for the development of wind energy (especially in the areas of the Dinaric Mountain range), solar energy (especially in the south), as well as hydropower (on which Albania, for example, completely relies today). In Serbia, a set of new laws in the field of energy has been adopted, including the first Law on Renewable Energy Sources and the new Law on Energy Efficiency and the Rational Use of Energy and Industrial Policy Strategy of the Republic of Serbia from 2021 to 2030 (Circular economy). In addition to the above-mentioned policies (which were adopted in 2021), when it comes to the pillar of Decarbonisation (Climate, Energy & Mobility), Serbia adopted other, also very important policies at the national level such as Integrated National Climate and Energy Plan for the period 2021 to 2030 with a vision until 2050 (INECP), The Law on the Spatial Plan of the Republic of Serbia, Draft of the Spatial Plan of the Republic of Serbia 2021–2035 (in procedure) (2021, July). Ministry of construction, traffic and infrastructure. Government of the Republic of Serbia), Long term building renovation strategy Republic of Serbia 2050, Sustainable Urban Development Strategy of the Republic of Serbia until 2030, Climate Strategy & Action Plan of the Republic of Serbia and Strategy for Development of the Energy Sector of the Republic of Serbia until 2025 with projections until 2030. Ministry of Mining and Energy is in the process of developing two important documents that would enable implementation of the green transition in a just way—Energy development Strategy and Integrated National Plan for Energy and Climate Up Until 2050. In general, the decarbonisation pillar went the furthest in terms of applying the postulates of the green transition. The adoption of these laws and policies in previous years initiated a series of activities and projects related to JGT. Nationally important are EU for Green Agenda in Serbia, advancing medium- and long-term adaptation planning in the Republic of Serbia, ALTERENERGY—Energy sustainability for Adriatic small communities (projects), Roadmap for circular economy in Serbia, Initiative for a Just Green Transition and Decarbonisation in Serbia, ECO SYSTEM Programme, Zero Waste Municipalities (initiatives and practices), etc. The Republic of North Macedonia is significantly exposed to the effects of climate change and has undertaken to implement an ambitious plan moderated to mitigate and adapt to those changes, with the aim of reducing emissions (Generalisation of emissions) to the effect of 0% by 2030, compared to the 1990 level. The key element of the decarbonisation process in the energy sector of North Macedonia is the gradual shutdown of outdated coal-fired thermal power plants, which have a harmful impact on the environment. This step is necessary to achieve the goal of reducing emissions and supporting the country's green transition. RNM aspires to make decarbonisation a priority in its policy through the development of a comprehensive strategy for the gradual withdrawal of coal, which will be well designed and integrated into various horizontal policies. The main strategic documents governing energy and climate policies include the Energy Strategy (2020–2040) and the National Energy and Climate Plan (NECP), which is currently under revision to ensure flexibility regarding the deadlines for decommissioning thermal power plants. Although these documents have been adopted, implementation has been slow, and there is a lack of a clearly defined date for coal phase-out, especially as coal-fired power generation has recently increased due to rising energy prices. Montenegro employs a comprehensive suite of laws, strategies, and plans to guide its climate actions.[3] The Directorate for International Cooperation, European Integration, and Climate Change, which operates within the Ministry of Ecology, Spatial Planning, and Urbanism, plays a key role in the coordination of international cooperation in the field of air conditioning and integration in Montenegro. The involvement of various stakeholders, including multiple ministries and specialised agencies, ensures that climate considerations are systematically integrated across sectors and at all administrative levels. Nevertheless, Montenegro faces certain challenges that threaten the effective implementation of climate policies. One of the key problems is the delay in the finalisation and adoption of the National Energy and Climate Plan (NECP). This plan is vital to establishing a clear and integrated strategy for achieving energy and climate goals, and its absence leaves a gap in climate policy. Montenegro faces challenges in the implementation and operationalisation of existing laws in the field of climate change. Although the Law on Protection from the Harmful Effects of Climate Change has been adopted, its effective implementation and compliance monitoring will require intensified efforts and additional resources. Also, Montenegro's reliance on international aid in achieving the goals of reducing greenhouse gas (GHG) emissions indicates a certain vulnerability in its climate strategy. To achieve ambitious climate goals, it will be crucial to ensure sufficient and sustainable international support, as well as to develop strong domestic mechanisms for financing climate activities. Montenegro has established a solid foundation for

3

- National Climate Change Strategy 2030: Provides a comprehensive approach to mitigation and adaptation efforts.
- National Sustainable Development Strategy 2030: Aligns with the UN Agenda 2030, promoting sustainable development goals.
- National Energy Development Strategy: Details the path for sustainable energy development.
- Revised Nationally Determined Contribution (NDC): Sets specific targets for GHG emissions reduction.
- Roadmap towards the Circular Economy: Guides the transition towards a more sustainable economic model.

Among the others: National Energy and Climate Plan (NECP): Currently in progress, expected to be finalized soon; National Adaptation Plan: To be prepared under the Law on Protection against Climate Change.

climate change management through comprehensive laws, strategies, and an inclusive institutional framework. However, to fully realise its climate goals, the country must overcome challenges related to the implementation of existing measures, finalise key plans such as the National Energy and Climate Plan (NECP) and ensure stable financing mechanisms. With continued commitment and the strengthening of international cooperation, Montenegro can advance its climate ambitions and contribute to global efforts towards sustainability. The energy transition in the WB countries is an inevitable and necessary process. Still, it is burdened with numerous challenges, such as dependence on coal, outdated infrastructure, financial constraints, and social risks.

3.4.6 Waste Management

Most countries in the region, such as Serbia, North Macedonia, Albania, and Montenegro, have adopted waste management strategies that are aligned with EU goals. These strategic documents envisage measures to reduce waste generation, increase recycling and improve separate collection. However, many of these strategies remain at the planning level, without adequate implementation in practice. Insufficient infrastructure is one of the biggest problems in waste management in the WB. The lack of sanitary landfills, recycling facilities, and systems for the collection and treatment of special types of waste, such as electronic waste and batteries, significantly limit the ability of countries to effectively manage waste in accordance with EU directives. This is one of the biggest challenges facing countries in the region, leading to low levels of recycling and waste separation. For example, in Serbia, Montenegro, Bosnia, and Herzegovina, some measures have been implemented for better management of special types of waste, such as electronic waste, textiles, and batteries, but these measures are still not sufficiently developed to significantly reduce the harmful effects of these types of waste. Despite some progress in alignment with EU waste management legislation, the implementation and enforcement of existing regulations in the WB remains a significant challenge. The strategies and roadmaps envisaged by the Action Plan for the WB should enable the development of waste management plans, including the definition of financial support models, such as subsidies, financial instruments, and public-private partnerships. These mechanisms aim to support the construction and maintenance of waste management infrastructure at the local and regional level. In this sense, these plans must be harmonised with the revised EU legislation and aimed at reducing waste generation, increasing separate collection and recycling, and encouraging circular business models. The economies of the region should define targets for the reuse and recycling of municipal waste by 2035, which is in line with EU targets, and set ambitious targets for packaging waste, scrap vehicles, and batteries, thus contributing to the overall goals of the circular economy. For example, the institutional and legislative framework in Serbia includes waste management issues and represents a significant part of the country's environmental policy. Adopted national policies that are under the other four pillars of the GAWB in Serbia are the Law on Waste Management, the Law on Packaging and Packaging Waste, and the Waste Management Program of the Republic of Serbia for the period 2022–2031. In general, the implementation of the Green Agenda is delayed due to various challenges, including lack of financial resources, political will, and coordination, so accelerating the implementation is necessary to achieve the goals of the green transition. The concept of a just transition is at an early stage of development, and more efforts are needed to ensure that the transition is inclusive and fair for all citizens. For example, Serbia receives support from the EU, but it is necessary to make better use of available funds and increase the capacity to absorb funds. On the other hand, reliance on EU funds and international financial institutions can create possible dependencies and uncertainty in the implementation of projects. Therefore, it is necessary to create more favourable conditions for attracting private investments and developing public–private partnerships. Continuous monitoring and transparent reporting are key to monitoring progress and adapting strategic documents, i.e. different types of plans at all planning levels as needed.

3.5 Conclusion

This overview of key international level and WB just transition regulations, guidelines, and policies showcases how the concept has evolved in relation to core JGT dimensions. Early references to just transitions focused on the environmental and social dimensions highlighting the need to reduce greenhouse gas emission while protecting jobs. The temporal dimension is clearly recognised in the development of time sensitive climate targets. As the JGT became more focused on implementation, territorial, democratic, governance and communication dimensions have become more prominent as policymakers have stressed the need for place-based JGT plans based on open and inclusive policymaking processes. It is vital JGT policies are developed on the basis of a well-defined transition concept that considers a range of different JGT dimensions. The broad nature of the just transition notion means it is interpreted differently by stakeholders across governance levels. While acknowledging the rights of the EU, nation states, and regions to interpret and implement the JGT concept in different ways, some conceptual coherence can provide important guidance in ensuring that plans strike a balance in addressing a range of different JGT dimensions. When looking closer at the WB region, we see that countries are learning to navigate the complex institutional and policy framework of JGT. None of the countries have direct institutions or policies on JGT and this is predominantly tackled through sectorial approaches. The EU plays an important role in shaping these sectorial approaches and trying to integrate them. The path towards the Europeanisation of WB goes through JGT (Berisha, 2025b), and it is evident that this will require not only legal and policy adaptation but a rather broader and more comprehensive institutional and cultural change. The required changes are deep and run through the roots of the governance system. Often, as argued, they are dependent on political will, human capacities, and the ability to think outside of the box and to use existing and new financial schemes to support JGT. Lastly, it is important that academic research also plays a role in JGT. As it was seen, there is a growth in research in the topic. Nevertheless, this needs to be integrated into policymaking as well as go beyond the identification of challenges or sectorial analysis in JGT. As such, the presence of WB institutions (academic, NGOs, and public authorities) in research programs such as Horizon, Erasmus, and COST supported by national research grants becomes very important in terms of pushing forward the JGT agenda.

References

Allkja, L., Toto, R., Dhrami, K., Toska, M., Sokoli, I., Delli, G., Muka, R., Kull, D. W., Milutin, D., Sultanov, O., Skalon, T., Gjika, K., & Mirzoyev, V. (2022). *Toolbox for Integrating disaster risk management into Albanian municipal planning and budgeting: Technical Annex—Guideline to create, unify and manage data on the local level for disaster risk management (English)*. World Bank Group. http://documents.worldbank.org/curated/en/099145009022241355

Balz, V., Verbeek, T., Qu, L., Baugh, R., Dąbrowski, M., Jungsberg, L., Ferry, M., Georgieva, N., Serdült, U., Trifonova, M., Amin, S., & Clouet, Y. (2023), DUST theoretical and conceptual framework, Deliverable 1.1. Horizon Europe DUST project.

Berisha, E. (2025a). Moving towards an inclusive Green Agenda in the Western Balkans. Foundation for European Progressive Studies (FEPS). https://doi.org/10.13140/RG.2.2.12092.76162

Berisha, E. (2025b). Green EU integration—The role of the green agenda in accelerating just green transitions in the Western Balkans. In E. Berisha, J. Moodie, L. Allkja, & M. Jeftić (Eds.), *The Palgrave handbook of just green transitions in the Western Balkans and beyond*. Palgrave Macmillan.

Berisha, E., Colic, N., Cotella, G., & Nedović-Budić, Z. (2018). Mind the gap: Spatial planning systems in the western Balkan region. *Transactions of the Association of European Schools of Planning, 2*(1), 47–62. https://doi.org/10.24306/TrAESOP.2018.01.004

Berisha, E., Colic, N., Cotella, G., & Nedović-Budić, Z. (2024). Potential future enlargement—Spatial planning in the Western Balkans. In V. Nadin, G. Cotella, & P. Schmitt (Eds.), *Spatial planning systems in Europe: Comparison and trajectories* (pp. 216–245). Edward Elgar. https://doi.org/10.4337/9781839106255.00022

Caka, F. (2021). Municipal capacity gaps and needs on integrating climate change aspects in spatial planning into Kosovo. UN-Habitat. Accessed on 20th August 2025. https://unhabitat-kosovo.org/wp-content/uploads/2021/11/Spatial-Planning-Climate-Change-Survey.pdf

Davoudi, S. (2020). Resilience, uncertainty, and adaptive planning. *Annual Review of Territorial Governance in the Western Balkans, 1*, 120–128.

Dietachmair, J., Duske, P., & Pfefferkorn, W. (2024). *Climate governance analysis and capacity development strategy Western Balkans, EUKI*. Accessed on 14th February 2025. https://www.euki.de/en/euki-publications/climate-governance-analysis-and-capacity-development-strategy-western-balkans/

European Commission (2019). Communication from the commission—The European green deal, COM, 2019, 640 Final.

European Commission (2021). The just transition mechanism: Making sure no one is left behind.

Gatarić, D., Đerčan, B., Živković, M. B., Ostojić, M., Manojlović, S., Sibinović, M., Lukić, T., Jeftić, M., Lutovac, M., & Lutovac, M. (2022). Can Depopulation Stop Deforestation? The Impact of Demographic Movement on Forest Cover Changes in the Settlements of the South Banat District (Serbia). *Frontiers in Environmental Science, 10*, Article 897201. https://doi.org/10.3389/fenvs.2022.897201

Georgieva, N., & Ferry, M. (2023), Identification and assessment of participatory processes in sustainability transition measures in case study regions. Deliverable 3.1. Horizon Europe DUST Project.

Georgieva, N., & Ferry, M. (2024), Factors impacting the scale and quality of participation of least engaged communities in deliberative governance of transition policies. Deliverable 3.1. Horizon Europe DUST Project.

ILO. (2015). *Guidelines for a just transition towards environmentally sustainable economies and societies for all* (p. 2015). ILO Publications.

Jeftić, M. R., Šećerov, V., & Tošić, D. (2021). Development of the urban systems in Serbia—Towards new spatial plan of the Republic of Serbia. *In Collection of Papers of the 5th Serbian congress of geographers "Innovative approach and perspectives of the applied geography"* (pp. 194–202). Novi Sad: University of Novi Sad—Faculty of Sciences, Department of Geography, Tourism and Hotel Management. https://hdl.handle.net/21.15107/rcub_gery_1120

Jeftić, M., Tošić, D., & Nikolić, T. (2019). Resilient functional urban regions: Spatial planning in the light of climate change. In W. Leal Filho, G. Trbic, & D. Filipovic (Eds.), *Climate change adaptation in Eastern Europe. Climate change management*. Springer, Cham. https://doi.org/10.1007/978-3-030-03383-5_8

Johansson, V. (2023). Just transition as an evolving concept in international climate law. *Journal of Environmental Law, 35*(2), 229–249.

Moodie, J., Tapia, C., Löfving, L., Gassen, N. S., & Cedergren, E. (2021). Towards a territorially just climate transition—Assessing the Swedish EU territorial just transition plan development process. *Sustainability, 13*(13), 7505.

Pinnavaia, L., & Berisha, E. (2021). The role of cross-border territorial development. Evidences from Albania. In E. Berisha, G. Cotella, & A. Solly (Eds.), *Governing territorial development in the Western Balkans. Advances in spatial science*. Springer, Cham. https://doi.org/10.1007/978-3-030-72124-4_15

Roessler, C. (2016). "Just Transition"—Just what is it?—Labor network for sustainability.

Rosemberg, A. (2017). *Strengthening just transition policies in international climate governance; policy analysis brief*. The Stanley Foundation

Shaker, Y., & Berisha, E. (2025, forthcoming). Just green transitions: Between terminological inexactitudes, conceptual fragmentations, and the exigency for a theoretical framework. *European Spatial Research and Policy*.

Spatial Plan of the Republic of Serbia 2021 to 2035 (Draft). (2021, July). *Ministry of construction, traffic and infrastructure. Government of the Republic of Serbia*. https://www.mgsi.gov.rs/sites/default/ f iles/PPRS%20Nacrt.pdf.

Solly, A., & Berisha, E. (2021). Towards the territorialisation of EU Cohesion Policy? The Case of EUSAIR. In E. Berisha, G. Cotella, & A. Solly (Eds.), *Governing Territorial Development in the Western Balkans. Advances in Spatial Science*. Springer, Cham. https://doi.org/10.1007/978-3-030-72124-4_16

Tapia, C., Giacometti, G., Moodie, J. R., Cedergen, E., & Shaneh, A. (2025–in this volume). From mitigation to transformation: An evolutionary perspective on the Just Green Transition. In E. Berisha, J. Moodie, L. Allkja, & M. Jeftić (Eds.), *The Palgrave handbook of just green transitions in the Western Balkans and Beyond*. Palgrave Macmillan.

Tavares, M. (2022). *A just green transition: Concepts and practice so far* (Future of the World, UN-DESA Policy Brief, No141).

Thomas, A. (2021). Framing the just transition: How international trade unions engage with UN climate negotiations. *Glob Environ Change, 70*, Article 102347.

United Nations (2015). *Transforming our World: The 2030 agenda for sustainable development, A/RES/70/1, 21st October 2015*. Accessed on 1 July 2021. https://sdgs.un.org/2030agenda

Živanović, Z., Popović, V., Protić, B., Djordjević, A., Šećerov, V., & Jeftić, M. (2025–in this volume). "Signs by the roadside" on the road to Just Green Transition in Serbia. In E. Berisha, J. Moodie, L. Allkja, & M. Jeftić (Eds.), *The Palgrave Handbook of Just Green Transitions in the Western Balkans and Beyond*. Palgrave Macmillan.

PART II

Theorising Just Green Transitions

CHAPTER 4

The Unknown Knowns of the Just Green Transitions: A Meta-Governance Approach for a Possible Enlargement of the European Union

Yahya Shaker

4.1 Introduction

As the future of Europe undergoes crucial transitions (European Commission, 2019), this chapter delves into the pivotal role of Just Green Transitions (JGT) at the nexus of European Union enlargement and the accession of the Western Balkans. These expected transitions are vital for integrating the Western Balkans into the European Union (Berisha, 2025a). Anchored by the European Green Deal (EGD) and the Green Agenda for the Western Balkans (GAWB) (European Commission, 2020a, European Commission, 2020c; European Commission, 2023), the chapter highlights the intricate challenges of conceptualising, operationalising, and advancing the JGT within a region marked by diverse governance structures and varying institutional capacities. Through a Meta-Governance perspective (Jessop, 2011), the chapter examines what the EU can offer the Western Balkans and how the region serves as a critical pivot in realising a unified European vision (Berisha, 2025a, b). The aims of this chapter are threefold. Firstly, it explores the intersection between the Western Balkans' transition towards carbon neutrality and their paths to EU accession. Secondly, it argues for the notion of a Meta-Governance approach to navigate the complexities of a macro-regional transition, ensuring no state is left behind. Finally, the chapter sheds light on the geopolitical dimension of the JGT within the broader context of EU enlargement, emphasising the

Y. Shaker (✉)
Interuniversity Department of Regional and Urban Studies and Planning, Politecnico Di Torino, Turin, Italy
e-mail: yahya.shaker@polito.it

E. Berisha et al. (eds.), *The Palgrave Handbook of Just Green Transitions in the Western Balkans and Beyond*,
https://doi.org/10.1007/978-3-031-95075-9_4

interconnectedness of institutional reforms inherent in aligning the Western Balkans' accession with the European Green Deal. The chapter is structured as follows: the first section establishes the linkage between the EU's enlargement strategies and the European Green Deal. The second section focuses on the geopolitical dimensions of the Just Green Transitions and their impacts on possible EU macro-regional integration. The third section introduces a Meta-Governance framework as a strategic tool for advancing both the JGT and EU accession goals. By addressing these interconnected themes, the chapter provides a holistic perspective on the transformative potential of the Just Green Transitions in shaping the future of the Western Balkans in the European Union.

4.2 A European Deal or a European Union Deal

The European Union—and not only—has been facing waves of compound crises. In particular, the Balkans is one of the most affected European regions, not only when it comes to the impacts of climate change (European Commission, 2020c). Simultaneous geopolitical, climatic, and economic challenges have affected the European Union's strategic partnerships (Bonomi, 2021) and its enlargement plans (European Commission, 2022). Simultaneously to the global pandemic, the European Commission introduced its European Green Deal, a once-in-a-lifetime transnational roadmap (IEA, 2021) and a geopolitical and geo-economic continental agenda (Leonard et al., 2021) pivoting on a transnational twinned transition which ought to be green, just, inclusive, fair (European Commission, 2019), and now also became competitive (European Commission, 2024b).

This continental agenda aims to forge institutional reform, which not only the Member States are required to undertake but also the Candidate States,[1] including the Western Balkans[2] (European Commission, 2020c) and is aligned with the New Growth Plan for the Western Balkans (European Commission, 2023) and the long-standing interest in enlarging the EU towards the East.

[1] See the Council conclusions on Enlargement, 17 December 2024, the 2024 Communication on EU Enlargement Policy (Council of the European Union, 2024). See also the speech (2024b) at the Interparliamentary Conference for CFSP and the CSDP—Europe and the countries wishing to join the EU, 09.09 2024. Candidate Countries in 2024: Albania, Bosnia and Herzegovina, Georgia, Kosovo, Moldova, Montenegro, North Macedonia, Serbia, Türkiye, and Ukraine.

[2] On the Spectre of the Balkans, see the reading of Žižek (1999). In this contribution, the term Western Balkans refers to a geo-political neologism used by the European Union referring to six countries formed during the collapse of The Socialist Federal Republic of Yugoslavia, located in Southern and Eastern Europe in the Balkan area: Republic of Albania (*Republika e Shqipërisë*), Bosnia and Herzegovina (*Bosna i Hercegovina*), Republic of Kosovo (*Republika e Kosovës; Република Косово*), Montenegro (*Црна Гора*), Republic of North Macedonia (*Република Северна Македонија*), and Republic of Serbia (*Република Србија*).

Its Reform Agenda has been aligned with the objectives of both the European Green Deal and the Green Agenda for the Western Balkans (European Commission, 2020a; European Commission, 2020c; European Commission, 2023), mobilising European funds for a "Green Transition" and a "Just Transition" (European Commission, 2019). On July 1st, 2021, the European Commission put into action the Just Transition Mechanism (JTM), aiming at leaving no person nor region behind and accompanying the Member States in preparing their Territorial Just Transition Plans targeting the most vulnerable carbon-dependent regions (European Commission, 2021). Within the framework of the JTM, both the Member States and the Candidate Countries—with their different spatial governance and planning systems (Berisha et al., 2021; Faludi, 2012; Nadin et al., 2018)—are still trying to conceptualise, operationalise, and implement the so-called Just Green Transitions while facing the predicament of not having completely agreed upon this conceptual compound terminology (Shaker & Berisha, 2024, 2025).

To achieve these transitions, the European Commission has been intensively investing and coordinating between the Member States, financing the transitions of the Candidate Countries, and creating agreements with the EU Neighbours included in its Enlargement Packages of the European Neighbourhood Policy and Enlargement Negotiations (European Commission, 2020b; European Commission, 2024c). Accordingly, several economic packages, political measurements, and reform support are in course, including the NextGenerationEU, the JTM, the Fit-for-55 package, the REPowerEU plan, the EU Reform Agenda, and the New EU Strategy for the Potential Enlargement of the European Union (European Parliament, 2021). Yet, there is still no common agreement on the definition of the Just Green Transitions (Shaker & Beirsha, 2024; Shaker & Persico, 2024). This deal, thus, represents an ambitious masterplan for the WB not only to address long-lasting socio-economic issues while trying to achieve carbon neutrality through the utilisation of the Just Transition Fund but, most importantly, it presents an opportunity both for macro-regional cohesiveness and a strategy for an institutional reform possibly leading to the accession to the EU. Meaning that this "deal" could be strategically utilised as a path towards a macro-regional integrative social vision, advocating for fairness and inclusivity to ensure that all vulnerable populations within the WB could truly benefit together from the transitions as one macro-region while *leaving no one nor region behind* (European Commission, 2019). As no one-size-fits-all, the Green Agenda of the Western Balkans (European Commission, 2020a) not only represents one foundational step towards the region's prospective integration but also a complementary strategic chess piece to the accession through the alignment with the EGD. Consequently, the Western Balkans are adapting to the European Green Deal objectives while tailoring a more suitable fit for their societal, political, and geopolitical contexts (Tahirović, 2024). The intersection of the two objectives—achieving the so-called JGT while simultaneously advancing the EU's enlargement—brings to the fore the geopolitical complexities entangled within the EGD, which is expected to be just, inclusive, and without leaving no person or region behind (European Commission, 2019), especially in a region characterised by complex (geo)political interests and (geo)economic disparities (Bieri, 2015). Borrowing from critical theory and political philosophy, without delving into the ontological, philological, and axiological polylemma of what are the *Balkans* and what are the *Western Balkans*, this chapter is building upon Slavoj Žižek's theorisation on the "*unknown knowns*" (2006) and *The Spectre of the Balkans* (1999), highlighting the cultural identity, strategic, and geopolitical value of the Balkan region as one fundamental missing pillar—or a phantom limb of European Union—to the full integration and cohesion of the European Union; culturally, politically, geographically, and geopolitically. In an analogy to Kennedy's proposition (Mehltretter, 2009), this chapter looks at what the Western Balkans can offer for the EU or whether the EU can lead the Just Green Transitions in the Western Balkans (Prelec et al., 2023), but rather what the EU can offer the Western Balkans and whether the Balkan Region, in turn, is one pivotal as the geopolitical region of the European Green Deal. Following this introductory section, The following one aims to find an association between the transitions of the Western Balkans in terms of European integration and accession to the European Union and their paths towards the so-called Just Green Transitions, while the third one advocates for a Meta-Governance approach for the transitions both to a carbon-neutral and an EU future for the Western Balkans.

4.3 The Geopolitical Dimension of the Just Green Transitions

Transition is an ambiguous terminology in social and political sciences (Škuflić, 2010. The EU Twin Transitions (European Commission, 2019) pivot on the security of energy. Their realisation has become the nucleus of the current global geopolitical environment (Najetović, 2024) and depends on the formation of global energy alliances, which both are influencing the ongoing accession processes of the Western Balkans into the European Union (Tziarras, 2019). The process of accessing new Members to the European Union is determined by the candidate Countries' institutional capacities (Abramowitz et al., 2011) to reform (Stanicek et al., 2023). Following the 2019 global pandemic emergency and its consecutive Energy and Food crises of 2022, the European Green Deal became the EU First Agenda (Almeida et al., 2023) for the Member States and the Western Balkans to progress key institutional reforms (European Commission, 2020b) required for the enlargement (European Commission, 2024c) and possibly another aspiration to join the EU block. The idea of joining the Western block following the fall of Yugoslavia has been subject to sociopolitical, geopolitical, and territorial governance entanglements in the Balkans (Berisha et al., 2018, 2024; Oberschall, 1996), which all have been highly influenced by identity projections (Hayden, 2012), ideological impositions, and Westernisation metonymicity (Tsaroucha, 2019), which all might have contributed to counteractive and retroactive bifurcation of the Balkan region from the rest of the Western world (Jukić, 2013). It is worth mentioning that it has been over a quarter of a century since the great expectations for EU stability and the WB's eventual accession to the EU have been first discussed (Steil & Woodward, 1999). Fifteen years passed, and the Paris Climate Deal 2015 (Ratha, 2019) offered room for new opportunities for EU accession through common Climate Change Policies tying the Western Balkans to the EU (Cierco Gomes, 2019). The term *Just Green Transitions* (JGT) remains an emerging and not yet widely established notion within either European Union policy discourse or academic literature. Positioned as a *meta-concept*—a synthesis of existing concepts and a concept of concepts (Shaker & Persico, 2024)—it is intended to capture the multi-perspectival and pluralistic nature of transitions across diverse socio-spatial-temporal contexts (Levin-Keitel et al., 2018). Unlike the terminological and conceptual paradigms of the *Green Transition* and the *Just Transition*, the Just Green Transitions, as an umbrella term, extends far beyond the separate yet connected policy constructs, offering a Meta-Governance (Jessop, 2011) perspective that acknowledges the interconnected, multi-layered, and simultaneous dynamics of change. JGT, therefore, emerges as an adaptive and integrative paradigm capable of addressing the complexities and inequities inherent in the ongoing processes of ecological and societal transitions/transformations (Shaker & Berisha, 2024, 2025). Putting into perspective both the possible geopolitical implications of the EGD and the Green Agenda for the Western Balkans (GAWB) (European Commission, 2020a), it is crucial to encompass the potential for the EU enlargement towards the East and the multifaceted dimensions of integrating the WB into the EGD ambitions. However, conceptualising, operationalising, and advancing the JGT present major challenges (Ignjatović et al., 2024) due to the sociopolitical, economic, and territorial governance specificities and varying levels of institutional capacities within the region (Berisha & Cotella, 2021). In a context where the EU strategies and deals go beyond the EU borders (Lavenex &Schimmelfennig, 2009), it could be argued that geopolitics is momentous in the attainment of the European Green Deal (EGD) (Official Journal of the European Union, 2022) and thus, current internal and external geopolitical interests in the Eastern Regions could shuffle the geopolitical landscapes and put governance models, frameworks, and decision-making processes to the test (European Commission, 2022). Both the European Green Deal and its twin transitions—inevitably—are forecasted to have a sociopolitical and geopolitical influence on the enlargement of the European Union (Official Journal of the European Union, 2022), not only at a transnational scale but at a macro-regional one as well.

The EU's rivalry in the clean energy space race (Official Journal of the European Union, 2022) aiming to make Europe the first climate-neutral continent (European Commission, 2019) is retroactively, retrospectively, and consequently re-establishing not only the speed, the geographic and [geo]political direction of the transitions (Scholten et al., 2019) but rather the level playing field (Borrell, 2023) of the global governance and its transitions. Viewing the EGD from an EU foreign and regional development perspective (Teevan et al., 2021), the geopolitical dimension of the JGT (Shaker & Persico, 2024) is thus becoming almost undetachable from the concepts of European territorial fragmentation (Sharpe, 1989) and the territorial phantom limbs (Billé, 2014) meaning that both the climatic and the geo-economic spatiality of the Deal goes beyond the territoriality of the European Union. In the case of the Balkan region—where a definition is beyond linguistic complexity, but both geographically and even more so geopolitically (Tahirović, 2024)—the path to the European Union (Dabrowski & Myachenkova, 2018) has been a constant topic on the European Council's agenda since 2009 (Anghel, 2020). It is, however, somehow interesting to observe the paradigm shift from Balkanisation towards Europeanisation (Jano, 2008), especially when, for the first time in European politics, adding the term *Western* to a political terminology counteracts the notion of Westernisation, the Western world, and Western Europe (Liridon, 2023). Semantically, one might argue that the term Western, in the Balkan case, might have been deliberately acting as a geopolitical identity inhibitor, identifying the last six states that have not yet joined the union away from Western Europe's cultural and political identity. Instead, in this context, the term Western indicated literally where these states are geographically located in the Balkan Peninsula. Logically, one might as well wonder, but what about Slovenia and Croatia, since their accession in 2004 and the latter in 2013, are no longer conceptually associated with the WB (Hayden, 2012; Liridon, 2023). These quasi-esoteric geo (graphical, political, social, economic) realities and their definition denote an issue of ideological antagonism (Žižek, 1999). One could also question this neologism "Western Balkans" (Vienna European Council, 1998), with its terminological implications and identity contrast from the term Balkans, which might present not only a strategic geopolitical chessboard predicament (Bieber &Tzifakis,, 2019) for various local realities having been reshaped over and over by global players' interests making the geopolitical game in this region particularly intricate. Mainly, the coining of this term (Dabrowski & Myachenkova, 2018) and its continuous use is paradoxically one main issue that might have been causing more avulsion than twinning, for it might have ideologically differentiating the Western Balkans from the rest of the Balkan Region (Liridon, 2023).

4.4 Towards a Meta-Governance Approach to the Just Green Transitions

The most pressing issue regarding the Just Green Transitions is how to govern these transitions, not only at the level of the Member States but also at the level of the Candidate Countries that are aiming to join the European Union within its Enlargement Plan (European Commission, 2024c) within its Western Partnership (European Commission, 2020b). Within the EU Multi-level Territorial Governance model (Faludi, 2012), the operationalisation processes can barely have all the necessary socio-territorial and institutional policy mixes (Zepa & Hoffmann, 2023) and (macro)regional capacities to govern energy transitions (Van Dijk et al., 2022) for advancing such ambitious transitional territorial strategies across various vertical and horizontal multi-level, multi-actor, and multi-sectoral institutional dynamics (Cotella et al., 2021) that reflect the various realities experienced by individuals, societies, communities, and regions. Moreover, the coordination and decision-making of the JGT at the transnational, national, and subnational levels are quite challenging (Pisano et al., 2014) due to the variety of the existing socio-territorial models, institutional and administrative traditions and technical and political cultures of the Member States. This issue becomes even more complex when added to the enlargement of the Eastern Neighbours (European Commission, 2024b) and their various Territorial Governance and Planning Systems (Cotella & Toto, 2021). Since the EGD extends beyond EU borders (Walker, 2024), emphasising partnerships to achieve climate neutrality for the Western Balkans, beyond the place-sensitive governance complexities (Cotella &Toto, 2021), a Meta-Governance approach (Jessop, 2011) of holistic, adaptive, flexible, and inclusive intergovernmental structure (Fratianni, 2003) and multilateral, regional cooperation (Tahirović, 2024) for advancing the Western Balkans' accession to the European Union through the European Green Deal might offer a possible governance path both to realise the Just Green Transitions and achieve the full accession to the EU. This contribution is thus arguing that a Meta-Governance framework forging a real-time (Meuleman, 2022) (Fig. 4.1).

Intergovernmental macro-regionalisation (Gänzle &Kern, 2016) framework corresponding to the New Growth Plan for the Western Balkans and bridging both the EGD and the GAWB through strengthening vertical and horizontal coordination across scales, sectors, actors, and levels of decision-making, and homogenising a strategic alignment on the accession of the rest of the Balkan countries to the EU (European Commission, 2023), could offer a more holistic governance mechanism for a macro-regional accession project. This meta approach could offer room to align the conceptualisation of the GAWB at various operationalisation levels:

- Environmental and climate directives and regulations (Knez et al., 2022),

Fig. 4.1 The Balkan, the Western Balkans, and Europe. (*Source*) author's own elaboration

- Trans-sectoral decarbonisation targets in the different WB countries (Jovanović et al., 2024), (c) Transnationally coordinating macro-regional energy governance (Cotella & Toto, 2021; Valkenburg & Cotella, 2016),
- Coordinating and monitoring the essential institutional reforms required for the alignment with the EU (European Commission, 2023),
- Ensuring that no country nor Balkan community should be left behind during the accession.

In other words, a macro-regional governance twinning process includes all the WB states under one regional accession to the European Union.

4.5 Conclusions

This contribution started with deconstructing the European Green Deal and its relevance to the European Regions, especially the Balkans. Followed by framing and debugging the undefined meta-concept of the so-called Just Green Transitions and their interconnectivity to both the Green Agenda for the Western Balkans and their relevance for possible accession to the

European Union. The chapter then examined the geopolitical dynamics underpinning both the EU's JGT and its enlargement policy, particularly towards the Western Balkans. This chapter discussed how the Meta-Governance could influence the political, social, and economic integration of the Western Balkans through a macro-regional Just Green Transition and advocating for a meta-conceptual approach, which recognises the multiplicity, simultaneity (Shaker & Berisha, 2024), diversity, and complexity of the transitions and emphasises the role of Spatial Planning and Territorial Governance (Berisha & Cotella, 2021; Berisha et al., 2021; Nadin et al., 2018) in achieving both institutional integration and accession and carbon neutrality. The chapter concludes that the Just Green Transitions, as envisioned in the European Green Deal, offer a viable framework for the Western Balkans' integration into the EU, but only if they are accompanied by a tailored meta-governance approach that addresses the macro-regional challenges and opportunities included in the Western Balkan' Green Agendas. By adopting a meta-governance approach into the broader framework of institutional reform, the Western Balkans would successfully navigate the internal challenges and take a step towards the expected EU enlargement (European Commission, 2024c).

Acknowledgements This research is funded by the research project European HORIZON-WIDERA-2021-ACCESS-02-01 | Foster Research Excellence for Green Transition in the Western Balkans-GreenFORCE Twinning of the Western Balkans, grant number: 101059411.

References

Abramowitz, M., Bieber, F., Jović, D., Manchin, R., Mungiu-Pippidi, A., Ordanoski, S., Radulović, M., Rupnik, J., Sarajlić-Maglić, D., Štiks, I., Surroi, V., & Teokarevic, J. (2011). The Western Balkans and the EU: 'The hour of Europe. In J. Rupnik (Ed.), *Chaillot papers*. The European Union Institute for Security Studies. ISBN 9789291981878. ISSN 1017-7566. https://www.iss.europa.eu/sites/default/files/EUISSFiles/cp126-The_Western_Balkans_and_the_EU.pdf

Almeida, D. V., Kolinjivadi, V., Ferrando, T., Roy, B., Herrera, H., Gonçalves, M. V., & Van Hecken, G. (2023). The "Greening" of Empire: The European green deal as the EU first agenda. *Political Geography. 105*(102925). ISSN 0962-6298. https://doi.org/10.1016/j.polgeo.2023.102925

Anghel, S. E. (2020). *Western Balkans on the European council agenda overview of discussions since the Lisbon treaty*. European Council in Action. European Parliamentary Research Service. European Council Oversight Unit PE 642.819—April 2020. https://www.europarl.europa.eu/RegData/etudes/BRIE/2020/642819/EPRS_BRI(2020)642819_EN.pdf

Berisha, E. (2025a). Moving towards an inclusive green agenda in the Western Balkans. Foundation for European Progressive Studies (FEPS). https://doi.org/10.13140/RG.2.2.12092.76162

Berisha, E. (2025b). Green EU integration—The role of the green agenda in accelerating just green transitions in the Western Balkans. In E. Berisha, J. Moodie, L.

Allkja, & M. Jeftić (Eds.), *The Palgrave handbook of just green transitions in the Western Balkans and beyond*. Palgrave Macmillan.

Berisha, E., Colic, N., Cotella, G., & Nedović-Budić, Z. (2018). Mind the gap: Spatial planning systems in the western Balkan region. *Transactions of the Association of European Schools of Planning, 2*(1), 47–62. https://doi.org/10.24306/TrAESOP.2018.01.004

Berisha, E., Colic, N., Cotella, G., & Nedović-Budić, Z. (2024). Potential future enlargement—Spatial planning in the Western Balkans. In V. Nadin, G. Cotella, & P. Schmitt (Eds.), *Spatial planning systems in Europe: Comparison and trajectories* (pp. 216–245). Edward Elgar. https://doi.org/10.4337/9781839106255.00022

Berisha, E., & Cotella, G. (2021). Territorial development and governance in the Western Balkans. In E. Berisha, G. Cotella, & A. Solly (Eds.), *Governing territorial development in the Western Balkans. Advances in spatial science*. Springer, Cham. https://doi.org/10.1007/978-3-030-72124-4_2

Berisha, E., Cotella, G., Janin Rivolin, U., & Solly, A. (2021). Spatial governance and planning systems in the public control of spatial development: A European typology. *European Planning Studies, 29*(1), 181–200. https://doi.org/10.1080/09654313.2020.1726295

Bieber, F., & Tzifakis, N. (2019). *The Western Balkans as a Geopolitical Chessboard? Myths, realities and policy options*. https://www.biepag.eu/wp-content/uploads/2021/07/The_Western_Balkans_as_a_Geopolitical_Chessboard.pdf

Bieri, M. (2015). *The Western Balkans Between Europe and Russia*. CSS Analyses in Security Policy, N0. 170. Centre for Security Studies (CSS), ETH Zurich. https://css.ethz.ch/content/dam/ethz/special-interest/gess/cis/center-for-securities-studies/pdfs/CSSAnalyse170-EN.pdf

Billé, F. (2014). Territorial Phantom pains (and other cartographic anxieties). *Environment and Planning d: Society and Space, 32*(1), 163–178. https://doi.org/10.1068/d20112

Bonomi, M. (2021). *In Search of EU Strategic Autonomy: What Role for the Western Balkans?* IAI—Istituto Affari Internazionali. ISBN 978-88-9368-200-8. https://www.iai.it/sites/default/files/9788893682008.pdf

Borrell, J. (2023). *Geopolitics of the green transition and improving EU's economic security*. The European External Action Service (EEAS). Published on 28.03.2023. https://www.eeas.europa.eu/eeas/geopolitics-green-transition-and-improving-eu's-economic-security_en

Cierco Gomes, T.M.R. (2019). The European Union accession and climate change policies in the Western Balkan countries. In T. Sequeira & L. Reis (Eds.), *Climate change and global development. Contributions to economics*. Springer, Cham. https://doi.org/10.1007/978-3-030-02662-2_8

Cotella, G., &Toto, R. (2021). Foreword: Territorial governance in the Western Balkans: Multi-scalar approaches and perspectives. *European Spatial Research and Policy, 28*(2), 5–20. https://doi.org/10.18778/1231-1952.28.2.01

Cotella, G., Janin Rivolin, U., Pede, E., & Pioletti, M. (2021). Multi-level regional development governance: A European typology. *European Spatial Research and Policy, 28*(1), 201–221. https://doi.org/10.18778/1231-1952.28.1.11

Council of the European Union. (2024). *Council conclusions on enlargement*. Brussels, 17 December 2024. 16983/24 ELARG 189 COWEB 218 COEST 750. https://data.consilium.europa.eu/doc/document/ST-16983-2024-INIT/en/pdf

Dabrowski, D., & Myachenkova, Y. (2018). *The Western Balkans on the road to the European Union*. Policy Contribution Issue 04. Bruegel. https://euagenda.eu/upload/publications/untitled-133335-ea.pdf

European Commission. (2019). *The European green deal*. COM (2019) 640 final. Brussels, 11.12.2019. https://eur-lex.europa.eu/resource.html?uri=cellar:b828d165-1c22-11ea-8c1f-01aa75ed71a1.0002.02/DOC_1andformat=PDF

European Commission. (2020a). *Guidelines for the implementation of the green agenda for the Western Balkans*. SWD (2020) 223 final. https://neighbourhood-enlargement.ec.europa.eu/system/files/2020-10/green_agenda_for_the_western_balkans_en.pdf

European Commission. (2020b). *Eastern partnership policy beyond 2020. Joint staff working document. Structured consultation on the future of the Eastern partnership*. JOIN (2020) 7 final. https://www.eeas.europa.eu/sites/default/files/jswd_structured_consultation_on_the_future_of_the_eastern_partnership.pdf

European Commission. (2020c). *An economic and investment plan for the Western Balkans. guidelines for the implementation of the green agenda for the Western Balkans. Commission staff working document*. SWD (2020) 223 final. Brussels, 6.10.2020. https://neighbourhood-enlargement.ec.europa.eu/system/files/2020-10/green_agenda_for_the_western_balkans_en.pdf

European Commission. (2021). *Commission staff working document on the territorial just transition plans*. SWD (2021) 275 final. https://ec.europa.eu/regional_policy/sources/funding/just-transition-fund/swd_territ_just_trans_plan_en.pdf

European Commission. (2022). *2022 Communication on EU enlargement policy*. COM (2022) 528 final. October 2022. https://data.consilium.europa.eu/doc/document/ST-13494-2022-INIT/en/pdf

European Commission. (2023). *New growth plan for the Western Balkans*. COM (2023) 691 final. https://neighbourhood-enlargement.ec.europa.eu/document/download/8f5dbe63-e951-4180-9c32-298cae022d03_en?filename=COM_2023_691_New%20Growth%20Plan%20Western%20Balkans.pdf

European Commission. (2024a). *On pre-enlargement reforms and policy reviews*. Brussels, COM (2024) 146 final. https://commission.europa.eu/document/download/926b3cb2-f027-40b6-ac7b-2c198a164c94_en

European Commission. (2024b). *The clean transition dialogues—Stocktaking A strong European industry for a sustainable Europe*. COM (2024) 163 final. Brussels, 10.4.2024. https://eur-lex.europa.eu/legal-content/EN/TXT/PDF/?uri=CELEX:52024DC0163

European Commission. (2024c). *2024 Communication on EU Enlargement Policy*. COM (2024) 690 final. Brussels, 30.10.2024. https://neighbourhood-enlargement.ec.europa.eu/document/download/7c67aed6-e7c2-47de-b3f8-b3edd26a3e26_en?filename=COM_2024_690_1_EN_ACT_part1_v11.pdf

European Parliament. (2021). Regulation (EU) 2021/1056 establishing the just transition fund. *Official Journal of the European Union*. https://eur-lex.europa.eu/legal-content/EN/TXT/?uri=CELEX%3A32021R1056

Faludi, A. (2012). Multi-level (Territorial) governance: Three criticisms. *Planning Theory and Practice, 13*(2), 197–211. https://doi.org/10.1080/14649357.2012.677578

Fratianni, M. (2003). EU Enlargement and flexible integration. In *EU economic governance and globalization*. Edward Elgar Publishing. https://doi.org/10.4337/9781781009864.00008

Gänzle, S., & Kern, K. (2016). Macro-regions, 'Macro-regionalization' and Macro-regional Strategies in the European Union: Towards a new form of European governance? In S. Gänzle & K. Kern (Eds.) *A 'Macro-regional' Europe in the making. Palgrave studies in European Union politics*. Palgrave Macmillan. https://doi.org/10.1007/978-1-137-50972-7_1

Hayden, R. (2012). *From Yugoslavia to the Western Balkans. Studies of a European disunion, 1991–2011 Balkan Studies library* (Vol. 7). Brill. ISBN: 978–90–04–24191–6. https://doi.org/10.1163/9789004241916

Ignjatović, J., Filipović, S., & Radovanović, M. (2024). Challenges of the green transition for the recovery of the Western Balkans. *Energy, Sustainability and Society, 14*, 2. https://doi.org/10.1186/s13705-023-00421-4

Ternational Energy Agency (IEA). (2021). *Net zero by 2050: A roadmap for the global energy sector*. OECD Publishing. https://doi.org/10.1787/c8328405-en

IRENA. (2018). *Global energy transformation: A roadmap to 2050*. International Renewable Energy Agency. ISBN 978-92-9260-059-4. https://www.irena.org/-/media/Files/IRENA/Agency/Publication/2018/Apr/IRENA_Report_GET_2018.pdf

IRENA. (2023b). *World ENERGY TRANSITIONS outlook 2023: 1.5°C Pathway—Preview*. International Renewable Energy Agency. https://www.irena.org/Publications/2023/Jun/World-Energy-Transitions-Outlook-2023

Jano, D. (2008). From 'Balkanization' to 'Europeanization': The stages of Western Balkans complex transformations. *L'europe En Formation, 349–350*(3), 55–69. https://doi.org/10.3917/eufor.349.0055

Jessop, B. (2011). Metagovernance. In *The SAGE handbook of governance* (pp. 106–123). SAGE Publications Ltd. https://doi.org/10.4135/9781446200964

Jovanović, B., Zlatanović, A., Kluge, S., Zec, A., Ibrahimi, M., Brašanac, M., Vasić, J., Šelmić, R., Mima, R., Stavreska, A. M., Nikolova, S., Rajič, A., Alihodžić, B., Lalović, O., Tomić, V., & Ćulafić, B. (2024). *Transforming the Western Balkans through Near-shoring and Decarbonisation*. Western Balkans WB6 Chamber Investment Forum. https://www.wb6cif.eu/wp-content/uploads/2024/09/Report_2024_v3_web.pdf

Jukić, T. (2013). The man who knew too much Žižek and the Balkans. *European Journal of English Studies, 17*(2), 160–175. https://doi.org/10.1080/13825577.2013.797198

Knez, S., Štrbac, S., & Podbregar, I. (2022). Climate change in the Western Balkans and EU green deal: Status, mitigation and challenges. *Energy, Sustainability, and Society, 12*, 1. https://doi.org/10.1186/s13705-021-00328-y

Lavenex, S., & Schimmelfennig, F. (2009). EU rules beyond EU borders: Theorizing external governance in European politics. *Journal of European Public Policy., 16*(6), 791–812. https://doi.org/10.1080/13501760903087696

Leonard, M., Pisani-Ferry, J., Shapiro, J., Tagliapietra, S., & Wolf, G. (2021). The geopolitics of the European green deal. *International Organisations Research Journal, 16*(2), 204–235. https://doi.org/10.17323/1996-7845-2021-02-10

Levin-Keitel, M., Mölders, T., Othengrafen, F., & Ibendorf, J. (2018). Sustainability transitions and the spatial interface: Developing conceptual perspectives. *Sustainability., 10*(6), 1880. https://doi.org/10.3390/su10061880

Liridon, L. (2023). The meaning of the Western Balkans concept for the EU: Genuine inclusion or polite exclusion? *Southeast European and Black Sea Studies*. https://doi.org/10.1080/14683857.2023.2170204

Mehltretter, S. A. (2009). John F. Kennedy, "Inaugural Address" (20 January 1961). *Voices of Democracy. The US Oratory Project*, 41–59. ISSN 1932-9539. https://voicesofdemocracy.umd.edu/wp-content/uploads/2010/08/mehltretter-kennedy.pdf

Meuleman, L. (2022). A Metagovernance Approach to Multilevel Governance and Vertical Coordination for the SDGs. In *Governing the interlinkages between the SDGs*. ISBN 9781003254683. https://doi.org/10.4324/9781003254683-5

Nadin, V., Maldonado, A. M., Zonneveld, W., Stead, D., Dąbrowski, M., Piskorek, K., Sarkar, A., Schmitt, P., Smas, L., Cotella, G., Rivolin, U. J., Solly, A., Berisha, E., Pede, E., Seardo, B. M., Komornicki, T., Goch, K., Bednarek-Szczepańska, M., Degórska, B..., Śleszyński, P. (2018). *COMPASS—Comparative analysis of territorial governance and spatial planning systems in Europe*. Applied Research 2016–2018: Final Report. ESPON and TU Delft. https://pure.tudelft.nl/ws/portalfiles/portal/51431091/20_11_18_COMPASS_Final_Report.pdf

Najetović, D. (2024). The Influence of a new geopolitical environment on the peace and security of the Western Balkans. *Studia Europejskie—Studies in European Affairs* (3), 57–71. https://doi.org/10.33067/SE.3.2024.3

Oberschall, T. (1996). The fall of Yugoslavia. Társadalom És Gazdaság Közép- És Kelet-Európában / *Society and Economy in Central and Eastern Europe. 18*(3), 78–105. http://www.jstor.org/stable/41468290

Official Journal of the European Union. (2022). *Opinion of the European Economic and social committee on geopolitical impact of the energy transition*. Document 52022AE1398. [JOIN (2022) 23 final]. EESC 2022/01398. Published on 21.12.2022. https://eur-lex.europa.eu/legal-content/EN/TXT/PDF/?uri=CELEX:52022AE1398

Pascual, C. (2015). *The new geopolitics of energy. Columbia SIPA*. Centre on Global Energy Policy (CGEP) Columbia University in the City of New York. https://relooney.com/NS3040/000_New_2395.pdf

Pisano, U., Lepuschitz, K., & Berger, G. (2014). *Sustainability transitions at the international, European, and national level: Approaches, objectives and tools for sustainable development governance. ESDN Quarterly Report 33*. European Sustainable Development Network (ESDN). https://www.esdn.eu/fileadmin/ESDN_Reports/2014-July-Sustainability_transitions_at_the_international__European_and_national_level.pdf

Prelec, T., Tzifakis, N., & Bechev, D. (2023). *External Actors and energy transition in the Western Balkans*. The Balkans in Europe Policy Advisory Group (BiEPAG). https://www.biepag.eu/wp-content/uploads/2023/12/Green-Power-Politics-External-Actors-and-Energy-Transition-in-the-Western-Balkans.pdf

Ratha, K. C. (2019). Paris climate deal: A bumpy road ahead. I*ndian Journal of Asian Affairs, 32*(1/2), 67–90. https://www.jstor.org/stable/26902686

Scholten, D., Criekemans, D., & Van de Graaf, T. (2019). An energy transition amidst great power rivalry. *Journal of International Affairs*, *73*(1), 195–204. https://www.jstor.org/stable/26872789

Shaker, Y., & Berisha, E. (2024) Transizioni Verdi Giuste': tra imprecisioni semantiche e necessità di un inquadramento teorico. In E. Formato & F. Vingelli (Eds.), Strumenti per il governo del valore dei suoli, per un progetto equo e non-estrattivo, Atti della XXV Conferenza Nazionale SIU "Transizioni, giustizia spaziale e progetto di territorio", Cagliari, 15–16 giugno 2023 (vol. 09, pp.193–203). Planum Publisher e Società Italiana degli Urbanisti, Roma-Milano. https://hdl.handle.net/11583/2989395

Shaker, Y., & Berisha, E. (2025). Just green transitions: Between terminological inexactitudes, conceptual fragmentation, and the exigency for a theoretical framework. *European Spatial Research and Policy, 32*(1), 121–138. https://doi.org/10.18778/1231-1952.32.1.05

Shaker, Y., & Persico, S. (2024). Mapping the Just green transitions in the socio-political virtual space. *European Journal of Spatial Development, 21*(1), 61–82. https://doi.org/10.5281/zenodo.13817780

Sharpe, L. J. (1989). Fragmentation and territoriality in the European state system. *International Political Science Review, 10*(3): 223–238. http://uniset.ca/microstates2/fragmentation_territoriality.pdf

Škuflić, L. (2010). Transition process in the Western Balkans: How much successful is this story? *Intelligent Information Management., 2*, 243–252. https://doi.org/10.4236/iim.2010.23029

Stanicek, B., Przetacznik, J., & Roman, A.A. (2023). *Enlargement policy: Reforms and challenges ahead*. Members' Research Service PE 757.575. Briefing December 2023. European Parliamentary Research Service (EPRS). European Parliament. https://www.europarl.europa.eu/RegData/etudes/BRIE/2023/757575/EPRS_BRI(2023)757575_EN.pdf

Steil, B., & Woodward, S. L. (1999). A European, "New Deal" for the Balkans. *Foreign Affairs, 78*(6), 95–105. https://doi.org/10.2307/20049535

Tahirović, M. (2024). The main Security challenges and threats in the Western Balkans and possible model for solving them in the process of European integration. *Studia Europejskie, 28*(3). ISSN 1428-149X. https://journalse.com/pliki/pw/3-2024-Tahirovic.pdf

Teevan, C., Medinilla, A., & Sergejeff, K. (2021). *The green deal in EU foreign and development policy*. Briefing Note 133. Making Policies Work. European Centre for Development Policy Management (ECDPM). ISSN1571-7577. https://ecdpm.org/application/files/2516/5546/8575/Green-Deal-EU-Foreign-Development-Policy-ECDPM-Briefing-Note-131-2021.pdf

The Oxford Institute for Energy Studies. (2021). *The geopolitics of energy: Out with the old, in with the new?*. Forum. February 2021: Issue 126. Oxford. https://www.oxfordenergy.org/wpcms/wp-content/uploads/2021/02/OEF-126.pdf

Tsaroucha, E. (2019). The Metonymicity of the Greek Deictic Adverbs εδώ [Here] and εκεί [There] in Politics. *Philosophies, 4*(3), 51. https://doi.org/10.3390/philosophies4030051

Tziarras, Z. (2019). *The new geopolitics of the Eastern Mediterranean: Trilateral partnerships and regional security*. Reimagining the Eastern Mediterranean Series: PCC Report 3/2019. Friedrich Ebert Stiftung. PRIO Cyprus Centre. ISBN: 978-82-7288-967-7. https://library.fes.de/pdf-files/bueros/zypern/15662.pdf

Valkenburg, G., & Cotella, G. (2016). Governance of energy transitions: About inclusion and closure in complex sociotechnical problems. *Energy, Sustainability, and Society., 6*, 20. https://doi.org/10.1186/s13705-016-0086-8

Van Dijk, J., Wieczorek, A. J., & Ligtvoet, A. (2022). Regional capacity to govern the energy transition: The case of two Dutch energy regions. *Environmental Innovation and Societal Transitions, 44*, 92–109. ISSN 2210-4224. https://doi.org/10.1016/j.eist.2022.06.001

Vienna European Council. (1998). *Presidency conclusions. Vienna, 11 and 12 December*. European Parliament. https://www.europarl.europa.eu/summits/wie1_en.htm

Walker, C. (2024). Global impacts of the European green deal. *Nature. Plants, 10*, 1439. https://doi.org/10.1038/s41477-024-01835-4

Yang, Y., Xia, S., & Qian, X. (2023). Geopolitics of the energy transition. *Journal of Geographical Sciences, 33*, 683–704. https://doi.org/10.1007/s11442-023-2101-2

Zepa, I., & Hoffmann, V. H. (2023). Policy mixes across vertical levels of governance in the EU: The case of the sustainable energy transition in Latvia. *Environmental Innovation and Societal Transitions, 47*(100699). ISSN 2210–4224. https://doi.org/10.1016/j.eist.2023.100699

Žižek, S. (1999). The spectre of Balkan. *The Journal of International Institute, 6*(21). ISSN: 1558–741X. http://hdl.handle.net/2027/spo.4750978.0006.202

Žižek, S. (2006). Philosophy, the "Unknown Knowns" and the public use of reason. *Topoi, 25*, 137–142. https://doi.org/10.1007/s11245-006-0021-2

CHAPTER 5

From Mitigation to Transformation: An Evolutionary Perspective on the Just Green Transition

Carlos Tapia, Alberto Giacometti, John Moodie, Elin Cedergen, and Aida Shaneh

5.1 Introduction

The Just Green Transition (JGT) concept helps us to understand the logic of long-term change and the engineering of new stable configurations for society and economies that are within the limit of the planet's boundaries (Cedergren et al., 2022). The notion of a JGT is often invoked in connection to other tenants of climate and energy policies (Evans & Phelan, 2016; McCauley &

C. Tapia (✉) · A. Giacometti · J. Moodie · E. Cedergen
Nordregio Research Institute for Regional Development and Planning, Stockholm, Sweden
e-mail: carlos.tapia@nordregio.org

A. Giacometti
e-mail: algia5534@uis.no; alberto.giacometti@nordregio.org

J. Moodie
e-mail: john.moodie@nordregio.org

E. Cedergen
e-mail: elin.cedergren@nordregio.org

A. Shaneh
Interuniversity Department of Territorial Sciences, Planning and Policies, Politecnico Di Torino, Torino, Italy
e-mail: aida.shaneh@polito.it

A. Giacometti
School of Business and Law, University of Stavanger, Stavanger, Norway

E. Berisha et al. (eds.), *The Palgrave Handbook of Just Green Transitions in the Western Balkans and Beyond*,
https://doi.org/10.1007/978-3-031-95075-9_5

Heffron, 2018). In this chapter we provide an overview of this co-evolutionary process by looking at the development of the concept of a JGT over time, from its origins to its current interpretation in practice, policy, and research. We show how the JGT has developed in a rather peculiar way, expanding its theoretical and practical scope and ambition as it moved from a corrective to a more transformative conceptualisation. The remainder of the paper is structured as follows: Sect. 5.2 provides a conceptual framework to guide the analysis and interpretation of the JGT under an evolutionary perspective; Sect. 5.3 puts the JGT concept in historical perspective and elaborates on the main social, economic, and environmental events that conditioned the evolution of the notion over time; Sect. 5.4 discusses the relevance of such historical evolution of the JGT concept and elaborates on the practical interpretation of current and future environmental policies at the global level.

5.2 Conceptual Framework

Societies are facing vast challenges in tackling climate change, biodiversity loss, and unsustainable levels of resource deployment. This requires large-scale systemic changes to mitigate and adapt societies to environmental change. The JGT is one of the strategies available to advance sustainability goals. In this section, we dissect the notion and propose a theoretical perspective to explore the concept and its evolution over time.

5.2.1 Deconstructing the Just Green Transition Concept

Research on transitions in relation to sustainable development and climate mitigation is expanding rapidly, with contributions from many fields, using a variety of approaches and terminology. The JGT represents a societal transformational paradigm grounded on two aspirational elements, namely social justice (Just) and environmental sustainability (Green), which are simultaneously enabled or become achievable through social and technological change (Transition). The term 'just' denotes a philosophically and politically contested concept that often defines positions of how green transitions should be designed. It is rooted in the notion of 'social justice' emerged during the nineteenth century, reflecting on the development of a justice-based political discourse, often in the context of the distribution of income and wealth (Jackson, 2005). This emphasises the importance of distributional perspectives in how impacts from policy decisions spread across different regions, groups or segments of society. As transitions can be understood as a policy-engineered change, recognitional and procedural justice have gained traction in academic research, elaborating on which groups of society are recognised or neglected and in what ways forms of decision-making processes entail the involvement of such groups or interests (Cedergren et al., 2022). Researchers are also interested in the meta-norms or commonly agreed tenets of justice in the green transition. Not least in a global setting where climate agreements and global goals are set for change-making, is not clear how such agreements are deployed in practice at different cultures of governance (e.g. Lawless et al., 2020). Policy measures and actions aimed at environmental sustainability, and the transformations arising from such measures, are thought to have both positive and negative social consequences for different parts of society (Cambou, 2020). This environmental perspective is captured by the second term in the concept: 'green'. However, the environmental sustainability priority encapsulated in it is an ambiguous concept with definitions that depend on political stands and disciplinary applications. Few sources in literature or policymaking specify the types of environmental outcomes associated with the JGT, and its inherent meaning may change over time and depend on the institutional context (Wøien Meijer & Peter, 2021). Simultaneous progress towards social justice and environmental sustainability requires reconfigurations in socio-technical systems, understood as nested systems that encompass market structures, consumer patterns, institutional, regulatory or governance models and infrastructure at various societal levels (Geels, 2010). This systems shift is the notion embodied in the last term in the JGT construct ('transition'), which denotes processes of social, technological, and economic change explicitly designed to enable a range of environmental goals.

5.2.2 Theoretical Frameworks for Social and Technical Transformations

As a normative agenda aiming for society-wide transformation, the JGT can be analysed building on the concepts developed within two main theoretical frameworks, namely the Social-Ecological Resilience and the socio-technical systems frameworks (Gillard et al., 2016). Social-Ecological Resilience is an interdisciplinary research perspective that provides a comprehensive framework for understanding and managing the resilience of coupled human-environment systems. Resilience management is about making it possible to deliberately alter the system's fundamental properties and about undertaking a transformation process to better cope with emergent conditions (Nelson et al., 2007). A social-ecological analysis of resilience enables the study of people-environment transactions across varying dimensions, time periods, and scales (Stokols et al., 2013). The resilience perspective emphasises non-linear dynamics, thresholds, uncertainty, and surprise. It explores how periods of gradual change interplay with periods of rapid change and how such dynamics interact across temporal and spatial scales. Resilience thinking focuses on understanding social processes such as social learning and social memory, agents and actor groups, social networks, institutional and organisational inertia and change, adaptive capacity, transformability and systems of adaptive governance (Folke, 2006). However, the assumption that short-term resilience, understood as bouncing-back to previous state, leads to long-term transformations has been questioned. A traditional understanding of resilience—as a single equilibrium framework—has been criticised for its conservatism, deterministic thinking, and political naivety. Critics argue that such an understanding of resilience can depoliticise issues such as climate change and reinforce neoliberal governance, exacerbating socio-economic inequality and vulnerability (Gillard et al, 2016). Contributions over the past two decades by scholars drawing on ideas from evolutionary economics, neo-institutional theory, and sociology of innovation have led to a more holistic and dynamic understanding of resilience and socio-ecological systems (Simmie & Martin, 2010). This evolution in scholarly thinking has shed light on the co-evolution of technologies, markets, institutions, social networks, and learning processes as components of socio-technical systems change (Fuenfschilling & Truffer, 2014; Geels, 2010). This framework incorporated the cultural, political, institutional, economic and spatial dimensions to understand the role of conflict, power struggles and cultural contexts in transition processes (Kern & Rogge, 2018; Köhler et al., 2018; Loorbach et al., 2017; Raven et al., 2012). The Multi-Level Perspective (MLP) provides a theoretical model to explain socio-technical transitions as socio-technical systems change. The MLP understands socio-technical change as a process where emerging innovations (e.g. low-carbon energy systems) and established socio-technical systems (e.g. carbon-based energy systems) compete over time (Geels, 2002; Geels et al., 2016; Smith et al., 2005). This framework helps to explain why change is difficult to achieve even when emerging innovations may have substantial technical or economic advantages over established practices and technologies. And that is because existing socio-technical systems are deeply entrenched due to institutional, economic and cultural factors, often creating the conditions for path-dependence and technological locks-in (Klitkou et al., 2021; Seto et al., 2016).

5.2.3 *Analytical Framework*

Important discrepancies can be identified in terms of how previous research conceptualises the notion of a JGT. Divergences may emerge in relation to the entity that is subject to transformation. This can be broadly associated to the normative agenda ('what ought to be done') that is implicitly or explicitly advocated for. Such normative agenda can be characterised through a set of aspirational and operational elements including (1) societal goals; (2) intervention logics; (3) change depth and; (4) prevailing social justice domain. Societal goals refer to the scope of the transformation envisioned by the JGT agenda. This may imply maintaining the status-quo or normal operation of landscape under a business-as-usual principle, or rather advocating a more systemic transformation of the economy and society, with all the intermediate stages (mitigate and adapt to changing environments). Each of these goals are subject to different intervention logics. For instance, under a palliative normative agenda seeking to preserve the status-quo, the intervention logic would be to prevent major changes on production systems and, when unavoidable, identify their potential impacts to alleviate their social and economic consequences as far as possible. In contrast, under a more transformational normative agenda, the intervention logic moves beyond palliative views to place the emphasis on helping society and the economy co-evolve and transform towards more sustainable production and consumption systems. The depth and breadth of change embedded in the transition can be used to classify the transition framework into incremental versus disruptive. Pelling (**2010**) found three visions of adaptation in terms of goals, scope and policy priorities:

- Resilience: Focuses on functional persistence in a changing environment, through shifts in technology, management practice and organisation.
- Transition: Focuses on introducing incremental change in the governance systems, including the exercise of rights within the established regime and changing practices of governance to secure procedural justice.
- Transformation: Focuses on re-configuring the structures (trajectories) of development, through change on overarching political-economy regimes.

Finally, the normative agendas for a JGT may differ in how they embed the social justice component in the rationale and which of the various expressions of social justice are promoted. This refers to which of the social justice tenants are prioritised in terms of distribution of assets (burdens and benefits) and power (recognition and engagement). Table 5.1 summarises different perspectives related to the social agendas and scope of socio-technical change advocated for by the JGT, according to the conceptual building blocks identified in the above-mentioned theoretical frameworks.

In the following section we review the historical development of the JGT concept, the context and debates surrounding its development, as well as its practical implementation in the form of targeted policy frameworks and measures.

Table 5.1 Social perspectives embedded in theories advocating system's change for sustainability

Normative agenda	*Societal goal*	*Intervention logic*	*Change level*	*Prevailing social justice component*
Palliative	Maintain status-quo	Identify and alleviate	Conservative	Restorative
Preventive	Mitigate impact, bounce back	Anticipate and recover	Gradual	Distributive
Adaptive	Adapt, bounce forward	Adapt and adjust	Incremental	Procedural
Transformative	Reconfigure, transform	Co-evolve and transform	Disruptive	Recognitional

5.3 The Just Green Transition in Historical Perspective

A fluid understanding of the scope of the JGT is motivated by its broad and flexible definition and long history of evolution. In this section, we review the history of the JGT against the analytical elements summarised in Table 5.1. With this, we aim to shed light on the conceptual evolution of the JGT and its practical operationalisation over time.

5.3.1 *First Phase: Emergence*

The concept of a Just Green Transition emerged from the struggles of North American workers in the 1970s and 1980s. The postwar economic boom in the U.S. began to decline in the 1960s, culminating in the 1973 oil crisis and stock market crash. This recession hit the manufacturing and mining sectors hardest, with manufacturing employment in the U.S. dropping from 34% in 1948 to 20% in 1987 (Milkman, 1990). The Rust Belt, a heavy manufacturing region, was particularly affected, losing significant population and income. From 1970 to 2006, Cleveland, Detroit, Buffalo, and Pittsburgh lost about 45% of their population and median household incomes fell: in Cleveland and Detroit by about 30%, in Buffalo by 20%, and Pittsburgh by 10% (Hartley, 2013). The fall was driven by the reduced competitiveness of traditional American industries compared to emerging low-wage manufacturing areas globally. However, while traditional industries struggled, sectors like information and communication technologies thrived in the 1980s. This period also saw a transformation in working conditions, with deregulation and a sharp decrease in unionisation leading to more temporary, part-time, and low-paying jobs (Kollmeyer, 2018). Neoliberal economic policies further exacerbated these issues, prioritising short-term profits over long-term competitiveness, worsening living conditions for many workers (Lazonick,2013; Newman, 1990). In response, American trade unions demanded investments in new capabilities to create well-paid jobs (Lazonick, 2013). Considering the limited success of these demands, their focus shifted to fighting the offshoring of manufacturing jobs, which later on influenced their stance on climate change mitigation agreements during the 1990s (Thomas, 2021). The idea of a just transition was developed in this context. Tony Mazzocchi, an American trade union leader, played a key role in merging union and environmental movements (Roessler, 2016). In the late 1970s, he proposed that workers whose jobs were threatened by disarmament should receive support similar to the GI Bill of Rights, which helped veterans transition to peacetime employment. This idea was also influenced by earlier programmes like the 1960s Canadian uranium mining dispute and the 1951 European Coal and Steel Community's Fund. In the early 1990s, Mazzocchi revived his original idea, calling it a "superfund for workers", which would provide financial support and education opportunities for workers displaced by environmental policies. The name of the plan was shortly changed to "just transition fund" on the pledge that the term "superfund" might have negative connotations for environmentalists. Mazzocchi's ideas sought to compensate workers affected by environmental policies, offering them support to start anew. While rooted in welfare theories, Mazzocchi emphasised that this support was not welfare but a means to transition from one type of job to another. His idea focused on mitigating the effects of policies designed by others without seeking structural change. It did not address indirect effects like cost of living or involve broader participation beyond the unions. Nonetheless, Mazzocchi's efforts were notable for bridging the gap between environmental and labour concerns, moving beyond the 'environment vs. jobs' dichotomy to address broader societal issues (Thomas, 2021).

5.3.2 *Second Phase: Formalisation*

The United Nations (UN) Conference on the Human Environment in 1972 in Stockholm marked the beginning of global climate governance, leading to the creation of the United Nations Environmental Program (UNEP). However, it wasn't until the 1990s that 'sustainability' became a key term in policy discussions. The 1992 Rio Conference, known as the Earth Summit, resulted in the Rio conventions on climate change, biodiversity, and desertification. The first Conference of Parties (COP) summit followed in Berlin in 1995, and in 1997, the Kyoto Protocol was introduced, setting the first treaty for countries to limit greenhouse gases. The global debate that preceeded the Kyoto Protocol led to a shift in the international union movement from opposing to supporting climate policies, driven by a growing societal consensus on the need to address environmental threats. Trade unions faced the challenge of balancing societal needs with protecting their members' interests. European unions, like the German Trade Union Confederation and the British Trades Union Congress, supported binding carbon emission targets, while the American Federation of Labor and Congress of Industrial Organizations opposed them, fearing job losses and industrial delocalisation to Asia and Mexico (Thomas, 2021). To break the deadlock, international trade unions needed to change their consensus-based decision-making approach, which often blocked proactive voices. The merger of the International Confederation of Free Trade Unions and the Christian World Confederation of Labour into the International Trade Union Confederation in 2006 aimed to strengthen their influence on climate policies. Unions began to recognise their role in climate policies, leading to broader political agreements like the Copenhagen Accord in 2009, which set a cap on global temperature rise of 2 °C by the end of the century. But it was not until 2015 with the Paris Agreement that commitments for limiting temperature rise became legally binding. The Paris Agreement emphasised the concept of just transitions, aligning with the UN 2030 Agenda for Sustainable Development's pledge to "leave no one behind"

(UN, 2022). In 2013, the International Labour Conference adopted a resolution on sustainable development, decent work, and green jobs "putting forward a policy framework for a just transition". This led to the development of the International Labour Organisations (ILO) 2015 Guidelines for a Just Transition towards Environmentally Sustainable Economies and Societies for All. These guidelines emphasise social dialogue, social protection, rights at work, and employment (ILO, 2015). Following the Paris Agreement, several international commitments and strategies were developed. The just transition gained attention at COP24 in Katowice in 2018. The COP24 event "Just Transition for All" gathered various stakeholders and resulted in a policy brief highlighting the need to link the just transition to the Nationally Determined Contributions (NDCs), which are the individual commitments that countries make to reduce their greenhouse gas emissions (Jenkins et al., 2020). However, the most important outcome from the COP24 was the "Solidarity and Just Transition Silesia Declaration" signed by 50 countries (Council of the European Union, 2018). This declaration emphasised creating decent work and quality jobs while addressing environmental goals, moving beyond compensatory measures. By 2022, 65 of the 170 countries that submitted updated NDCs explicitly reference the just transition (Lee & Baumgartner, 2022). In 2019, at the UN Climate Action Summit, 46 nations committed to developing Just Transition strategies. Countries like Canada, Chile, the Czech Republic, and Germany established commissions to plan coal transitions, while South Africa and Indonesia formed international Just Energy Transition Partnerships. In 2021, just transition was a central theme at COP26 in Glasgow, resulting in the "Just Transition Declaration". Developed countries committed to supporting just transitions internationally through the Global Accelerator on Jobs and Social Protection for Just Transitions. The EU, US, UK, Germany, and France pledged €8.5 billion to aid South Africa's just transition. The Just Transition Mechanism was launched as part of the European Green Deal (EUGD) investment plan for 2021–2027. In 2022, the G20 Sustainable Finance Working Group developed a Just Transition Framework. During this phase, the concept of a just green transition was formalised as a consensus policy priority at the international level. The just transition concept gained recognition not only as a tool to mitigate the impacts of climate policies on workers but also increasingly as a framework to anticipate the social externalities of climate policies. It aimed to plan strategies that reconcile economic and environmental goals under a coherent policy framework. However, the approach remained broadly focused on specific groups and industries, with the advocated changes being more incremental and adaptive rather than transformative and disruptive.

5.3.3 *Third Phase: Maturation & Policy Implementation*

Recently, the Just Green Transition (JGT) has become an umbrella term for various policy themes related to climate and environmental sustainability and justice. Strategies for JGT development and implementation have emerged at both international and national levels, alongside sectoral and organisational roadmaps promoting carbon-neutral processes. At the supranational level, the UN 2030 Agenda for Sustainable Development, including the 17 Sustainable Development Goals (SDGs), and the EUGD advocate for transformative climate and environmental changes, made explicit and implicit links to the JGT. Key JGT themes include the circular economy, energy-efficient buildings, biodiversity, nature protection, and sustainable farming. Both the SDGs and the EUGD aim to balance economic growth and social justice, especially for regions and groups most vulnerable to JGT impacts. However, the SDGs have been criticised for lacking a clear implementation plan and sufficient political support, limiting their transformative impact. In contrast, the EU Green Deal, expected to mobilise over a trillion Euros in the 2020–2030 period, has been criticised for prioritising sectoral actors and economic growth over societal challenges and citizen needs.

The EU's Just Transition Mechanism (JTM), part of the Green Deal, is the most comprehensive effort to deliver a JGT. It focuses on supporting Europe's most affected regions, particularly those dependent on fossil fuels and energy-intensive processes, to transition to a low-carbon economy. The JTM balances socio-technical elements, such as investments in carbon-neutral technologies, and social elements, like worker protection and reskilling opportunities. A core instrument of the JTM is the Just Transition Fund (JTF), which mobilises around €55 billion to facilitate socio-economic diversification in regions most affected by the climate transition. Access to the JTF requires EU member states to develop Territorial Just Transition Plans (TJTPs). However, these plans have been dominated by national and sectoral actors, with limited involvement from regional and local actors and limited citizen participation. As a result, many TJTPs focus on specific sectors while neglecting territorial socio-economic challenges (Moodie et al., 2021). National and regional JGT strategies vary in scope and ambition. Chile's Strategy for a Just Energy Transition has a clear sectoral focus, while programmes in Colorado, USA, and Alberta, Canada, support communities affected by coal mine closures (UN DESA, 2022). These programmes are mainly compensatory, with some emphasis on industry adaptation and diversification. Greece's Just Transition Development Plan (ibid) focuses on "transforming the productive model in the lignite areas of the country, promoting clean energy, industry and trade, smart agricultural production, sustainable tourism, technology and education" (ibid, p. 5). In parallel to the initial implementations of the just transition agenda at a practical policy level, the international climate debate shifted towards the systemic nature of the transformation and the need for more active participation from a broader group of stakeholders. The Sharm el-Sheikh Implementation Plan, adopted after COP27 (UNCCC, 2022), affirms that "sustainable and just solutions to the climate crisis must be founded on meaningful and effective social dialogue and participation of all stakeholders" and notes that "the global transition to low emissions provides opportunities and challenges for sustainable economic development and poverty eradication" (UNCCC, 2022, p. 6). This represents a significant expansion of the scope of the JGT, which is increasingly seen as a tool to advance both environmental and social priorities beyond labour markets. In response to the potential risk of "jargonisation", where broadening the just transition concept might dilute its meaning, the Climate Action Network, an alliance of Non-Governmental Environmental Groups, released a newsletter in which it was stressed that "workers have always been, and need to remain, at the heart of agreements on just transition" (Climate Action Network, 2022, p. 4). These examples highlight the challenges in achieving transformative JGT strategies, particularly balancing technical and social dimensions and engaging different stakeholders in the process.

5.4 Discussion

In this chapter we have examined the evolution of the JGT notion of JGT from a historical perspective, connecting its major conceptual and operational milestones to the development of the international climate policy agenda. The historical timeline presented shows how the JGT paradigm has not evolved in the same way as other strategic frameworks for sustainable development. Often, these frameworks are first developed on a theoretical level, to be then applied on a more practical level. The JGT is a peculiar example of a sustainable appraisal concept that has evolved in exactly in the opposite direction. Originally proposed as a targeted policy measure focusing on workers affected by industrial reconversion, the notion developed towards a more comprehensive and strategic policy framework applicable to a global policy agenda. As we have seen, the JGT concept was originally formalised by the international labour union movement in response to a growing concern that new policies (and technologies) introduced as a response to growing global environmental concerns would lead to layoffs and undermine working conditions in various industries. This interpretation conditioned the practical implementation of the first generation of JGT policies, which were deployed as a rather reactive set of mitigation and preventative measures addressing those concerns. During the three decades following the adoption of the Kyoto Protocol, the JGT concept broadened its scope and gained in international recognition as a comprehensive and enabling framework defining the boundaries for a fair application of climate policies. In this process, concepts and principles related to societal fairness were increasingly mainstreamed to the environmental agenda, and social criteria implicit in the notion of a 'just transition' challenged the first generation of climate change mitigation policies. Simultaneously, as scientific evidence about the anthropocentric nature of climate change consolidated, climate policies increasingly expanded their scope beyond mitigation towards climate change adaptation. This process made the JGT evolve in a similar direction, that is, from mitigation of the adverse consequences of climate policies on workers (via compensation mechanisms), towards a more adaptive perspective where the emphasis was on proactively dealing with the unavoidable consequences of the green transition. Today, the JGT is framed as a transformative paradigm. This is characterised by a global policy agenda for sustainability, which requires an overhaul of the interlinked social, economic, and institutional systems besides technological innovations. Figure 5.1 below showcases this evolutionary process of JGT concept at different stages of maturity of climate policies.

Figure 5.1 provides a broad overview of the evolution of the JGT policies over time, highlighting how its focus moved from a predominantly preventative and mitigative perspective to a more adaptive and transformational approach. The linear progression outlined in the table helps us to isolate and zoom in on specific elements of this JGT evolutionary process; however, we recognise that this oversimplifies a more complex reality in the sense that as the JGT focus and approach moves forward, the approaches outlined in previous phases remain relevant and do not become obsolete. At the same time, these different approaches are not mutually exclusive but complementary, as JGT policies may require a combination of mitigative, adaptive and/or transformative measures. Furthermore, the territorial specific and wicked nature of JGTs mean that different policy measures might be needed depending on the industry, region or social groups affected nature of the transition. For instance, closing a power plant may have different repercussions on employment in a region than changing farming practices, therefore, different JGT measures need to be tailormade. In sum, the JGT emerges as a non-linear and non-incremental process at both policy and practice levels.

There is also a clear dissonance between academic and policy discourses and reality. The increased understanding of the complexity of socio-technical systems has led to the evolution of environmental, political-economic, and

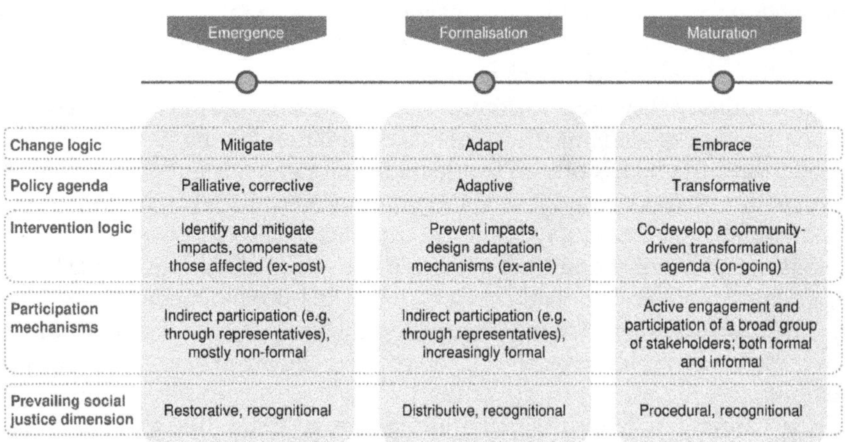

Fig. 5.1 Tentative characterisation of the just green transition concept at different stages of maturity

social theories, as well as the definition of policy goals and strategies. However, implementing JGT strategies and policies have proved demanding in practice. This is evident in the design and practical implementation of the European TJTPs, which are specifically meant to address the socio-economic consequences of climate policies. However, the social implications surrounding the technical questions are narrowly formulated, and the top-down policy processes largely disregard place-based leadership. This denotes the difficulty of addressing technical issues such as a specific energy mode at the same time as issues around jobs and skills development, which may demand deeper transformation of political and economic systems. As often these issues are discussed at different levels of abstraction, the socio-economic perspectives tend to water down the more we zoom into the specific sectoral and technological questions of transitions.

Furthermore, it is important to acknowledge that transformational processes are inevitably disruptive, which often requires painful trade-offs between conflicting economic, ethical, societal, and environmental priorities and objectives. For instance, the creation of alternative 'green' jobs for those who will become unemployed in polluting industries in one area may be challenged by practical and economic considerations, as the new jobs may follow different economic and location logics and thus be created in other regions at other points in time. Therefore, space and time become vital dimensions to consider during the design of JGT policy agendas. While policymakers and citizens are impatient for a rapid implementation of JGTs, industrial transformations take time. While this may not be good news for the environment, it may be more favourable for allowing the labour market to adapt. While novel industries and practices within existing industries emerge, new generations gain the necessary skills to work in such industries while polluting industries gradually phasing out continue to sustain a portion of the labour market. At the same time, regions and municipalities that are losing economic activities have the time to invest in the diversification of their economies before their labour force flees out to other regions.

As in other policy areas, institutional and social actors have often clashed in the process of defining the scope and ambition of the JGT agenda. A paradigmatic example is offered by the trade unions. While international union organisations have been positive and even sponsored the just transition claims by environmental activists and indigenous rights groups, the position of local unions at the national and local levels is more varied. As Thomas (2021) put it, "unions representing workers from fossil fuel-dependent industries have mostly opposed stringent mitigation policies, while unions in industries with a lower carbon footprint have developed proactive climate policies". This further demonstrates the wicked nature of the JGT, where the position of conflicting interests among and across societal groups are generally based on legitimate claims.

Whatever JGT approach policymakers choose to take, it is essential that the policies are 1) place- and time-based, addressing regional specificities and meeting the needs of local citizens; and 2) strike a balance between the different technical, economic, and social dimensions of JGTs. This is vital as the more JGTs shift towards more adaptive and transformational approaches, there is a risk that the social justice element of transitions become neglected, with JGT strategies becoming a tool for cultivating technical and structural change and promoting economic growth. At the same time, there is a growing risk that the JGT agenda is co-opted by interest groups to keep the status quo and prevent the green transition from even taking place or to slow it down substantially, on the claim that such transformations are 'socially unfair'.

Another interesting paradox emerges where the early preventative and mitigative measures dealt more directly with the socio-economic impacts of JGTs on citizens and workers, while not engaging them directly in the policy process; in contrast, adaptive and transformational approaches advocate for wide ranging citizen engagement in the policy process, while focusing more on the technical and economic dimensions of JGTs. Indeed, citizen engagement is essential for building public support and acceptance for JGTs. This is particularly important during a period where there is growing climate-scepticism championed by far-right political parties lifting their voices against stringent environmental policies. The growing tension between EU members, particularly around the Nature Restoration Law, has already forced EU regulators to weaken the rules and overall ambitions. Against this backdrop, raising citizen awareness, understanding and participation in JGTs becomes vital, especially in regions that have been left behind by the forces of globalisation.

5.5 Relevance to Western Balkans

The Western Balkans is in its relative infancy when it comes to JGTs, as the region is still overwhelmingly dependent on energy-intensive industries, with coal power accounting for 70% of electricity produced. Many countries and regions across the WBs must still undergo substantial structural changes in the shift to carbon neutrality, transformations that will create both significant socio-economic challenges and opportunities (Berisha, 2025a, 2025b). Policymakers across the WBs must, therefore, carefully consider the potential impacts of the green transformation in their region and whether to adopt and apply mitigative, adaptive, or transformative measures, meticulously tailoring these different JGT approaches to meet the needs of heterogenous regional contexts, characteristics, and specificities. In embarking on this selection process, policymakers can learn from best and worst practice examples of JGT strategies and policies from across Europe and the globe. Currently, a combination logic focusing on mitigation, adaptation, and transformation applies when it comes to JGTs in the WBs. This logic is most visible in the support for the 2016 Paris Agreement, which was ratified by five out of six Western Balkan countries, thereby committing signatories to climate change mitigation and adaptation. This global commitment was given more regional specificity and relevance at the Western Balkan Summit held in Sofia on the

10th November 2020 under the framework of the Berlin Process initiative, where WB leaders signed the Sofia Declaration. The declaration acknowledges "the need to set the basis for a major transformation of our region to turn sustainability and resilience challenges into opportunities and transpose elements of the EUGD in all interrelated priority sectors" (Regional Cooperation Council, 2020). The declaration was complemented by the Green Agenda for the Western Balkans (GAWB) and corresponding GAWB Action Plan. The GAWB outlines transformative actions designed to support and accelerate JGT changes and processes in the Western Balkan Region. The GAWB largely represents an extension of the EUGD in both scope and focus, but it is also considered a tool for enhancing WB integration and cohesion with the EU through increased policy alignment. The GAWB is embedded in the EU Economic and Investment Plan for the Western Balkans, which entails that as conditionality for funding and for the accession process to the EU. WB countries must implement EU directives and transpose regulation across policy areas and on a broad range of JGT policy themes also reflected in the GAWB, including climate, energy, air, soil, and water pollution, waste management and circularity, and biodiversity. While the GAWB and GAWB Action Plan outline plans for transformative change, it has faced several implementation challenges. Similarly to UN SDGs, the GAWB has been difficult to enforce, lacks fundings, received limited high-level political support, and resulted in little structural or policy change. These obstacles highlight tension between the development and implementation of transformative JGT policies in the WBs and globally. As with the EU Just Transition Mechanism, the GAWB has also been criticised as being a top-down down process dominated by national level and sectoral actors, which prioritises economic growth over socio-economic impacts and challenges. Furthermore, so far there has been no direct citizen engagement in the development and implementation of the GAWB. Communicating JGTs to citizens effectively is particularly important in a WB setting where the term 'transition' has become loaded with negative connotations. Toto et al (2023) argue that the WBs has been a loop of post-socialism transition since the early 1990s, with the slow transition towards liberal democracies, free economies, and EU integration, creating disillusionment and "transition fatigue" among citizens across the WBs. The notion of JGT is viewed as being enforced from abroad drawing on a deliberate set of expectations and their implementation relying on foreign aid and political steering. However, the justice component is largely watered down in the GAWB compared to the EUGD. Therefore, a key concern is that in prioritising the fulfilment of EUs demands, measures in place will disregard the local context and the social consequences. Implementing environmental policies can potentially exacerbate the tensions and the precarious economic situation in the region. This is particularly critical in the transition of the energy sector, where numerous households and regions risk losing their main source of income.

References

Berisha, E. (2025a). Moving towards an inclusive green agenda in the Western Balkans. Foundation for European Progressive Studies (FEPS). https://doi.org/10.13140/RG.2.2.12092.76162

Berisha, E. (2025b). Green EU integration—The role of the green agenda in accelerating just green transitions in the Western Balkans. In E. Berisha, J. Moodie, L. Allkja, & Jeftić, M. (Eds.), *The Palgrave handbook of just green transitions in the Western Balkans and beyond*. Palgrave Macmillan.

Cambou, D. (2020). Uncovering injustices in the green transition: Sámi rights in the development of wind energy in Sweden. *Arctic Review on Law and Politics, 11*(0), 310. https://doi.org/10.23865/arctic.v11.2293

Cedergren, E., Tapia, C., Gassen, N. S., & Lundgren, A. (2022). *Just green transition—Key concepts and implications in the Nordic Region*. Nordregio.

Climate Action Network. (2022). *ECO—NGO newsletter No. 16. No climate justice without human rights*. https://climatenetwork.org/wp-content/uploads/2022/11/ECO-16.11.2022.pdf

Council of the European Union. (2018). *Silesia declaration on solidarity and just transition*. Pub. L. No. 14545/1/18 REV 1 PL/bsl 1 TREE.1.B EN.

Evans, G., & Phelan, L. (2016). Transition to a post-carbon society: Linking environmental justice and just transition discourses. *Energy Policy, 99*, 329–339. https://doi.org/10.1016/j.enpol.2016.05.003

Folke, C. (2006). Resilience: The emergence of a perspective for social-ecological systems analyses. *Glob Environ Change, 16*(3), 253–267. https://doi.org/10.1016/j.gloenvcha.2006.04.002

Fuenfschilling, L., & Truffer, B. (2014). The structuration of socio-technical regimes—Conceptual foundations from institutional theory. *Research Policy, 43*(4), 772–791. https://doi.org/10.1016/j.respol.2013.10.010

Geels, F. W. (2010). Ontologies, socio-technical transitions (to sustainability), and the multi-level perspective. *Research Policy, 39*(4), 495–510. https://doi.org/10.1016/j.respol.2010.01.022

Geels, F. W. (2002). Technological transitions as evolutionary reconfiguration processes: A multi-level perspective and a case-study. *Research Policy, 31*, 1257–1274. https://doi.org/10.1016/S0048-7333(02)00062-8

Geels, F. W., Kern, F., Fuchs, G., Hinderer, N., Kungl, G., Mylan, J., Neukirch, M., & Wassermann, S. (2016). The enactment of socio-technical transition pathways: A reformulated typology and a comparative multi-level analysis of the German and UK low-carbon electricity transitions (1990–2014). *Research Policy, 45*(4), 896–913. https://doi.org/10.1016/j.respol.2016.01.015

Gillard, R., Gouldson, A., Paavola, J., & Van Alstine, J. (2016). Transformational responses to climate change: Beyond a systems perspective of social change in mitigation and adaptation. *Wires Climate Change, 7*(2), 251–265. https://doi.org/10.1002/wcc.384

Hartley, D. (2013). *Urban decline in rust-belt cities*. Econ Comment.

ILO. (2015). *Guidelines for a just transition towards environmentally sustainable economies and societies for all*. ISBN 978-92-2-130628-3.

Jackson, B. (2005). The conceptual history of social justice. *Political Studies Review, 3*(3), 356–373. https://doi.org/10.1111/j.1478-9299.2005.00028.x

Jenkins, K., Sovacool, B., Błachowicz, A., & Lauer, A. (2020). Politicising the just transition: Linking global climate policy, nationally determined contributions and targeted research agendas. *Geoforum, 115*, 138–142. https://doi.org/10.1016/j.geoforum.2020.05.012

Kern, F., Rogge, K. (2018). Harnessing theories of the policy process for analysing the politics of sustainability transitions: A critical survey. https://doi.org/10.1016/j.eist.2017.11.001

Klitkou, A., Capasso, M., & Hansen, T. (2021). Understanding conditions for path development after path exhaustion. *European Planning Studies, 29*(8), 1538–1555. https://doi.org/10.1080/09654313.2021.1875995

Köhler, J., de Haan, F., Holtz, G., Kubeczko, K., Moallemi, E., Papachristos, G., & Chappin, E. (2018). Modelling sustainability transitions: An assessment of approaches and challenges. *Journal of Artificial Societies and Social Simulation, 21*(1), 8. https://doi.org/10.18564/jasss.3629

Kollmeyer, C. (2018). Trade union decline, deindustrialization, and rising income inequality in the United States, 1947 to 2015. *Research in Social Stratification and Mobility, 57*, 1–10. https://doi.org/10.1016/j.rssm.2018.07.002

Lawless, S., Song, A. M., Cohen, P. J., & Morrison, T. H. (2020). Rights, equity and justice: A diagnostic for social meta-norm diffusion in environmental governance. *Earth System Governance, 6*, Article 100052. https://doi.org/10.1016/j.esg.2020.100052

Lazonick, W. (2013) The Financialization of the U.S. Corporation: What has been lost, and how it can be regained. *Seattle Univ Law Rev 36*(2), 857

Lee, S., & Baumgartner, L. (2022). *How just transition can help deliver the Paris agreement*. K. J. Smith (Ed.). UNDP.

Loorbach, D., Frantzeskaki, N., & Avelino, F. (2017). Sustainability transitions research: Transforming science and practice for societal change. *Annual Review of Environment and Resources, 42*(1), 599–626. https://doi.org/10.1146/annurev-environ-102014-021340

McCauley, D., & Heffron, R. (2018). Just transition: Integrating climate, energy and environmental justice. *Energy Policy, 119*, 1–7. https://doi.org/10.1016/j.enpol.2018.04.014

Milkman, R. (1990). 7. Labor and management in uncertain times: Renegotiating the social contract. In *7. Labor and management in uncertain* times: Renegotiating the social contract (pp 131–151). University of California Press.

Moodie, J., Tapia, C., Löfving, L., Gassen N. S., & Cedergren E. (2021). Towards a territorially just climate transition—Assessing the Swedish EU territorial just transition plan development process. *Sustainability, 13*(13), 7505. https://doi.org/10.3390/su13137505

Nelson, D. R., Adger, W. N., & Brown, K. (2007). Adaptation to environmental change: Contributions of a resilience framework. *Annual Review of Environment and Resources, 32*(1), 395–419. https://doi.org/10.1146/annurev.energy.32.051807.090348

Newman, K. S. (1990). 6. Uncertain seas: Cultural turmoil and the domestic economy. In *6. Uncertain seas: Cultural turmoil and the domestic economy* (pp. 112–130). University of California Press

Pelling, M. (2010). *Adaptation to climate change: From resilience to transformation*. Routledge.

Raven, R., Schot, J., & Berkhout, F. (2012). Space and scale in socio-technical transitions. *Environ Innovation and Societal Transitions, 4*, 63–78. https://doi.org/10.1016/j.eist.2012.08.001

Regional Cooperation Council. (2020). *Sofia declaration on the green agenda for the Western Balkans*. https://www.rcc.int/download/docs/Leaders%20Declaration%20on%20the%20Green%20Agenda%20for%20the%20WB.pdf/196c92cf0534f629d43c460079809b20.pdf

Roessler, C. (2016). *"Just Transition"—Just what is it?—Labor Network for sustainability*. https://www.labor4sustainability.org/uncategorized/just-transition-just-what-is-it/. Accessed 9 April 2024

Seto, K. C., Davis, S. J., Mitchell, R. B., Stokes, E. C., Unruh, G., & Ürge-Vorsatz, D. (2016). Carbon lock-in: Types, causes, and policy implications. *Annual Review of Environment and Resources*, 41(1), 425–452. https://doi.org/10.1146/annurev-environ-110615-085934

Simmie, J., & Martin, R. (2010). The economic resilience of regions: Towards an evolutionary approach. *Cambridge Journal of Regions, Economy and Society, 3*(1), 7–43, 2010. SSRN: https://ssrn.com/abstract=1570002 or https://doi.org/10.1093/cjres/rsp029

Smith, A., Stirling, A., & Berkhout, F. (2005). The governance of sustainable socio-technical transitions. *Research Policy, 34*, 1491–1510. https://doi.org/10.1016/j.respol.2005.07.005

Stokols D, Lejano RP, Hipp J (2013) Enhancing the resilience of human–environment systems: A social ecological perspective. *Ecology and society, 18*(1).

Thomas, A. (2021). Framing the just transition: How international trade unions engage with UN climate negotiations. *Glob Environ Change, 70*, Article 102347. https://doi.org/10.1016/j.gloenvcha.2021.102347

Toto R., Imami F., Poro E., et al. (2023) Just Green Transition Conceptualisation in the Western Balkans. GreenFORCE Horizon project, Deliverable 4.1

UN DESA. (2022). *A just green transition: Concepts and practice so far* (POLICY BRIEF NO. 141).

UNFCCC. (2022). *"Sharm El-Sheikh Implementation Plan."* Climate Change Conference (COP27). Sharm el-Sheikh, Egypt. https://unfccc.int/sites/default/files/resource/cop27_auv_2_cover%20decision.pdf?download.

Wøien Meijer, M., & Peter, O. (2021). Nordic industrial green transition: A literature review and Nordic benchmark. NORCE Samfunn report number: 20–2021. ISBN: 978–82–8408–169–4

CHAPTER 6

Epistemic Work and Epistemic Justice in Just Green Transitions

Govert Valkenburg

6.1 Introduction

Just Green Transitions (JGTs) pose complex policy challenges (Shaker & Berisha, 2025). Changes of technical, social, cultural, economic, and institutional nature need to be accomplished simultaneously. These policy aims require input of a broad variety of types of knowledge. Traditionally, this need for an extended knowledge base has been pursued by attempts at cultivating an *extended peer community* (Funtowicz & Ravetz, 1993, 2003), bringing together many kinds of formal expertise as well as different sorts of knowledge from the broader public that has other authorisations and authentications than formalised expertise: experiential and professional knowledge, practical knowledge, local and cultural knowledge, etc. This enrolment is ideally done through *transdisciplinary research*. Such research offers a specific sociality that is required to take a problem-based approach and generate the required comprehensive bodies of knowledge (Silvast & Foulds, 2022). However, what has so far been poorly addressed, both in policy and in practical approaches, is the inevitable emergence of issues of justice related to the nature of different sorts of knowledge. Different sorts of knowledge are not necessarily equally and fairly adopted, and it is a matter of *epistemic justice* to ensure that any inequitable reception of knowledge is repaired. I propose the notion of *epistemic work* to capture the effort to bring together and build bridges between

G. Valkenburg (✉)
Department of Interdisciplinary Studies of Culture, Norwegian University of Science and Technology NTNU, Trondheim, Norway
e-mail: govert.valkenburg@ntnu.no

E. Berisha et al. (eds.), *The Palgrave Handbook of Just Green Transitions in the Western Balkans and Beyond*,
https://doi.org/10.1007/978-3-031-95075-9_6

different sorts of knowledge and to ensure that no knowledge is excluded only because it is not rooted in formal expertise. Such epistemic work is particularly pertinent to the development of JGTs, and I henceforth refer to it as *sustainable transition epistemic work* or STEW. It is vital to shape STEW as a work practice with longevity, which is not a given as transition work is mostly organised as projects, pilots, and living labs. This ad-hoc structure entails that innovation arenas, as well as professionals with relevant knowledge-related abilities, remain in place only temporarily, and capturing the routines and building persistent structures to ensure epistemic justice remains challenging. Thus, an underlying challenge is to make STEW a standard part of all the work in transitions, tied to actors and institutions with more permanence than project-based scopes. This chapter explores how STEW can be made durable. What challenges related to epistemic justice emerge in transitions? What skills do knowledge workers need to warrant a proper treatment of all knowledges involved? How can such skills be translated from one transition arena to the next? How can initiatives of public engagement be shaped with epistemic integrity? And what are the different sorts of knowledge that matter in the first place? I first explore the knowledge content of transitions and why we need to pay particular attention to it. Then, the argument is made that the project-based organisation of much of what goes on in transitions poses challenges to stabilising practices of proper dealing with knowledge. This entails challenges for knowledge justice. The chapter concludes that there are two core value sets that STEW would need to uphold. First, the notion of 'just transitions' is all around. Considerations of justice, including specific fields such as energy justice and environmental justice, need to be part and parcel of STEW, as the knowledge work involved is, by definition, not neutral but conducive to promoting some configurations and making others less likely. Second, and more in the centre court of the current chapter, the idea of epistemic justice needs to be similarly central. Addressing this within the remit of STEW helps to improve the justice of technological societies and the quality of their democratic governance.

6.2 Transitions as a Knowledge Object

Because of their heterogeneity, sustainable transitions are too complex to manage from within traditional scientific disciplines, and even cross-disciplinary collaborations may not suffice to accomplish the tasks. Therefore, novel modes of producing and circulating knowledge are needed. In particular, modes are needed that traverse technical, scientific, social-scientific, and cultural fields of knowledge. Only with the aforementioned extended peer community (Funtowicz & Ravetz, 1993, 2003) and transdisciplinary research (Silvast & Foulds, 2022) can the breadth of all the normative and factual uncertainties be addressed.

Within such strategies towards knowledge production for transitions, learning is a crucial element. Transitions are not structured puzzles where the problem is clearly defined at the start and then resolved by recourse to known strategies and existing knowledge bases. Rather, both the questions and their answers are developed and learned-by-doing in the transition projects themselves (Grin et al., 2010; Raven et al., 2008). Thus, the knowledge developed in projects somehow needs to be captured, generalised and transported to other situations. Knowledge needs to be ported to novel contexts: translations need to be made, connections need to be forged, boundary objects should be tended to so as to help them retain meaning, novel normative frameworks must be adapted to, etc. Moving between these different epistemic worlds requires creativity and engagement, and it is hardly ever sufficient to distribute a report after a pilot project. However, learning and incorporating knowledge has remained challenging still today (Bremer et al., 2022; Foulds et al., 2023; Raven et al., 2008; Ryghaug & Skjølsvold, 2021, p. 48). Often, knowledge is tied to persons who develop the knowledge in the form of contracted experience. There is little or no possibility to codify the knowledge, if only because limited resources and working time are allocated to this codification. Also, important parts of the knowledge are difficult to codify and disseminate, as they are tied to local insights. If a person leaves a project, the knowledge leaves with them. Thus, the curation of knowledge in sustainable transitions calls for a particular kind of work. I will henceforth refer to this work as *epistemic work*, defining it as all those actions that engage with the origin, production, validation, and social connection and distribution of knowledges.[1] Epistemic work is also about translation, or bridging the epistemological gaps between various knowledges. Furthermore, it needs to reflect on the origins of the gaps and address possible hierarchical and hegemonic schemes that produce them. This chapter focuses specifically on epistemic work in sustainable transitions and the difficulties engendered by the projectified nature of transitions. I will therefore refer to this work as *sustainable transition epistemic work*, henceforth STEW. I explore the challenges this kind of work faces, as well as the possible ways of conducting it. Some examples of STEW have already been provided. Examples include the articulation of actors' expectations and visions and the discursive space they have to revise those (Pesch, 2014), helping actors to connect socio-cultural and techno-scientific aspects of innovations (Naber et al., 2017), workshops to facilitate stakeholders in devising future visions (Soneryd, 2016), or comprehensive approaches to *transition arenas* (Hyysalo et al., 2019) consisting of activities such as stakeholder co-design of systems, operational activities of learning-by-doing, and workshops where knowledge positions of actors are respected and discussed. In *just transitions*, these activities are usually considered part of *procedural justice*, meaning that citizens should be involved in decision-making procedures in such a way that they can recognise the outcomes as theirs (Jenkins et al., 2016, 2020). Indeed,

[1] This notion of 'epistemic work' is radically different than the one from Cook and Brown (1999), who use the term to refer to a distinctly action-based and interactional understanding of knowing, in contrast to 'knowing as the static possession of knowledge'.

procedural justice is a prerequisite for achieving epistemic justice. Yet, I hold that procedural justice needs additional stipulations to also achieve epistemic justice. Measures need to ensure that the inclusion of knowledge is genuine and fair. The looming tokenism must be overcome of simply inviting people to the discussion table for the sake of inviting them rather than ensuring meaningfulness to their participation, for example, by exerting influence on decisions or by establishing genuine learning points that will be taken on board in the ensuing process (Chilvers & Kearnes, 2016; Pandey & Sharma, 2021). In a broader sense, epistemic justice needs to be part of notions of just transitions and energy justice themselves (Valkenburg, 2024).

6.3 The Transience of Transition Knowledge Work

One important problem with STEW is that it is caught in structures of exceptionality: projects, pilots, living labs, and all those other purpose-built temporary arenas and other institutionalisations that perish once the project period is over. As I will continue to argue, this transience is one of the important reasons why it remains difficult to learn from projects and reap lessons, upscale them, and transfer lessons learned to other situations. This chapter will explore what it takes to position STEW as normal work: a regular part of organisations that find themselves in societies where sustainable transitions occur. Most transition-related changes are a 'one-of' event, even though they may serve long-term goals. They involve many actors with diverging interests at once and are connected to other systemic structures. Also, they are path-dependent, meaning that definitions of problems change along the way, as do the possible and preferable strategies for solutions. With all these properties, the governance of transitions is essentially a *wicked problem,* as defined by Rittel and Webber (1973). This ephemeral nature of transition work is reflected by the fact that a large part of what happens in transitions is shaped as pilot projects, living labs, and other experimental settings (Ryghaug & Skjølsvold, 2021). It is in such complex spaces of decision-making that knowledge workers are set to task. Significantly, this project basis is different from continuous, long-standing tasks and jobs. I refer to this project-based and time-limited organisation of STEW as *exception work.* It stands in contrast to *normal work,* which I understand as work that is a common part of continuous and longer-term work processes. The exceptionality of energy transition project work poses specific challenges to STEW, and it is high time to explore what it takes for STEW to be done responsibly and how this can be sustained over time. As transitions are here to stay, STEW needs to become a part of modern societies' *normal work.* Social sciences have attended to the combination of the complexity of transition issues and broader concerns about how expertise should function in a healthy democracy. This discourse has sparked interest in the question of how processes of learning can be made to be less dominated by experts and more welcoming to a broader range of sources of relevant knowledge. Under the theme of *responsible research and innovation*

(Owen & Pansera, 2019; Owen et al., 2013), the exchange of knowledge and values between the public and practices of science and innovation are a central concern. Work on participation (Chilvers & Kearnes, 2016, 2020) has rekindled the long-standing engagement with the democratisation of technological societies and the role of expert knowledge in it (Collins & Evans, 2007; Grundmann, 2016; Wynne, 1991), as well as the long-standing literature on transdisciplinarity (Silvast & Foulds, 2022) and the extended peer community (Funtowicz & Ravetz, 1993, 2003). Dealing with knowledge not only requires making a serious effort to include valuable knowledge in policymaking. It is also about liberating newly generated knowledge from its confines so that the broader democratic constituency can make use of it. Knowledge needs to be disseminated, communicated, and attributed with proper credit to its origin. Not only do originators of knowledge deserve this credit, but it is also about framing the knowledges such that many actors can see them as their own rather than 'stamped and appropriated' by experts in core sites of transition work. It is also about recognising and acknowledging the particularities of the situations in which lessons were learnt and the costs that were borne for these lessons by local stakeholders. Moreover, it must be ensured that these remain visible when we discuss upscaling and transfer to other situations. These questions bring the concept of STEW necessarily under the ambit of *epistemic justice* or knowledge justice. As will be reviewed below, epistemic justice is about giving people due credibility as sources of knowledge and about providing them with the proper means to develop knowledge and make sense of the issues they are confronted with. One core task for STEW then follows: to critically assess the hierarchies in which knowledge contributions are caught, not least the dominant understanding of innovation as connected to markets and linear narratives of technological progress. This understanding is needed for a clear view of the knowledges that potentially get lost in transitions and to build epistemic justice in practical work in complex situations.

6.4 Justice Also Means Knowledge Justice

Justice aspects of sustainable transitions have been addressed widely (Heffron & McCauley, 2018). The notion of *just transition* is even part of the European Commission's strategy, according to (then) EC Executive Vice President Frans Timmermans.[2] The theme connects to notions such as energy justice (Jenkins et al., 2016, 2020; Sovacool & Dworkin, 2015) and environmental justice (Menton et al., 2020). Still, the challenges that the exception activities described above pose to justice have not been addressed sufficiently. On the one hand, transition activities are fluid, and hence, they are important

[2] https://commission.europa.eu/strategy-and-policy/priorities-2019-2024/european-green-deal/finance-and-green-deal/just-transition-mechanism_en (Accessed 30 January 2023).

sites of politics and innovation. Yet, they are also often couched in the conceptions that reproduce existing social structures, celebrate technology-centred paradigms, and rely on market mechanisms for defining value (Ryghaug & Skjølsvold, 2021). As I will argue, this is not the best frame to achieve justice, especially if we zoom in on its knowledge aspects. These aspects should be a particular concern for STEW. The literature on *epistemic justice* (Fricker, 2007) has thematised how the transaction and production of knowledge can happen in just and unjust ways. In terms of epistemic justice, fairly treating knowledge means extending existing notions of inclusion from mere *testimonial justice*—i.e. justice with respect to being heard—towards *hermeneutic justice*—i.e. justice with respect to being enabled to develop knowledge. First, testimonial injustice is done when a person is denied the possibility to present their knowledge as valid, rational, reasonable, or relevant. One classic example is that women, throughout history, have been taken less seriously than men when presenting knowledge for the sole reason that they were not men. Similar exclusions have happened along racist, classist, colonial, and other faultlines. Also, such misrecognition might happen because people are situated in unfavourable social, political, or institutional arrangements. Second, hermeneutic injustice is done when people are denied the intellectual means to make sense of a socially relevant phenomenon. Such injustice is, for example, done when schooling is curtailed, as the Taliban do to Afghan girls, leading to the girls having less access to knowledge and concepts that help them make sense of their situation and which would ultimately empower them. Another example is the fact that large parts of the corpus of medical science have been grafted on the male body, which entails that women have relatively less access to relevant medical-scientific knowledge to make sense of their own bodies. Thus, epistemic justice is a comprehensive matter in that it pertains to both the content of knowledge and the procedures and interactions through which knowledge is produced and shared. Process and substance are inextricable, and a sensitivity to this complexity is a crucial part of the STEW toolkit. Epistemic justice, a virtue for individuals and institutions, includes recognition of difference and acknowledging the existence of other epistemologies. It involves attending to sources of knowledge other than the ones championed by conventional science and expertise. Without proper recognition, the dominance of science and expertise may lead to the exclusion of other knowledges and the actors holding those knowledges. For example, indigenous and traditional knowledges are not straightforwardly included at the same level as Western, certified expert knowledge. Also, practical knowledge from professionals, from care and education to agriculture, is not always included in a fair way in processes of decision-making, despite the many attempts at inclusion. In view of the high heterogeneity of transition processes, the incapacity to include other knowledge is utterly problematic. Epistemic justice involves recognition of one's own potentially privileged position in hierarchies of epistemologies. At the same time, not all knowledge should be uncritically included, and critical assessment remains a necessity. Yet, what matters is how we do so and how we steer clear of the risk of unjustly excluding knowledge. There is a moral duty to assess the grounds on which exclusion happens or the reasons why it is not sufficiently prevented. The recognition that there are other epistemologies with a right to existence is a first step. The topic has been connected to various practices of innovation and governance (Valkenburg, 2020, 2022, 2024; Valkenburg et al., 2020). Thus, epistemic justice extends existing notions of just transitions, where principles of *procedural justice* are mostly geared towards including knowledge but not necessarily nurturing the capacity to produce and present knowledge. Epistemic justice should be a guiding principle for STEW. Capitalist and technology-centred paradigms that characterise thinking in sustainable transitions need to be assessed in view of epistemic justice. These paradigms relate to the rationality of modern science, which is an important driver of innovation and an important asset to what makes modern life liveable. Yet, it also retains dominance over other types and sources of knowledge. It needs to be recognised that there are large domains of knowledge that are meaningful yet not compliant with the epistemologies of modern science and, therefore, risk exclusion. For example, through art, we learn about what various realities mean to our lives. And our normative and moral frameworks derive from the cultures we grew up in. These are not validated by markets or measurements, yet they are crucially important in shaping our lives, and hence to such uprooting changes as sustainable transitions. Epistemic justice requires an effort to bring these ways of knowing to the places where they matter. Regarding just transitions, there is a broad literature available already (Heffron & McCauley, 2018; Jenkins et al., 2016; Jenkins et al., 2020; Menton et al., 2020; Sovacool & Dworkin, 2015). Its explicit connection with epistemic justice is still developing (Valkenburg, 2024), and in the next section, I explore how it can be shaped more concretely.

6.5 Knowledge Justice as a Task for Transition Knowledge Workers

As knowledge work is indispensable in the governance of transitions, it is a straightforward concern how this knowledge work can be conducted in a way that is both practically feasible and compliant with the moral principles of epistemic justice. The question, then, is how STEW can promote epistemic justice. It is sensible to focus on knowledge justice, given that knowledge is the primary object of engagement in the work. The question of epistemic justice is then how knowledge workers in transitions can grant that people are not by any unjust means impeded from developing knowledge and getting their knowledge heard. What sets transitions apart from other social situations is the fact that transitions are explicitly geared towards uprooting change. This change may destabilise both people's means of knowledge-making and the frameworks through which their knowledge is assessed as valid and relevant. In consequence, the circulation of knowledge and the fairness thereof are destabilised. As transitions are at once the object of knowledge and a destabiliser of knowledge circulations, epistemic justice in transitions is a highly reflexive affair.

Epistemic justice as a guiding principle for STEW seems to call attention to three main points of engagement. First, in its professional transaction of knowledge, STEW must recognise different knowledges and how these differences matter. Given their technical, spatial, and socio-cultural ramifications, transitions protrude into many different cultural situations and, hence, different ways of knowing the transitions and their consequences. What is a neutral switch from natural gas-based to electrical heating of homes? For some, it might be a matter of ideology or impediment to individual freedom for others. For some, it may be an insurmountable financial cost. None of these is more correct than the others, and no ultimate truth can be established. Yet, thinking of the move as neutral is most compatible with how formal experts are likely to frame it. Hence, this position has a higher likelihood of being taken into account in decision-making. However, this may amount to testimonial injustice, and ultimately, it may lead to citizens rejecting the decisions made and the transition failing because of a lack of support. Hence, the responsibility of STEW includes taking responsibility for the vulnerabilities of knowledges and those who own them. In this first sense, epistemic justice is an *instrument* in STEW and its aim to secure public support for transitions.

Second, STEW must develop sensitivity to how processes are sound in view of epistemic justice. When facilitating any learning or knowledge development in transitions, it is often up to STEW workers to make sure that processes of public engagement are conducive to this. These processes institutionalise ways of going about knowledge and prioritisations between those ways of knowing. With 'institutionalising', I refer to the fact that dynamics may emerge beyond the actions and intentions of individual people (whether STEW workers or others) and that these dynamics may stay in place over time. Thus, STEW workers have a responsibility to shape such institutionalisations so that all actors see their epistemic agency and respect granted. In this second sense, implementing epistemic justice is also the *substance* of STEW. Third, STEW needs to attend to the mentioned reflexivity, as categories and conceptual frameworks are destabilised. Epistemic justice needs to be related constantly to the uprooting changes that transitions incur and the reflexive nature that follows from that. While the technical changes in transitions are the most prominent, the ensuing economic, social and cultural ones are just as significant. In consequence, the economy of knowledge might also change. For example, people may need to change jobs because of transitions, and this may impede their economic means to contribute to governance processes, or it may kill their motivation to do so. It may also change their relative epistemic agency, as people's ability to establish and share knowledge is dependent on social context. Thus, their ability to frame and share knowledge is compromised, and STEW workers must be aware of the consequences of these issues and creatively seek to mitigate them. While implementing emancipatory interventions is beyond the scope of this chapter, one example one could think of consists of articulating the (not least social) context of the origin of all knowledge, such that their hierarchy becomes less self-evident. One could also engage with how non-dominant forms of knowledge are valuable and useful even if they are not cast in typical expertise forms. Epistemic justice as a philosophical-normative construct thus requires considerable translation to be applied concretely in a work situation. What epistemic skills are needed to build epistemic justice through STEW? What specific transition-related challenges are identified through the lens of epistemic justice? What epistemic challenges appear through a lens of just transitions? We need an inventory of the epistemic challenges that emerge in sustainable transitions and the epistemic skills needed to 'get just transitions done'. These skills need to afford an inclusion of the relevant knowledge along the lines of epistemic justice and just transitions. Finally, this raises the question of *who* is going to perform STEW. While the work necessarily involves reflexive knowledge production that is characteristic of academic research in social sciences and humanities (SSH), it is not straightforward that this work needs to be left to academics as a permanent task. Not only does it need to be rooted in society and, hence, to some extent, placed outside the confines of academia, but it would also be problematic that current academic research has, in fact, exactly the project-based organisation structure that hampers the continuity that is needed here (Felt et al., 2016). It is, therefore, probably a better strategy to implement STEW as a permanent portfolio in governmental agencies, corporate actors (even though the required investment is likely to clash with their market logic), and dedicated agencies that offer this work as a service to governmental and corporate actors.

6.6 Doing Transitions Knowledge-Work Well: An Agenda

It is high time to develop this agenda further and shape STEW as a set of skills and a recognised kind of work that helps knowledge circulation needed for sustainable transitions, does so in an epistemically just way, and is capable of securing a place in institutional and governmental structures over time. A lot of knowledge work in transitions is currently done, and dedicated training programmes exist. However, this training mostly focuses on the technical and social interventions needed (see Köhler et al., 2019). It lacks a systematic perspective on the epistemic content of the work, let alone a perspective integrating the normative perspective of epistemic justice.

The challenges presented in this chapter are particularly pressing in the context of the Western Balkans. As reported by (Toto et al., 2023, p. 53ff), cross-sectoral cooperation and accountability have generally been poor over the past decades. Also, parts of society have developed a certain fatigue towards EU-driven policy initiatives and considerable scepticism towards transitions at large (p. 10). In a general sense, the principles discussed in this chapter should be expected to contribute to resolving these barriers. Promoting epistemic justice through STEW is a double-edged sword: not only does it make people feel taken seriously as sources of knowledge and contributors to governance processes, but it also enforces reflection on the dominant position of those in power and the (possibly unjust) control over discourses and knowledge validations that they hold. As sketched in this chapter, epistemic justice needs to be a consideration that runs through the veins of STEW. Epistemic justice is to be implemented at three levels.

First, it needs to be pursued and promoted by the people conducting STEW. Considerations of epistemic justice must be tabled when interacting with people who need to be recognised as knowledgeable and where there might be social, cultural, or institutional structures that potentially impede such recognition. STEW workers must shape their work in line with principles of epistemic justice to ensure that their work contributes value to society instead of perpetuating existing hegemonies, not least knowledge hegemonies. What is more, they bear responsibility for the institutional structures to which they contribute and which need to be arranged according to the same principles. Through the material with which these professionals are to work, i.e. the knowledge circulations that happen in the design and implementation of transitions, they ideally add societal value in the form of epistemic justice.

Second, epistemic justice must be ingrained in and safeguarded by the organisations that govern transitions and that are responsible for their proper governance. It is not only in the individual interactions that STEW professionals engage in but also in the broader arrangements they put in place that facilitate knowledge circulations in better or worse ways. STEW workers build situations and institutionalisations, and that comes with the responsibility of installing epistemic justice as a virtue of those institutionalisations. For example, corporate actors such as energy and building companies need to have permanent staff to conduct STEW as part of all their projects, and it is up to STEW workers to continually point out the need for this work. Also, STEW needs to be incorporated into educational programmes, meaning regular professional and higher education, as well as dedicated courses and training in which transition workers participate.

Third, STEW must itself be experimental and permanently open for debate and revision, as this is the only way to accommodate its internal reflexivity and dynamics. One way in which change can be seen as radical is if not only a certain reality changes but also the more fundamental understandings through which we make sense of that reality. It is a general principle in technological societies that new technologies not only challenge the norms of moral frameworks but also require that we revise those very norms in view of novel technological opportunities and hazards. In view of such challenges, a certain moral prudence is required to achieve moral resilience (Swierstra, 2013).

In relation to the norms by which we assess the fair circulation of knowledge, this raises three principal questions. First, people may become positioned differently by transitions, bearing in specific ways on their opportunities to establish knowledge and voice their concerns and knowledges. Second, the norms by which we validate knowledge may change. A concrete phenomenon is that expertise has, over the past decades, become much more critically assessed and sceptically approached in the public sphere. This critical approach is likely to continue, and therefore, we need to be aware that disenfranchised groups may be further disadvantaged if expertise, or whatever source of knowledge they tap into that can help them is driven out by such forces. It is up to STEW to counteract such impediments. Third, the changes may mean that our norms of fairness themselves become revised, meaning that we come to think differently about how and when unfair knowledge circulations must be corrected. Also, it is STEW's important responsibility to identify and articulate such changes and offer solace where possible. A further challenge is that STEW with an eye on epistemic justice is typically a skill powered by insights from social sciences and humanities (SSH), and it is not always easy to gain ground for this kind of professional style amidst corporate cultures that are by and large governed by commercial interests. Even though companies may care for values of justice and fairness, the interests of shareholders usually pose a competing interest. These interests are predicated on the logic of efficiency and profitability and narratives of linear technological progress, which is intimately connected to the dominant styles of knowledge production in Western academic style. My argument has been that, on top of serving these managerial needs, STEW has a normative content that is, in part, to offer critical mass against exactly those logics.

References

Bremer, S., Wardekker, A., Baldissera Pacchetti, M., Bruno Soares, M., & van der Sluijs, J. (2022). Editorial: High-quality knowledge for climate adaptation: Revisiting criteria of credibility, legitimacy, salience, and usability. *Frontiers in Climate, 4*(905786), 1–3.

Chilvers, J., & Kearnes, M. (Eds.). (2016). *Remaking participation. Science, environment, and emerging publics*. Routledge.

Chilvers, J., & Kearnes, M. (2020). Remaking participation in science and democracy. *Science, Technology, & Human Values, 45*(3), 347–380.

Collins, H. M., & Evans, R. (2007). *Rethinking expertise*. University of Chicago Press.

Cook, S. D. N., & Brown, J. S. (1999). Bridging epistemologies: The generative dance between organizational knowledge and organizational knowing. *Organisation Science, 10*(4), 381–400.

Felt, U., Igelsböck, J., Schikowitz, A., & Völker, T. (2016). Transdisciplinary sustainability research in practice: between imaginaries of collective experimentation and entrenched academic value orders. *Science, Technology & Human Values.*

Foulds, C., Valkenburg, G., Ryghaug, M., Suboticki, I., Skjølsvold, T. M., Korsnes, M., & Heidenreich, S. (2023). Implementing mission-oriented experiments: Recommendations on epistemic inclusion for city stakeholders working in climate change initiatives. *Journal of City Climate Policy and Economy.*

Fricker, M. (2007). *Epistemic injustice*. Oxford University Press.

Funtowicz, S. O., & Ravetz, J. R. (1993). Science for the post-normal age. *Futures, 25*(7), 739–755.

Funtowicz, S. O., & Ravetz, J. R. (2003). *Post-normal science*. The International Society for Ecological Economics.

Grin, J., Rotmans, J., & Schot, J. (2010). *Transitions to sustainable development: New directions in the study of long term transformative change*. Routledge.

Grundmann, R. (2016). The problem of expertise in knowledge societies. *Minerva, 55*(1), 25–48.

Heffron, R. J., & McCauley, D. (2018). What is the 'Just Transition'? *Geoforum, 88*, 74–77.

Hyysalo, S., Lukkarinen, J., Kivimaa, P., Lovio, R., Temmes, A., Hildén, M., Marttila, T., Auvinen, K., Perikangas, S., Pyhälammi, A., Peljo, J., Savolainen, K., Hakkarainen, L., Rask, M., Matschoss, K., Huomo, T., Berg, A., & Pantsar, M. (2019). Developing policy pathways: Redesigning transition arenas for mid-range planning. *Sustainability, 11*(3).

Jenkins, K. E. H., McCauley, D., Heffron, R., Stephan, H., & Rehner, R. (2016). Energy justice: A conceptual review. *Energy Research & Social Science, 11*, 174–182.

Jenkins, K. E. H., Stephens, J. C., Reames, T. G., & Hernández, D. (2020). Towards impactful energy justice research: Transforming the power of academic engagement. *Energy Research & Social Science, 67.*

Köhler, J., Geels, F. W., Kern, F., Markard, J., Onsongo, E., Wieczorek, A., Alkemade, F., Avelino, F., Bergek, A., Boons, F., Fünfschilling, L., Hess, D., Holtz, G., Hyysalo, S., Jenkins, K., Kivimaa, P., Martiskainen, M., McMeekin, A., Mühlemeier, ...Wells, P et al. (2019). An agenda for sustainability transitions research: State of the art and future directions. *Environmental Innovation and Societal Transitions, 31*, 1–32.

Menton, M., Larrea, C., Latorre, S., Martinez-Alier, J., Peck, M., Temper, L., & Walter, M. (2020). Environmental justice and the SDGs: From synergies to gaps and contradictions. *Sustainability Science, 15*(6), 1621–1636.

Naber, R., Raven, R., Kouw, M., & Dassen, T. (2017). Scaling up sustainable energy innovations. *Energy Policy, 110*, 342–354.

Owen, R., & Pansera, M. (2019). Responsible innovation and responsible research and innovation. In D. Simon, S. Kuhlmann, J. Stamm, & W. Canzler (Eds.), *Handbook on science and public policy*. Edward Elgar Publishing.

Owen, R., Stilgoe, J., Macnaghten, P., Gorman, M., Fisher, E., & Guston, D. (2013). A Framework for responsible innovation. In R. Owen, J. Bessant, & M. Heintz (Eds.), *Responsible innvoation. Managing the responsible emergence of science and innovation in society* (pp. 27–50). John Wiley & Sons.

Pandey, P., & Sharma, A. (2021). Knowledge politics, vulnerability and recognition-based justice: Public participation in renewable energy transitions in India. *Energy Research & Social Science, 71.*

Pesch, U. (2014). Tracing discursive space: Agency and change in sustainability transitions. *Technological Forecasting and Social Change, 90-B*(January), 379–388.

Raven, R. P. J. M., Heiskanen, E., Lovio, R., Hodson, M., & Brohmann, B. (2008). The contribution of local experiments and negotiation processes to field-level learning in emerging (Niche) technologies. Meta-analysis of 27 new energy projects in Europe. *Bulletin of Science,Technology & Society, 28*(6), 464–447.

Rittel, H. W. J., & Webber, M. M. (1973). Dilemmas in a general theory of planning. *Policy Sciences, 4*(2), 155–169.

Ryghaug, M., & Skjølsvold, T. M. (2021). *Pilot society*. Palgrave Macmillan.

Shaker, Y., & Berisha, E. (2025). Just Green Transitions: Between terminological inexactitudes, conceptual fragmentation, and the exigency for a theoretical framework. *European Spatial Research and Policy*, 32(1), 121–138. https://doi.org/10.18778/1231-1952.32.1.05

Silvast, A., & Foulds, C. (2022). *Sociology of interdisciplinarity*. Palgrave Macmillan.

Soneryd, L. (2016). Technologies of participation and the making of technologised futures. In J. Chilvers & M. Kearnes (Eds.), *Remaking participation. Science, environment and emergent publics* (pp. 144–161). Routledge.

Sovacool, B. K., & Dworkin, M. H. (2015). Energy justice: Conceptual insights and practical applications. *Applied Energy, 142*, 435–444.

Swierstra, T. E. (2013). Nanotechnology and technomoral change. *Etica & Politica / Ethics & Politics, XV*(1), 200–219.

Toto, R., Imami, F., Poro, E., Garvanlieva, V., Jeftić, M., Gjoka, R., & Bejko, A. (2023). *Report on Western Balkans just green transition conceptualisation* (Vol. D4.1). The GreenFORCE project.

Valkenburg, G. (2020). Consensus or contestation: Reflections on governance of innovation in a context of heterogeneous knowledges. *Science, Technology and Society, 25*(2), 341–356.

Valkenburg, G. (2022). Temporality in epistemic justice. *Time & Society, 31*(3), 437–454.

Valkenburg, G. (2024). Energy justice as epistemic justice. Ethics, Policy & Environment, 1–22. https://doi.org/10.1080/21550085.2024.2418789

Valkenburg, G., Mamidipudi, A., Pandey, P., & Bijker, W. E. (2020). Responsible innovation as empowering ways of knowing. *Journal of Responsible Innovation, 7*(1), 6–25.

Wynne, B. (1991). Knowledges in context. *Science, Technology, & Human Values, 16*(1), 111–121.

CHAPTER 7

Participatory Planning Culture and Its Role and Potential for Fostering Just Green Transitions

Matej Jaššo and Milan Husár

7.1 Introduction

Spatial planning and all the processes related to spatial development on different scales have recently faced numerous shifts, changes and modifications concerning the dynamics and depths of the planning process, the role of the planners as well as participation of various actors and stakeholders. The role of planners has been modified from the position of the designers of the future through the position of technical experts toward the role of mediators and leaders of the development processes (Jaššo & Finka, 2019). This process requires special attention to be paid to cultural phenomena, communication processes and behavioral variables. Understanding of cultural dynamics, the ability to develop a clear vision of the desired planning culture based on democratic leadership and its active fostering, should be an essential competence of any planner (Jaššo & Finka, 2019). Just green transition is a phenomenon concerning all societal layers and planning as a major managerial and steering activity plays an undisputable role within this process.

Just green transition like every major paradigmatic change in human history brings changes in cultural frameworks. This concerns various cultural subsystems: national cultures, professional cultures, planning culture as well as

M. Jaššo · M. Húsár (✉)
Spectra Centre of Excellence of the EU, Slovak University of Technology in Bratislava, Bratislava, Slovakia
e-mail: milan.husar@stuba.sk

M. Jaššo
e-mail: matej.jasso@stuba.sk

E. Berisha et al. (eds.), *The Palgrave Handbook of Just Green Transitions in the Western Balkans and Beyond*,
https://doi.org/10.1007/978-3-031-95075-9_7

corporate culture of organizations active on the market. The evolutionary and gradual change of each culture often stands in a different rhythm with the pace of technological change, and this imbalance can then produce tension in society. Culture, as an accumulation of best practices from the past, often stands in opposition to the required actions; it can act as a braking, questioning or retarding element. Just green transition is a process involving a plethora of actors with different positions, motivations and competencies. This process is impacting many cultural realms and frameworks and is not externalized in vacuum. Underlying assumptions, beliefs, attitudes, values and expectations play a major role in the effectiveness and smoothness of this process although their role might be not visible and obvious on the first sight. Many important variables related to leadership and culture (individual initiative, decision-making, risk acceptance and communication styles) may contribute to overall success or failure of entire just green transition. An important factor is public trust (or distrust) toward the whole process, which is also determined by cultural and human resource related determinants. Attitudes toward the goals of just green transition, resentments and fears related to abrupt changes in many accepted and proven actions and proceedings as well as commitment to values of sustainability and innovative behavioral patterns are also inseparable parts of cultural background. Just green transition needs to lean on large societal consensus and acceptance of wide array of stakeholders in democratic decision-making process. The aim of the chapter is to analyze and emphasize very delicate and multidimensional relation between the process of just green transition and cultural elements accompanying this process within the field of spatial planning and spatial development. The just green transition is a very complex and highly dynamic process that transcends sectoral boundaries and must involve all relevant societal sectors and actors. Many of these actors are both principal actors in this process as well as recipients of its deliverables (e.g. companies, universities, communities, etc.). The success of this highly delicate process depends not only on the technological and managerial aspects but also on the commitment and positive attitudes reflected in the planning culture of communities, cities or regions. Thus, we predominantly focused on participatory planning elements and participatory planning culture which is supposed to have best chances to help accomplish just green transition as smooth, democratic and participatory process. Our contribution introduces challenges of just green transition with regard to the process of democratic public participation, explains its supposed interlinkages toward planning culture and introduces our model of evaluation of planning cultures. In the final phase, we briefly analyze two examples of recent planning practice. Concluding remarks take into the consideration regional specifics of Western Balkans.

7.2 Challenges of the Just Green Transition in Relation to Public Participation

The challenges and problems related to just green transition are becoming a significant part of national, regional or municipal strategies. They are undertaken for various strategic reasons like achievement of territorial cohesion or synergy benefits arising from sustainable measures made in different fields (Shaker & Berisha, 2025). On the other hand, introducing a new paradigm heavily influenced by perceived risks and threats is one of the most difficult and complex tasks even for the most experienced and successful leaders active within the planning practice. Recent discussions indicate that most challenging obstacles are not only in technological dimensions but are present also due to the lack of systematic, strategic and highly sensitive approach to public participation, education and cooperation of various actors. Managing green transition also means managing fears and distrust. It has been verified in many similar historical situations that a larger number of transitional processes fail due to inadequacies in the communication process rather than to any fundamental failure of strategic concept. For these reasons, a considerable part of our effort must be aimed at developing and elaborating public participation strategy and its implementation in all phases. Real public participation means a mature polyvocal discussion and readiness to accept different opinions and point of views, which goes far beyond pure information exchange.

Energy governance is a multifaceted issue facing a variety of technological, economic, environmental, political and institutional challenges (Burke & Stephens, 2018; Stevis & Felli, 2020). These also include the overall transition to renewable energy sources as the EU is the leader in this process, where integration of various renewable energy sources into the existing grid, decentralization of energy generation and synergies between different sectors require considerable additional investments or phasing out fossil fuels while ensuring energy stability (Dobravec et al., 2021; White, 2020). Achieving consensus on EU energy policy is difficult due to a complex EU regulatory framework and the challenges of its transposition to national legislations. Achieving consensus here is challenging due to differing national interests and different energy profiles of member states (Glynn et al., 2020). Scientific literature focusing on national and local levels is often rich, regional level remains often forgotten, as Hoppe and Miedema (2020) argue. However, regional governance level plays a crucial role as in many countries it includes competences in energy governance (Melica et al., 2018). While the overall costs accompanying the energy transition is high, EU is providing a series of funding and investment opportunities focusing on financing this transition, incentivizing green technologies and managing the economic impact on citizens and industries (Hughes & Hoffmann, 2020). In this regard, the greatest challenge remains to improve the use of these technologies by communities and citizens (Wang & Lo, 2021). Besides financial benefits, many citizens increasingly consider the ethics of using renewables and focus on environmental impact of energy sources rather than looking strictly at the cost side (Stevis & Felli, 2020). Public support and resilient policies are crucial for energy transition, requiring a strong political will and public support, the policies need to be resilient to political changes and public acceptance of new energy infrastructure is crucial (White, 2020).

To ensure just green transition, the challenges remain similar and diverse. For participatory planning culture, many questions occur regarding social equity, environmental sustainability and community involvement (Wang & Lo, 2021). Just energy transition needs to ensure inclusivity involving all segments of the society, actively including vulnerable communities, minority groups and economically disadvantaged populations in planning and decision-making processes to prevent exacerbating existing inequalities (Stevis & Felli, 2020). As the communities are culturally diverse including their attitudes toward environmental conservation and sustainability, participatory approaches to just green transition necessitate tailoring approaches that respect and integrate local cultural values (Hisschemöller & Sioziou, 2013; Hughes & Hoffmann, 2020). For maintaining long-term engagement and participation and motivation of stakeholders in green transition processes, it is important to keep in mind that interest and incentives may decrease over time, particularly if the immediate benefits are not apparent or if the initiatives are not effectively communicated (White, 2020). To support enduring engagement of communities, there must be mechanisms for measuring progress with integrated feedback procedures for transparent monitoring and evaluation processes (Revez et al., 2020). Due to the high costs associated with implementing green technologies, particularly for economically less-developed regions, it is crucial to ensure that infrastructure for such transitions is accessible, affordable and sustainable (Green & Gambhir, 2020). It requires balancing the diverse interests of various stakeholders, including private companies, local communities, governmental bodies and environmental activist groups (Galgóczi, 2020). To foster effective and efficient participatory planning, all participants need to have a high level of understanding of the issues at hand and this includes educating the communities about the benefits and challenges of green transitions and the importance of sustainable practices (Stevis & Felli, 2020). When successful local participatory cases are achieved, another challenge may arise in scaling them to wider and different areas regarding their transferability due to differences in economic conditions, social structures or environmental factors (Hughes & Hoffmann, 2020; Lowitzsch et al., 2020). Policies dealing with just green transitions require robust regulatory frameworks that are flexible enough to accommodate local and regional variations, while being stringent enough to enforce essential guidelines and standards (White, 2020). Effectively addressing these aspects involves an integrated approach combining engagement, policy innovation, educational outreach and technological advancements to ensure that the green energy transition is both just and sustainable (Revez et al., 2020).

7.3 Participatory Planning Culture and Just Green Transition

Culture is present in every layer of human activity and plays a significant role in the development of any territory, shaping also the dynamics of the entire society. Every culture has a proven and elaborate way of dealing with the unknown based on collectively shared experience. Spatial development is a process combining many processes from the technological, economic, social and cultural spheres. Every culture is interpretative and has various connotations—knowledge of its interpretation codes distinct the insiders from outsiders. Hofstede (2001) defines culture as collective programming of the mind, consisting of two poles: human nature and individual personality. Each culture has diffused borders, and this is essentially valid also for planning culture. Planning culture might be interpreted as one of the cultural subsystems (similarly like corporate culture, culture of various professions, regional culture). Planning culture is always embedded in society, as well as within the physical territory. According to Knieling and Othengrafen (2009) it refers to the different planning systems and traditions, institutional arrangements of spatial development and the broader cultural context of planning. In our considerations, planning culture is specific, unique as well as typical, non-casual set of approaches (based on underlying values) toward the factors playing fundamental roles in the process of spatial development within certain territory.

Other definitions of planning culture are rather operational: Keller et al. (in Sanyal, 2005, p. 30) concentrate on the tasks and subject matter of planning, the organization, structures and legal frameworks in which planning practices are embedded, and the background assumptions, values and general orientation that guide planning endeavors, Rode (2006) considers planning culture an umbrella for situations when different social contexts and values are ultimately generating a particular planning approach, without further analysis of these contexts and Dollinger (in Fürst, 2007) is referring to the ways how the decision-makers are dealing with planning processes. According to Stead et al. (2015) planning culture affects planners' worldview, cognitive frames, planning practice and planning outcomes.

Participatory planning culture largely dominates planning practice in all democratic societies. We are witnessing fundamental revision of the role and competence of the spatial planner. The role of today's planner is significantly shifting to the field of so-called soft skills. Planner's main task is to identify interests, mediate dialogue and communication, propose a fair solution and ensure its acceptance across various stakeholder groups (Stiftel, 2000). The role of spatial planner lost its mere technocratic character (engineer, designer) and moved toward leadership, facilitation, moderation and negotiation. Postmodernism as a whole has sometimes been accused of contributing to the loss of the meaning of the entire society and the subsequent loss of 'security' provided by modernist concepts, especially in science and politics (see

Váně, 2008). Participatory planning, however, accepted for the first time that society is in a fluid stance. Acceptance and encouragement of diversity, plurality of positions that are constantly changing—these are all necessary prerequisites for a successful planning process today. Participatory/communicative planning is a direct illustration of Habermas' basic social situation, where intersubjective action will always be more productive than a rigid systemic approach. The passive actor (rational-predictive planning) is replaced by the idea of a concrete, active actor whose actions are not determined by social structures, but who creates and transforms these structures himself/herself (Balon, 2008).

Since communicative/collaborative/participatory planning originated 'from below', as a reaction to dysfunctional (often theoretically well-developed) rationalist or incrementalist approaches, it cannot even be subjected to a systemic analysis and does not have a uniform methodology (as was the case by rational-predictive planning). Participatory planning is thus more a set of procedures than a concise theory or planning school. Harris (2002, p. 37) believes that 'collaborative planning is most suitably interpreted as an element in a longer-term program of research and theoretical development focused upon a concern with the democratic management and control of urban and regional environments and the design of less oppressive planning mechanisms'.

Healey´s (1997) model of participatory planning is quite broad and integrates a broad range of topics from community, the relationship to the distribution of power, global economic restructuring, culture and governance, institutional design and the nature of expertise to mediation, conflict resolution and spatial planning as a whole (Harris, 2002). The evaluation of the planning process is no longer restricted to evaluation of outcome: during the process (participation), new relations might be established, and new values might appear. Participatory planning highlights the educative and learning character of such interaction. Suddenly, the central focal point in the planning discourse is occupied by language and communication. Language and, in a broader sense, communication are the key to every social phenomenon or process: for Foucault, language and meaning were the means by which power and power configurations in society can be analyzed, for Giddens, communication is the key to inclusion in social structures (structural theory) and for Habermas (communicative action theory) language and communication are the key to understanding the dominance of instrumental rationality and to the 'unfinished project of modernity'.

7.4 Our Model of Planning Cultures

Every model of culture is a largely simplified concept of reality used to describe the key traits and features of the cultural milieu. Each planning culture has both universal variables and unique traits that can be interpreted differently, as well as a set of unique, deeply engraved characteristics which might be diversely interpreted. Just green transition tackles cultural traits such as shared

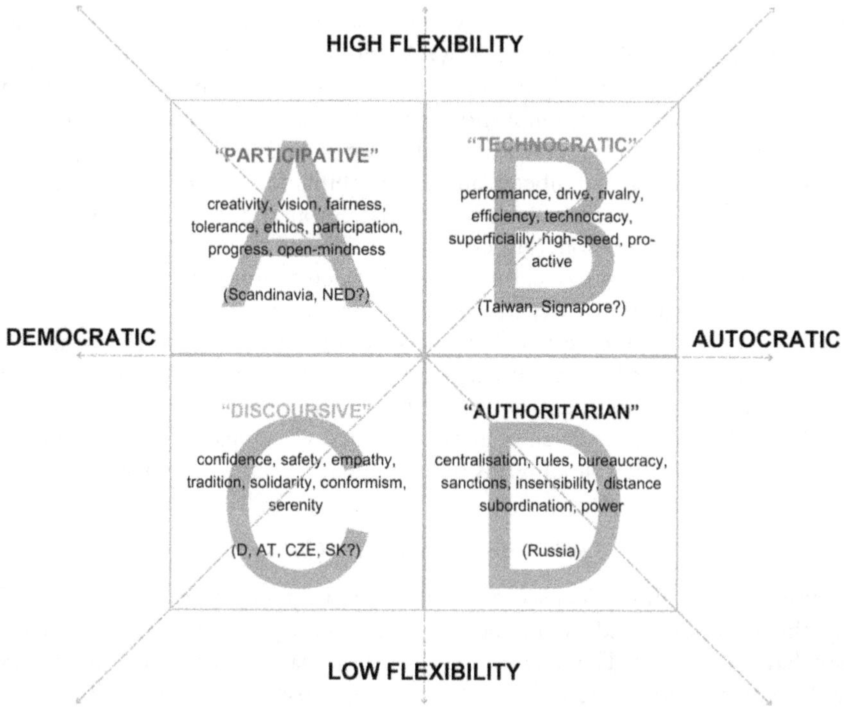

Fig. 7.1 Typology of planning cultures (adapted from Jaššo & Finka, 2019)

values, leadership style and risk aversion. Different types of planning culture might deliver different results in the process of just green transition and might be important preconditions of overall success of just green transition. Many, often hidden, cultural determinants might hinder, slow down or transform important elements and steps of just green transition.

Research and survey of every planning culture thus should combine both standardized, normative methods (e.g. questionnaires, comparative profiles) as well as individual and interactive methods (e.g. workshop, interview, desk research, storytelling, analysis of written texts and maps etc.). Our model of planning culture is based upon horizontal dimension, mapping out the degree of democratic versus autocratic decision-making in various levels and vertical dimension, mapping out the flexibility of the system, its ability to react in proper ways to changes in external and internal environment (Fig. 7.1).

7.4.1 *A—Participatory Culture*

Participatory planning culture is focused on wide participation of different actors of spatial development and the mutual consensus is the key milestone of its effectiveness. This type evolves through feedback and consensus and is present in societies built upon strong cohesion and balance. The power hierarchy in this culture is rather weak and the culture is inclusive, especially toward 'small players' and marginalized groups within the society. Main values include creativity, innovation, progress, shared vision, fair competition, tolerance, flexibility, ethics, individual freedom and the main competitive advantage is support of liberal and open thinking, proper conditions for creative development, agreeable social climate, honest and friendly social relationships and strong vision. Successful leaders are often creative, visionary and team-oriented. This type of culture is highly open to innovation and flexibility in facing new challenges. Regarding just green transition, we assume that this type of culture has the best odds to be successful in long-term horizons and large spatio-temporal scales. Participatory planning culture is most effective when utilizing commitment of communities, their sound interpersonal relations and democratic leadership. Systems located within participatory planning culture are highly flexible in terms of learning and optimizing their capacities.

7.4.2 B—Technocratic Culture

Technocratic culture is focused on results and progress, its success is measured by the processual and economic efficiency and the small stakeholders do not have much say. This culture leans on pragmatic and measurable factors and elements. Sometimes it is more 'evidence-based planning' rather than 'planning-for-people'. Values include competitiveness, rivalry, high efficiency, performance, individuality, pro-activity and a certain superficiality. Successful leaders are often individualistic, calculating and ambitious. Overall target-oriented work climate, quick (especially quantitative) feedback and a certain flexibility in the decision-making process make this culture attractive for performance-oriented individuals. A successful leader is often an individualistic, calculating, tactician, closed, calm, inscrutable, professional, ambitious and resistant person. This culture might be helpful in the process of a green transition by generating leadership based upon visionary technological perfection, explaining new technological tools and approaches, and illustrating them with a plethora of data and quantitative indicators. On the other hand, when dealing with soft factors and looking for fragile and delicate compromise, this type of culture is often ineffective.

7.4.3 C—Discursive Culture

This culture, focused on integration and evolution, has arisen by the modification of the rather obsolete, normative, rational-predictive culture of organized modernity into a more communicative and sensitive one. The transition retained features like bureaucracy and inertia. Key values include integrity, safety, confidence, solidarity and conformism. Lower flexibility and lower efficiency of processes is, on the other hand, the main hindrance. In stable systems, it offers social safety and cohesion. It is backed by data and research, stability, efficient performance of routine tasks and risk aversive decision-making and management style. Successful leaders are responsible, empathetic and loyal. This culture struggles with abrupt changes of the outer environment that require strong personal leadership. Its advantage in green transition lies in institutional and methodological robustness, though it may lack managerial leadership.

7.4.4 D—Authoritarian Culture

Authoritarian culture is strongly related to national culture realms and the tradition of the current or former totalitarian regimes. It is a bureaucratic culture with numerous restrictions, rules and norms. Participation is rather low and small stakeholders are sidelined. Various power alliances rule the system, resisting change. Typical values are centralization, authority and strict rules. Unified behavior and structural stability can be seen as advantages. Successful leaders are authoritarian, powerful, formal, insensible and strict. In terms of just green transition, this type of culture is in democratic milieu largely ineffective.

7.5 Case Studies

7.5.1 Case Study 1: Güssing Case

The town of Güssing of about 3,700 inhabitants is located in the Burgenland region of Austria. It was considered one of the poorest districts in Austria with heavy dependence on agriculture and was suffering from high unemployment rates and emigration (Safian, 2014). Since the 1990s, local leaders have made a decision to improve the economy by focusing on the town's energy self-sufficiency utilizing locally available renewable resources. The turning point for Güssing came with the establishment of the Güssing Renewable Energy GmbH in 1990 (Hermann et al., 2002). Its goal was to develop and implement technologies producing energy from local biomass resources, mainly wood from the surrounding forests. The project began with a biomass district heating plant and expanded to include biogas and biodiesel production facilities (Hofbauer, 2015). Güssing became a pioneer in developing and implementing technologies for sustainable energy.

One crucial aspect of Güssing's success was the active involvement of the local community. The projects were designed to benefit the local population both economically and environmentally. Local businesses thrived by supplying raw materials to the energy facilities, and many new jobs were created in the renewable energy sector. A particular focus was put on education and public outreach, annually hosting thousands of visitors and professionals who are eager to learn from its success, influencing energy policies and projects worldwide. Community involvement grew as the benefits became apparent, leading to broader participation in subsequent projects (De Waal & Stremke, 2014).

Participatory planning aspects were built on strategic vision from local leaders who set the vision and framework (ibid.). Güssing invested heavily in education and training centers, which not only supported local technology development but also kept the community informed and engaged (Nay, 2012). Such participatory approach led to substantial local economic development, reduced energy costs and a high degree of technological innovation that attracted international attention and tourism.

7.5.2 *Case Study 2: Integration of Solar and Wind Power in North Macedonia*

North Macedonia has historically used mostly coal for its energy production, particularly from the Bitola coal plant, which supplied the majority of the country's electricity. North Macedonia has made efforts toward energy diversification, focusing intensively on renewable energy sources such as solar and wind, aiming to reduce carbon emissions, foster local economic development and improve energy security. An important milestone in this transition took place in 2014 with the construction of the Bogdanci Wind Farm, the country's first large-scale wind power project. Located close to the southern border, the farm has an installed capacity of 50 MW, producing sufficient energy to supply over 60,000 households annually. The project received funding from the European Bank for Reconstruction and Development and the EU, with a second phase of the project in preparation (Serbia Energy, 2023). Following the Bogdanci project, North Macedonia implemented the Oslomej Solar Plant, expanding its focus on solar power. Built on the site of a former coal mine, the plant has a capacity of 10 MW (WeBalkans, 2022). As part of an energy transition strategy, the government supported and implemented policies encouraging household and business investments in rooftop solar panels by offering incentives to local residents, which led to job creation and innovation in the renewable energy sector (WBIF, 2022). An important aspect and one of the key drivers of North Macedonia's energy transition has been the active involvement of various stakeholders, including local communities, businesses and public authorities (Petrova & Dimitrov, 2020). The renewable energy projects were designed to economically and environmentally benefit local communities. Extensive support for these initiatives was fostered through community inclusion via educational programs, workshops and outreach efforts (WeBalkans, 2022). Participatory aspects of planning surrounding these projects have been crucial to their success. The North Macedonian government worked with local communities and international stakeholders to set strategic goals and create a framework for the expansion of renewable energy (European Commission, 2019). Locals were incentivized to adopt small-scale solar installations and participate in discussions about energy policies (EBRD, 2022). Additionally, this participatory planning approach has been fostered through various public-private partnerships (Balkan Green

Energy News, 2022). These investments have also contributed to substantial economic development, not only reducing dependence on coal but also attracting green technology companies. It is expected to contribute to long-term economic stability by creating sustainable energy sources that can help buffer against global energy price fluctuations.

7.6 Discussion—Comparison of Case Studies and Recommendations

The presented case studies highlight how participatory spatial planning can be leveraged to support a just green transition and illustrate how inclusive planning approaches can drive successful energy transitions. From a participatory planning framework perspective, Güssing's renewable energy project was initiated by local leaders and expanded with community involvement. The municipality was focusing on the maximalization of local biomass resources, with the leadership emphasizing long-term sustainability and economic benefits. The strategic vision was set by the local government, which actively engaged the community in subsequent stages of the projects through education, training and building partnerships with local businesses (De Waal and Stremke, 2014; Hofbauer, 2015). This gradual approach empowered residents to become integral parts of the renewable energy supply chain by providing raw materials and eventually contributing to broader projects. In contrast, North Macedonia case study required a more complex, multi-stakeholder planning framework, as it operated at a national scale. The government of North Macedonia was working with international stakeholders to implement large-scale wind and solar energy projects. A participatory framework was developed by encouraging household and business investments in small-scale renewable energy installations. Strategic partnerships between local authorities, international investors and communities ensured the planning process included the views of all stakeholders, and it incentivized local participation in renewable energy projects, especially in small-scale rooftop solar systems (EBRD, 2022; Petrova Dimitrov, 2020). Looking at community empowerment, the Austrian case shows that the role of the community was gradual but crucial to its renewable energy initiatives. The local population participated initially by supplying biomass materials, but as the benefits of the energy transition became more visible, community involvement deepened and allowed for continuous innovation, integrating new technologies such as biogas and solar power. The participation of local businesses in supplying raw materials and services for renewable energy facilities fostered local economic growth and job creation, further encouraging community involvement (Nay, 2012; Safian, 2014). Community participation in North Macedonia in the planning process was driven by educational workshops and outreach efforts aimed at informing the public about the benefits of renewable energy and the environmental importance of reducing coal dependence. The participatory nature of planning was evident in the way local businesses provided construction services for the renewable energy projects, while also directly benefiting from job creation (Petrova & Dimitrov, 2020). Güssings's participatory planning culture allowed for continuous innovation and flexible approach to planning. This fostered its position as an international model for renewable energy success (De Waal and Stremke, 2014; Hofbauer, 2015). North Macedonia's participatory planning model emphasizes long-term sustainability by focusing on both large-scale infrastructure projects and local household participation in renewable energy initiatives. The participatory framework introduced by the government incentivised small-scale solar installations, creating a more decentralized and flexible energy system.

7.7 Conclusions

Just green transition should cover not only the integration of strategies, systems, know-how and management approaches. Major attention should also be paid to the smooth integration of social systems concerning communities, citizens, employees, clients and other important individual stakeholders. Attitudes and opinions of the public are important factors that should be taken into consideration as a cornerstone of public debates of green transition. Green transition is a unique, highly complex process generating ultimate requirements and challenges not only in terms of new competencies and decision-making capacity of planners, planning bodies and institutions but also tackling the deepest emotional and identification mechanisms of participating actors. The collective perception of what is just and what is promising is changing, and reaching consensus as the cornerstone of the participatory planning paradigm might be a more and more arduous effort. The complexity of just green transition represents a big challenge for rational comprehension, and some individuals might tend to abandon rational reasoning and start to use defensive mechanisms. This is often resulting in denial or rejection of any new approach, new technology or new idea influencing/forming social systems. Some people tend to play an opposing role in the whole process and hamper various initiatives. Nostalgia for 'good old times' is understandable (particularly pronounced in some parts of Western Balkans, see Velikonja, 2009), though an irrational escapist mechanism of the human psyche which is not a passive recorder of the past, but it is (sometimes confused) a reconstructor and interpreter of our past. Western Balkan society, with its turbulent past, fragmentation and record of many culturally influenced disputes and confrontations, might be highly sensitive terrain in this aspect. On the other hand, cooperative patterns of behavior and tolerant traditions always present in certain layers (especially in the urban milieu) of Western Balkan societies might also play a very important supportive role in the process.

All this is happening during the current highly ambivalent global transition period when societies are facing a wide range of threats and risks and are much more fragile concerning their capacities to retain the most talented human resources and innovative leaders. Successful and professional mastering of just green transition is decisive for delivering value and synergies expected by stakeholders. Just green transition represents one the most complex and demanding combinations of challenges for the entire society, and the participation of all relevant actors, enhancement of their motivation and commitment, as well as respect for their right to agree or disagree, is a fundamental precondition for the general success of the whole process. This could become a formative experience for society in generating a unique planning culture.

References

Balon, J. (2008): *Teorie komunikativního jednání Jürgena Habermase [Theory of Communicative Action of Jurgen Habermas]*. In Šubrt a kol. (Ed.): Soudobá sociologie II (Teorie sociálního jednání a sociální struktury). Univerzita Karlova v Praze. Nakladatelství Karolinum. Praha. ISBN 978-80-246-1413-7.

Balkan Green Energy News. (2022). North Macedonia completes first solar park on abandoned coal pit. [ONLINE] https://balkangreenenergynews.com/north-macedonia-completes-first-solar-park-on-abandoned-coal-pit/ [accessed 05/09/2024].

Burke, M. J., & Stephens, J. C. (2018). Political power and renewable energy futures: A critical review. *Energy Research & Social Science, 35*, 78–93. https://doi.org/10.1016/j.erss.2017.10.018

De Waal, R. M., & Stremke, S. (2014). Energy transition: Missed opportunities and emerging challenges for landscape planning and designing. *Sustainability, 6*(7), 4386–4415. https://doi.org/10.3390/su6074386

Dobravec, V., Matak, N., Sakulin, C., & Krajačić, G. (2021). Multilevel governance energy planning and policy: A view on local energy initiatives. *Energy, Sustainability and Society, 11*, 1–17. https://doi.org/10.1186/s13705-020-00277-y

EBRD. (2022). EBRD and donor-funded solar plant starts operating in North Macedonia. Electric Energy Online. [ONLINE] https://electricenergyonline.com/article/energy/category/solar/142/954101/ebrd-and-donor-funded-solarplant-starts-operating-in-north-macedonia.html [accessed 05/09/2024].

European Commission. (2019). North Macedonia Energy Transition Report. [ONLINE] https://www.ebrd.com/content/dam/ebrd_dxp/assets/pdfs/office-of-the-chief-economist/transition-reportarchive/transition-report-2019/country-assessments-2019-20/transition-report-201920-north-macedonia.pdf [accessed 05/09/2024].

Fürst, D. (2007). Planungskultur. Auf dem Weg zu einem besseren Verständniss von Planungsprozessen?, PND Online III/2007 [ONLINE] www.planung-neu-denken.de. Accessed January, 06, 2018.

Galgóczi, B. (2020). Just transition on the ground: Challenges and opportunities for social dialogue. *European Journal of Industrial Relations, 26*(4), 367–382. https://doi.org/10.1177/0959680120951704

Glynn, P. J., Błachowicz, A., & Nicholls, M. (2020). Incorporating just transition strategies in developing countries nationally determined contributions. In *Handbook of climate change management: Research, leadership, transformation* (pp. 4449–4469). Cham: Springer International Publishing. https://doi.org/10.1007/978-3-030-22759-3_221-1.

Green, F., & Gambhir, A. (2020). Transitional assistance policies for just, equitable and smooth low-carbon transitions: Who, what and how? *Climate Policy, 20*(8), 902–921. https://doi.org/10.1080/14693062.2019.1657379

Harris, N. (2002). '*Collaborative planning—From theoretical foundations to practice forms*'. In P. Allmendinger & M. Tewdwr-Jones (Eds.), *Planning futures: New directions in planning theory* (pp. 21—43). Routledge. ISBN: 9780203996195.

Healey, P. (1997). *Collaborative planning: Shaping places in fragmented societies*. Macmillan. https://doi.org/10.1007/978-1-349-25538-2

Hermann, H., Reinhard, R., Klaus, B., Reinhard, K., & Christian, A. (2002). Biomass CHP plant Güssing–A success story. In *Proceedings of the expert meeting on pyrolysis and gasification of biomass and waste, Strasbourg, France* (Vol. 30).

Hisschemöller, M., & Sioziou, I. (2013). Boundary organisations for resource mobilisation: Enhancing citizens' involvement in the Dutch energy transition. *Environmental Politics, 22*(5), 792–810. https://doi.org/10.1080/09644016.2013.775724

Hofbauer, H. (2015). Energy from biomass via gasification in Güssing. In *Transformations to effective use biomass power for the world* (pp. 343–373). Pan Stanford Publishing. ISBN: 9814613894.

Hofstede. G. (2001). *Cultures consequences, comparing values, behaviors, institutions and organizations across nations*. (2nd Ed). Sage Publications. ISBN 9780415577762.

Hoppe, T., & Miedema, M. (2020). A governance approach to regional energy transition: Meaning, conceptualization and practice. *Sustainability, 12*(3), 915. https://doi.org/10.3390/su12030915

Hughes, S., & Hoffmann, M. (2020). Just urban transitions: Toward a research agenda. *Wiley Interdisciplinary Reviews: Climate Change, 11*(3), Article e640. https://doi.org/10.1002/wcc.640

Jaššo, M., & Finka, M. (2019). The role of leadership in the current contexts of central Europeal planning culture. In M. Jaššo, M. Finka, & M. Husár (Eds.), *The role of public sector in local economic and territorial development. Innovation in central, eastern and South Eastern Europe = EAI/Springer innovations in communication and computing* (pp.131–159). 1. vyd. Cham, Switzerland: Springer, 2018. https://doi.org/10.1007/978-3-319-93575-1. ISBN: 978-3-319-93574-4.

Knieling, J, & Othengrafen, F. (2009). *Planning cultures in Europe*. (HafenCity University). https://doi.org/10.4324/9781315246727. ISBN 9781315246727.

Lowitzsch, J., Hoicka, C. E., & van Tulder, F. J. (2020). Renewable energy communities under the 2019 European clean energy package-governance model for the energy clusters of the future? *Renewable and Sustainable Energy Reviews, 122*, Article 109489. https://doi.org/10.1016/j.rser.2019.109489

Melica, G., Bertoldi, P., Kona, A., Iancu, A., Rivas, S., & Zancanella, P. (2018). Multilevel governance of sustainable energy policies: The role of regions and provinces to support the participation of small local authorities in the covenant of Mayors. *Sustainable Cities and Society, 39*, 729–739. https://doi.org/10.1016/j.scs.2018.01.013

Nay, E. (2012). Disruptive green technologies and the gussing model. *The International Journal of the Constructed Environment, 2*(1), 81. https://doi.org/10.18848/2154-8587/CGP/v02i01/37516

Petrova, G., & Dimitrov, N. (2020). *Community involvement in North Macedonia's energy transition*. Balkan Green Energy News.

Revez, A., Dunphy, N., Harris, C., Mullally, G., Lennon, B., & Gaffney, C. (2020). Beyond forecasting: Using a modified delphi method to build upon participatory action research in developing principles for a just and inclusive energy transition. *International Journal of Qualitative Methods, 19*, 1609406920903218. https://doi.org/10.1177/1609406920903218

Rode, P. (2006). *City design—A new planning paradigm?*. Discussion paper on Urban Age. London School of Economics. Accessed February, 2, 2014. [ONLINE] http://www.urban-age.net/0_downloads/archive/Philipp_Rode-City_Design_A_New_Planning_Paradigm-DiscussionPaper.pdf

Safian, F. (2014). The synergies of community ownership, renewable energy production and locality–The cases of Güssing and Samsø. *Geographical Locality Studies, 2* (1), 386–410. ISSN: 2053-3667.

Sanyal, B. (Ed.) (2005). *Comparative planning cultures*. Routledge. ISBN 9780415951357.

Serbia Energy. (2023). Macedonia Bogdanci Wind Farm project enters second phase. [ONLINE] https://serbiaenergy.eu/macedonia-bogdanci-wind-farm-project-enter-second-phase/ [accessed 05/09/2024].

Shaker, Y., & Berisha, E. (2025). Just Green Transitions: Between terminological inexactitudes, conceptual fragmentation, and the exigency for a theoretical framework. *European Spatial Research and Policy*, 32(1), 121–138. https://doi.org/10.18778/1231-1952.32.1.05

Stead, D., de Vries, J., & Tasan-Kok, T. (2015). Planning cultures and histories: Influences on the evolution of planning systems and spatial development patterns. *Eur Plann Stud., 23*(11), 2127–2132. https://doi.org/10.1080/09654313.2015.1016402

Stevis, D., & Felli, R. (2020). Planetary just transition? How inclusive and how just? *Earth System Governance, 6*, Article 100065. https://doi.org/10.1016/j.esg.2020.100065

Stiftel, B. (2000). Planning theory. In R. Pelaseyed (Ed.) *The national AICP examination preparation course guidebook* (pp. 4–16).

Váně, J. (2008). *Postmoderna—předsudky, pomluvy, nedorozumění aneb pokus o typologii [Postmodern - prejudices, rumors, misunderstandings or attempt of typology]*. In Šubrt a kol. (Ed.): Soudobá sociologie III (Diagnozy soudobých společností). Univerzita Karlova v Praze. Nakladatelství Karolinum. Praha. ISBN: 978E80E246E1486E1.

Velikonja, M. (2009). *Titostalgia-a study of nostalgia for Josip Broz*. ISBN: 365980309X.

Wang, X., & Lo, K. (2021). Just transition: A conceptual review. *Energy Research & Social Science, 82*, Article 102291. https://doi.org/10.1016/j.erss.2021.102291

WeBalkans. (2022). An ex-coal mine reworked as North Macedonia's first large solar plant. [ONLINE] https://webalkans.eu/en/stories/an-ex-coal-mine-reworked-as-north-macedonias-first-large-solar-plant/ [accessed 05/09/2024].

Western Balkans Investment Framework (WBIF). (2022). EU-funded solar plant starts operating in North Macedonia. [ONLINE] https://www.wbif.eu/news-details/eu-funded-solar-plant-starts-operating-north-macedonia[accessed 05/09/2024].

White, D. (2020). Just transitions/design for transitions: Preliminary notes on a design politics for a green new deal. *Capitalism Nature Socialism, 31*(2), 20–39. https://doi.org/10.1080/10455752.2019.1583762

PART III

Just Green Transitions Policy Themes

CHAPTER 8

Landscape Design as a Tool to Foster Spatial Transformations in Light of Just Green Transitions in the Western Balkan

Dea Buza and Ledio Allkja

8.1 Introduction

The Green Agenda for the Western Balkan (GAWB), ratified in 2020, opens up a new perspective for the region in fostering the Just Green Transition (JGT) (Berisha, 2025a, 2025b). By agreeing with the GAWB the region entered into the path of the EU Green Deal for becoming carbon neutral by 2050. The Agenda focuses on five main pillars such as climate action, circular economy, biodiversity, pollution, and food. Meanwhile, digitalisation (innovation) is seen as a horizontal aspect which crosses all 5 pillars (European Commission, 2020). Besides the GAWB, five of the WB countries have signed in 2016 the Paris Agreement for fighting climate change, while Kosovo, due to its status within the UN, is applying it voluntarily without a formal agreement. The main goal of the agreement is to "limit temperature increase well below 2° above before pre-industrial levels", with the main aim of keeping temperatures at 1,5° above pre-industrial levels within this century (UN, 2015). Not being able to hold the temperatures at these adequate levels, the risks of

D. Buza (✉)
Agricultural University of Tirana, Tirana, Albania
e-mail: dbuza@ubt.edu.al

L. Allkja
Co-Plan Institute for Habitat Development, Tirana, Albania

Polis University, Tirana, Albania

L. Allkja
e-mail: ledio_allkja@universitetipolis.edu.al

E. Berisha et al. (eds.), *The Palgrave Handbook of Just Green Transitions in the Western Balkans and Beyond*,
https://doi.org/10.1007/978-3-031-95075-9_8

climate change-induced disasters will increase with higher magnitudes while the impact on territories will be devastating. Thus, the implementation of the Paris Agreement, requires countries to make meaningful efforts and take active actions in mitigating climate change but at the same time to adapt to climate change. While on the one hand, it is highly important to reduce Green House Gas (GHG) emission levels and decarbonize our economic systems (mitigation), on the other, it is a well-recognized fact that we are now living through climate changes, and the need for adaptation is high as well. Therefore, countries need to coordinate efforts and create synergies between mitigation and adaptation. When looking at the WB, the region is highly affected by climate changes, also due to the low capacity to respond and adapt. Climate changes will produce increased risks in terms of flooding, heatwaves, droughts, increased presence of urban heat islands, etc., which in turn will create knock-on effects and multiple risks leading to other social and territorial stresses such as health problems due to heat, economic problems due to increased cost of energy production, landslides, higher pressure on existing infrastructure and agricultural produce, etc. The increased frequency and magnitude of impact can be seen also through recent events such as the flooding in Bosnia & Herzegovina in 2024 which not only destroyed physical infrastructure but also had several casualties. Similarly, in June 2023, after 54 mm of rainfall within just one hour, lead to flash flooding in Western Kosovo, creating several territorial impacts as well as casualties. In January 2021, extreme weather conditions lead to flooding across multiple countries in WB including Albania, Kosovo, and Serbia as well as parts of Bulgaria and Greece. Meanwhile, extreme heat levels in the summer pose higher stress on population and cities. Droughts and water stress are one issue, while health, especially impacting citizens above a certain age create other problems. Thus, the pressure for adaptation is quite high.

On the other hand, the pressure to reduce/cut GHG by 55% requires creative thinking in terms of transitioning the economy, especially through the decarbonization of energy and transport systems as well as careful use of resources, towards a greener and socially just system. This transition cannot be done by taking singular or sectorial measures which are not coordinated with each other. The costs of non-coordination and lack of integrated and comprehensive measures are quite high, especially when considering the status of WB countries which have to operate under limited budgets and capacities. Thus, every investment, besides careful thinking, needs to maximize its impact.

Spatial planning is seen as one of the tools which governments use in achieving policy coordination and integration among different sectorial policies as well as provides the overall policy framework for territorial transformations and achieve resilience (Davoudi et al., 2013). Thus, planning has a high potential for fostering JGT and be a driver for climate change mitigation and adaptation. Planning is one of the main tools for achieving resilience not just from a territorial/spatial point of view but also for fostering resilient governance of urban and spatial transformations. Spatial planning as an inter-disciplinary tool is composed of several different aspects. As said, it is a governance tool, it is a tool for policy development, it is a tool for land-use change management as well as a tool for designing space. Spatial planning is composed of different disciplines, including landscape planning and landscape design. The aim of this chapter is to look into landscape planning and landscape design in the WB and see the potentials that these two disciplines within the wider planning framework have to support JGT and resilience building. Initially, a theoretical discussion of what is landscape planning and landscape design will be conducted. Secondly, an analysis of the potential of landscape planning and design in terms of supporting the five main pillars of the GAWB will be done. Thirdly, an analysis of the planning system in Albania, with reference and comparison to other WB countries, will be made in order to understand the role that landscape plays in the systems and the wider policy framework. This will allow to understand the gaps and offer some recommendations as to how landscape should be better used within the wider framework of planning in WB countries, thus supporting the conclusion and recommendations sections. The paper is based on desk studies and literature review. Different cases from the WB and EU are used as illustrations of the landscape policies or design which can be used to achieve different components of the GAWB.

8.2 Landscape Planning and Landscape Design

Spatial planning focuses on "*the problem of coordinating or integrating the spatial dimension of sectoral policies through a territory-based strategy*" (Nadin, 2006, p. 91). It is not simply a regulation of land use, but addresses tensions and contradictions between sectoral policies, e.g. conflicts between economic development, environmental development and social cohesion policies (Albrechts, 2004). The key role of spatial planning is to promote the most rational distribution of activities and to find common ground between competing policy objectives (Duhr et al., 2010). The purpose of spatial development varies widely across countries, but many share some similar basic principles. Spatial planning is a key instrument in defining long-term and sustainable frameworks for social, economic and territorial development, both within and between countries. Its main role is to promote integration between sectors such as housing, transport, energy and industry, and to improve national and local urban and rural development systems, taking into account environmental issues. The aim of spatial planning is to create a more rational territorial organization of land use and the relationship between different uses (Berisha et al., 2021, 2023; Janin Rivolin, 2012), in order to balance development requirements with the need to protect the environment and to achieve social and economic objectives (Nadin, 2006). Spatial planning also includes measures to coordinate and improve the spatial impacts of other sectoral policies, in order to guarantee a more equitable distribution of economic development in a given territory, which would otherwise be controlled by market forces. Spatial planning is thus an important driver in promoting sustainable development and improving the quality of life.

As mentioned above, over the years, the concept and approach of planning has evolved. While it started as an urban and sectorial approach, mainly a regulatory tool, the paradigm of planning has shifted and expanded. When one speaks about spatial planning now, it can mean various things and approaches, such as strategic planning, mobility planning, resilience planning, adaptive planning, etc. As an overarching framework, planning can also include issues of landscape. Landscape often in planning is reduced to auxiliary issues of urban design. It is seen as a tool for beautification, often limited to urban greenery, rather than a functional aspect of territorial development. This lack of awareness demises the role that landscape can have in spatial development and green transition. However, landscape has emerged across different disciplines as a *model* and a *medium* for the contemporary city. Landscape has thus acquired crucial importance as a "lens through which the contemporary city is represented and a medium through which it is constructed" (Waldheim, 2006). Thus, the planning of landscape can be considered as a comprehensive and systematic process that involves the design, management, and organization of outdoor spaces, with the goal of achieving specific objectives while considering the natural environment, human needs, and aesthetic preferences (Thorbjorn, 2010). Thus, landscape planning is related to different processes of planning which impact development such as spatial planning, regional planning, urban planning and design, change and planning of land-use as well as creating a regulatory framework of development. Therefore, landscape planning by definition has a multidisciplinary approach which aims to plan and design outdoor spaces which are sustainable and not only focus on the aesthetic or form of development. Landscape planning acts across different territorial scales, albeit regional, urban, rural, or areas in between. In fact, landscape planning is seen as highly important in the transition between urban and rural areas such as peri-urban or suburban. Focusing on the planning of the outdoor spaces, landscape has a strong environmental focus, often borrowing ecological planning approaches, to foster territorial changes and development. Meanwhile, landscape design involves creating plans that create synergies and combine the aesthetic, functional and sustainability factors in different spatial settings. As such it can include vegetation, public spaces, streets and roads, buildings but also greater regional and national infrastructures albeit transport, energy or for the protection from disasters. New approaches in landscape design aim to create resilient territorial frameworks which are designed to withstand and recover from various shocks and stresses, such as natural disasters, economic crises, and public health emergencies. As such it is a powerful tool in addressing climate change challenges. Landscape planning and design also plays an important role in terms of community engagement and involvement in sustainable spatial change. Landscape planning processes require the engagement of with various stakeholders, the local community, government agencies, environmental organizations, in order to build on the diverse perspectives and interests.

8.3 Landscape Planning and the GAWB

The Green Agenda for the Western Balkans has five main pillars:

1. climate action, including decarbonization, energy, and mobility
2. circular economy, addressing in particular waste, recycling, sustainable production, and efficient use of resources
3. biodiversity, aiming to protect and restore the natural wealth of the region
4. fighting pollution of air, water, and soil
5. sustainable food systems and rural areas.

In 2023 Regional Cooperation Council (RCC) prepared a monitoring report for the GAWB implementation. The findings are quite interesting and also related to this chapter. Due to the geopolitical situation and major global events such as the war in Ukraine or the war between Israel and Palestine, attention to this issue was rather shifted by WB6 to also mitigate the impact of these external shocks. The RCC acknowledges the fact that the WB6 have made progress in terms of legal and institutional efforts to implement the GAWB. However, one of the main backdrops and gaps remains the on-the-ground implementation. Here it is difficult to see real impactful actions on issues such as energy, transport, Nature-based Solutions (NbS), land use, land-use change and forestry (LULUCF), circular economy, and climate adaptation (RCC, 2023, p. 6). Additionally, coordination across the different sectors continues to remain weak and poses challenges to the implementation of the GAWB. Meanwhile, concepts such as nature-based solutions are not integrated in the wider policy framework in the WB6. Another issue identified by the RCC monitoring report is that the GAWB goals are not reflected in spatial planning at the different level of governance. Meanwhile, in terms of Biodiversity there is a serious lack in data and information which makes the whole monitoring process highly difficult and thereof also the implementation of the GAWB for this pillar. In this sense, one of the main actions for biodiversity also requires the preparation of The Regional Forest Landscape Restoration Plan. Referring to Action 52, it is also important to prepare a landscape restoration plan which differs from an afforestation plan and requires a different approach. Another important challenge for the implementation of the GAWB is the financing of actions.

Thus, based on the finding of the report it can be said that each of the pillars of the GAWB can be addressed in different manners and require policies, plans, projects, financing, and engagement of different actors. Without a doubt there is a role for landscape planning and design to support these pillars. Now let's look in more detail how landscape planning and design can contribute to the five main pillars of the GAWB.

Adaptation to climate change can be quite costly, especially when considering the adaptation of existing infrastructures to cope with higher levels of precipitation for example. Thus, the use of nature-based solutions can be a rewarding approach. Landscape planning and design can help in transforming public spaces into areas which can support this challenge and offer multiple benefits to the community. Developing multi-purpose spaces through careful landscape design can be a win-win approach and a cost-efficient one for climate-proofing WB cities. This approach was used, for example, in the River Irwell, in the UK, flood control scheme. After a careful risk assessment, the authorities decided to convert the area into a multi-purpose public space so that during heavy rain it is used as a flood storage basin while when it is not flooded, it is an open public space with several functions such as a sport center, playing fields, walking areas etc. (CABE, 2008). Another successful example comes from the city of Leiden in the Netherlands. The transformation of the Noorderkwartier-Oost neighborhood is a good case study which combines nature-based solutions with citizen engagement in order to enhance climate proofing of urban areas. The main aim of the project was to improve sewerage systems and water management in the area through the use of different nature-based solutions. Citizens were involved from the early stages of the design which allowed the authorities to transform the area even with a limited budget (Urban Green- Blue Grids, 2025).

Often, heavy rainfalls, besides the flooding disasters, also serve as catalyst of landslides. The protection from landslides usually requires engineering measures such as retention walls which do have negative impacts on the environment and sometimes denaturalize the areas. Hence, with careful landscape design, the use of appropriate vegetation can help in restoring landslides or in their prevention Arte-Mojica et al. (2019). Additionally, landscape planning is a powerful tool for combating erosion. The latter is a typical factor which contributes to both landslides and flooding; thus by addressing this phenomenon through landscape design and nature-based solutions or by enhancing eco-system services, it can support in achieving a higher territorial resilience.

Rising sea-levels or sea erosion is too an important factor for the coastal areas in the WB, especially for countries such as Albania and Montenegro. The advancement of sea towards land is evident along the coast of both countries, and often this has been exacerbated by deforestation on the justification of increased touristic activities. Sand Dunes have been used for centuries as a tool/ natural barrier for stopping sea advancement (Strodt, 2023). Until few decades ago, Dunes were considered as unproductive and unnecessary areas. Meanwhile now, coastal agents are working to restore these areas due the higher understanding that they have on the environment and the protection they provide (Freitas, 2020). Sea advancement creates not only problems of land loss and flooding, but also increases salinity of nearby agricultural lands thereby making them unusable for food production. Salinity is one such abiotic stress that hampers plant physiological functioning in a number of ways: resulting in lower crop production (Ahmad et al, 2022, p. 2). Use of different types of vegetation, such as for example typha plants, can help in reducing land salinity (Jesus et al., 2014). Another option, as suggested by Ahdmad et al., the use of plant extracts to combat salinity is a green, ecofriendly, and sustainable approach. Also in this case, landscape planning and design can offer different solutions which are not only functional but also esthetical.

Meanwhile, urban heat islands are becoming more and more present in WB cities. The increase of impervious surfaces, combined with rapid urban growth and the increase of concrete and asphalts in cities is serving as a main factor for urban heat islands. Combined with extreme hot temperatures during the summer they manifest in conditions which make cities unlivable and unhealthy, especially for vulnerable groups. Thus, the use of urban greenery can be a strategic tool for the decrease and cooling of urban areas. While it is common for cities to increase urban greenery by planting various trees, which can offer shade and are aesthetically pleasing, it is also true that the unplanned planting of trees can also create negative effects in health. The increase of pollution, ie higher presence of CO_2 in urban settings, is increasing also blooming of trees thus leading to higher pollen production (Gisler, 2021). Hence, higher concentrations of pollen can lead to higher levels of allergies such as hay fever, creating problems in the respiratory tracts of humans. Thus, landscape planning and design can help in reducing these risks by planting appropriate trees which not only reduce the impact of urban heat islands but also increase health in the city.

The use of renewable energies is very important in the JGT, and this is especially important for some of the WB countries which heavily rely on fossils. Often, solutions include wind and solar energy, and in some cases, it can also be biomass. Wind and solar plans while offer a viable solution in energy production can be problematic when are not carefully planned. Besides the environmental issues such as the disruption of habitats, which need to be evaluated and mitigated, from an aesthetic point of view, they are not always pleasing to the eye. Hence, landscape planning and design can help in mitigating some of these issues.

Lastly, when referring to decarbonisation of transport, one of the common approaches especially for urban settings is the increase of cycling routes with the aim of getting people to use more bikes and reduce car use. Many cities have taken up the race of building cycling lanes; however, in most cases they are reduced to adding a bike lane on the side of the road, removing parking for the bike lanes, and in some cases reducing walkways for the sake of bikes. In some cases, where infrastructure and urban settings allow, they have been quite successful, but in others, they produce sometimes paradoxical results. Imagine riding your bike right next to a road where vehicles and their exhaust pipes are. While on the one hand you think you are healthy, on the other you are getting all of the emissions in your face, which can be quite discouraging for cyclists. Hence, landscape design can be used for improving the experience of cyclists in the city, but it can also be used more strategically to promote long-distance cycling. For example, in Germany and the Netherlands, the concept of bike highways is becoming quite common. Developing bike highways in areas with beautiful landscapes that are environmentally friendly is offering a new attraction to cyclists to travel longer distances and also commuting to work. The Ruhr Valley Cycle Path is a 230 km cycle route which passes across some of the main cities in North Rhine Westphalia. The cycle path is developed along the Emscher River and passes 50% of it passes on separate cycle paths, 30% on forest roads and about 20% on local secondary roads offering an impressive diversity and high level of environmentally friendly cycling (Eco-Cycle, 2020). This makes it quite attractive for cyclists to use not only for recreational activities but also as a daily commute to work. Thus, landscape planning can offer different solutions from local to regional scales also in terms of mobility and decarbonization.

Landscape design can play an important role in terms of promoting circular economy, although at a rather small scale. One of the main problems nowadays is the treatment of plastic waste. Through landscape design, recycled plastic or other types of waste can be used as tools to improve urban furniture or public art. There are many cases around Europe where, for example, art in public spaces is developed through recycled plastic, glass, or other types of waste. Bristol Whales by Sue Lipscombe is one example of a sculpture using over 100,000 plastic bottles to depict a giant whale. A similar example is the "World of Litter" sculpture in Amsterdam developed by artist Peter Smith in 2012. Sometimes, this type of art can also be created by community engagement, therefore increasing the awareness of local actors in terms of recycling and re-using. Although, as said, these are initiatives of small scale, and they have a rather low impact in the overall quantity of waste which needs to be redesigned, it can still be a powerful tool for awareness raising.

Accession in the EU also requires the improvement of management of the protected areas, a prudent and sustainable use of resources. Following the fall of the dictatorial regimes, WB countries have found severe limitations in managing and preserving biodiversity. This is often due to lack of planning or adequate attention to this issue. Several areas have been destroyed or compromised due to the tentative of increasing tourism. On the other hand, landscape planning, as part of the management plans of environmentally sensitive areas can be an adequate approach in planning in order to enhance biodiversity but also to support and increase eco-system services.

Urban forests can play an important role for preserving biodiversity while at the same time increasing green spaces in dense urban areas. Landscape planning in this case is important in terms of creating the right conditions for these forests to flourish and be of an added value to urban systems. These can be used in different settings from derelict former industrial areas, to abandoned infrastructure. The New York Highline, for example, is a typical approach that through a project of landscaping a derelict infrastructure running through the city has been turned into a functional green space which offers cleaner air, attractive recreational areas and a good opportunity to increase biodiversity within urban areas (Highline, 2025). A key component here is not just creating "islands" of greenery within the city, but rather to create a network of interlinked green spaces which communicate with each other. Only if they are connected they can offer enhanced eco-system services but also offer a space for biodiversity to flourish within urban areas and in order to achieve that careful landscape planning is necessary. Another example is the River Quaggy in the UK, where the riverbed was restored to its natural course with more meanders and areas of deep and shallow water. This allows to better connect the riverbed with the flood plains but also enhance biodiversity. Additionally, where engineering measures such as hard defenses were necessary, they were combined with natural features such as bird nesting or planted boxes which allow for biodiversity to flourish (CABE, 2008).

Landscape planning can help in terms of reducing air, water, and land pollution. Especially in former industrial areas or brownfields the use of landscaping can lead to major transformations and offer different urban services to the city. A typical example is the Ruhr region in Germany, where by using landscape planning and different environmentally friendly techniques, including nature-based solutions, many of the city services, such as offering drinking water or sewerage treatment, are being done. These interventions create attractive open spaces which have a high level of functionality and are crucial for the city. They can use different types of plants for filtering water and cleaning them, thus reducing the pressure on typical gray infrastructure. Additionally, different plants can be used for absorbing heavy metals from soils which after some years can be restored to their natural state and re-purposed even for agriculture. Similarly, for the cleaning of polluted waters, different types of plants can serve as highly efficient and effective filters and thereof offer a natural purification of these waters.

Landscape design can play an important role in food production and in increasing food security. Concepts such as "agritecture", linking landscape architecture with agriculture are emerging and are being used as an approach by many cities. Urban farming can serve as a tool to further support the greening of cities with functional plants. At the same time, it can also serve to combat food poverty and can be used as social policies to support vulnerable groups. While urban farming, or urban allotments, are quite common for European cities, there are also some emerging initiatives in the WB countries. For example, the city of Korca, in Albania, recently has developed an urban farming project. An area of the city, opened by the municipality, has been developed as an urban farm by the local authority. The farm is cared by different citizens while its produce is freely available to vulnerable groups. Additionally, green roofs too can serve as a basis for increasing urban farming as well as providing green spaces in over densified areas. Urban farming can be an interesting approach also in terms of community engagement and the increase of the sense of community in different urban areas. The Augustenborg district in the city of Malmo in Sweden was re-developed under a similar principle. From the beginning of the initiative in 1998, citizens were actively involved in the transformation of the area. The scheme includes an open storm water system, green roofs, renewable energy, and transport initiatives (CABE, 2008). This approach not only offered a better living environment, involved the community and transformed the area, but also it decreased unemployment and increased property values.

Besides the city level, landscape planning can help in addressing issues of food productivity at the city/regional level. By carefully planning the surrounding/suburban areas for food production, it can help in increasing food security as well as productivity. This approach can help in preventing sprawl but at the same time offer food with a minimal distance from the main consumers. After the fall of dictatorial regimes in the WB, the state run cooperatives failed and land remained unused. By creating a network of food production combined with markets at the city-regional level, it can open up interesting perspectives for landscape design. Many European countries are using the concepts of market spaces as areas not only of selling agricultural/food products but also as attractions for tourism and creating a more polycentric city. Gastronomy plays an important role for attracting tourists and promoting and strengthening local produce. Food markets have been historically a place for the economic development of cities, and therefore they should be seen as such also nowadays with a higher attention on sustainability. This type of policy with adequate landscape design can serve as a transformative tool for the cities. One of the successful case studies for the transformation of historical market spaces in new and attractive areas for the city is the market of Ribeira in Lisbon, Portugal (Madeira et al., 2023). Now, the Ribeira market besides offering food products from the region, it is also house to over 20 restaurants, different bars, and it is also a space for cultural activities. Thus, now the market serves as a reference for the promotion of indigenous ingredients and cuisine, as well as local culture.

8.4 Planning and Policy Framework

Countries in the WB have gone through different planning reforms following the end of the dictatorial regimes (Berisha et al., 2018, 2024). Mostly, after the 2000s, with the signature of the Stabilization and Association Act, supported by the EU or other international donors, all of the WB countries have changed their planning system. The change has not only included a change in the institutional and legal frameworks but also in the planning approach moving away from traditional urbanistic and regulatory approaches (inherited from the old governance systems) towards more open, comprehensive, and integrated approaches of spatial planning (Allkja, 2021). Through these changes, planning has become a central element of territorial governance and serves as a tool to coordinate the different sectorial policies and plans.

Although there are significant differences among the planning systems in the Western Balkan countries, one can say that there is also a high degree of similarity (Berisha et al, 2024). As such, the Albanian planning system will be analyzed to demonstrate some of the main institutional and instrumental features of the systems.

In Albania there are two distinct levels of governance and thereof planning, the central level represented by the government of Albania and the local level represented by local authorities. This is also typical in other WB places, for example, in Kosovo, Montenegro, and North Macedonia where there is a lack of the meso-level of regional planning. Considering also the size of the countries it has been seen optimal to plan at two distinct levels rather than 3. Meanwhile, Bosnia & Hercegovina has a far more complex federal system with multiple levels of planning (Berisha et al., 2021). What is typical in the institutional framework in the WB for planning is that there is usually a Ministry (responsible for planning policies) and a (semi) independent institution in the form of an Agency (see National Territorial Planning Agency in Albania) or Institute (Institute of Spatial Planning in Kosovo). These agencies or institutes are responsible for drafting national plans but also coordinating the work with local authorities and ensuring that the national and the local level are harmonized.

In terms of planning instruments at the national level, in Albania there are three. The National Spatial (Territorial) Plan, sectorial or cross-sectorial plans which are drafted by different ministries, and detailed plans for areas of National Importance. This is a similar setting to other WB such as in Kosovo, North Macedonia, and Serbia. So, there is one instrument which is of the highest level, comprehensive and integrated, which harmonizes the impact of different sectorial policies in the territory and provides a development vision for a period of 15–20 years. This is afterward detailed/operationalized through other instruments such as sectorial plans (i.e. for energy, transport, environment, tourism, etc.) or through plans of areas of national importance such as protected areas, priority areas for tourism or economic development.

At the local level, municipalities or communes are the main institutions with planning competences. In Albania, municipalities have three main instruments for direct planning such as the General National Territorial Plan (GLTP), Sectorial Plans for different issues (mobility, environment, etc.), and Detailed Local Plans. The GLTP is composed of a Strategic and Policy Instrument which defines the development vision and the main policies of development for a period of 15 years, the land-use instrument, and the planning and development regulation. Similarly, in Kosovo, North Macedonia, and Serbia, the local level uses more or less the same type of instruments. A general and comprehensive plan, and then sectorial plans or detailed plans for specific issues or areas. As it can be seen, the structure of planning is quite similar, also due to the fact that the reforms in planning have been guided by similar types of international actors operating in the region in the same timeframe. Although, as previously mentioned, there are differences in how spatial planning works, it is put in place in practice and the way the different countries implement it.

Meanwhile, when it comes to landscape planning, it can be said that there is no distinct legal framework which defines and structures it. In Albania, for example, landscape is considered an element/dimension of territorial planning and policies, rules, or projects of landscape should be developed under this umbrella. In Kosovo it is a similar situation where landscape is developed as part of the overall strategic planning, and there is no instrument nor an institution which is dedicated to landscape planning albeit at the national or the local level. Sometimes, landscape is considered as a purely environmental issue at the national level or an urban furniture issue at the local level rather than a comprehensive approach.

When looking at the National Territorial Plan of Albania, we see that there are some ambitious goals and policies in terms of climate change adaptation, which in turn could also foster JGT. The plan speaks about polycentric development with regions that strive under endogenous growth and cooperation with their cross border counterparts. The GNTP proposes a wider network of protected areas which need to be better planned to enhance biodiversity and at the same time be integrated as part of the Natura 2000 network. It goes to the extent of proposing a national "green" cycling network across Albania as well as to develop the European Green Belt on the eastern border of the country. Meanwhile, in terms of agriculture and food production it proposes different policies which aim at the innovation in agriculture and in meeting local needs through different products through the development of agricultural poles of development (NTPA, 2017a).

Similarly, the Integrated Cross Sectorial Plan for the area Tirane-Durres, develops on detailing the vision and the policies of the GNTP. It offers a very ambitious vision for turning the Tirane-Durres area into one of the most competitive areas in the WB (NTPA, 2017b). As such, it proposes different policies for improving the livability of the metropolitan area such as improved public transportation systems, development of cycling routes, adaptation to climate change. Although the plan in itself has a strong emphasis on competitiveness and economic development, it still offers some interesting proposals in terms of just green transitions, although not necessarily related to the GAWB as the plan precedes the agenda in time. The plan for example offers a perspective for developing a regional park called the "Green Crown of Tirana". Additionally, it supports the implementation of renewable energy systems, primarily solar, at different scales. It supports local agricultural production by proposing different collection points and different regional markets that work together in an intertwined network.

As it can be seen from the above, both planning instruments offer a myriad of policies related to green transition which can help in the implementation of the Green Agenda. However, what is noted is that as for the GAWB, while there is advancement in terms of policy, there is a gap in terms of implementation. Both plans lack processes for their operationalization and not often are met with financial means to foster implementation. In most cases the plans are considered as reference/framework documents that need to be implemented by the local level. However, when looking at the municipalities, their main focus somewhat remains on using the local plans as a tool for issuing building permits and fostering urban growth rather than achieving sustainable spatial transformations. This is typical of former eastern planning systems where there is a gap between policy objectives and implementation in practice. From an instrumental point of view, they are well aligned with EU policy objectives and wider global goals such as the Sustainable Development Goals (SDGs); however, in practice, they manage to achieve very little of these objectives.

8.5 Conclusions/Recommendations

This chapter looked into the prospect of using landscape planning and landscape design as a tool to support the implementation of the Green Agenda for the WB. Currently, while the WB6 are making efforts to adapt the legal and policy framework to align with the EU goals and the pillars of the GAWB there are some evident gaps in the real-life and in the field implementation of the agenda. These gaps are evident in terms of policy coordination among the different sectors but also evident in the implementation of projects that contribute to the Agenda (Berisha, 2025a, 2025b).

The current planning systems and planning policies while they have policies and aims related to climate change, decarbonization, or green transition, although not directly related to the process of the GAWB, fail to achieve these policy objectives. As it was also discussed in the Albanian case, there are several policies at national, regional, and local level which could contribute to the JGT, nevertheless their implementation in practice remains quite low. There is a gap between policy objectives and real territorial development which needs to be fulfilled. While policies at different level point to climate adaptation and green transition, the implementation in the territory points more towards urban growth rather than qualitative territorial transformations. In this sense, landscape planning and landscape design can offer an alternative solution to reducing the gaps and foster the implementation of the GAWB. Landscape planning and design can offer solutions at different scales albeit national, regional, local, or even at the community scale. As demonstrated by the different case studies, focusing on landscape can offer multi-purpose solutions which have added value to spatial development but also guarantee positive changes. One of the main challenges of implementing the GAWB is also related to financing. In this case, the landscape planning and design approach can be quite supportive. Through the implementation of different "pilot"/small scale projects which can serve as best practice solutions with low levels of investments WB6 can enhance community transformations. Another important factor is that these small scale projects can help in engaging communities and making them the real drivers of producing the required change.

Additionally, landscaping can also offer social and economic benefits as a result of the production of new types of jobs which are environmentally friendly.

The development of landscape planning policies would be a recommended step to complement planning policies in the WB6. These can be at different levels of governance, starting from national, regional, or local. Landscape policies would offer a strong framework for future actions, designs or projects at the respective level of interventions. These policies should be aligned with EU directives and policies. Landscape planning can serve as a tool to harmonize different sectorial policies which sometimes can be conflicting with each other, for example, the dichotomies between environment–tourism, environment–energy, agriculture–urban development, infrastructure–tourism–environment to name a few. Landscape planning at the local level should also be seen as a priority, and municipalities should be encouraged to develop landscape plans. These would offer interesting prospects of harmonious spatial development at the local level and increase the quality of space. Landscape plans can also be supported by landscape design guidelines with simple rules to be followed by different authorities, thereby enhancing the power of transformation of space.

Finally, municipalities can start with different small scale projects of landscape which include the community. These types of projects can serve as best practice, which afterward can be replicated at different territorial scales. Additionally, the involvement of the community, would be beneficial in terms of raising awareness and pushing JGT forward. The advantage of these pilots is that usually they can be done with limited budgets, sometimes with very low budgets if the community is involved in a coherent and systematic manner.

References

Ahmad, A., Blasco, B., & Martos, V. (2022). Combating salinity through natural plant extracts based biostimulants: A Review. *Frontiers in Plant Science, 13*, Article 862034. https://doi.org/10.3389/fpls.2022.862034

Albrechts, L. (2004). Strategic (spatial) planning reexamined. *Environment and Planning B: Planning and design, 31*(5), 743–758.

Allkja, L. (2021). The Europeanisation of Spatial Planning in Albania: Opportunity or Threat?," Advances in spatial science. In E. Berisha, G. Cotella, & A. Solly (Eds.), *Governing territorial development in the Western Balkans, chapter 0* (pp. 173–193). Springer.

Arce-Mojica Th., Nehren, U., Sudmeier-Rieux, K., Miranda, P., & Anhuf, D. (2019). Nature-based solutions (NbS) for reducing the risk of shallow landslides: Where do we stand?. *International Journal of Disaster Risk Reduction, 41*, 101293

Berisha, E. (2025a). Moving towards an inclusive green agenda in the Western Balkans. Foundation for European Progressive Studies (FEPS). https://doi.org/10.13140/RG.2.2.12092.76162

Berisha, E. (2025b). Green EU integration—The role of the green agenda in accelerating just green transitions in the Western Balkans. In E. Berisha, J. Moodie, L. Allkja, & M. Jeftić (Eds.), *The Palgrave handbook of just green transitions in the Western Balkans and beyond*. Palgrave Macmillan.

Berisha, E., Čolić Marković, N., Cotella, G., & Nedović-Budić, Z. (2024). Potential future enlargement: Spatial planning in the Western Balkans. *Spatial Planning Systems in Europe*, 216–245. https://doi.org/10.4337/9781839106255.00022

Berisha, E., Colic, N., Cotella, G., & Nedović-Budić, Z. (2018). Mind the gap: Spatial planning systems in the western Balkan region. *Transactions of the Association of European Schools of Planning*, *2*(1), 47–62. https://doi.org/10.24306/TrAESOP.2018.01.004

Berisha, E., Cotella, G., Janin Rivolin, U., & Solly, A. (2023). Spatial governance and planning systems *vis-à-vis* land consumption in Europe. *European Planning Studies*, *32*(3), 553–568. https://doi.org/10.1080/09654313.2023.2207605

Berisha, E., Cotella, G., & Solly, A. (2021). *Governing territorial development in the Western Balkans. Challenges and prospects of regional cooperation* (p. 381). Cham, Springer Nature. ISBN: 978-3-030-72123-7

CABE. (2008). *Public space lessons—Adapting public space to climate change,* (online). https://www.designcouncil.org.uk/fileadmin/uploads/dc/Documents/adapting-public-space-to-climate-change1.pdf last Accessed March, 20, 2025.

Davoudi, S., Brooks, E., & Mehmood, A. (2013). Evolutionary resilience and strategies for climate adaptation. *Planning Practice and Research, 28*(3), 307–322.

Dühr, S., Colomb, C., & Nadin, V. (2010). '*European spatial planning and territorial cooperation*'.

Eco-Cycle. (2020). *The Ruhr valley cycle path—RuhrtalRadweg*, (online). https://www.interregeurope.eu/good-practices/the-ruhr-valley-cycle-path-ruhrtalradweg last Accessed March, 20, 2025.

European Commission. (2020). *Guidelines for the implementation of the green agenda for the Western Balkans,* (online). https://enlargement.ec.europa.eu/system/files/2020-10/green_agenda_for_the_western_balkans_en.pdf last Accessed March, 20, 2025

Freitas, J. G. (2020). Making a case for an environmental history of dunes. *Anthropocenes—Human, Inhuman, Posthuman, 1*(1), 5. https://doi.org/10.16997/ahip.4

Gisler, A. (2021). Allergies in urban areas on the rise: The combined effect of air pollution and pollen. *International Journal of Public Health, 66*, 1604022. https://doi.org/10.3389/ijph.2021.1604022

https://doi.org/10.1016/j.ijdrr.2019.101293

Janin Rivolin, U. (2012). Planning systems as institutional technologies: A proposed conceptualization and the implications for comparison. *Planning Practice & Research*, *27*(1), 63–85. https://doi.org/10.1080/02697459.2012.661181

Jesus, J. M., Calheiros, C. S., Castro, P. M., & Borges, M. T. (2014). Feasibility of Typha latifolia for high salinity effluent treatment in constructed wetlands for integration in resource management systems. *International Journal of Phytoremediation, 16*(4), 334–346. https://doi.org/10.1080/15226514.2013.773284. PMID: 24912235.

Madeira, A., Rodrigues, R., Palrão, T., & Mendes, A. S. (2023). Tourists' fascination with urban food markets: The successful case of time out market Lisbon. *Foods, 12*(9), 1795. https://doi.org/10.3390/foods12091795

Nadin, V. (2006). *Town and country planning in the UK* (14th ed.). Routledge. https://doi.org/10.4324/9780203004258

NTPA. (2017a). *National territorial plan of Albania*, (online). https://planifikimi.gov.al/index.php?id=252 last Accessed March, 20, 2025.

NTPA. (2017b). *Integrated cross sectorial plan for the economic area Tirane- Durres*, (online). https://planifikimi.gov.al/index.php?eID=dumpFile&t=f&f=1310&token=79845b4945ed987ac47dd693d35b72da0a680b92 last Accessed March, 20, 2025

RCC. (2023). *Report on the implementation of the green agenda for the Western Balkans action plan (GARI)*, (online). https://www.rcc.int/pubs/202/2023-report-on-the-implementation-of-the-green-agenda-for-the-western-balkans-action-plan-gari last Accessed March, 20, 2025.

Strodt, S. (2023). *Sand dunes offer clues to coastal erosion and how to prevent it*, (online). https://projects.research-and-innovation.ec.europa.eu/en/horizon-magazine/sand-dunes-offer-clues-coastal-erosion-and-how-prevent-it last Accessed March, 20, 2025.

The Highline. (2025). The history, (online). https://www.thehighline.org/history/ last Accessed, March, 20, 2025.

Thorbjörn, A. (2010). Landscape Urbanism versus Landscape Design The potential of design must not be neglected. European Landscape Magazine 71:80

UN. (2015). Paris agreement, (online). https://unfccc.int/sites/default/files/english_paris_agreement.pdf last Accessed March, 20, 2025.

Urban Green- Blue Grids. (2025). *Leiden Noorderkwartier-Oost, climate-proof public space*. https://urbangreenbluegrids.com/projects/leiden-noorderkwartier-oost-climate-proof-public-space/ last Accessed March, 20, 2025.

Waldheim, Ch. (2006). Landscape as Urbanism. In C. Waldheim (Ed.), *The landscape Urbanism, reader*. Princeton Architectural Press.

CHAPTER 9

Green Energy Transition in the Western Balkans: Challenges and Opportunities for Bosnia and Herzegovina

Aliaksandr Novikau and Mustafa Krupalija

9.1 Introduction

Anthropogenic global climate change is an existential threat to individual countries and the entire planet. The United Nations recognizes global climate change as a severe threat through its Sustainable Development Goals (SDGs) and, specifically, SDG13—Climate Action, and calls for urgent action to combat climate change and its impacts (United Nations, 2015). Energy use is responsible for about 80% of total greenhouse gas (GHG) emissions in developed countries. The current global energy system is dominated by fossil fuels such as coal, oil, and natural gas, the combustion of which releases GHGs such as carbon dioxide into the atmosphere and causes the greenhouse effect. Therefore, the decarbonization of the global energy system is one of the current priorities of all humankind. Consequently, the "green" or "clean" energy transition aims to replace fossil fuels responsible for pollution and the emission of greenhouse gases with low-carbon renewable energy sources such as solar, wind, hydro, and bioenergy (Arent et al., 2017; Tian et al., 2022). In line with such a call, many countries and organizations are taking concrete actions to combat global climate change. For instance, to address this challenge, the European Union has committed to turning the union into the

A. Novikau (✉) · M. Krupalija
Department of Political Science and International Relations, International University of Sarajevo, Ilidža, Bosnia and Herzegovina
e-mail: anovikau@ius.edu.ba

M. Krupalija
e-mail: mkrupalija@ius.edu.ba

E. Berisha et al. (eds.), *The Palgrave Handbook of Just Green Transitions in the Western Balkans and Beyond*,
https://doi.org/10.1007/978-3-031-95075-9_9

first climate-neutral continent by 2050, and as a first step, by 2030, it will have at least 55% less net greenhouse gas emissions compared to 1990 levels (European Council, 2023). Transitioning to green technologies and practices offers several specific environmental benefits. First, the adoption of green technologies can lead to a reduction in GHG emissions. Second, decarbonizing energy systems can significantly reduce air pollution, resulting in improved air quality and a decrease in respiratory and cardiovascular diseases. Third, the shift to renewable energy sources can enhance a country's energy security by diversifying its energy mix and reducing dependence on foreign fossil fuels. Renewable energy resources are usually locally available, mitigating the risks associated with supply chain disruptions or geopolitical tensions that can affect the availability of imported energy sources.

However, transitioning to green energy can also bring some challenges, primarily affecting regions where the economy relies on the fossil fuel industry. Therefore, transitioning towards green energy in coal-dependent industries can lead to social tensions due to the dilemma between environmental protection and job opportunities. Additionally, some of the proposed methods for transitioning to cleaner energy systems, such as the construction of small hydroelectric power plants, are subject to debate, especially in cases where the construction and operation of such plants are not adequately regulated.

Recently, justice has emerged as a vital component in governing the green energy transition. This concept, increasingly referred to as a "just energy transition," lacks a universally agreed-upon definition. However, it usually includes rigorous analysis and assessment of the impacts of green energy transition policies on the social dimensions of energy. The just transition concept initially focused on ensuring workers have good jobs during the shift away from fossil fuels, but over time, it included protecting vulnerable communities and emphasizing fairness and support. Therefore, just transition promotes the engagement of all parties affected by energy policies in the decision-making process, with a commitment to acknowledging their perspectives and necessities (Johansson, 2023).

Considering all the challenges that communities face during the green energy transition, a just energy transition represents a safeguard that would make this process more acceptable, fairer, and less painful for all relevant parties. For instance, the mechanisms can include social protection, training, and skill development programs for workers potentially impacted by the transition, such as those employed in coal regions, and economic diversification. Achieving a just energy transition can require a more active role for the state within the economy. This role can focus on fostering job creation in low-carbon sectors of the economy to offset job losses in declining fossil fuel industries. Additionally, it requires shared responsibility between the state and business to internalize the negative social externalities associated with the transition, for instance, through mechanisms such as increased taxes on industries responsible for GHG emissions. Ultimately, the goal is to establish a robust social safety net and provide adequate compensation for individuals and communities affected by the shift towards a green energy system (Graaf & Sovacool, 2020).

Although the green energy transition is widely focused on academic literature, little attention has been paid to regional energy transitions (Loewen, 2022). The region of the Western Balkans is relatively underrepresented in the scholarly literature on green energy transition. Yet, all the Western Balkan countries that started the green energy transition have faced issues typical to that process, including but not limited to the just transition.

The purpose of the chapter, therefore, is to analyze the green energy transition in Bosnia and Herzegovina (BiH), a Western Balkan country with a high share of coal in energy balance with a focus on its just energy transition. The chapter is structured as follows. The next section covers the energy landscape in BiH. The following sections explore the benefits and challenges BiH is encountering as it moves towards a greener and more sustainable future. This examination will shed light on both the positive outcomes and the obstacles that arise in the context of green transition. The final section discusses some lessons learned from the case and offers some suggestions for further research.

9.2 Bosnia and Herzegovina's Energy Paradox

The structure of the energy sector in Bosnia and Herzegovina is complex, and the competencies in the sector are divided between the federal state, two entities—Federation of Bosnia and Herzegovina (FBiH) and Republika Srpska (RS), and a self-governing administrative unit—the Brčko District. BiH is characterized by abundant energy resources, at least for electricity generation—mostly coal, hydropower, and forest biomass. As a result, BiH is the net exporter of 27% of its electricity generation (mainly from coal), which is one of the country's primary export goods (OECD, 2022). The electricity sector in BiH is dominated by three independent vertically integrated power utility companies that reflect the ethnic division of the country (IRENA, 2023).

Like other Western Balkan countries, coal is one of the primary energy sources in Bosnia and Herzegovina, with 5 coal-fired power plants with an installed capacity of 2 GW. Their share in total power generation constantly remains above 50% (OECD, 2022). The total geological coal reserves in the country account for 5.6 billion tons, with exploitation reserves of 1.9 billion tons (Council of Ministers of BiH, 2018). However, most of the existing power plants are old and inefficient. Even with the newest Stanari power plant opened in 2016, BiH's coal plants are, on average, 44 years old (Ciuta et al., 2024). That results in high emissions of air pollutants (mostly particulate matter and sulfur dioxide) and GHGs (OECD, 2022). In addition, land pollution from coal plants results in the degradation of meadows, arable land, and forests (OECD, 2021).

Because of domestic and international pressure, these coal plants, therefore, need either upgrade or decommissioning. As a result, several new coal power plants have been planned in Bosnia during the last decade with the potential to double the country's generation capacity. Yet, all of them have been either canceled or put on hold (Ciuta et al., 2024).

9.3 Green Energy Transition in Bosnia and Herzegovina

BiH has had close relationships with the European Union since the 1990s and has a strong desire to join the union. The country was given the prospect of EU membership through the Thessaloniki Summit Declaration in 2003. It applied for EU membership in 2016, gained official candidate status in 2022, and in March 2024, EU leaders approved opening accession talks with Bosnia and Herzegovina. In 2005, Bosnia and Herzegovina joined the Energy Community Treaty with the EU. By signing the treaty, the country has been obliged to gradually take over parts of the EU acquis in such a way as to transpose into its legislation the requirements and rules of the corresponding directives and regulations of the EU in the areas of energy (Energy Community Secretariat, 2020). The transition to green energy in BiH is considered part of a more considerable effort to comply with the European Green Deal, aiming to make the EU carbon-neutral by 2050. Despite not being an EU member, these actions are significant for BiH due to its location and political alignment with EU standards. Therefore, achieving a just transition in Bosnia and Herzegovina involves balancing two seemingly incompatible realities—an aspiration to join the EU and its heavy reliance on coal (Novikau, 2022). In 2018, a Framework Energy Strategy until 2035 was published. The document prioritizes the integration of renewable energy resources under the EU Renewable Energy Directive, which aims for a share of over 50% of renewable energy in energy consumption by 2040 (IRENA, 2023). The document provides four different scenarios for the planned phase-out of all existing coal-fired power plants. Yet, according to all the scenarios, at least some of the power plants will be operational even after 2035 (Council of Ministers of BiH, 2018). BiH adopted its Nationally Determined Contribution (NDC) in 2021 and submitted it to the UN Climate Change Secretariat. While outlining the objective to reduce emissions of GHGs and ultimately reach carbon neutrality, it lacks a concrete phase-out date for coal-fired power plants. Furthermore, the singular measure addressing a just transition appears insufficient for managing the social and economic impacts of phasing out coal in Bosnia and Herzegovina (Spasić, 2024). BiH adopted the National Renewable Energy Action Plan, resulting from the transposition of the EU Renewables Directive. BiH pledged to achieve a share of 40% in total final consumption by 2020—the highest target among all Energy Community members. In 2014, FBiH adopted a similar Action Plan for the use of renewable energy in which it committed itself to achieve a share of 41% in total final consumption by 2020, while RS commited to 48% (Council of Ministers of BiH, 2018; IRENA, 2023). As part of its obligations to comply with the Large Combustion Plants Directive under the Energy Community Treaty, BiH has drawn up a National

Emission Reduction Plan (NERP) covering the period from 2018 to 2027. Yet the country's coal power plants have not adhered to the pollution ceilings for any of the criteria air pollutants. In 2023, sulfur dioxide (SO_2) emission was more than eight times as much as allowed under NERP—181,807 tons, with Kakanj 7 and Ugljevik power plants highest contribution. Particulate matter (PM) emission was 4,647 tons, with the highest contribution from the Gacko powerplant. Nitrogen oxide (NOx) emissions from BiH's coal power plants in 2023 totaled 11,752 tons, with Gacko and Kakanj 7 power plants as the highest contributors. Because of its non-compliance with the NERP ceilings, in 2021, the Energy Community Secretariat initiated a dispute settlement case against BiH, which remains open due to the continued breaches (Ciuta et al., 2024). The Energy Community Secretariat stated that the FBiH decision to extend the operation of Tuzla 4 (installed capacity of 200 MW) and Kakanj 5 (installed capacity of 110 MW) until 2028 is a breach of obligations to phase-out these plants (Todorović, 2022). The controversial project, the Tuzla 7, a 450 MW coal plant, was initiated in 2014 and intended to replace aging coal-fired power plants and improve the country's energy security. Yet, in 2023, the project was canceled due to a lack of funding and environmental concerns (Spasić, 2023).

9.4 Just Energy Transition in BiH

The entire economy of BiH heavily depends on coal, which is used in electricity generation and specifically in iron, steel, and cement manufacturing (Ignjatović et al., 2024). In addition, there are social aspects related to the coal industry. BiH has a rich coal mining heritage. During the industrialization era, the coal mining sector emerged as a pivotal industrial branch of Bosnia and Herzegovina. During Socialist Yugoslavia times, miners were elevated to a unique position within the societal consciousness, being portrayed as the embodiment of the industrious socialist worker whose labor and commitment were integral to the revolutionary cause. This perception was reinforced through various mediums, including currency, popular music of the period, and even street names that commemorate legendary miners. One such individual, Alija Sirotanović, whose statue still stands at the Breza mine, achieved international recognition in 1949 by breaking the world record for coal extraction, a record previously held by a Soviet miner. This earned him a state honor and positioned him as a symbol of Yugoslav strength and pride, promoting a spirit of competition and asserting the superiority of Yugoslav workers over their socialist counterparts in the Soviet Union. The mining tradition was further expanded during this period with the introduction of new mines and technologies. Miner's Day, celebrated in Bosnia and Herzegovina on December 21, receives significant media attention every year, highlighting this profession with recognition of miners' integrity, harsh working conditions, and hard-earned bread (Anadolija, 2023). In addition, the symbol of miners, the hammer and chisel, is featured on the official coat of arms of the fourth-largest city in the country, Zenica, indicating the deep-rootedness of this profession in almost all social strata.

In today's BiH, coal miners have once again become a central figure in the symbolic narrative that seeks to address the challenges of the modern era. If the mining sector can successfully adapt to the demands of the green transition, the miner could potentially be recast as a symbol of the contemporary European individual who has made a significant contribution to the advancement of his nation and society on the European path. The process of green transition, which ultimately requires the closure of coal mines and coal-fired power plants, represents a particular challenge in regions where this sector is a vital industrial branch. Currently, the country has 11 coal mines (8 in FBiH and 3 in RS) with annual production of 14 million tons of coal. The coal-related industry directly employs almost 17,000 workers (14,500 in mining and 2,500 in power plants) (Somers & Mandras, 2021). Thus, phasing out coal could potentially result in the loss of 1.3% of all jobs in the country. In addition, the number of indirect workers related to coal activities is estimated to be more than 13,000 (Somers & Mandras, 2021). Moreover, coal-related jobs are not equally distributed across the country and are concentrated near key coal deposits located in the basins of Tuzla, Central Bosnia (Kakanj and Zenica), Bugojno, Livno-Duvno, Gacko, and Doboj-Banja Luka (Council of Ministers of BiH, 2018). Thus, some regions would be disproportionately affected by the potential phase-out of coal, and without considering the just dimension, the social effects of the green energy transition could be devastating. For instance, Banovići is a municipality located in the Tuzla Canton with 6,000 residents in the urban area and 27,000 in the entire municipality (Perks, 2023a). Approximately half of the population is employed in the local mine, and about a quarter depends on the mine indirectly (A. K., 2022). Considering the seriousness of the difficulties in transition and Bosnia and Herzegovina's obligation to shut down coal mines and fossil fuel power plants by 2050, local authorities in coal-dependent areas heavily rely on support from higher levels of government and close cooperation with international organizations. Considering the potential impact on jobs, skills development, and the overall well-being of affected communities in 2019, the World Bank launched a global initiative entitled to support just energy transition in coal regions (Perks, 2023b). The European Commission launched the Initiative for coal regions in transition in the Western Balkans and Ukraine to support their transition to a carbon-neutral economy. The initiative, which ran from December 2020 to December 2023, offered technical assistance, capacity building, and access to financing for coal regions in Bosnia and Herzegovina, Kosovo*,[1] Montenegro, North Macedonia, Serbia, and Ukraine. The initiative established an open platform facilitating region-wide, multi-stakeholder dialogue. It created a forum for sharing experiences, knowledge, and best practices on

[1] This designation is without prejudice to positions on status, and is in line with UNSCR 1244 and the ICJ Opinion on the Kosovo declaration of independence.

transition-related issues. It fostered connections between coal regions in the Western Balkans and Ukraine and their EU counterparts through coal region-to-region exchanges. The selected pilot regions received technical assistance through expert support to develop transition roadmaps for relevant public authorities (European Commission, 2024).

The Federal Ministry of Energy, Mining and Industry of BiH unveiled its "Strategy for a Just Transition of the Energy Sector in FBiH." This strategy aims to address challenges associated with the green transition, with a particular focus on vulnerable areas. It includes action plans to reduce production and close coal mines in FBiH, alongside worker resettlement through workforce requalification programs in mining regions.

The Zenica Coal Mine, employing 615 people (out of a total of 5,120 across Elektroprivreda BiH's seven mines) in 2024 (Zenicainfo, 2024), exemplifies the complexities of this transition. Despite government loans, production reached only 34% of the target, with insufficient revenue to cover wage costs (S. H., 2024). Minister Vedran Lakić announced a pilot project in collaboration with the World Bank. This project would involve the closure of Zenica's coal mines, resetting a select number of workers, and developing a plan for closing other facilities within a specified timeframe, ultimately leading to the cessation of coal production in Zenica (TIP, 2023). However, the closure appears driven by the mine's financial unsustainability rather than successful project management. Consequently, this pilot project's success as a model for BiH's transition is questionable. Sinan Husić, president of the Independent Union of Miners of the Federation of BiH, proposed reassigning workers to other mines in need, highlighting the need for more comprehensive solutions (Dajić, 2024). Currently, approximately 200 employees maintain equipment while 400 await overdue salaries despite legal obligations of €320 per month. These workers have not received wages for months, nor have their health and pension contributions been paid since 2010. Although the government secured funds for one outstanding salary, further payments are contingent on selling non-essential assets. The law for official closure remains in limbo. Kadrija Hodžić, head of the House of Representatives Energy, Mining, and Industry Committee, argues that existing bankruptcy and financial consolidation legislation suffices. Regardless, the Federal government seeks credit-based financing and asset sales to manage the collapse (Katić, 2024).

9.5 Conclusion

Bosnia and Herzegovina's green energy transition presents a complex challenge, requiring a careful balancing act between environmental necessity and social well-being. While transitioning away from fossil fuels like coal is crucial for reducing greenhouse gas emissions and aligning with the European Green Deal, the economic and social risks caused by dependence on coal mining necessitate not only green but a just energy transition.

The absence of a comprehensive strategy to address the social and economic consequences of coal phase-out remains a significant hurdle. A rigid "comply or close" approach without accompanying region-specific transition plans for displaced workers and coal-dependent communities is not a viable solution. Reskilling initiatives, infrastructure development, and the cultivation of alternative industries in these areas are crucial for ensuring a just transition that leaves no one behind.

The Zenica mine closure exemplifies the critical challenges BiH faces in its green transition. While closing unsustainable mines is inevitable, an unplanned and worker-unprepared approach exacerbates economic and social hardships in affected communities. A just transition necessitates more than mere mine closures—it demands a comprehensive strategy encompassing long-term economic sustainability, worker social security through retraining, the development of new industrial sectors, and environmentally friendly alternatives. Without such measures, the green transition risks leaving workers and communities unsupported, undermining public confidence in the energy sector's future and the overall sustainability of the process.

BiH's green energy transition demands a multi-faceted approach that prioritizes both environmental sustainability and social justice. By fostering collaboration between government agencies, international bodies, private investors, and local communities, the country can navigate the complexities of just transition and emerge as a leader in renewable energy development within the Western Balkans. Further research is needed to explore best practices for decommissioning coal plants, fostering alternative industries, and developing sustainable, community-supported renewable energy projects. Through such efforts, BiH can unlock the full potential of its green future.

References

A. K. (2022) Može li BiH postati "zelena" država i kako ugasiti termoelektrane i rudnike bez kolapsa sistema? [Can BiH become a "green" state and how to phase out thermal power plants and mines without collapsing the system?]. In: Klix. https://www.klix.ba/biznis/privreda/moze-li-bih-postati-zelena-drzava-i-kako-ugasiti-termoelektrane-i-rudnike-bez-kolapsa-sistema/221108110. Accessed 22 Sep 2024

Anadolija (2023) Dan rudara: Čuvari digniteta i tradicije ove teške i časne profesije [Miner day: Guardians of dignity and tradition of this hard and honorable profession]. In: Buka Magazin. https://6yka.com/bih/dan-rudara-cuvari-digniteta-i-tradicije-ove-teske-i-casne-profesije. Accessed 18 May 2024

Arent, D, Arndt, C., & Miller, M., et al (2017) Introduction and synthesis. In D. Arent, C. Arndt, & M. Miller et al. (Eds.), *The political economy of clean energy transitions* (1st ed, pp 3–15). Oxford University PressOxford.

Ciuta, I., Gallop, P., & Pehchevski, D. (2024). Comply or close: Six years of deadly legal breaches by Western Balkan coal plants. CEE Bankwatch Network

Council of Ministers of BiH. (2018). Framework energy strategy of Bosnia and Herzegovina until 2035

Dajić, M. (2024). Elektroprivreda BiH obustavila radove u Rudniku uglja Zenica [Elektroprivreda BiH suspended works in Zenica coal mines]. In Al Jazeera. https://balkans.aljazeera.net/teme/2024/4/4/elektroprivreda-bih-privremeno-obustavila-radove-u-rudniku-uglja-zenica. Accessed 19 May 2024

Energy Community Secretariat (2020) Policy guidelines by the Energy Community Secretariat on small hydropower projects in the Energy Community

European Commission (2024) Initiative for coal regions in transition in the Western Balkans and Ukraine. https://energy.ec.europa.eu/topics/carbon-management-and-fossil-fuels/coal-regions-western-balkans-and-ukraine/initiative-coal-regions-transition-western-balkans-and-ukraine_en

European Council. (2023). Fit for 55. In *European green deal*. https://www.consilium.europa.eu/en/policies/green-deal/fit-for-55-the-eu-plan-for-a-green-transition/. Accessed 22 Sep 2024

van de Graaf, T., & Sovacool, B. K. (2020). *Global energy* Politics.

Ignjatović, J., Filipović, S., & Radovanović, M. (2024). Challenges of the green transition for the recovery of the Western Balkans. *Energy Sustainability and Society, 14*, 2. https://doi.org/10.1186/s13705-023-00421-4

IRENA (International Renewable Energy Agency). (2023). Renewables readiness assessment: Bosnia and Herzegovina. nternational Renewable Energy Agency, Abu Dhabi, UAE

Johansson, V. (2023). Just Transition as an evolving concept in international climate law. *Journal of Environmental Law, 35*, 229–249. https://doi.org/10.1093/jel/eqad017

Katić, G. (2024). Zenički rudnik s milionskim dugovima ne radi, rudari bez svjetla na kraju tunela [Zenica mine with millione debt does not work, miners without light at the end of the tunnel]. In Radio Free Europe. https://www.slobodnaevropa.org/a/rudnik-uglja-zenica-zatvoren-kasne-plate/33066461.html. Accessed 29 Sep 2024

Loewen, B. (2022). Coal, green growth and crises: Exploring three European Union policy responses to regional energy transitions. *Energy Research & Social Science, 93*, Article 102849. https://doi.org/10.1016/j.erss.2022.102849

Novikau, A. (2022). Energy security in Bosnia and Herzegovina: Will the EU energy acquis be helpful? In D. Šiljak & K. L. Nielsen (Eds.), *Bosnia and Herzegovina and European integration: Obstacles and challenges* (pp. 93–111). International University of Sarajevo.

OECD. (2022. Clean energy transition in the Western Balkans

OECD. (2021). *OECD development pathways multi-dimensional review of the Western Balkans: assessing opportunities and constraints*. OECD Publishing.

Perks, R. B. (2023a). *Concept environmental and social review summary (ESRS)—Pilot approaches to a just transition in select coal regions of BiH*. World Bank Group.

Perks, R. B. (2023b). *Concept project information document (PID) - pilot approaches to a just transition in select coal regions of BiH*. World Bank Group.

S. H. (2024) Elektroprivreda žestoko o rudniku “Zenica”: Imaju prihod od 3,2 miliona KM, a na plate im treba 10 miliona [Elektroprivreda fiercely about the “Zenica” mine: have revenue of 3.2 million KM, and they need 10 million on their salaries]. In: Klix. https://www.klix.ba/biznis/privreda/elektroprivreda-zestoko-o-rudniku-zenica-imaju-prihod-od-3-2-miliona-km-a-na-plate-im-treba-10-miliona/240312101. Accessed 19 May 2024

Somers, J., & Mandras, G. (2021). *Recent trends in coal and peat regions in the Western Balkans and Ukraine*. Publications Office of the European Union.

Spasić, V. (2024). *BiH's NECP lacks clear set of policies, measures, investment needs–Energy community*. In Balkan Green Energy News. https://balkangreenenergynews.com/bihs-necp-lacks-clear-set-of-policies-measures-investment-needs-energy-community. Accessed 18 May 2024

Spasić, V. (2023). *Tuzla 7 coal plant project in BiH is dead, entity prime minister says*. In Balkan Green Energy News. https://balkangreenenergynews.com/tuzla-7-coal-plant-project-in-bih-is-dead-entity-prime-minister-says/. Accessed 18 May 2024

Tian, J., Yu, L., Xue, R., et al. (2022). Global low-carbon energy transition in the post-COVID-19 era. *Applied Energy, 307*, Article 118205. https://doi.org/10.1016/j.apenergy.2021.118205

TIP. (2023). *Ministar Lakić otkrio koji rudnik ide u zatvaranje i proces tranzicije [Minister Lakić uncovered which mine goes in closing and the transition process]*. https://tip.ba/2023/12/10/ministar-lakic-otkrio-koji-rudnik-ide-u-zatvaranje-i-proces-tranzicije/. Accessed 19 May 2024

Todorović, I. (2022). *Energy community: Prolonging operation of coal plants in BiH is breach of treaty*. In Balkan Green Energy News. https://balkangreenenergynews.com/energy-community-prolonging-operation-of-coal-plants-in-bih-is-breach-of-treaty/. Accessed 18 April 2022

United Nations. (2015). Transforming our world: The 2030 agenda for sustainable development.

Zenicainfo. (2024). *Nakon Odluke Elektroprivrede, gasi li se rudnik u Zenici? [After the decision of Elektroprivreda, if a mine in Zenica is shutting down?]*. https://zenicainfo.ba/2024/04/06/nakon-odluke-elektroprivrede-gasi-li-se-rudnik-u-zenici/. Accessed 19 May 2024

CHAPTER 10

Cross-Border Dimension of Just Green Transitions in the Western Balkans—Case Study of the Republic of Serbia

Ana Nikolov

10.1 Introduction

The chapter focuses on identifying the cross-border dimension of the Just Green Transition in policy documents, strategic documents, and programmes targeting the Western Balkan countries, particularly the cross-border area between Serbia and Bosnia and Hercegovina and between Serbia and Croatia. In addition, the comprehensive analysis of the data sets of the Archive of Ratified International Agreements and Other Acts of the Republic of Serbia elaborate on legal frameworks for supporting cross-border cooperations and cross-border public services that can contribute to the implementation of Just Green Transition in the IPA-IPA (Serbia—Bosnia and Hercegovina) and INTERREG-IPA (Croatia and Serbia) cross-border areas. The main questions of the chapter are: How is the cross-border dimension of the Just Green Transition conceptualised in policy documents, strategic documents and studies? Are the needs of cross-border areas to mitigate climate change and transition to a low-carbon society reflected in legal frameworks for cross-border cooperation? Which legal frameworks can be used to establish cross-border public services to support the Just Green Transition in the cross-border areas of the Republic of Serbia?

A. Nikolov (✉)
ICBC Institute, AEBR Balkans, Belgrade, Serbia
e-mail: ananikolov@gmail.com

E. Berisha et al. (eds.), *The Palgrave Handbook of Just Green Transitions in the Western Balkans and Beyond*,
https://doi.org/10.1007/978-3-031-95075-9_10

10.2 Background

The European Green Deal, the new growth strategy for the EU that supports the just green transition to a sustainable, fair, and prosperous society, can only be fully achieved if the EU joins forces with its neighbours and partners, especially considering that the transition must be just and inclusive. (European Commission, 2019). Building on the Lisbon Treaty that introduced territorial cohesion, seeking to reduce disparities between the levels of development of various regions, with a particular focus on least developed regions, addressing, among others, cross-border regions, making sure that no areas or citizens should be "left behind" in the just transition to a climate-neutral, sustainable, and prosperous European Union (European Commission, 2023). In addition, the EU Territorial Agenda 2030 sets two main objectives liked by the Green Deal, namely Just Europe and Green Europe, and already in the second article, it encourages neighbouring countries to join forces to implement its objectives at European, transnational, macro-regional and cross-border levels (EU, 2020). Border regions' participation in cross-border, interregional, and transnational cooperation initiatives under the Interreg programme or Macro-Regional Strategies addresses joint challenges, and leverages shared resources, promotes more integrated and cohesive development to help align policies, share best practices, and facilitates cross-border cooperation for more effective implementation of the Territorial Agenda (Lindblad, 2023; Pinnavaia & Berisha, 2021; Solly & Berisha, 2021). The challenges linked to climate change and the environment do not recognise human-made borders, and it includes action to step up cooperation across borders in several sectors, for example, setting up ecological corridors, measures related to river basin management, disaster risk management, flood risk management and forest protection plans, or national and regional climate change adaptation strategies (European Commission, 2021). The EU supports Western Balkan countries' paths towards a Just Green Transition, particularly the Western Balkans Green Agenda (WBGA) (European Commission, 2019). All the activities envisaged with the Sofia Declaration and WBGA Action Plan can have a cross-border dimension if implemented in cross-border areas, as each activity of the WBGA is interlinked and interconnected with others. Activities of the WBGA that have cross-border dimensions can be grouped based on the nature of the activities as follows:

- Legal Activities: Ratification of the Conventions and Conclusions of the Agreements and their implementation
- Strategic Activities: Development and Implementation of Regional Strategic and Action Plans, Road Maps, and Impact Assessments
- Monitoring Activities: Establishment of the regional monitoring and reporting systems
- Educational Activities: Capacity building and knowledge exchange

For their coordinated, sustainable, and efficient implementation, the Sofia Declaration envisaged the parties' active participation in current pan-European networks, Horizon 2020, regional organisations, macro-regional strategies, and other relevant initiatives such as the Covenant of Mayors (RCC, 2020). Moreover, European Commission SWD (2020) 223—Guidelines for the Implementation of the Green Agenda for the Western Balkans emphasises that EUSDR and EUSAIR *should continue to support innovation and cross-border cooperation on a medium-term vision for the regions* (European Commission, 2020) and Action Plan for the Implementation of the Sofia Declaration on the Green Agenda for the Western Balkans recognised that the '*external instruments under the EU Multiannual Financial Framework (2021–2027), IPARD with its 11 measures and Instrument for Pre-Accession Assistance (IPA) III will be the key funding instruments to kick-start actions and initiate a comprehensive green transition in the Western Balkan region*' (RCC, 2021). Also, 43 cross-border structures have been established in the Balkan peninsula, primarily formed in cross-border areas with shared environmental and cultural resources and the need for joint actions for protection, preservation, mitigation, and sustainable and economic development. (AEBR Balkans, 2021) Cross-border areas are often called "laboratories of European integration" as new concepts and interventions are tested on how they contribute to the European Integration process (European Berisha, 2025a, 2025b; Commission, 2021), in this case, to test cross-border frameworks that can be used to achieve a Just Green Transition in the Western Balkans.

10.3 Case Study of the Cross-Border Area of Serbia with INTERREG-IPA (Croatia) and IPA-IPA (Bosnia and Herzegovina) Borders

For this chapter, the cross-border area overlapping with cross-border programmes is selected, namely cross-border Programmes INTERREG-IPA Croatia—Serbia and IPA CBC Serbia—Bosnia and Hercegovina, comparing the results on EU external border with the ones on IPA-IPA borders. The programme area of the INTERREG-IPA Programme Croatia and Serbia includes four counties on the Croatian side and five districts on the Serbian side, covering a total of 25.505 km2 with around 2.14 million inhabitants (European Commission, 2021b), and the programme area territory of the IPA-IPA Programme Serbia-Bosnia and Herzegovina extends across four regions in the Republic of Serbia and 67 local self-government units in Bosnia and Hercegovina, covering in total 32.014,18 km2 with around 2,66 million inhabitants (European Commission, 2021a). The state border between Serbia and Croatia is not determined, and it is more than 260 km long, of which more than 150 km are hydrographic elements (rivers and canals). The terrain along the border line is of a plain type, with the developed road network that provides relatively good connections between regional, municipal and rural settlements in the border area. The blue border between Serbia and Croatia is represented by the Danube, the internationally navigable river, where the Convention on the Danube Navigation Regime regulates the navigation regime. The state border between Serbia and Bosnia and Herzegovina is not determined, and it is more than 370 km long, of which more than 260 km are hydrographic elements (rivers, lakes, and streams). Natural barriers are represented by the mountain relief on the southern part of the border, the Drina River with reservoirs and the Sava River. Permeability of the border is limited to communications. The blue border between Serbia and Bosnia and Hercegovina is represented by the Sava and Drina rivers, where the Sava is the international navigable river, where the navigation regime is regulated by the Framework Agreement on the Sava River Basin with annexes and the Protocol on navigation regime, and Drina River, that is navigable from 0.00 to 15.00 river kilometre (RS, 2022). The main challenges that are faced in the cross-border area of Croatia and Serbia are a decline in population, regional imbalances and disparities, economic transition processes, climate change, the modest expenditure on research and innovation, relatively unfamiliar concepts of the European Green Deal, and the long-term socio-economic consequences of the COVID-19 pandemic (European Commission, 2021b). Cross-border areas of Serbia and Bosnia and Hercegovina are exposed to the following challenges: negative migration trends, high level of unemployment, narrow skills of the unemployed, focus on traditional industries, gaps in the infrastructure and accommodation possibilities, lack of diversification and innovativeness and low quality of services in the tourism sector, and long-term socio-economic consequences of the COVID-19 pandemic (European Commission, 2021a). In addition, the priorities of the cross-border structures established in the case study's territorial scope also reflect the area's challenges (Table 10.1). Six cross-border structures have been established, two of which are very active, namely Drina-Sava and Drina-Tara. In addition, the Five-country Biosphere Reserve Mura-Drava-Danube is also partly covering the case study area.

The cross-border structures in the Balkan area are often established in areas with environmental qualities that need to be preserved and improved, which is reflected in their priorities. It is also visible that the priorities of the cross-border structures in the case study areas are established as priorities that align with the WBGA pillars. Comparing the results from the cross-border structures, it can be concluded that the dominant pillars are (2) Circular economy and (5) Biodiversity, followed by (3) Depollution, and then (1) Climate, energy, mobility, and (4) Sustainable agriculture and food production. Two cross-border programmes cover the case study area: INTERREG-IPA Programme Croatia-Serbia 2021–2027, where influence and interconnection of the European Green Deal and Western Balkans Green Agenda can be observed, and IPA CBC Programme Serbia-Bosnia and Herzegovina 2021–2027, where cross-border cooperation to achieve WBGA can be observed.

Table 10.1 Cross-border structures' priorities in line with WBGA pillars (*Source* Authors' interpretation of the data from AEBR Balkans database on cross-border structures in the Balkans)

Cross-Border structure	*Priorities in line with WBGA pillars*	*WBGA pillars*
Dunav-Drava-Sava	Promoting joint spatial planning, Nature conservation and environmental protection, Transport and communications, Economy and tourism	(1) Climate, energy, mobility (2) Circular economy (5) Biodiversity (3) Depollution text
Drina-Sava-Majevica	Traffic and transportation, Economic development, tourism and agricultural development, Environmental and nature protection issues, Culture, sports and health care, Energy issues, waste management, Innovation and technology transfer, Education and professional training	(1) Climate, energy, mobility (2) Circular economy (5) Biodiversity (3) Depollution (4) Sustainable agriculture and food production
Drina Euroregion	Protection of the Drina River and coordination of water management activities, Economic development and tourism Environmental protection,	(5) Biodiversity (1) Climate, energy, mobility (2) Circular economy (3) Depollution
Drina-Sava	Green and smart economy, Multifunctional use of space and natural resources, Increase areas cohesion, Strengthen social capital and improve the quality of life	(2) Circular economy (5) Biodiversity

(continued)

Table 10.1 (continued)

Cross-Border structure	*Priorities in line with WBGA pillars*	*WBGA pillars*
Drina-Tara	Rural development and agriculture, SMEs and entrepreneurship Tourism Environment	(4) Sustainable agriculture and food production (2) Circular economy (5) Biodiversity (3) Depollution
Adria Alpe Pannonia	Promotion of traditional producers of organic food, Innovative patents and geographical brand protection, Tourism, Environmental protection	(4) Sustainable agriculture and food production (2) Circular economy (5) Biodiversity

The INTERREG-IPA Programme Croatia-Serbia 2021–2027 have four priorities: (1) Cooperating for smarter programme area; (2) Cooperating for a greener and climate change resilient programme area; (3) Cooperating for healthier and more inclusive programme area; and (4) Cooperating for more sustainable and socially innovative tourism and culture (European Commission, 2021b)—Priority 1—Cooperating for smarter programme area will contribute to the (2) Circular economy and pillar (1) Climate, energy, and mobility, as it will support the activities that contribute to the smart specialisation: nature-based solutions, green and digital transitions; Industry 4.0; technology transfer for bio, green and circular economy; transfer of knowledge and skills for social development, economic growth, environment and climate change. Moreover, Priority 2—Cooperating for greener and climate change resilient programme area will support the activities that contribute to the sustainability and resilience, environmental protection and climate change adaptation; renewable energy and energy efficiency; joint climate change adaptation, disaster prevention and first response plans; joint emergency centres and systems for monitoring, prevention and management of potential risks; green infrastructure; circular, innovative and climate-neutral solutions; and ecosystem services, contributing to the (1) Climate, energy, mobility and (2) Circular economy. Finally, Priority—4 Cooperating for more sustainable and socially innovative tourism and culture of the Programme will contribute to the (2) Circular economy, as it will support the activities that contribute to the diversification of tourism, place-based and nature-based solutions, creating smart destinations; and support job creation including up-skilling and re-skilling of vulnerable groups.

Overall, the INTERREG-IPA Programme Croatia-Serbia 2021–2027 will directly contribute to (1) Climate, energy, and mobility and (2) Circular economy, and indirectly to (3) Depollution; (4) Sustainable agriculture and food production; and (5) Biodiversity.

The IPA CBC Programme Serbia-Bosnia and Herzegovina 2021–2027 have three thematic priorities: (1) Technical Assistance; (2) Investing in youth, education and skills; and (3) Encouraging tourism and cultural and natural heritage (European Commission, 2021a). Thematic Priority—Investing in youth, education and skills will contribute to the (2) Circular economy as it will support the activities that contribute to the development of the joint training for specific skills and competencies; re-training of workers from large traditional industrial systems; and enhancement of first-job experiences. Moreover, Thematic Priority—Encouraging tourism and cultural and natural heritage of the Programme contributes to (4) Sustainable agriculture and food production and (5) Biodiversity by supporting the activities that contribute to the development of niche tourism for sport, adventure, culinary, rural tourism; connecting tourism with agriculture and food processing industry; implementation of preservation action plans; prevention of risks for users of natural and cultural sites. From a broader viewpoint, the IPA CBC Programme

Table 10.2 Bilateral agreements between Serbia and Croatia and between Serbia and Bosnia and Herzegovina that can support the implementation of WBGA pillars (*Source* Authors interpretation of the data from the Archive of Ratified International Agreements and Other Acts, Legal Information System of the Republic of Serbia, Official Gazette RS [2023])

Border area	*WBGA pillars*	*No. of bilateral agreements*[1]
Serbia-Croatia	(1) Climate, energy, mobility;	6
	(2) Circular economy;	5
	(3) Depollution;	2
	(4) Sustainable agriculture and food	1
	production;	2
	(5) Biodiversity	16
	Not relevant for WBGA	
Serbia-Bosnia and Herzegovina	(1) Climate, energy, mobility;	20
	(2) Circular economy;	6
	(3) Depollution;	3
	(4) Sustainable agriculture and food	3
	production;	2
	(5) Biodiversity	33
	Not relevant for WBGA	

Serbia-Bosnia and Herzegovina 2021–2027 will directly contribute to (2)—Circular economy, (4) Sustainable agriculture and food production; and (5) Biodiversity and indirectly to (1) Climate, energy, mobility, and (3) Depollution. Comparing the results from the two Programmes, it can be concluded that the dominant pillar that can be supported through the CB projects is (2) Circular economy, followed by (1) Climate, energy, mobility, then (5) Biodiversity, (4) Sustainable agriculture and food production and (3) Depollution. Besides the CB Programmes, bilateral agreements also strengthen and specify cross-border cooperation. There are, in total, 84 bilateral agreements covering the case study area, namely 26 signed by relevant institutions of Serbia and Croatia (Table 10.2) and 58 signed by relevant institutions of Serbia and Bosnia and Hercegovina (Table 10.3). Evaluation of the bilateral agreements between Serbia and Croatia and between Serbia and Bosnia and Herzegovina categorise them following the objectives to determine if they can influence the implementation of the pillars of the Western Balkans Green Agenda (Table 10.4).

Of the 10 relevant agreements that can be used to support WBGA pillars in the cross-border area between Serbia and Croatia, 37% can address (1) Climate, energy, mobility; 31%—(2) Circular economy; 12%—(3) Depollution; 12%—(5) Biodiversity; and 6%—(4) Sustainable agriculture and food production. Of the 25 relevant agreements that can be used to support WBGA

[1] Some of the Bilateral Agreements can support more than one WBGA pillars.

Table 10.3 Bilateral agreements between Serbia and Croatia and between Serbia and Bosnia and Herzegovina that can support CPS policy area (*Source* Authors interpretation of the data from the Archive of Ratified International Agreements and Other Acts, Legal Information System of the Republic of Serbia, Official Gazette RS [2023])

CPS policy area	*No. of bilateral agreements*[2]	*No. of bilateral agreements*[3]
	Between Serbia and Croatia	*Between Serbia and Bosnia and Herzegovina*
1. Transport	6	10
2. Spatial planning, economic development, tourism and culture	2	8
3. Healthcare, long-term care and social inclusion	4	1
4. Education and training	1	2
5. Labour market and employment	3	5
6. Communication, broadcasting and information society	1	1
7. Environmental protection, natural resources management and climate change action	2	5
8. Civil protection and disaster management	2	4
9. Citizenship, justice and public security	2	4
Not relevant for CPS	8	28

pillars in the cross-border area between Serbia and Bosnia and Hercegovina, 80% can address (1) Climate, energy, mobility; 24%—(2) Circular economy; 12%—(3) Depollution; 12%—(4) Sustainable agriculture and food production; and 8%—(5) Biodiversity. Comparing the results' evaluation of the bilateral agreements between Serbia and Croatia and between Serbia and Bosnia and Herzegovina, it can be concluded that the most legal bases for influencing the implementation of the WBGA have pillars: (1) Climate, energy, mobility; and (2) Circular economy; followed by (3) Depollution; (5) Biodiversity; and (4) Sustainable agriculture and food production. Additional analysis is done to elaborate on legal frameworks that can support establishing and providing cross-border public services that can contribute to implementing Just Green Transition in the case study area.

[2] Some Bilateral Agreements can support more than one CPS policy area.

[3] Some Bilateral Agreements can support more than one CPS policy area.

Table 10.4 Bilateral agreements between Serbia-Croatia and Serbia and Bosnia and Herzegovina that can support WBGA pillars through CPS fields of intervention (*Source* Authors interpretation of the data from the Archive of Ratified International Agreements and Other Acts, Legal Information System of the Republic of Serbia, Official Gazette RS [2023])

CPS field of intervention	*WBGA pillars RS-CRO*	*No. of bilateral agreements RS-CRO*[4]	*WBGA pillars RS-BiH*	*No. of bilateral agreements RS-BiH*[5]
1.1. Public transport services	(1) Climate, energy, mobility; (2) Circular economy; (3) Depollution;	4	(1) Climate, energy, mobility; (2) Circular economy;	6
1.2. Services for transport infrastructure maintenance	(1) Climate, energy, mobility;	2	(1) Climate, energy, mobility;	3
1.3. Services at border crossing points	(1) Climate, energy, mobility; (2) Circular economy; (3) Depollution;	5	(1) Climate, energy, mobility;	5
1.4. Other transport services	(1) Climate, energy, mobility;	1	(1) Climate, energy, mobility; (2) Circular economy;	2
2.1. Spatial planning or sector policy planning	/	/	(1) Climate, energy, mobility; (3) Depollution;	3
2.2. Services supporting economic development	(2) Circular economy;	2	(1) Climate, energy, mobility; (2) Circular economy;	3
2.3. Services for culture and cultural heritage	/	/	(1) Climate, energy, mobility;	1

[4] Some Bilateral Agreements can support more than one CPS Field of Intervention, which can influence the implementation of WBGA pillars.

[5] Some Bilateral Agreements can support more than one CPS Field of Intervention, which can influence the implementation of WBGA pillars.

CPS field of intervention	*WBGA pillars RS-CRO*	*No. of bilateral agreements RS-CRO*	*WBGA pillars RS-BiH*	*No. of bilateral agreements RS-BiH*
2.5. Other services for spatial planning, economic development, tourism and culture	/	/	(4) Sustainable agriculture and food production;	1
5.3 Services for further qualification and life-long learning	(1) Climate, energy, mobility; (4) Sustainable agriculture and food production; (5) Biodiversity	2	/	/
6.1 Mail delivery, telephone or mobile phone services	(1) Climate, energy, mobility; (2) Circular economy;	1	/	/
6.3 Digital services	/	/	(1) Climate, energy, mobility; (2) Circular economy;	1
7.1. Protecting / restoring and managing terrestrial freshwater water bodies (blue infrastructures), estuaries and coastal waters	(2) Circular economy; (3) Depollution; (5) Biodiversity	1	(1) Climate, energy, mobility; (3) Depollution;	2

(continued)

Table 10.4 (continued)

CPS field of intervention	*WBGA pillars RS-CRO*	*No. of bilateral agreements RS-CRO*	*WBGA pillars RS-BiH*	*No. of bilateral agreements RS-BiH*
7.2. Restoring, protecting and managing valuable terrestrial ecosystems or landscapes and for developing green infrastructures, incl. services for risk prevention and climate change resilience	(2) Circular economy; (3) Depollution; (5) Biodiversity	1	(1) Climate, energy, mobility; (3) Depollution; (5) Biodiversity	3
7.3. Improving resource efficiency and for promoting a low-carbon economy or a greening of the society	(2) Circular economy; (3) Depollution; (5) Biodiversity	1	/	/
7.4. Solid waste	(2) Circular economy; (3) Depollution; (5) Biodiversity	1	/	/
7.5. Sewage water collection / treatment & drinking water	(2) Circular economy; (3) Depollution; (5) Biodiversity	1	/	/
7.7. Other services for environmental protection, natural resources management and climate change action	(1) Climate, energy, mobility; (2) Circular economy; (3) Depollution; (5) Biodiversity	2	(1) Climate, energy, mobility; (3) Depollution; (5) Biodiversity	4

CPS field of intervention	*WBGA pillars RS-CRO*	*No. of bilateral agreements RS-CRO*	*WBGA pillars RS-BiH*	*No. of bilateral agreements RS-BiH*
8.1. Services for firefighting and for assistance in accidents	(1) Climate, energy, mobility; (5) Biodiversity	1	(1) Climate, energy, mobility; (3) Depollution; (5) Biodiversity	3
8.2. Services for flooding management	(1) Climate, energy, mobility; (5) Biodiversity	1	(1) Climate, energy, mobility; (3) Depollution; (5) Biodiversity	3
8.3. Managing large-scale incidents and major disasters	(1) Climate, energy, mobility; (5) Biodiversity	1	(1) Climate, energy, mobility; (3) Depollution; (5) Biodiversity	3
8.4. Other services for civil protection and disaster management	(1) Climate, energy, mobility; (5) Biodiversity	2	(1) Climate, energy, mobility; (3) Depollution; (5) Biodiversity	4

ESPON CPS Study created the conceptual framework for cross-border public service (CPS) provision, as no analytical concept has existed. Flexible criteria rather than a formal definition for CPS have been developed to remain flexible and adhere to a large variety of possible CPS in Europe. In general, all criteria have to comply, but it is suggested that the criteria be applied less strictly for services at EU external borders.[6] (ESPON, 2018) The main Policy Areas of CPS provisions are (1) Transport; (2) Spatial planning, economic development, tourism and culture; (3) Healthcare, long-term care and social inclusion; (4) Education and training; (5) Labour market and employment; (6) Communication, broadcasting and information society; (7) Environmental protection, natural resources management and climate change action; (8) Civil protection and disaster management; and (9) Citizenship, justice and public security (ESPON, 2022). Each Policy Area, together with its specific Fields of Interventions,[7] needs a normative base that allows the establishment of the CPS and defines the rules for services provision. Evaluating the bilateral agreements between Croatia and Serbia and between Serbia and Bosnia and Herzegovina, they were categorised according to whether they provide the legal bases for the establishment and/or provision of cross-border public services, assigning specific Policy Areas and Fields of Interventions to each agreement. Each field of intervention is then evaluated to determine if it can influence the implementation of the pillars of the Western Balkans Green Agenda.

Of the 18 relevant agreements that can be used to support CPS Policy Areas between Serbia and Croatia, 33% can address—(1) Transport; 22%—(3) Healthcare, long-term care and social inclusion; 16%—(5). Labour market and employment; 11%—(2) Spatial planning, economic development, tourism and culture; 11%—(7) Environmental protection, natural resources management and climate change action; 11%—(8) Civil protection and disaster management; 11%—(9) Citizenship, justice and public security; 5%—(4) Education and training; and 5%—(6) Communication, broadcasting and information society. Of the 30 relevant agreements that can be used to support CPS Policy Areas between Serbia and Bosnia and Herzegovina, 33% can address—(1) Transport; 27%—(2) Spatial planning, economic development, tourism and culture; 17%—(5) Labour market and employment; 17%—(7) Environmental protection, natural resources management and climate change action; 13%—(8) Civil protection and disaster management, 13%—(9) Citizenship, justice and public security; 7%—(4) Education and training; 3%—(3) Healthcare, long-term care and social inclusion; and 3%—(6) Communication, broadcasting and information society. Comparing the results evaluation of the

[6] Revised the list of CPS characteristics you can find in ESPON Cross-Border Public Services 2.0.

[7] Overview policy areas and fields of interventions in which CPS may exist you can find in ESPON Cross-Border Public Services Targeted Analyses—Final report—Scientific report.

bilateral agreements between Serbia and Croatia and between Serbia and Bosnia and Herzegovina, it can be concluded that the most legal bases for supporting the establishment or provision of the CPS have the Policy Areas of (1). Transport; (2). Spatial planning, economic development, tourism and culture; and (5). Labour market and employment; followed by (7). Environmental protection, natural resources management and climate change action; (8). Civil protection and disaster management; and (9). Citizenship, justice and public security; followed by (3). Healthcare, long-term care and social inclusion; (4). Education and training; and (6). Communication, broadcasting and information society.

There are 17 relevant CPS Fields of Interventions that can influence WBGA pillars in the cross-border area between Serbia and Croatia; 11 address (1) Climate, energy, mobility; 11—(5) Biodiversity; 10—(2) Circular economy; 7—(3) Depollution; and 1—(4) Sustainable agriculture and food production.

There are 16 relevant CPS Fields of Interventions that can influence WBGA pillars in cross-border areas between Serbia and Bosnia and Herzegovina; 15 address (1) Climate, energy, mobility; 8—(3) Depollution; 6—(5) Biodiversity; 4—(2) Circular economy; and 1—(4) Sustainable agriculture and food production.

Some of the examples of possible CPS that have legal bases in the bilateral agreements and can be used for supporting the implementation of the WBGA are Protection and response in emergency situations, Planning and implementation of measures for protection against floods, earthquakes, fires, and environmental pollution; Hydrotechnical and dredging works on the cross-border waterway with technical documentation, environmental impact assessment, construction, and use permit; Synchronised missile action on hail clouds along the borderline.

Comparing the results of the evaluation of the bilateral agreements between Serbia and Croatia and between Serbia and Bosnia and Herzegovina, it can be concluded that the most legal bases that can support WBGA Pillars through CPS Fields of Intervention have (1) Climate, energy, mobility; followed by (5) Biodiversity; (3) Depollution; and (2) Circular economy; followed by (4) Sustainable agriculture and food production.

10.4 Conclusions

Cross-cutting analyses of the results elaborated in the Case study show that Cross-Border Structures, Cross-Border Programmes, and Bilateral Agreements set the framework that should be utilised by national, regional, and local authorities as well as all other relevant stakeholders involved in or able to be involved in implementing the Western Balkan Green Agenda. Cross-border cooperation can play a critical role in supporting a Just Green Transition in the Western Balkans alongside the EU integration process (Berisha, 2025a, 2025b), especially considering cross-border public services that can be crucial in facilitating and supporting the process of achieving the goals of the

Western Balkans Green Agenda. Moreover, to implement the Sofia Declaration, the RCC is invited to be the regional CPS provider, namely, develop the Action Plan for the Western Balkans Green Agenda and monitor its implementation. The Action Plan also assigned to existing regional bodies the provision of regional CPS, for example, to the Transport Community Permanent Secretariat, Regional Working Group on Environment, Southeast Europe Biodiversity Taskforce, and RCC, and envisaged the establishment of regional CPS bodies like Regional Working Group, Expert Advisory Working Group on GAWB, and Decarbonisation Committee. Regional CPS provision in the Western Balkans needs to be further analysed in more detailed way to capitalise on existing regional bodies. An inventory of the operational cross-border and transnational CPS and in-depth analyses of the Regional Agreements that provide the normative basis for regional CPS provision should be done to support the Just Green Transition in the Western Balkans fully.

References

AEBR Balkans. (2021). *AEBR Balkans database on cross-border structures in the Balkans*. Belgrade.

Berisha, E. (2025a). *Moving towards an inclusive Green Agenda in the Western Balkans*. Foundation for European Progressive Studies (FEPS). https://doi.org/10.13140/RG.2.2.12092.76162

Berisha, E. (2025b—in this volume). Green EU integration—the role of the green agenda in accelerating just green transitions in the Western Balkans. In E. Berisha, J. Moodie, L. Allkja & M. Jeftić (Eds.), *The Palgrave handbook of just green transitions in the Western Balkans and beyond*. London.

ESPON. (2018). *Cross-Border Public Services (CPS) targeted analysis, final report, scientific report*. www.espon.eu

ESPON. (2022). *Outreach and uptake of territorial evidence // Cross-Border Public Services 2.0 (CPS 2.0)*. www.espon.eu

EU. (2020). *Territorial agenda 2030—A future for all places*. Germany. https://territorialagenda.eu/wp-content/uploads/TA2030_jun2021_en.pdf.

European Commission. (2019). Communication from the Commission to the European Parliament, the European Council, the Council, the European Economic and Social Committee and the Committee of the Regions—The European Green Deal—COM/2019/640 Final. https://eur-lex.europa.eu/legal-content/EN/TXT/?qid=1588580774040&uri=CELEX:52019DC0640

European Commission. 2020. 'Commission Staff Working Document—SWD(2020) 223 Final—guidelines for the implementation of the green agenda for the Western Balkans—Accompanying the communication from the commission to the European Parliament, the council, the European economic and social committee and the committee of the regions—An economic and investment plan for the Western Balkans {COM(2020)641final}'. https://www.rcc.int/news/383/rcc-publishes-study-on-climate-change-in-the-western-balkans-alarming-increase

European Commission. (2021). Action plan for the implementation of the Sofia declaration on the green agenda for the Western Balkans 2021–2030.

European Commission. (2021). Report from the commission to the European Parliament, the council, the European economic and social committee and the committee of the regions—EU border regions: Living labs of European integration—COM(2021) 393 Final. Brussels. https://ec.europa.eu/regional_policy/en/information/publications/reports/2021/eu-border-regions-living-labs-of-european-integration

European Commission. (2021a). EN annex to the commission implementing decision on the financing of the cross-border cooperation programme Serbia-Bosnia and Herzegovina for 2021–2027 action document for: Cross-Border cooperation programme Serbia-Bosnia and Herzegovina for 2021–2027.

European Commission. (2021b). Interreg-IPA Programme Croatia—Serbia 2021–2027.

European Commission. (2023). *Territorial cohesion*. https://ec.europa.eu/regional_policy/policy/what/territorial-cohesion_en.

Lindblad, Sverker. (2023). *Territorial cohesion-the story*.

Official Gazette, R. S. (2023). Archive of ratified international agreements and other acts, legal information system of the Republic of Serbia.

Pinnavaia, L., & Berisha, E. (2021). The role of cross-border territorial development. Evidences from Albania. In E. Berisha, G. Cotella, & A. Solly (Eds.), *Governing Territorial development in the Western Balkans*. Advances in Spatial Science. Springer, Cham. https://doi.org/10.1007/978-3-030-72124-4_15

RCC. (2020). Sofia Declaration on the Green Agenda for the Western Balkans.

RS. (2022). *Стратегија интегрисаног управљања границом у Републици Србији за период 2022—2027. године*. Belgrade. RS. https://rsjp.gov.rs/upload/1660915406108_Strategija_IBM_SRP_20.%2004.2022_.pdf

Solly, A., & Berisha, E. (2021). Towards the territorialisation of EU cohesion policy? The case of EUSAIR. In E. Berisha, G. Cotella, & A. Solly (Eds.), *Governing territorial development in the Western Balkans*. Advances in Spatial Science. Springer, Cham. https://doi.org/10.1007/978-3-030-72124-4_16

CHAPTER 11

Just Green Transition of Coal-Mining Cities

Tamara Maričić and Marijana Pantić

11.1 Introduction

The concern for the future of our planet has grown worldwide in recent decades, especially since 1987 and the Brundtland Report, which introduced the concept of sustainable development as inevitable for ensuring a long-term equilibrium between humankind's progress and the environment, thus protecting future generations. This strategy includes efforts to reduce the use of non-renewable energy resources and their various negative environmental impact. The "just green transition" (JGT) lies among recent pro-environmental and sustainable concepts that pursue new development directions—a shift towards carbon–neutral economies and societies.

The European Green Deal (EGD) (European Commission, 2019) is an ambitious and comprehensive plan set forth by the European Commission to address climate change and environmental sustainability. It encompasses various policies and initiatives aimed at transforming the European Union (EU) into a climate-neutral economy by 2050. This includes reducing greenhouse gas emissions, investing in clean energy technologies, promoting sustainable agriculture and biodiversity, and fostering green innovation and job creation. The transition towards low-carbon development in Serbia slowly intensifies, as the country moves forward to the EU membership and full implementation of climate acquis (Pantić et al., 2024). This should reduce the economic competitiveness of carbon-intensive energy production, which

T. Maričić (✉) · M. Pantić
Institute of Architecture and Urban & Spatial Planning of Serbia, Belgrade, Serbia
e-mail: tamara@iaus.ac.rs

E. Berisha et al. (eds.), *The Palgrave Handbook of Just Green Transitions in the Western Balkans and Beyond*,
https://doi.org/10.1007/978-3-031-95075-9_11

dominates in Serbia. Nevertheless, some of the leaders in renewables—such as Germany—show how the uncertainties in the global market and gas shortages due to geopolitical factors, e.g. the current war in Ukraine, can result in a return to investments in coal and shelving climate goals.

Green transition requires significant efforts, especially in fragile coal-mining regions focused on the mono-functional production and processing of non-renewable energy resources. Since most citizens are employed in and dependent on the mining industry, a sharp transition to other economic activities could result in unemployment and poverty. A gradual transition is therefore crucial. The key challenges revolve around: (1) political (the decision and commitment to change the energy policy strategically), (2) financial (available resources, loans etc.), (3) technological & organizational (the possibility of a quick and easy shift to renewable energy sources), (4) societal (the community's willingness to move on from traditional employment models; brain drain) and (5) implementation obstacles (legislation, strategic plans and other development documents). After analysing the main aspects of JGT in mining regions, we will present different examples of regions with a long tradition in coal mining—Germany, the UK, and Poland. The study will then focus on Serbia as a representative case of a post-socialist transitional Western Balkans society, that largely relies on coal as an energy resource, and has many economic (mono-functional industry) and environmental problems in coal-mining regions. Interviews of selected stakeholders helped in identifying the most suitable paths to JGT of coal-mining cities in Serbia.

11.2 Literature Review

11.2.1 European Green Deal and Just Green Transition

Mining regions and cities play a crucial role in JGT, as they often significantly rely on non-renewable resources, thus generating large pollution. To ensure a successful green transition, sustainable mining practices including minimizing resource extraction should become a priority (Nurdiawati & Urban, 2021; Zheng et al., 2020). Their development should focus on economic diversification by encouraging non-mining related industries, attracting investments in renewable energy projects, and supporting the retraining and employment of local workers in green industries (Huang et al., 2020). This will help reduce dependency on mining and create a more resilient and sustainable economy for the region. The JGT shift holds economic, environmental and social risks, in which we should try to eliminate inequalities existing in the current energy regime (Carley & Konisky, 2020; Healy & Barry, 2017; Jasanoff, 2018; Winkler, 2020). As Pianta and Lucchese (2020) have noticed, the EGD already shows some weaknesses such as lacking a vision of a just, post-carbon economy; and overambitious goals that require more resources than available—a problem recognized also by Claeys et al. (2019) and Storm (2020); and insufficient implementation tools. A wider range of policies for the "green" industry is needed to achieve a balance between environmental quality, development and social satisfaction. According to Sharpe and Martinez-Fernandez (2021), the transition to an ecologically sustainable economy brings significant changes in the world of work. These changes encompass the emergence of new green jobs and industries, as well as varying degrees of transformation in existing roles, alongside the gradual phasing out of jobs linked to carbon-intensive activities. In essence, achieving a just transition requires the provision of social protection for workers impacted by decarbonization, alongside the investments in their communities. Filipović et al. (2022) observe that the social aspect is not adequately considered in the EGD and JGT. ITUC (2019) underlines the implementation of measures such as income security and job transition support for affected workers, and their requalification. The shift from the mining to the renewable energy sector is considered as one option (Sharpe & Martinez-Fernandez, 2021). However, it remains uncertain whether the same region is suitable for both sectors.

11.2.2 Just Green Transition in Mining Regions

The energy transition is both an opportunity and a threat to the mining regions in Europe (Janikowska & Kulczycka, 2021). The current challenges facing mining regions stem from divergent processes that result in long-term instability and hinder efficient resource utilization (Danciu et al., 2023). According to Danciu et al. (2023) and Rising et al. (2021), mining areas lack coordination, cooperation, and comprehensive spatial planning, particularly in regions that have lost economic significance despite being the focus of numerous national and international support initiatives. Spasić et al. (2009) also emphasize the importance of synchronizing activities in planning the development of mining regions. García-García et al. (2024) estimated the effects on net employment and land availability of restructuring towards renewable energy sources in the rural mining areas of León (Spain). Their results point out the decline and ageing of mining-induced depopulation, loss of industrial fabric, and unemployment. If the economy and overall population size of a town are diverse, the impacts on economic, social and demographic aspects are more subtle. While analyzing post-mining regions in Upper Silesia (Poland) Jonek-Kowalska & Turek have also concluded that cities with more diversified economic activity and sources of income were in a better economic condition. Evans and Phelan (2016) investigated Australian coal-mining regions, noticing that the pressure to transition and restructure mining regions does not come only from climate change-induced policies but also from the decline of global coal markets. The example from the USA also shows that the coal industry has plunged due to the increasing extraction of oil and gas and for fast development of technologies for renewable energy production (Mayer, 2018). China still lacks a JGT as a policy tool, leading to powerful stakeholders safeguarding their own interests at the expense of the least powerful actors who bear the costs of transition. According to Zhang and Zhang (2024), JGT tools could mitigate the injustices associated with phasing out excess coal production. Kolde and Wagner (2022) investigated on Rhineland in Germany. The results pointed out that JGT must be secured through different action areas, from supporting labour in the mining industry, and citizen initiatives to preservation of land that is easily consumed. Yarkova et al. (2021) conducted another study based on interviews, focusing on examples from Bulgaria, specifically the Marishki Basin, to identify determinants of energy transition. This case highlighted the necessity of incorporating the concept of a JGT into local spatial plans.

11.3 Methods and Materials

This research attempts to contribute to the vast body of literature on JGT of mining regions, especially highlighting challenges posed to insufficiently analyzed coal-mining examples in the post-socialist societies in the Western Balkans. It starts with the comparative method as one of the fundamental analytical tools (Collier, 1993) by presenting the three transition examples of large coal regions in the UK, Germany (Ruhr) and Poland (Silesia). Then, our research concentrates on the case of Serbia, and especially the Kolubara and Kostolac coal-mining regions. A case-study approach that has been applied is understood as a multi-perspectival analysis (Tellis, 1997). It should capture the complexity of a single case, investigated with a multitude of methods and be contemporary, thus bridging the gap between quantitative and qualitative methods in the social sciences (Johansson, 2007). As a source of quantitative data for the case of Serbia, we have used national statistical documents, spatial plans for those regions and the latest reports. For a better understanding of the possibilities and obstacles for the JGT of mining regions in Serbia, we conducted structured interviews (Firmin, 2008) with the key stakeholder representatives in April 2024. All foreseen questions were asked to all respondents without exception; this achieves high reliability, with the limitation that the planned questions do not entirely reveal the phenomenon. Due to the distinguishing characteristics of each interviewee, some questions have been slightly tailored to suit each of their sectors and fields (spatial planning, education, energy, NGO, and local government).

11.4 Results and Discussion

11.4.1 Examples of Transition

There are examples of transitions in mining areas that occurred long before the term "JGT" was introduced, i.e. Germany and the UK. The primary driver in those cases was the closure of mining sites due to the exhaustion of extraction resources and/or saturation of the mining market that led to the profitability decline (i.e. cheap imports; substitutes such as crude oil and natural gas). The improvement of the deteriorated environments gained importance only in the second transitional phase. More recent examples are related to the European Green Deal (EGD) policy. Both types of examples are beneficial: the former for providing insights into realistic outcomes after many decades of transitioning performance, and the latter for offering measures that are globally considered within the European context under the EGD policy. Germany and the UK have adopted ambitious net-zero emissions targets for 2050, and a target of 65–68% GHG emissions reductions by 2030. Poland is, however, more cautious about making commitments as it relies much more on fossil fuels compared to other EU Member States. "Poland's Energy Policy until 2040" targets a 30% reduction of GHG emissions by 2030, while further negotiations under the EU pressure to raise its rate of decarbonization are still ongoing.

11.4.2 Germany—The Ruhr District

The Ruhr District is a densely populated urban area in western Germany comprising 11 cities and four counties. It spreads over 4,435 km^2, with a population of 5.3 million (Regionalverband Ruhr, 2006). Historically, this region was a prominent industrial hub with coal mines and steel mills. The peak of hard coal production was in 1957, with over 473,000 (4/10 of Ruhr's entire workforce) workers in the coal-mining sector, and it came to zero in December 2018 (Arora & Schroeder, 2022). The region underwent a successful transition over the past decades from coal, iron and steel to a knowledge economy (ibid.), which positions the Ruhr area as a beacon of success, serving as a potential model for other regions worldwide (Kretschmann & Nguyen, 2020; Lee et al., 2014). According to Arora and Schroeder (2022), Ruhr received great support for economic diversification and usage of renewables already in the 1980s. The government provided large financial support for the protection of individuals, communities and industry from such a huge socio-economic shock (ibid.). As a result of de-industrialisation, cities in the Ruhr District managed to offset job losses in the mining, coal, and steel sectors by attracting employment opportunities in the tertiary sector (Bronny et al., 2004). Instead of providing subsidies to those who lost their jobs in the industry, they focused on nurturing architectural and industrial heritage to develop the cultural, tourism, and hospitality sectors. In the high education sector, TH Georg Agricola University of Applied Sciences in Bochum stands out. Established in 1816, it offers practice-oriented education in the field of mining. Over time, it focused on transitioning mining regions to non-mining economies (FZN, 2024) with the ambition to find ways to remediate abandoned mines and repurpose them. The challenge of landscape restoration has been successfully completed. The mining waste heap Hoheward stands as one of the largest in the Ruhr area. It has been transformed into a landscape feature seamlessly integrated into its surroundings, meeting ecological and recreational needs (Hofmann & Winter, 1991). Additionally, efforts have been made to enhance leisure opportunities by incorporating pathways, viewpoints, and artistic installations. Another example of landscape fitting is Lake Lanstrop in Dortmund (stemming from mining activities in a coal mine that closed in 1985), which has been reimagined as a nature reserve and popular recreational spot (Bell et al., 2000). Various initiatives have emerged in the domain of renewable energy production. These include the installation of photovoltaic systems on mining dumps and factory roofs, harnessing heat from mine water for building heating or enhancing biomass production for energy generation and installation of wind turbines atop dump hills. Furthermore, there is the extraction of methane from coal beds for energy production. Pump-storage power plants utilize the topography of dump hills and underground mine structures, while geothermal energy production further diversifies the energy portfolio in the area (Kretschmann & Hegemann, 2012). Coal mining has left behind a plethora of architectural landmarks, including extraction facilities and production plants, which are integral to the industrial heritage and culture. Striving to preserve this heritage, mitigating the effects of ageing on these structures through material research and conservation methods was conducted. The Zollverein Coal Mine Industrial Complex in Essen stands as a prime example, designated as a UNESCO World Heritage Site (UNESCO, 2002). Its renovation revived regional identity (Lee et al., 2014). Museums, restaurants and entertainment venues of this complex attract visitors worldwide. The complex includes amusement parks, swimming pools, ice-skating fields, restaurants, cafes, and museums. A pivotal initiative in showcasing the industrial heritage of the region is the "Route der Industriekultur" (Industrial Heritage Trail), a 400 km circular route encircling the Ruhr area (Route Industriekultur, 2020).

11.4.3 United Kingdom

The United Kingdom (UK) is an example of a country that went through a coal phase-out some decades ago, which was motivated by economic reasons. Coal mining was a significant driver of economic development—before WWI, around 1.1 million miners worked in more than 3,000 coal mines in the UK (Tahseen et al., 2022). The shift away from coal started already in the late 1970s, and the majority of coal mines were closed during the 1980-90s. In 2023 only 360 workers remained in eight coal mines. The transition was difficult for local mining communities and resulted in high levels of poverty and unemployment, higher mortality rate, health issues retired miners due to a contaminated working environment, less youngsters going to higher education, outmigration increase (Fothergill et al., 2024; Tahseen et al., 2022). The country is still dealing with the consequences of the phase-out period. A report on 40 years of just transitions in the UK (Rising et al., 2021) outlines mistakes from the 1980s when interventions in deprived regions focused mainly on economic recovery (when a mine was about to close, they offered workers early retirement or shifting to another mine). In the 1990s, regeneration policies were developed on a regional scale, including private, public and third sector partnerships, plus community engagement. From 1998 significant coalfield policies developed to tangle serious social problems: National Coalfields Programme (1996), Coalfields Regeneration Trust (1999) and Coalfields Enterprise Fund (2004). However, since 2010 the government backed away and devoluted regional development to local actors (ibid.). The latest analysis by Fothergill et al. (2024) highlights the major statistical evidence on the situation in former coal regions of England, Scotland and Wales: the population is older and slower growing compared to other UK regions and of poor health conditions; although the number of jobs increased significantly in the last decade—there are still not enough quality jobs and businesses; extensive deprivation; many people receive out-of-work benefits. Decarbonisation agendas lacked social protection of impacted workers and communities. Settlements on coalfields have now taken on new roles. Yorkshire coalfield has become the capital of warehousing, along with Lancashire and Nottinghamshire (over 175,000 people in the UK employed in warehousing—similar to the number of miners before the 1984–85 strikes). Many other coalfields became dormitories for nearby cities (Lothian coalfield for Edinburgh, Welsh Valleys for Cardiff). Although there has been substantial growth in jobs, their quality is not satisfying enough to prevent brain drain, so the worklessness and reliance on welfare benefits endure.

11.4.4 Poland Silesia Region

Poland presents a good example of an economy still reliant on coal, rooted deeply in strong mining traditions (Krzywda et al., 2021). Despite the imperative to shift from coal, Poland remains the largest hard coal producer among current EU member states (Szpor & Ziółkowska, 2018). The Silesian Voivodeship houses 4,333,100 residents and is emblematic of mining together with Katowice, the regional capital (279,100 inhabitants in 2023) (Statistical Office of Katowice, 2023). The restructuring of the coal-mining sector, intertwined with broader political and economic reforms since 1989, has been marked by challenges. Poland's coal production experienced steady growth until the late 1970s. However, this growth was largely driven by incentives within the Eastern Bloc, leading to inefficiencies in the sector due to central planning and nationalization (Blaschke, 2002 in Szpor & Ziółkowska, 2018). By the 1980s, declining efficiency and limited access to global markets left the coal sector in a catastrophic situation. The transformation from 1989 to 2015 marked a shift towards a free market economy, with the restructuring of large industrial companies and the privatization of weaker entities. This restructuring involved reducing the workforce, liquidating unprofitable collieries, and aligning production with demand. The 1998 Mining Law redefined "mining communes" and provided them with higher financial incentives, including a percentage of personal income tax and preferential loans for economic development. However, approximately 70,000 miners remain (Janikowska & Kulczycka, 2021). Despite private–public partnerships (Szpor & Ziółkowska,

2018), challenges remained as many communes lacked the expertise and financial resources to utilize these funds effectively. Some mining communes successfully revitalized, creating new job opportunities through initiatives like the "New Gliwice" Education and Business Centre. Poland's accession to the EU enabled additional financial opportunities, leading to the establishment of industrial and technology parks in Silesia (ibid.). The rapid development of this region was also intertwined with the advancement of social rights and the rise of workers' movements, fostering the emergence of a new cultural identity deeply rooted in mining-related labour (Swadźba, 2001). Throughout the transition period, families with longstanding mining traditions residing in housing estates owned by mines witnessed the erosion of traditional social bonds, accompanied by economic hardships and the fragmentation of group identity (Mandrysz, 2011). Local authorities introduced measures such as miners' early retirement, redundancy payments, welfare allowances, and retraining courses. However, the economic situation of households receiving these benefits steadily deteriorated after leaving their jobs, largely due to a lack of knowledge or inclination to invest the received funds (Szpor & Ziółkowska, 2018).

Urbanization surged around burgeoning industrial hubs, notably the Silesian conurbation, which saw the rapid modernization of existing villages (such as Gliwice and Bytom) and the transformation of new settlements into cities (like Katowice and Zabrze) (Gwosdz, 2013 in Szpor & Ziółkowska, 2018). Nowadays, one of the most prominent suburban neighbourhoods in Katowice is Nikiszowiec—renewed as an example of the industrial and architectural heritage. With the sunsetting of fossil fuels and heavy industry, many relics of the city's industrial past have been transformed into galleries, cafes and cultural spaces. Remnants of the past—abandoned mining shafts, shuttered porcelain factories and other industrial edifices—have been transformed into contemporary museums, galleries and cultural spaces. The Silesian Museum architectural complex marries revamped heritage brick buildings with ultra-modern glazed cubes. In the 1950s, Silesian Park—a Provincial Park of Culture and Recreation in Chorzów—was established on the severely degraded post-industrial areas of Chorzów and Katowice. Through extensive earth-moving efforts, millions of cubic metres of soil were rearranged, and nearly 3.5 million trees and shrubs of 70 different species were planted.

11.4.5 Serbia

Similar to mining regions in the Western Balkans, coal-mining regions in Serbia are hindered by pollution, outdated technologies and weak rule of law. They are of great significance, as Serbia's energy production and consumption rely on coal. Fossil fuels participated in the country's primary energy production with 73.67% in 2022 (SORS, 2024a). The coal alone had the largest share, of as much as 61.68%. Most of Serbia's coal resources are, however, low-calorie lignite (6–12.5 MJ/kg), predominantly used for electricity production. The important indicator for analysis is workers' salaries. In the mining sector, including the subsector of coal mining, average monthly earnings have been almost 60% higher than the national average in 2019 (SORS, 2024b). This steadily decreased and dropped to 36% higher for mining and 28.5% higher than the national average for coal mining in 2023. A similar trend is observed for the electricity supply sector, where income dropped from 64% higher to around 28% higher than the national average in the same period (ibid.). In the meantime, some other sectors, such as information and communication, became much more profitable. Kolubara and Kostolac are the largest lignite open-pit coal-mining and carbon-intensive industrialized regions in Serbia. Along with mining, both regions have related sectors of mining supply and power generation systems (thermal power plants). Kolubara coal region spreads over 585 km2, with 83.259 inhabitants (SPKLB, 2017). In the Kostolac coal region lives 76,300 people on around 543.5 km2 (SPSPAKCB, 2013). In both mining regions lives 2.2% of the national population, and they occupy 1.45% of Serbia's territory. Strategic documents dealing with the planning and arrangement of these areas are the Spatial plan of the exploitation area of the Kolubara lignite basin (SPKLB 2017, amended in 2024) and the Spatial plan of the special purpose area of the Kostolac coal basin (SPSPAKCB 2013, amended in 2018). Relevant to note, amendments proposed more usage of renewables and provided legal bases for constructing wind and solar power plants. The majority of municipalities in Kolubara and Kostolac regions are among the top-ranked in terms of income in the country. The municipalities with the dominant share of jobs in coal mining and/or electricity supply: Lajkovac, Lazarevac and Obrenovac in Kolubara region; Požarevac and Kostolac in Kostolac region are also the ones whose whole territory or its major part belong to a coal region (elaborated based on SORS, 2023). Data presented in Table 11.1 indicate that those municipalities, in general, have a more favourable trend of significant indicators (employment, unemployment, salaries). All dominantly mining municipalities (in terms of spatial coverage), except Kostolac, have higher employment than the average of the country. Unemployment is lower than the national average (64/1,000) in all municipalities except Aranđelovac (78/1,000) and Kostolac (98/1,000). Požarevac has the best unemployment ratio (29/1,000) and Lazarevac has the best employment ratio (422/1,000). UNDP (2020) has identified their local communities as potentially the most vulnerable to the effects of the green transition process, mostly because the majority of employed works in mining and related jobs.

Coal production and electricity generation in both regions are concentrated in Elektroprivreda Srbije (EPS), which was transformed in 2023 from a public company to a non-public joint stock company. Between 2019 and 2023, the number of workers in organizational unit Open cast mines decreased by 7.8%, but the number of workers in organizational unit Thermal power plants (TPPs) rose by 25% (EPS, 2024). A total of 15,769 workers were employed at

Table 11.1 Key population and employment data for the principal coal municipalities (2022) (*Source* SORS [2023])

			Kolubara coal region					*Kostolac coal region*		
		Serbia	*Aranđelovac*	*Lajkovac*	*Lazarevac*	*Obrenovac*	*Ub*	*Požarevac (entire)*	*Kostolac (entire)*	*Veliko Gradište*
Population		6,664,449	41,301	13,846	55,366	69,079	25,807	56,754	12,117	15,521
Employed per 1000 inhabitants		347	334	**347**	**422**	**383**	312	**350**	337	245
Unemployed per 1000 inhabitants (31.12.2022)		64	78	**39**	**34**	**38**	**33**	**29**	94	**43**
Average net salaries per employee (RSD)	2022	74,933	64,296	74,609	**80,854**	71,909	63,092	70,147	**77,603**	56,457
	2021	65,864	57,545	**70,488**	**76,856**	65,472	58,208	**64,546**	**75,663**	49,546
	2020	60,073	52,641	**67,763**	**73,717**	**60,197**	53,743	**60,311**	**73,269**	46,009

the coal mines and TPPs in the Kolubara and Kotolac coal regions in 2023 (a decrease of 2.2% compared to 2019). Transition to a low-carbon economy has a direct impact on those workers. As the average household in Serbia has 2.55 members, the existence of more than 40,200 people will be threatened due to the potential loss of mining-related jobs. However, the number of affected is much higher and includes other workers and the communities around the mines.

Research has been carried out for some time on the possibility of using wind and solar energy for electric power generation on recultivated land of Serbian coal mines (Pavlovic et al, 2021). In Kostolac, solar plant "Petka" with 9.75 MW was completed and released into trial operation in July 2025, and the wind park with 66 MW should start producing electricity in 2025. Amendment to Spatial Plan of the Kostolac Coal Basin (SPSPAKCB, 2018) foresaw three more wind parks (in Drmno, Ćirikovac and Klenovnik) and another solar power plant on "Srednje Kostolačko Ostrvo". Serbian Government is trying to keep up with the new decarbonisation tendencies and has officially decided to construct a solar power plant instead of coal-fired TPP Kolubara B (along with another one in the area of Kolubara A) by adopting an amendment to the Spatial Plan of Kolubara Coal Basin in March 2024. That is a significant change, as EPS started with the construction of TPP Kolubara B of 2 × 350 MW already in the mid-1980s, paused it from 1992–1996 due to sanctions by the UN Security Council and the disintegration of SFR Yugoslavia (Maričić, 2006) and invested USD 380 million by 2012 (EPS, 2012). After EPS signed a contract with Power Construction Corp. of China Ltd. for this investment worth 385 million euros in 2020 (to start electricity production in 2024), the final twist happened in 2021 (before the prepared Spatial plan of specific purpose area for construction of TPP Kolubara B has been adopted) with canceling construction of TPP Kolubara B as an attempt to solve the numerous problems plaguing Serbia's energy sector.

Some important strategic documents regulating energy sector have been adopted in 2024: Energy development strategy of the Republic of Serbia 2040 with projections up to 2050, Integrated National Energy and Climate Plan (INECP) for the period up to 2030 with a vision to 2050, and the Law on amendments to the Law on Energy; while Spatial plan of the Republic of Serbia 2035 (public display ended in May 2021, public hearing held in June 2021) and "Just transition diagnostics" with an Action Plan are still on-hold. Those documents concentrate much more on achieving energy independence, and implementation of the green energy transition while ensuring the stability of the energy system and security of supply, mainly by increasing the share of renewable sources in electricity production to 45.2%, reducing emissions of harmful gases by 40.3% in 2030 compared to 1990, as well as a significant increase in energy efficiency (INECP, 2024). Amendments to the Law on Energy lifted the moratorium on the construction of nuclear power plants in Serbia, which was introduced in 1989. INECP and Energy development strategy of RS both considered the possibility of using nuclear energy.

At the same time, the Government continues investing in traditional energy production, and new 350 MW Block 3 of TPP Kostolac has been opened in December 2024. The new Spatial plan for the Kostolac mining region is in preparation and foresees opening of another open-pit lignite mine.

11.5 Interview Results and Discussion

We conducted structured interviews with selected relevant stakeholder representatives in April 2024. This included one representative of (1) the local self-government Lazarevac (Kolubara mining region); (2) EPS, as the largest company in the field of electricity production, supply and trade; (3) the IAUS as an expert institution in the field of spatial planning of mining regions; (4) the University of Belgrade, Technical faculty in Bor (mining region in East Serbia) as educational sector; and (5) the Association of Young Researchers Bor as NGO representative. The interviews showed that the respondents were familiar with the JGT concept and the importance of decarbonisation, especially if they have been directly involved in related initiatives. They estimated that recently increasing political will, experienced through the adoption of relevant laws, strategies and plans, is an optimistic sign that decarbonisation might become more relevant in Serbia. When it comes to the obstacles to the implementation of this concept in Serbia, the interviewees recognized: (a) many relevant documents have not yet been adopted; (b) "invisibility", in the sense that the concept of JGT is not recognized as a subject of spatial planning in national legislation; (c) inherited power system highly dependent on fossil fuels, primarily coal; (d) insufficient participation at the local level, which suffers major consequences, as lack of capacity and inadequate structure of employees in the municipal administration (see also Toković et al. 2025); and (e) insufficient awareness of population.

Regarding the integration of decarbonisation in practice, positive experience has been noted in Lazarevac, where the municipality supports the transition from coal-fired to biomass boilers for family housing heating systems along with gasification of the whole territory. Regarding the education system, the Technical Faculty in Bor developed a study programme called "Recycling Technologies and Sustainable Development" and introduced the subject "Environmental Management" as a part of the study programme "Engineering Management". Study programmes for the empowerment of professionals outside of the mining sector have not been introduced. In the field of environmental protection, the major educational activity is connected with after-mining land reclamation and renewable energy resources. Melchers and Goerke-Mallet (2016) emphasize that future post-mining experts need knowledge and skills in geology, hydrogeology, hydrochemistry, rock mechanics and geotechnology, mine site rehabilitation, socio-economic issues, law, business administration, management, monitoring techniques, and digitalisation but also know-how in understanding mine maps and subsidence, etc. A transformation of the education system would be a long-term and complex process, primarily due to the lack of appropriate personnel and complicated administrative procedures. García-García et al. (2024) recommend diversifying the local economy through such avenues as rural tourism, agroindustry, and a "silver economy", which corresponds to the age structure of the Serbian population and long-lasting agricultural tradition.

A part of the respondents believe that a strategy of re-industrialization and restructuring is needed and propose the development of SMEs, services, tourism and agribusiness on reclaimed land. Representatives of EPS and spatial planning experts however consider that from an economic point of view, agriculture and other activities do not represent a satisfactory substitution for coal extraction. All of the respondents agreed that the population in mining regions identifies itself to a great extent with mining activities, mainly due to the huge effects that mining has on community development, as well as their dominant economic dependence on mining. Continuous retraining of miners is necessary, and JGT strategies should particularly address the situation of women in the future labour market (Janikowska & Kulczycka, 2021), but adequate strategies have not been developed yet in Serbia. According to García-García et al. (2024), environmental impacts, job quality, and land use implications, along with the landscape effects of coal phase-out, should not be treated as separate dimensions. In addition, Nikolić et al. (2021) stress the relevance of implementation processes. The interviewees were partially informed on realized positive examples of decarbonization in Serbia. EPS has been very active in this regard since recently (solar power plants, and wind farms). In the respondents' opinion, producing a part of the required energy from solar energy, wind, biomass and geothermal energy is a great opportunity for Serbia.

It is noteworthy to emphasize that the majority of the respondents (except spatial planning expert who thinks that the private sector and NGOs are the most important) agreed that the political-administrative sector has the major role in this process in Serbia, as it creates preconditions for JGT. In the second place, the majority put the private sector. The representative of the Technical Faculty in Bor considers that the private sector will be active if there is an appropriate incentive, and that the educational sector along with NGOs has a more significant role. The role of labour unions has also been emphasized by representative of EPS. According to Kretschmann and Nguyen (2020), shutting down mines has an impact on the socio-economic development of their surroundings. New jobs must be created, the old sites have to be reutilized, where JGT is of utmost importance for the region. An important factor of success is the explanation of measures to the public (Melchers & Goerke-Mallet, 2016; Pantić et al., 2021). The bottom line of such policies, according to Szpor and Ziółkowska (2018) should be a scientific evaluation of communities undergoing restructuring processes. To efficiently promote and implement JGT in Serbia, the interviewees pointed out in particular: coordination between relevant ministries, environment protection legislation, vertical cooperation between related administration levels, strengthening of administrative personnel, knowledge and experience transfer, multidisciplinary expert teams in the compilation of spatial plans, transparent deliberative public engagement in plan & policy preparation, informing and educating citizens. Respondents are aware that the decarbonisation process in Serbia has a slow pace, as energy production is dominantly based on coal. They doubt that Serbia's energy dependence on fossil fuels could end before 2050 and added that complete decarbonisation will be possible only with the introduction of nuclear power plants.

The literature confirms that the transition to a sustainable economy requires deep changes in technologies and production systems. Radically new technologies must be developed, and older ones replaced (Melchers & Goerke-Mallet, 2016). Economic structures have to evolve with a broader use of knowledge, a reduction of technology gaps, the development of new production capabilities, and economic and social activities. In Europe, this evolution should consider the divergence in economic performances and incomes that have emerged in Southern and Eastern European economies (Pianta & Lucchese, 2020). The managing of this demanding path of structural change envisages a key role for public policies to set the direction of technological development towards the creation and diffusion of new clean technologies, creating new market opportunities, and fostering green transition (Lamperti et al., 2018). García-García et al. (2024) go further and claim that the decline of coal began in the late twentieth century, but the push for a just transition gained momentum with the introduction of the Sustainable Development Goals in 2015. Melchers and Goerke-Mallet (2016) think that a positive result in the preservation and development of a mine site for new purposes can only be achieved with a consensus among the stakeholders concerned. Socioeconomic measures applied in Poland acknowledged as the most effective measures, failed (Szpor & Ziółkowska, 2018). Therefore, coal sector restructuring requires finding new pathways of development for the whole region. Our insight into Serbian and other cases led to the conclusion that there is no "one size fits all" strategy to achieve the JGT of coal-dependent communities. Each country and each coal-dependent region must develop a strategic approach based on its strengths and weaknesses. According to Rising et al. (2021), Arora and Schroeder (2022), Jonek-Kowalska and Turek (2022), Tahseen et al. (2022), some of the most important issues and aspects that need to be addressed to achieve JGT include: (1) long-term comprehensive planning; (2) development of specific policies to support transition of post-mining regions as general regional development policies are not sufficient; (3) cooperation between stakeholders—establishing a "just transition board" to coordinate transition plans with all stakeholders (academia, government, companies, labour unions, local community, citizens) and consultations with workers and local communities as early as possible (formal and informal participation); (4) strategies for new jobs, workforce reskilling and retraining; (5) maintaining or development of the supply chain; (6) investment in local community to mitigate social costs including youth emigration; (7) reuse of infrastructure and skills of local people; (8) supporting businesses and their partnerships with universities and research centres. Recognizing the importance of mitigating the negative impacts that mine closure can have on people and communities, the World Bank (2018) has developed guidelines based on main pillars: (1) clear policies and strategy development, large budget and, again, genuine stakeholder consultation; (2) people and communities—meeting the needs of workers, their families and wide community; and (3) timely planning of land use and environmental remediation.

11.6 Concluding Reflections

The JGT will contribute to the enhancement of the environment and life quality (Shaker & Berisha, 2025), but the risks of pauperisation and deprivation of post-coal-mining cities and regions are high. Both national and local governments must provide strong commitments and dedication to prevent unjust transition. The greatest barrier to the application and implementation of JGT in coal-mining communities is the likely tensions between different social actors (miner's trade unions, politicians, environmentalists, and mining companies) having diverse, often opposed goals and expectations. According to Krzywda et al. (2021), they create a diverse discourse, attempting to influence public opinion and gain support.

The lessons learned from analyzed transitions of post-mining cities and regions certainly provide valuable support for the ongoing decarbonisation processes. They outline the significance of timing, i.e. it is important to prepare well in advance to prevent numerous and diverse threads of decarbonisation in mining communities by ensuring financial resources and by developing thoughtful strategies. It is also of great importance to concentrate on developing and using the soft infrastructure, providing timely discussions and consultations with workers and locally impacted populations. A JGT will reshape the labour market by opening new opportunities, along with new risks, and creating new jobs while extinguishing the existing ones. A coordinated approach of all state authorities and participants in this process is required, along with transparent decision-making and broad dialogue with all relevant stakeholder groups. To transform Serbian coal-mining regions into more sustainable and diversified economies, the energy sector's dependence on fossil fuels must be reduced. Conducted interviews and analysis of relevant documents showed that the decarbonisation awareness in Serbia has been gradually growing, but it is still not embedded into policy systems and strategic thinking. Even more advanced economies than Serbian, with longer history of implementing JGT mechanisms (as Poland) are doubted to reach the 2050 goal. Some measures in Poland failed, and the example of Germany shows that positive outcomes are possible only after performing a clear restructuring policy for several decades. Interviews also showed that Serbia's education system takes small steps on the way to JGT, while most changes remain economy and decarbonization-oriented. Variations of goals in Serbian national policies show that alternatives to coal-based power generation are not consistently considered. A place-based response seems inevitable if we want a JGT process to go smoothly. The bottom-up policy approach implemented in the Ruhr region, as well as in the UK in later stages that involved local actors in the design and implementation of JGT policies and strategies proved to be successful and should be considered in other countries.

Acknowledgements This work was funded by the Ministry of Science, Technological Development, and Innovation of the Republic of Serbia (Grant No. 451-03- 136/ 2025-03/200006).

References

Arora, A., & Schroeder, H. (2022). How to avoid unjust energy transitions: Insights from the Ruhr region. *Energy, Sustainability and Society, 12*(1), 19.

Bell, F. G., Stacey, T. R., & Genske, D. D. (2000). Mining subsidence and its effect on the environment: Some differing examples. *Environmental Geology, 40*, 135–152.

Bronny, H. M., Jansen, N., & Wetterau, B. (2004). *The Ruhr Area. structural changes of an outstanding European conurbation.* Kommunalverband Ruhrgebiet, Essen.

Carley, S., & Konisky, D. M. (2020). The justice and equity implications of the clean energy transition. *Nature Energy, 5*(8), 569–577.

Claeys, G., Tagliapietra, S., & Zachmann, G. (2019). *How to make the European Green Deal work.* Policy Contribution Bruegel Institute. www.bruegel.org/2019/11/how-to-make-the-european-green-deal-work/ (accessed on 29th March, 2024).

Collier, D. (1993). The comparative method. In A. W. Finifter (Ed.), *Political science: The state of discipline II* (pp. 105–119). American Political Science Association.

Danciu, M., Domokos-Paşcu, G., & Bădescu, Ş. (2023). *Monitoring and control of just transition in mining regions: Relevant tools.* AIP Conf. Proc. 27 September; 2928 (1):210005.

EPS (2012). TE "Kolubara B", Procena uticaja na životnu sredinu za projekat izgradnje prve faze TE "Kolubara B", 2x350 MW, Studija—Deo I, Beograd.

EPS (2024). *2023 environmental report*, Belgrade.

European Commission (2019). The European Green Deal. Communication no. COM/2019/640. Brussels: European Commission. https://eur-lex.europa.eu/legal-content/EN/TXT/?uri=COM:2019:640:FIN (accessed on 1st April 2024).

Evans, G., & Phelan, L. (2016). Transition to a post-carbon society: Linking environmental justice and just transition discourses. *Energy Policy, 99*, 329–339.

Filipović, S., Lior, N., & Radovanović, M. (2022). The green deal—just transition and sustainable development goals Nexus. *Renewable and Sustainable Energy Reviews, 168*, Article 112759.

Firmin, M. W. (2008). *Structured interview*. SAGE Publications.

Fothergill, S., Gore, T., & Leather, D. (2024). *The state of the coalfields 2024: Economic and social conditions in the former coalfields of England.* Centre for Regional Economic and Social Research, Sheffield Hallam University, Sheffield.

FZN (2024). https://fzn.thga.de/en/fzn/ (accessed on 4th April, 2024).

García-García, P., Carpintero, Ó., & Buendía, L. (2024). Just transitions to renewables in mining areas: Local system dynamics. *Renewable and Sustainable Energy Reviews, 189*(A), 113934.

Healy, N., & Barry, J. (2017). Politicizing energy justice and energy system transitions: Fossil fuel divestment and a "just transition." *Energy Policy, 108*, 451–459.

Hofmann, W., & Winter, T. (1991). Steinkohlebergehalden als Landschaftsbauwerke. In H. Wiggering, & M. Kerth, (Hrsg.) (Eds.), *Bergehalden des Steinkohlenbergbaus. Beanspruchung und Veränderung eines industriellen Ballungsraumes*, pp. 33–46. Vieweg, Braunschweig, Wiesbaden.

Huang, J. Guo, D., & Hou, H. (2020). *An analysis of progress and models of local green mining*. IOP Conf. Ser.: Earth Environ. Sci. 558 022022.

INECP (2024). Integrated national energy and climate plan of the RS for the period up to 2030 with a vision to 2050.

ITUC. (2019). *Economic and social policy brief: The role of social protection in a just transition*. ITUC CSI IGB.

Janikowska, O., & Kulczycka, J. (2021). Just transition as a tool for preventing energy poverty among women in mining areas—A case study of the Silesia region. *Poland. Energies, 14*, 3372.

Jasanoff, S. (2018). Just transitions: A humble approach to global energy futures. *Energy Res. Social Sci., 35*, 11–14.

Johansson, R. (2007). On case study methodology. *Open House International, 32*(3), 48–54.

Jonek-Kowalska, I., & Turek, M. (2022). The economic situation of polish cities in post-mining regions. Long-Term analysis on the example of the Upper Silesian Coal Basin. *Energies, 15*(9), 3302.

Kolde, L., & Wagner, O. (2022). Governance policies for a "just transition"—A case study in the Rhineland lignite mining district. *Journal of Sustainable Development of Energy, Water and Environment Systems, 10*(1), 1–16.

Kretschmann, J., & Hegemann, M. (2012). *New chances from old shafts. Risk management in abandoned mine sites in Germany*. In Proceedings of the Annual Meeting of the Society for Mining, Metallurgy & Exploration. Red Hook, NY: Curran Associates, 153–158.

Kretschmann, J., & Nguyen, N. (2020). Research areas in post-mining areas—Experience from German hard coal mining. *Mineral Engineering, 1*(2), 255–262.

Krzywda, J., Krzywda, D., & Androniceanu, A. (2021). Managing the energy transition through discourse. *The Case of Poland. Energies, 14*, 6471.

Lamperti, F., Dosi, G., Napoletano, M., Roventini, A., & Sapio, A. (2018). Faraway, so close: Coupled climate and economic dynamics in an agent-based integrated assessment model. *Ecological Economics, 150*, 315–339.

Lee, H.-J., Lee, S.-J., & Lee, C. (2014). A study on the meme features appear in renewal space -focusing on the renewal project of the Zollverein coal mine industrial complex in Germany. *Korean Institute of Interior Design Journal, 23*(6), 60–69.

Magel, H. (2012). *Sustainable land management—new challenge for surveyors and land experts*. Keynote speech at the Conference Land Management days in Turkey, 15.-17.11.2012 in Istanbul. www.landentwick-lung-muenchen.de/master/news/Istanbul/Magel_LM_Istanbul.pdf (accessed on 17th February 2024).

Mandrysz, W. (2011). Restrukturyzacja przemysłu na terenie Górnego Śląska a przemiany tożsamości indywidualnej i zbiorowej jego mieszkańców: Na przykładzie osiedli przyzakładowych. *Górnośląskie Studia Socjologiczne. Seria Nowa, 2*, 103–112.

Maričić, T. (2006). Implementation of strategic environmental assessment in Serbia—case of Spatial plan of Kolubara Lignite basin. *Spatium, 13–14*, 16–20.

Mayer, A. (2018). A just transition for coal miners? Community identity and support from local policy actors. *Environmental Innovation and Societal Transitions, 28*, 1–13.

Melchers, C., & Goerke-Mallet, P. (2016). Research institute of post-mining. In J. Kretschmann, & C. Melchers (Eds.), *Done for good—Challenges of post-mining* (pp. 11–18). Anthology by the Research Institute of Post-Mining. THGA University.

Nikolić, D., Pantić, M., & Jokić, V. (2021). Urban and spatial planning: Pragmatic considerations for plan implementation improvements (A Case Study of the City of Bor). *SAGE Open, 11*(1).

Nurdiawati, A., & Urban, F. (2021). Towards deep decarbonisation of energy-intensive industries: A review of current status. *Technologies and Policies. Energies, 14*(9), 2408.

Pantić, M., Cilliers, J., Cimadomo, G., Montaño, F., Olufemi, O., Torres Mallma, S., & van den Berg, J. (2021). Challenges and opportunities for public participation in urban and regional planning during the COVID-19 pandemic—Lessons learned for the future. *Land, 10*(12), 1379.

Pantić, M., Maričić, T., & Milijić, S. (2024). Visualising the relevance of climate change for spatial planning by the example of Serbia. *Applied Sciences, 14*(4), 1530.

Pavlovic, N., Ignjatovic, D., & Subaranovic, T. (2021). Possibility of using wind and solar sources for electric power generation on Serbian opencast coal mines. *Materials Proceedings, 5*(1), 50.

Pianta, M., & Lucchese, M. (2020). Rethinking the European Green Deal: An industrial policy for a just transition in Europe. *Review of Radical Political Economics, 52*(4), 633–641.

Ruhr, R. (Ed.). (2006). *The Metropole Ruhr—Facts and figures*. Regionalverband Ruhr.

Rising, J., Dumas, M., Dicker, S., Propp, D., Robertson, M., & Look, W. (2021). *Regional just transitions in the UK: Insights from 40 years of policy experience*. Resources for the Future & Environmental Defense Fund.

Route Industriekultur (2020). www.route-industriekultur.ruhr/ (accessed on 4th February, 2020).

Shaker, Y., & Berisha, E. (2025, forthcoming). *Just green transitions: Between terminological inexactitudes, conceptual fragmentations, and the exigency for a theoretical framework*. European Spatial Research and Policy.

Sharpe, S. A., & Martinez-Fernandez, C. M. (2021). The implications of green employment: Making a just transition in ASEAN. *Sustainability, 13*, 7389.

Skoczkowski, T., Bielecki, S., Kochański, M., & Korczak, K. (2020). Climate-change induced uncertainties, risks and opportunities for the coal-based region of Silesia: Stakeholders' perspectives. *Environmental Innovation and Societal Transitions, 35*, 460–481.

SORS (2023). Municipalities and regions of the Republic of Serbia, SORS: Belgrade.

SORS (2024a). *Annual energy statistics and energy balances*. https://data.stat.gov.rs/Home/Result/040201?languageCode=en-US&displayMode=table (accessed on 27th February, 2024).

SORS (2024b). https://data.stat.gov.rs/Home/Result/2403040102?languageCode=en-US (accessed on 4th April, 2024).

Spasić, N., Jokić, V., & Maričić, T. (2009). Managing spatial development in zones undergoing major structural changes. *Spatium, 21*, 53–65.

SPKLB—Spatial plan of the exploitation area of the Kolubara lignite basin, "Sl. glasnik RS", 107/2017, 19/2024.

SPSPAKCB—Spatial plan of the special purpose area of the Kostolac coal basin, "Sl. glasnik RS", 1/2013, 20/2018.

Statistical Office of Katowice (2023). https://katowice.stat.gov.pl/en/ (accessed on 2nd April 2024).

Storm, S. (2020). *The EU's Green Deal: Bismarck's "What is possible" Versus Thunberg's "What is imperative."* INET Working Paper no. 117. New York: Institute for New Economic Thinking.

Swadźba, U. (2001). Śląski etos pracy. Studium socjologiczne. Wydawnictwo Uniwersytetu Śląskiego, Katowice.

Szpor, A., & Ziółkowska, K. (2018). *The transformation of the polish coal sector*. The International Institute for Sustainable Development, Winnipeg, Manitoba, Canada.

Tahseen, J., Watson, E., Mattar, S. D., & Mikulewicz, M. (2022). *Just transitions and structural change in coal regions: Central and Eastern Europe*. Glasgow Caledonian University.

Tellis, W. M. (1997). Application of a case study methodology. *The Qualitative Report, 3*(3), 1–19.

Toković, M., Milosavljević, D., Vuković, M., Pantić, M. & Maričić, T. (2025). Resistance and engagement in mining communities: Experiences from Bor and Majdanpek (Serbia). *Spatium, 3*, 11–21.

UNDP (2020). *Initiating the just transition in Serbia, Belgrade*. https://www.klimatskepromene.rs/wp-content/uploads/2021/03/08-03-2021_Report-on-a-Roadmap-for-Just-Transition-for-Serbia.pdf

UNESCO (2002). *Zollverein Coal Mine Industrial Complex in Essen*. https://whc.unesco.org/en/list/975/ (accessed on 2nd April 2024).

Wander-Lush (2023). wander-lush.org/things-to-do-in-katowice-poland/ (accessed on 2nd April, 2024).

Winkler, H. (2020). Towards a theory of just transition: A neo-Gramscian understanding of how to shift development pathways to zero poverty and zero carbon. *Energy Res. Social Sci., 70*, 101789–101800.

World Bank. (2018). *Managing coal mine closure: Achieving a just transition for all*. World Bank. https://documents1.worldbank.org/curated/en/484541544643269894/pdf/130659-REVISEDPUBLIC-Managing-Coal-Mine-Closure-Achieving-a-Just-Transition-for-All-November-2018-final.pdf

Yarkova, Y., Yarkov, D., & Tsacheva, S. (2021). *The just transition of the European Green Deal and the development of the regions with intensive coal mining (example of Marishki Basin—Bulgaria)*. SHS Web of Conferences "Business and regional development". 120:01002.

Zhang, G., & Zhang, B. (2024). Hidden dimensions of injustice in the green transition of China's coal mining industry. *Social Inclusion, 12*, 7588.

Zheng, Y., Xiao, J., & Cheng, J. (2020). Industrial structure adjustment and regional green development from the perspective of mineral resource security. *International Journal of Environmental Research and Public Health, 17*(19), 6978.

CHAPTER 12

Between Light and Shadow: Exploring Spatial and Multi-Dimensional Justice in the EU's Green Transition

Grazia Brunetta, *Ombretta Caldarice*, *and Danial Mohabat Doost*

12.1 Introduction: Exploring the Just Dimension of the European Union's Approach to Green Transition

Currently, the European Green Deal (EGD) is one of the most important and progressive European climate initiatives. Its primary objective is to achieve climate neutrality for the European Union (EU) by 2050. This forward-looking common goal requires reducing carbon emissions by a minimum of 50% by 2030 and progressing towards a 55% reduction by 2050 (Fit for 55: Delivering on the Proposals—European Commission). By implementing different environmental, climate, and energy policies for the EGD, the European Union (EU) aims to become the global leader in combating the impacts of climate change. In support of the EGD implementation, there are several areas of intervention that are introduced by the European Agenda which include climate action and emissions reduction; environment and the circular economy; skills, education, and training; greening public and private finances;

G. Brunetta · O. Caldarice (✉) · D. M. Doost
Interuniversity Department of Regional and Urban Studies and Planning (DIST), Responsible Risk Resilience Centre (R3C), Politecnico Di Torino, Turin, Italy
e-mail: ombretta.caldarice@polito.it

G. Brunetta
e-mail: grazia.brunetta@polito.it

D. M. Doost
e-mail: danial.mohabat@polito.it

E. Berisha et al. (eds.), *The Palgrave Handbook of Just Green Transitions in the Western Balkans and Beyond*,
https://doi.org/10.1007/978-3-031-95075-9_12

Just Transition; research and innovation (Directorate-General for Structural Reform Support, 2020). In addition, the EGD has two key pillars: first, the Green Transition and its financing, and second, the Just Transition to leave no one behind.

The first pillar, Green Transition, aims to fundamentally reshape Europe's economic model and represents a comprehensive strategy to combat climate change, promote sustainable development, and foster a resilient future for Europe. Through the use of EU financial tools, particularly InvestEU, the Sustainable Europe Investment Plan, part of the EGD, aims to unlock private money and mobilise public investment of at least €1 trillion (Financing the Green Transition—European Commission).

The other key pillar of the EGD is the Justice dimension of the transition. From a policy point of view, one important instrument to guarantee that the shift to a climate-neutral economy proceeds fairly and leaves no one behind is the Just Transition Mechanism (JTM) which offers focused assistance to help mobilise around €55 billion in the most affected regions between 2021 and 2027 in order to lessen the socio-economic effects of the transition (The Just Transition Mechanism—European Commission). According to the EU Commission, the JTM is a key tool to ensure that the transition towards a climate-neutral economy happens fairly, leaving no one behind.

This means that the European Commission recognises that the shift towards a climate-neutral economy will inevitably cause economic and social impacts, particularly in regions dependent on fossil fuel extraction and processing or where carbon-intensive industries are located. The European states and regions that are still heavily dependent on the extraction and production of fossil fuels are more likely to face economic decline as a consequence of the energy transition. This suggests that the transition must follow sustainable regional development on a macro scale. Furthermore, within the impacted regions dependent on fossil fuels, the economic and social impacts of the transition to renewable energies are not equally distributed among citizens. In other words, in the community scale, the cost of transition is not paid equally.

Hence, the JTM is the tool through which the European Commission aims to alleviate the situation and facilitate a fair and just Green Transition to support workers and citizens of the regions most impacted by the transition (Fleming & Mauger, 2021). As mentioned earlier, the EU's primary concern is with areas that depend on the extraction and processing of fossil fuels, such as refineries, or with extremely carbon-intensive businesses, such as cement factories. The Green Transition in these regions will impact a broader array of economic sectors, potentially leading to a larger number of workers facing job displacement. The JTM explains that three main groups will benefit from the JTM: People and citizens most vulnerable to the transition; Companies and sectors active in or comprising carbon-intensive industries; Member States and regions with high dependence on fossil fuel and carbon-intensive industries (The Just Transition Mechanism—European Commission). Additionally, it points out the manner in which each of these groups is supported, as summarised in Fig. 12.1.

Therefore, the JTM aims first to support individuals who are particularly susceptible to the impacts of transitioning away from carbon-intensive industries through various means, for instance, by facilitating their future employment and guaranteeing their access to energy resources. Businesses and industries heavily reliant on carbon-intensive processes are the following beneficiaries through economic and technological support. Finally, JTM aims to support certain regions and countries that have economies heavily reliant on fossil fuels by providing technical, digital, economic, and infrastructural assistance. Although this mechanism can significantly assist specific individuals, companies, and regions in managing the adverse effects of the Green Transition, there are several areas for improvement within the agenda that this chapter aims to address in the following sections. The aim is to discuss the following issues critically. Which dimensions of spatial justice green policies and actions need to be addressed? How do we address procedural justice

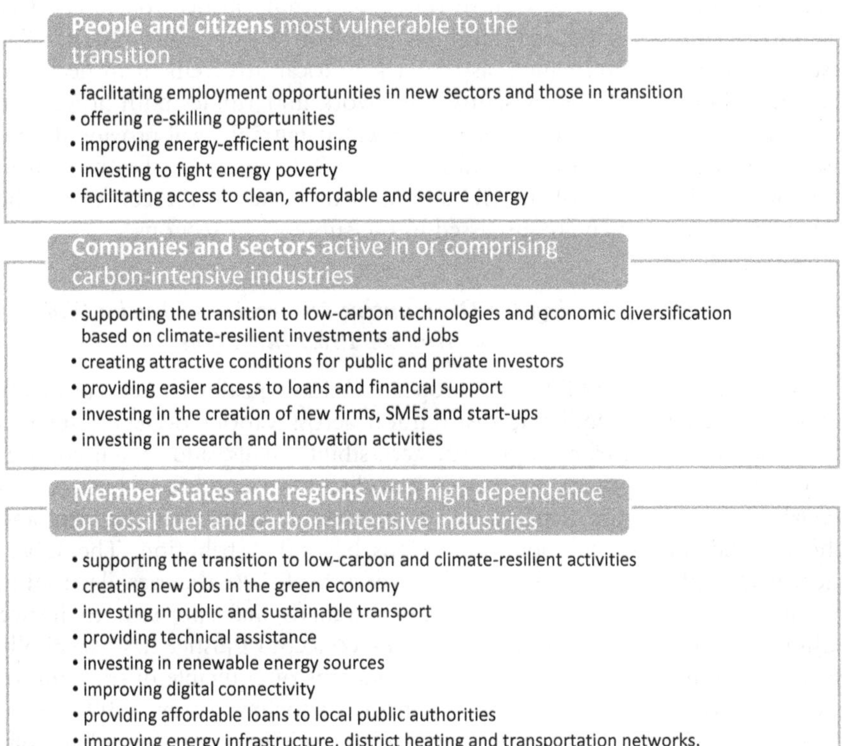

Fig. 12.1 Groups that will benefit from the JTM (*Source* authors elaboration adapted from The Just Transition Mechanism—European Commission)

in national policies and regional strategies? How do we support distributional justice in mitigation and adaptation actions at the local level? How can promoting the idea of spatial justice in the Green Transition help overcome the current limitations of the EU's approach to just transition?

12.2 Questioning the EU's Just Transition Strategy

The term Just Transition, originally proposed by global trade unions in the 1980s, recently became a mobilising concept for promoting green jobs as a necessary component of transitioning away from fossil fuels (Abraham, 2017). The incorporation of JTM's policy framework into the EGD shows the European Union's acknowledgement of the pivotal role of justice in mitigating the socio-economic impacts of transitioning to a climate-neutral economy. However, there remain unresolved debates and uncertainties regarding the EU's approach to addressing justice issues within this transition mechanism.

One critique refers to the insufficient emphasis on the procedural aspect of justice. The other critical aspect refers to the absence of consideration for the spatial dimension. Finally, the last issue refers to the fact that both policy discussions and academic discourse often undermine justice to a secondary role in the context of Green Transition (Heffron & McCauley, 2018). The first two issues are intertwined, as the lack of local involvement in decision-making within the GTM's legislative framework undermines both procedural justice and its spatial implications. Subsequent paragraphs will provide deeper insights into these debates. In addition, Fig. 12.2 represents the three main interconnected critiques of the EU approach to justice in the context of the Green Transition, which are discussed in the subsequent sections.

12.2.1 *Overemphasising the Distributive Dimension of Justice While Undermining the Procedural One*

In the context of the EGD, the European Union's justice strategy primarily focuses on achieving equitable distribution across various domains, such as jobs, resources, infrastructure, energy accessibility, skills, and digital connectivity. While this approach emphasises the distributive aspect of justice, it is crucial to acknowledge that justice, as posited by numerous theorists and scholars, encompasses multiple dimensions beyond distribution. The subsequent paragraphs will provide a concise overview of these theories, illustrating why it is necessary to delve into a multi-dimensional approach to justice within the context of Green Transition. The concept of justice has historically been predominantly associated with the concerns of equitable distribution of resources and opportunities among members of society (Barry, 1999; Rawls, 1971; Walzer, 1983). This perspective—which is widely adapted in urban policies and planning—focuses on the allocation of societal goods and is known as distributive justice. Distributive justice in societal governance pertains to the

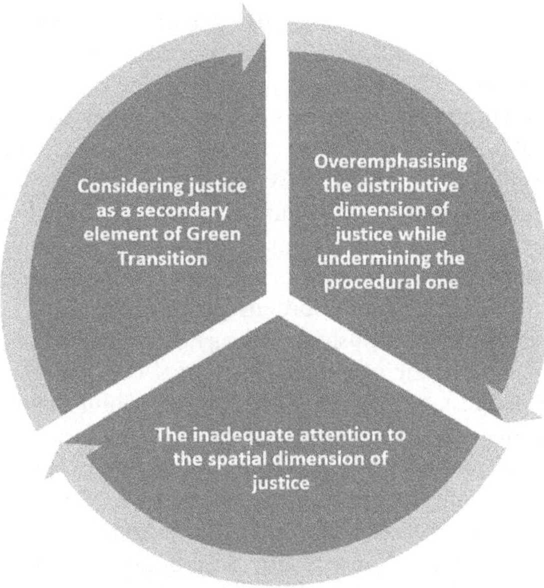

Fig. 12.2 The interconnected critiques of the EU approach to justice in the context of the Green Transition (*Source* authors elaboration)

equitable allocation of both benefits and burdens across populations (Brighouse, 2004). Benefits encompass resources, opportunities, and freedoms, while burdens encompass costs, risks, and limitations. Accordingly, distributive justice governs the fair allocation of societal advantages and disadvantages (Schlosberg, 2009), thereby striving for an equitable distribution of social goods (Rawls, 1971). The distributive dimension of justice from a spatial planning perspective concerns the geographical distribution of both resources and adversities across territories. Populations living in specific urban or rural areas might be more likely to face adversities and less likely to benefit from available resources. In this context, justice encompasses access to jobs, essential services, green spaces, infrastructure, recreational areas, and more. At the same distribution of adversities and burdens plays a key role, including the level of pollution in resources such as air, water, and soi, exposure to different natural hazards, the presence of urban decay, etc. Although the distributive dimension of justice is crucial, contemporary scholars (Fraser, 1999; Honneth, 2004; Schlosberg, 2004, 2009; Young, 1990) challenged the merely distributional understanding by proposing a more nuanced interpretation of justice that encompasses a broader array of interconnected elements. For instance, Schlosberg (2004, 2009) contends that justice, particularly as manifested in political activism such as environmental justice movements, comprises multiple intertwined components, including distribution, recognition, participation, and capability. Thus, while acknowledging the significance of distributive justice, these theorists advocate for a comprehensive approach that also considers individual and communal recognition, participation, and functioning. This perspective posits that alongside distributional concerns, it is important to address the processes that contribute to unequal distribution, as well as issues of respect and recognition, which are deemed essential prerequisites for procedural justice (Miller, 2001). In expanding the conceptual boundaries of justice, scholars such as Iris Young (1990) and Nancy Fraser (1999, 2014, 2020) have emphasised the need to focus on the underlying causes of unequal distribution rather than merely its outcomes. They highlight the significance of recognising marginalised groups within social and political spheres, as the absence of recognition contributes to disparities in the allocation of benefits and burdens (Schlosberg, 2004). Procedural justice concerns itself with the inclusivity and fairness of decision-making structures, advocating for democratic processes as essential components of social justice. Therefore, as suggested by IPCC (2022), procedural justice refers to who decides and participates in decision-making. The empirical study by Susan S. Fainstein (2009) is an excellent example of the importance of this dimension of justice, as it compares the justice issues in the three cities of New York, London, and Amsterdam. She argues that among the three cities, Amsterdam shows higher levels of justice not only because it fosters equality through policies like social housing but also through diversity and participation. In addition, a well-established approach to integrating the procedural dimension of justice is participatory planning. Although there are authors criticising certain aspects of it, participatory planning—if designed properly—can amplify the voices of marginalised groups and provide several additional advantages, such as transforming the planning process into a learning system which, in the long term, leads to a more equitable society. The first step to implementing procedural justice is the recognition of different voices. As Schlosberg (2009) suggests, if people are not recognised, they are less likely to participate, and if they do not participate, they are not recognised. Furthermore, if they do not participate, they are less likely to receive what you truly need. In this sense, the lack of a procedural dimension not only limits participation but also undermines distributive justice. According to this background, the focus of the EGD is on the distributive dimension of justice, and above all, the concern for jobs has become central to the concept of Just Transition (McCauley & Heffron, 2018; Wang & Lo, 2021). While it is very relevant to consider the unequal outcomes that the transition might cause, the question here is how to involve those who will be more affected by the transition's unintended consequences, such as job loss. The importance of the procedural or participative dimension arises from the fact that top-down processes, in which the just procedures for decision-making are not considered, might lead to outcomes that are not necessarily acceptable and fair for all. This is why, for a holistic inclusion of socially just policies towards carbon neutrality, effective governance tools must consider several factors to foster civic engagement, including active communication between citizens, policymakers, and stakeholders. In the context of the Green Transition, procedural justice can be addressed by ensuring that decision-making processes are fair, transparent, inclusive, and participatory. There are several steps to be taken to do so. First, it is necessary to engage a wide range of stakeholders, including citizens, non-profit organisations, businesses, and, above all, marginalised groups. While it may appear that the most vulnerable group during an energy transition consists solely of those directly impacted by job loss, it is crucial to acknowledge and address other marginalised groups that have historically been overlooked and excluded during transitional phases (Mohabat Doost et al., 2023). In this way, we can transform the transition phase into an opportunity to actively engage these groups in decision-making processes and amplify their voices, facilitating their meaningful participation. The other step is to enhance the transparency of the decision-making process. This will also contribute to creating trust and accountability. In addition, equitable capacity building for the whole society is needed to inform people about their procedural rights and help them understand how they can more effectively participate in making decisions that will affect their lives. To operationalize the participative dimension, the governance system plays a key role. A government that does not undermine bottom-up approaches to decision-making can create more just processes. Many practical steps can be taken, such as establishing open communication venues where different groups and stakeholders can express their opinions and concerns. Additionally, implementing multi-layered governance frameworks with multiple centres of authority is essential, as is considering the needs of marginalised or vulnerable groups at every stage of the decision-making process.

12.2.2 The Inadequate Attention to the Spatial Dimension of Justice

Examining the outlined agenda of GT, it becomes evident that the primary focus lies on specific groups, notably individuals facing job loss and workers, rather than addressing the broader community within a given locality. In terms of spatial dimension, solidarity under the JTM will be limited to the designated regions despite numerous justice concerns regarding the transition to climate neutrality on a local scale. The absence of local-level engagement among residents is notable within the mechanism, as no legal framework mandating such participation exists (Fleming & Mauger, 2021). Due to its ability to facilitate a more thorough analysis of the structural injustices underlying environmental issues, a spatial approach to justice is beneficial. Instead of only treating the symptoms, policymakers can discover and address the underlying causes of socio-economic interactions by examining how they manifest geographically. A spatial conception of justice can play a key role in overcoming certain place-based challenges and helps provide a holistic framework that addresses both the processes and outcomes associated with Green Transition. In brief, the spatial understanding of justice aims to consider social, economic, and environmental factors within specific geographical contexts. Many academics and philosophers from urban studies and planning fields have examined and elaborated on the idea of spatial justice. The French philosopher and social theorist Henri Lefebvre is considered one of the important figures involved in the development of the concept. In his writings, Lefebvre investigates the relationship between space, society, and power, most notably in his seminal book 'The Production of Space' (Lefebvre, 1991). He makes the case that space is a social product influenced by power relations rather than a mere physical place. Lefebvre's other book, 'The Right to the City' (Lefebvre, 1996), is another globally recognised work that expands upon the concept of spatial justice by incorporating a range of demands in cities, including housing, participation in urban governance, access to public spaces, and prevention of displacement and gentrification, to mitigate spatial inequalities within urban environments. The concept of spatial justice has also been explored in the context of urbanisation and capitalism by geographer and social theorist David Harvey. In works such as 'Social Justice and the City' (Harvey, 1973) and 'Spaces of Hope' (Harvey, 2000), Harvey suggests how to develop more equitable urban planning and development practices and examines how unequal development and spatial segregation contribute to societal inequities. Harvey defines justice within the framework of place and time, asserting that it is impossible to discuss justice without considering its contested nature as a product of power dynamics within specific locations and historical periods (Harvey, 1996). Several other authors are developing the concept of spatial justice. For example, when it comes to planning and policymaking, Fainstein (2014) has contributed significantly to the field by exploring the role of governance structures in shaping spatial inequalities. She suggests that three central concepts must be considered to create more inclusive and socially just cities: diversity, democracy, and equity. In addition, Neil Smith's (2008) theoretical insights on the concepts of uneven development, gentrification, and the relationship between nature and urbanisation emphasise the importance of considering spatial dimensions of inequality and injustice in cities. The theories mentioned above are very relevant in addressing current issues and challenges in the context of the Green Transition. The relevance of spatial justice theories in this context comes from the fact that Just transition is not only about employment and job creation. It is also important to guarantee that all segments of the society have equitable access to the new opportunities of any kind. These opportunities can be related to different actions that are part of the Green Transition, and the benefits vary, from cleaner air and water to more desirable urban spaces or more accessible infrastructures and services. All of these benefits have spatial dimensions, and their accessibility matters in terms of spatial equity and justice. Several studies have shown that so-called sustainable urban development projects do not always contribute to the transition towards a more spatially just future. In some cases, it can even undermine social equity in cities, exacerbating inequalities rather than addressing them.

As an example, a study by Long (2014) discusses that the city of Austin (Texas) is recognised as a sustainable city with strong economic growth, and its ability to address political tensions between development and environmental protection has made it a model for policymakers. However, the popular narrative overlooks key aspects of sustainable development, such as fair political representation, affordability, and social justice issues. The author shows that Austin, a newly segregated city, is emerging. The downtown area features a sustainable community that is attractive to the affluent creative class, while the outer areas are seeing a growing marginalised working class facing challenges such as limited access to city services, suburban poverty, increased traffic congestion, etc. Therefore, Long views the case as a 'sustainability fix,' related to the concept of the 'spatial fix,' originally introduced by Harvey (1982, 1996). Harvey's theory suggests that capitalism, due to its inherent instability, seeks a spatial fix during periods of cyclical over accumulation. This involves expanding, reconstructing, and investing in infrastructure to open new markets and resources—not to solve the core problem but to alleviate it temporarily. Thus, rather than facilitating a comprehensive green transition, this approach merely acts as a remedy that perpetuates existing economic crises. Ultimately, such an agenda is unlikely to achieve meaningful progress towards justice or a fundamental Just transition.

Another example is provided by Hoover et al. (2021), who analyse the issue from a planning point of view. Their research focuses on the investment in Green Infrastructure (GI) in 19 U.S. cities, showing that although GI is increasingly used to enhance sustainability and resilience, the issues of justice are usually missing. From a spatial planning perspective, investment in GI is a matter of distributional environmental justice, and their research identified numerous spatial siting criteria outlined in the plans, most of which are highly technical or economic in nature, with little emphasis on spatial and environmental justice.

In the same vein, a study by Shokry et al. (2021) showed that in the city of Philadelphia (Pennsylvania), Green Infrastructures—which are originally designed to enhance the resilience of communities to climate risks—might lead to gentrification processes. Through a spatial quantitative study, they observe that the GI interventions aimed at enhancing resilience are concentrated in the wealthy and already gentrified neighbourhood of Central Philadelphia. These areas are associated with real estate developments and economic reinvestments for greater profit. This article interestingly points out the critical issue of compliance between greening initiatives and housing policies. In more detail, it mentions that housing programmes are currently disconnected from greening initiatives, which prevents the planning of neighbourhoods that become greener without displacing economically disadvantaged populations. More on that, 'Greening, housing, and other community advocates must therefore work together to guarantee that when greening is negotiated into new developments, that affordable housing—whether through land value capture, inclusionary zoning, or other measures—as well as support for the kinds of social infrastructure discussed in this article—is a key part of the plan' (Shokry et al., 2021, p. 238).

In light of this, academics and decision-makers can create more effective strategies for incorporating spatial justice into the Green Transition agenda, guaranteeing that the transition is both environmentally sustainable and socio-economically just. Accordingly, future planning and policy responses could also favour a multi-scalar approach, taking into consideration the challenges and opportunities at different spatial levels (Brunetta et al., 2023). In addition, the approach to justice outlined here goes beyond the distributive dimension, which currently emphasises only job loss and economic decline in fossil fuel-dependent regions. Instead, the concept of spatial justice also encompasses the procedural dimension, highlighting the significance of civic participation, equitable capacity building, and bottom-up approaches to transition planning. Considering the justice element of the transition in a comprehensive, multidimensional way can help enhance the solidarity of the green transition towards climate neutrality, which is key to operationalising the foreseen strategies (Sikora, 2021).

12.2.3 Considering Justice as a Secondary Element of Green Transition: A Reductionist Approach

The last critique of the current EU approach to justice issues in the Green Transition refers to the fact that justice is usually considered as a remedy for the Green Transition's unwanted impacts. In other words, the understanding of justice within the context of the Green Transition often falls short, reducing it to merely an element of transition rather than recognising it as the primary objective of transformative change. Many recent authors promote this view when they aim to answer why justice is key to the Green Transition (Fleming & Mauger, 2021; Heffron & McCauley, 2018; McCauley & Heffron, 2018; Newell & Mulvaney, 2013; Wang & Lo, 2021). Although the need to transition towards a sustainable, green economy is clear, justice is often perceived in a reductionist manner, neglecting its significant implications for addressing the underlying causes of environmental challenges. Indeed, the current climate crisis not only exacerbates existing inequalities but is fundamentally rooted in unjust socio-economic relations. Consequently, the notion of justice as an element to be considered within the transition arises from a failure to acknowledge that the crisis itself partly stems from these inequalities.

According to the widely accepted view of Just Transition, the transition to a green economy is the key objective that inevitably brings about socio-economic challenges. These challenges might lead to regional inequalities, meaning that the dependence on fossil fuel industries in particular regions makes them more vulnerable to change. To alleviate this situation, the transition to a green economy must provide compensation for the most affected regions. However, framing justice solely as a reactionary measure overlooks its deeper significance. Justice should not merely serve as a remedy for immediate injustices but also as a lens through which we can address the structural inequities as root causes of social and environmental challenges. Failure to identify injustice as a structural issue risks causing cycles of inequality and hindering real progress towards sustainability.

Therefore, within the discourse of the Green Transition, justice must not be relegated to a secondary role since this reductionist view fails to grasp the essence of justice as the core objective of the transformation. While the urgency of transitioning to a sustainable, green economy is evident, justice must not be marginalised; rather, it should be recognised as integral to addressing the root causes of environmental crises. Therefore, what is suggested is that justice, instead of being an aftermath of the Green Transition, must become an objective hand-in-hand with it: a transition to both a just society and a green economy.

12.3 Concluding Remarks. Towards an Integrated Framework for Spatial Justice in the Green Transition

The concept of Green Just Transition is closely tied to sustainable development, which aims to balance economic growth, social equity, and environmental protection (Shaker & Berisha, 2025). Green Transition and growth, advocated by organisations like the World Bank and the UN, seek economic growth while protecting the environment (Mandelli et al., 2018). The Green Transition could lead to new green jobs and increased welfare. However, it also poses challenges such as job losses, particularly in fossil fuel-dependent sectors of the economy.

The Just Transition approach emphasises the need to address such issues, including employment and social concerns, in moving towards sustainability. The EGD aims to make green growth and transition work together to implement sustainable development effectively and holistically.

However, as illustrated in this chapter, the EU's approach to justice issues can be improved if some existing limitations are considered. The limitations analysed in this chapter are overemphasising the distributive dimension of justice while undermining the procedural one, the inadequate attention to the spatial dimension of justice, and considering justice as a secondary element of Green Transition. Here, we argue that a spatial conception of justice can help overcome the challenges outlined in this chapter by providing a comprehensive framework that addresses both the immediate and underlying challenges of the just and Green Transition. A spatial understanding emphasises the interconnectedness of social, economic, and environmental factors within specific geographical contexts. Therefore, during the whole transition process, it is important for the policymakers to:

- Assess how environmental policies regarding the Green Transition impact different geographies on a local scale. The analysis of these impacts can consider access to resources and basic services, socio-economic changes, and place-based assessments of communities' vulnerability to radical economic changes.

- Guarantee equitable geographical access to the new infrastructures, including urban green, renewable energy sources, new green jobs, etc., in different neighbourhoods.
- Consider that new green job creation policies and plans can be created in disadvantaged areas to support local development and provide opportunities to address socio-economic and spatial inequality.
- Provide fair access to training programmes, especially in disadvantaged areas, and empower citizens by informing them about their ways of involvement in decision-making processes. This can be complemented by additional skills training to prepare the population for emerging job market requirements.
- Enhance the bottom-up decision-making and procedural dimension of equity to ensure that decision-making processes are fair, transparent, inclusive, and participatory. This does not mean that only soft governance tools are crucial; also, hard governance tools are necessary to create more effective outcomes.
- Consider justice issues not only as a remedy for immediate impacts caused by transition but also as a lens through which we can address structural inequities as root causes of social and environmental challenges.

This chapter mainly aimed to study the existing gaps and limitations in the institutional discourse of Green Transition in the European context and proposed some strategies to overcome these limitations. However, it is essential also to consider the potential challenges and barriers to implementing the recommended strategies.

The first challenge refers to the political resistance of the dominant groups whose benefits are not in coherence with the redistributive policies or planning initiatives, and so these policies might threaten their interests. This once again points to the importance of recognising and amplifying the voice of the least powerful and most vulnerable groups in societies and empowering these groups so that they can contribute to the decision-making processes.

The second challenge refers to the distribution of public and private funding across different geographies. Green Transition aims to foster welfare, well-being, and social equity through a sustainable economic path in coherence with EGD. Therefore, to reach this goal, it is crucial to allocate resources based on place-based factors that can prioritise the allocation of resources.

The last challenge is the structural characteristics of governance systems, which can play a crucial role. As an example, rigid local planning laws might be a barrier to implementing justice-oriented policies. Traditional zoning laws that historically favour socio-economic groups over others are an important constraint, for instance. Another example of structural constraints to justice-oriented spatial policies is the institutional inefficiencies usually caused by dysfunctional bureaucratic systems. Spatial justice also has a temporal dimension, for example, due to the displacement of specific groups from one place to another at a certain point in time. Institutional structures that are not efficient, adaptive to changing conditions, and flexible might create critical delays in the implementation of spatial justice policies. Therefore, the lack of integrated planning creates serious problems. Justice is not a fragmented issue, and disadvantaged people usually suffer from issues that relate to different sectors, from housing to transportation and environmental planning. Finally, fragmented and purely sectoral approaches fail to address the root causes of existing injustices in various places.

In brief, a spatial approach to justice -at different geographical levels, from national to local- also makes it possible to look more closely at the systemic injustices that underlie environmental problems. Policymakers can address the underlying causes of socio-economic linkages instead of just managing their symptoms by examining how these relations emerge regionally. The significance of local knowledge and community views in forming policy responses is emphasised by a spatial definition of justice. Through grassroots stakeholder engagement, policymakers can provide solutions for specific circumstances, with a focus on social justice and environmental sustainability. Importantly, we argue that the existing EU framework for justice within the context of the Green Transition relegates it to a secondary role, primarily focusing on mitigating the transition's adverse effects on particular groups, such as workers in the fossil fuel sector. While acknowledging the importance of addressing these impacts, it is vital to recognise that the concept of a Just Transition extends beyond mere alleviation. It must align closely with the broader goals of energy and environmental transition, operating in parallel to ensure comprehensive and sustainable change.

A potential recommendation for upcoming research is to investigate the outcomes of the EU's JTM and implemented strategies in local contexts. This could entail evaluating their ability to fulfil the requirements of marginalised communities, the sufficiency of assistance for businesses and sectors undergoing transformation, and the consistency and inclusivity of policies designed to achieve a fair transition at a local level.

References

Abraham, J. (2017). Just transitions for the miners: Labor environmentalism in the Ruhr and Appalachian coalfields. *New Political Science, 39*(2), 218–240. https://doi.org/10.1080/07393148.2017.1301313

Barry, B. (1999). Sustainable and intergenerational justice. In D. Andrew (Ed.), *Fairness and futurity: Essays on Environmental sustainability and social justice*. Oxford University Press. https://doi.org/10.1093/0198294891.003.0005

Brighouse, H. (2004). Justice. *Malden, MA: Polity.* https://scholar.google.it/scholar?hl=en&as_sdt=0%2C5&q=Brighouse%2C+Harry+%282004%29.+Justice.+Cambridge%3A+Polity.&btnG=

Brunetta, G., Caldarice, O., Mohabat Doost, D., & Pellerey, F. (2023). Notes on spatial implications of COVID-19. Evidence from Piedmont Region, Italy. In *Post*

Un-Lock: From territorial vulnerabilities to local resilience (pp. 13–31). Springer International Publishing. https://doi.org/10.1007/978-3-031-33894-6_2

Directorate-General for Structural Reform Support. (2020). *Technical support for implementing the European Green Deal*. https://doi.org/10.2887/90157

Fainstein, S. (2009). Spatial justice and planning. *Justice Spatiale/Spatial Justice, 1*(1), 1–13. https://www.jssj.org/wp-content/uploads/2012/12/JSSJ1-5en1.pdf

Fainstein, S. (2014). The just city. *International Journal of Urban Sciences, 18*(1), 1–8. https://doi.org/10.1080/12265934.2013.834643

Financing the green transition. Retrieved April 24, 2024, from https://ec.europa.eu/commission/presscorner/detail/en/ip_20_17

Fit for 55: Delivering on the proposals—European Commission. Retrieved April 24, 2024, from https://commission.europa.eu/strategy-and-policy/priorities-2019-2024/european-green-deal/delivering-european-green-deal/fit-55-delivering-proposals_en

Fleming, R. C., & Mauger, R. (2021). Green and just? An update on the "European Green Deal." *Journal for European Environmental and Planning Law, 18*(1), 164–180. https://doi.org/10.1163/18760104-18010010

Fraser, N. (1999). Social justice in the age of identity politics: Redistribution, recognition and participation. *Iride*. https://doi.org/10.1414/11370

Fraser, N. (2014). *Justice interruptus: Critical reflections on the" postsocialist" condition*. Routledge. https://doi.org/10.4324/9781315822174

Fraser, N. (2020). From redistribution to recognition? dilemmas of justice in a "post socialist" age. *In Justice Interruptus*. https://doi.org/10.4324/9781315822174-8

Harvey, D. (1973). *Social justice and the city*. University of Georgia Press.http://www.jstor.org/stable/j.ctt46nm9v

Harvey, D. (1982) *The limits to capital*. Blackwell. https://files.libcom.org/files/The%20Limits%20to%20Capital.pdf

Harvey, D. (1996). *Justice, nature, and the geography of difference*. Wiley-Blackwell. https://www.wiley.com/en-us/Justice%2C+Nature+and+the+Geography+of+Difference-p-9781557866813

Harvey, D. (2000). *Spaces of hope*. Edinburgh University Press. http://www.jstor.org/stable/10.3366/j.ctvxcrcz8

Heffron, R. J., & McCauley, D. (2018). What is the 'just transition'? *Geoforum* (Vol. 88, pp. 74–77). Elsevier Ltd. https://doi.org/10.1016/j.geoforum.2017.11.016

Honneth, A. (2004). Recognition and justice: Outline of a plural theory of justice. *Acta Sociologica, 47*(4), 351–364. http://www.jstor.org/stable/4195049

Hoover, F. A., Meerow, S., Grabowski, Z. J., & McPhearson, T. (2021). Environmental justice implications of siting criteria in urban green infrastructure planning. *Journal of Environmental Policy & Planning, 23*(5), 665–682. https://doi.org/10.1080/1523908X.2021.1945916

IPCC. (2022). *Climate change 2022: Mitigation of climate change*. Contribution of Working Group III to the Sixth Assessment Report of the Intergovernmental Panel on Climate Change (P. R. Shukla, J. Skea, R. Slade, A. Al Khourdajie, R. van Diemen, D. McCollum, M. Pathak, S. Some, P. Vyas, R. Fradera, M. Belkacemi, A. Hasija, G. Lisboa, S. Luz, & J. Malley, Eds.). Cambridge University Press. https://doi.org/10.1017/9781009157926

Lefebvre, H. (1991). The production of space (D. Nicholson-Smith, Trans.). Blackwell. (Original work published 1974).

Lefebvre, H. (1996). The right to the city. *Writings on Cities, 63181.* https://theanarchistlibrary.org/library/henri-lefebvre-right-to-the-city.lt.pdf

Long, J. (2014). Constructing the narrative of the sustainability fix: Sustainability, social justice and representation in Austin, TX. *Urban Studies, 53*(1), 149–172. https://doi.org/10.1177/0042098014560501

Long, J. (2016). Constructing the narrative of the sustainability fix: Sustainability, social justice and representation in Austin. *TX. Urban Studies, 53*(1), 149–172. https://doi.org/10.1177/0042098014560501

McCauley, D., & Heffron, R. (2018). Just transition: Integrating climate, energy and environmental justice. *Energy Policy, 119*, 1–7. https://doi.org/10.1016/j.enpol.2018.04.014

Miller, D. (2001). *Principles of social justice.* Harvard university Press. https://books.google.it/books?hl=en&lr=&id=9SAwEAAAQBAJ&oi=fnd&pg=PP1&dq=Miller,+David+(1999a).+Principles+of+Social+Justice.+Cambridge,+&ots=oKL_8UFtsx&sig=_wxpAsIH93DGp-ikvK6XuKP-u7w

Mohabat Doost, D., Brunetta, G., & Caldarice, O. (2023). In search of equitable resilience: Unravelling the links between urban resilience planning and social equity. *Sustainability, 15*(18), 13818. https://doi.org/10.3390/su151813818

Newell, P., & Mulvaney, D. (2013). The political economy of the "just transition." *Geographical Journal, 179*(2), 132–140. https://doi.org/10.1111/geoj.12008

Rawls, J. (1971). A theory of justice. In *Essays and reviews: 1959–2002.* Oxford University Press. https://doi.org/10.5860/choice.37-4151

Sabato, S., & Mandelli, M. (2018). *The EU's potential for promoting an ecosocial agenda.* Report prepared for the project "Sustainable welfare societies: Assessing linkages between social and environmental policies", NOVA Norwegian Social Research. Brussels: European Social Observatory, December, 35p.

Schlosberg, D. (2004). Reconceiving environmental justice: Global movements and political theories. *Environmental Politics, 13*(3). https://doi.org/10.1080/0964401042000229025

Schlosberg, D. (2009). *Defining environmental justice: Theories, movements, and nature.* Oxford University Press. https://doi.org/10.1093/acprof:oso/9780199286294.001.0001

Shaker, Y., & Berisha, E. (2025, forthcoming). *Just green transitions: Between terminological inexactitudes, conceptual fragmentations, and the exigency for a theoretical framework.* European Spatial Research and Policy.

Shokry, G., Anguelovski, I., Connolly, J. J. T., Maroko, A., & Pearsall, H. (2021). "They didn't see it coming": Green resilience planning and vulnerability to future climate gentrification. *Housing Policy Debate, 32*(1), 211–245. https://doi.org/10.1080/10511482.2021.1944269

Sikora, A. (2021). European Green Deal—Legal and financial challenges of the climate change. *ERA Forum, 21*(4), 681–697. https://doi.org/10.1007/S12027-020-00637-3/METRICS

Smith, N. (2008). Uneven development: Nature, capital, and the production of space. *Uneven development: Nature, capital, and the production of space.* https://doi.org/10.2307/622014

The just transition mechanism—European commission. Retrieved April 24, 2024, from https://commission.europa.eu/strategy-and-policy/priorities-2019-2024/european-green-deal/finance-and-green-deal/just-transition-mechanism_en

Walzer, M. (1983). Spheres of justice: A defence of pluralism and equality. *Philosophy*, *59*(229), 413–415. https://scholar.google.it/scholar?hl=en&as_sdt=0%2C5&q=Walzer%2C+Michael+%281983%29.+Spheres+of+Justice.+Oxford%3A+Blackwell.&btnG=

Wang, X., & Lo, K. (2021). Just transition: A conceptual review. *Energy research and social science* (Vol. 82). Elsevier Ltd. https://doi.org/10.1016/j.erss.2021.102291

Young, I. M. (1990). *Justice and the politics of difference*. Princeton University Press. https://doi.org/10.7312/blau17412-114

PART IV

Just Green Transitions on the Western Balkans

CHAPTER 13

Just Green Transitions in North Macedonia: A Place-Based Case Study from Southwest Planning Region

Vesna Garvanlieva Andonova, *Elena Gotovska*, *and Marjan Nikolov*

13.1 Introduction

The Western Balkan (WB) countries, including Republic of North Macedonia (RNM), have signed the Green Agenda for the Western Balkans (GAWB) as a binding commitment to transition towards a more sustainable and environmentally conscious future. The agreement reflects a recognition of the urgent need to address climate change, promote renewable energy sources, and foster green economic growth in the wider region (Berisha, 2025a). RNM, is facing the challenges of transitioning towards a greener, more sustainable energy future. This endeavour entails large magnitudes of effort restrained by a myriad of limitations. Nevertheless, one of the larger steps involves the decommissioning of the coal-fired thermal plants. This chapter explores the challenges and opportunities presented by the shift from coal-based energy production to renewable energy sources (RES), with a focus on the territorial economic and social impacts of this transition. By aligning with the goals outlined in the GAWB, North Macedonia seeks to leverage its natural resources, innovative

V. Garvanlieva Andonova (✉) · E. Gotovska · M. Nikolov
CEA—Center for Economic Analyses, Skopje, North Macedonia
e-mail: vesna_nl@yahoo.com

E. Gotovska
e-mail: elena.gotovska@gmail.com

M. Nikolov
e-mail: makmar2000@yahoo.com

E. Berisha et al. (eds.), *The Palgrave Handbook of Just Green Transitions in the Western Balkans and Beyond*,
https://doi.org/10.1007/978-3-031-95075-9_13

capacity, and stakeholder collaboration to navigate the challenges and capitalize on the opportunities associated with the transition towards a greener and more resilient energy system yet, abiding to the principles of "just transitions". As North Macedonia embarks on its transition to a just energy future, it is important to adopt a comprehensive methodology that encompasses both financial and social dimensions. This chapter presents the summary of a study that employs a systematic mixed-method approach, integrating quantitative and qualitative analyses to assess the socioeconomic implications of phasing out coal-fired power generation, with a specific focus on the Southwest planning region (SWPR). We apply a Cost–Benefit Analysis (CBA) framework to quantify the costs and benefits of three distinct scenarios related to the retirement of the TPP Oslomej coal-fired plant, considering direct and indirect costs, environmental impacts, public health effects, and potential job losses. These costs and benefits are converted into net present value (NPV) to facilitate direct comparisons among the policy alternatives. Complementing this quantitative analysis, we utilize qualitative content analysis to understand the social impacts on local communities and economies. Additionally, input–output analysis is employed to examine the indirect economic effects, allowing us to quantify how potential job losses from the coal phase-out may reverberate through the local economy, impacting sectors such as manufacturing and services. We compare three scenarios: a baseline scenario where the plant is closed with no further interventions, a less ambitious scenario with labour assistance programs, and a more ambitious scenario that incorporates both labour support and investment in renewable energy. This selection of scenarios aims to explore various pathways for a just transition, taking into account differing political and economic feasibilities associated with each option. The chapter serves two main purposes. Firstly, it aims to provide an overview of RNM's current energy and climate policies, while also examining how the social-economic aspects of the energy transition policy plan fit into this framework. Secondly, it focuses on a territorial perspective by exploring the specific place-based effects in the Southwest planning region of the country. This section contextualizes the significance of energy transition and decarbonization in ensuring energy security, while emphasizing the importance of incorporating proper support for social justice and reaping benefits from a place-based approach. This study aims to assess the implications of the implementation of just transition principles in the Southwest planning region of North Macedonia during the transition towards a greener energy future.

13.2 Overview of the Energy Sector and Policies in North Macedonia

13.2.1 Overview of the Current Energy Landscape in North Macedonia

Fossil fuels continue to play a significant role in North Macedonia's energy mix, with electricity generation being the primary contributor to the country's greenhouse gas (GHG) emissions. The domestic energy production in RNM indicates a trend of decreasing domestic production and increased import dependency.[1] Recently, there are indications of gradual dependency decrease on coal-based production, nevertheless the contribution of the coal-fired thermal power plants (TPP) is still contributing to around 70% of the domestic electricity production (ERC, 2023). The production of electricity by the coal-fuelled TPP Oslomej, situated in the Southwest planning region (SWPR), contributes with less than 5% of the total domestic electricity production (ERC, 2023).[2] At the same time, RES contribute to around 29% of the energy production[3] predominantly from hydropower (25% hydropower, and 4% from solar, wind and bio sources cumulatively). Coal's heavy use in energy production, widespread reliance on wood and coal for heating, and elevated levels of pollution from transportation, all play a part in air quality problems and total emissions. Electricity production accounts for approximately 60% of total greenhouse gas emissions, while the transport, heating, and agricultural sectors each contribute with around 10% (MoEPP, 2023). In addition, North Macedonia has an aging and outdated energy generation infrastructure that heavily relies on fossil fuels. This poses a significant challenge, nonetheless also offers potential to improve energy efficiency and security. The country has set ambitious targets for reducing greenhouse gas emissions, with a strong emphasis on advancements in the energy sector. In 2021, the country pledged to reduce GHG emissions by more than 50% compared to 1990 levels by 2030. Meeting these goals will require substantial reductions in emissions across all economic sectors, particularly focusing on the energy industry. Following the European Green Deal (EGD), and as a contracting Party of the Energy Community (EnC, 2006), North Macedonia has committed to working towards achieving climate neutrality by 2050 while also matching current per capita GDP levels of some neighbouring EU countries (MoE, 2019, Energy Development strategy). Progress in reducing GHG emissions and air pollution however has been limited. Furthermore, the country, has

[1] The production of electricity of JSC ESM in 2021 compared to 2010 has decreased by 50%.

[2] Production contribution for Y2022, which has drastically increased in 2021 and 2022, after gradual decrease starting from 2012 to 2020; local coal reserves have been depleted and the increased production is fueled by externally procured (imported) coal. Energy and water services regulatory commission of RNM (ERC), Annual reports, (2023).

[3] RES electricity production is variable on an annual level, foremost due to the hydro power plants capacities.

inconsistently increased energy production by reviving older energy generation capacities in response to the recent energy crisis. In the onset of the energy crisis RNM derailed from past policy efforts and instead of decreasing it increased the coal-fuelled energy production. However, at the same time there has been accelerated private investments in RES, particularly in photovoltaics (PVs).[4] This last development is in line with the decarbonization strategy efforts for increased RES share, and the overall Enhanced National Determined Contributions' (MoEPP, 2021a, 2021b) objectives outlined in both the energy and climate strategies (MoEPP, 2021c).

13.2.2 Decarbonization Policy Landscape in North Macedonia

The national and sectorial strategic documents provide direction for decarbonization efforts in the energy sector (MoE, 2019), with the aim to create integrated internal energy market which is more efficient, environmentally friendly, and competitive in supporting the economic growth. Likewise, the country's climate policy and strategic documents (MoEPP, 2021c) outline the commitments to mitigation and adaptation.[5] These decarbonization plans hinge on the expansion of RES (mainly hydropower plants and PVs), to reach 38% of gross energy consumption from RES by 2030 and the goal for RES in the electricity sector production to reach 66% by 2030. Nonetheless, the funding required to achieve these targets is considerable, accompanied by an ambitious investment agenda. The two key strategies for climate action and energy are the Long-term Strategy on Climate Action and Action Plan (MoEPP, 2021c) and the Strategy for Energy Development of the RNM up to 2040 (MoE, 2019). These comprehensive documents delineate measures to address climate change, including plans for decarbonization and adaptation. They estimate the required capital investment—encompassing both public and private sources—to be between €18.5 billion and €34.1 billion through 2050, representing approximately 4.3% to 8.0% of the annual GDP for the year 2023. The viability for securing and leveraging funding of such a immense size[6] is questionable given the history of the outturns of such ambitions failing to effectuate fast or at all. For instance, it had been already set that the decommissioning of TPP Oslomej to have taken place in 2021, which did not occur and has been postponed yet again. There is no guarantee that the region will abide by the opting-out commitment on the legacy coal-fuelled power plants as in 2021, when the EnC initiated legal cases against

[4] According to ERC statements, during 2022 and 2023, 551 MW newly installed and operational RES PV capacity are added (399 MW in 2023, 152 MW in 2022).

[5] Although the National Energy and Climate Plan was adopted in June 2022, the Law on Climate Action is still pending. To make progress, it will be essential to make a coordinated effort to address the legislative gaps, involve stakeholders, and prioritize climate action.

[6] This raises another flag of caution as the PIM processes are weak and the past paints a picture of investment outlays have not been efficiently nor effectively been realized with significant capital budgeting bias.

four of the WB countries for breaching the emission ceiling (EnC, 2021). There is no straightforward definition of "just green transition" in the strategic policy documents of North Macedonia, nevertheless, these documents reference the concept primarily concerning the energy transition area. Specifically, the anticipated shift from coal to renewable energy sources is addressed within several key strategic frameworks including the Energy Strategy (2020–2040), the National Energy and Climate Plan (NECP), the Just Transition Roadmap (JTR), and the Accelerated Coal Transition Investment Plan (ACT IP). Collectively, these documents refer to the importance of a just transition to "mitigate the negative effects associated with job losses, redeploy workers, and foster new employment opportunities in low-carbon technologies and services." More specifically, the JTR delineates guidelines at addressing the socioeconomic impacts of the energy transition and the establishment of an institutional framework to facilitate these processes, noting the significance of fostering inclusivity among various stakeholders in coal-dependent regions and the necessity to "address and reflect the specific needs of municipalities and regions." It is undoubtedly that the success of decarbonization via coal phasing-out (as other climate adaptation endeavours as well) by large will be determined by the local adaptive capacity. The complex nature of the interplay between the adaptive capacity and the "place" calls for an inclusive unequivocally all-social efforts and engagement. Thus, the more mindful is the approach in which "places" shape the collective understanding of just transition, the better it will be for effective actions to be taken. The transition design or conceptualization of the overall green transitions, based on a specific place-based aspects are still lacking in the country. The advancement though on a policy level yet considerably determined by the top-down central level, can be noted by the national document Just transition roadmap (MoE, 2023b) and the Investment plan for accelerated coal transition (MoE, 2023a) which envisage the closure of the plant as most viable option coupled with intentions for replacement of energy with PV RES.

13.3 Fitting in the "Just" in the Policy for Regional Energy Transition

Government policies play a crucial role in shaping the socioeconomic impacts of the closure of the coal-fired power plants. However, each region and locality has its unique characteristics, thus tailored measures are necessary to facilitate a just and fair transition (Shaker & Berisha, 2025). The presence or absence of suitable government policies and regulations significantly influences the social outcomes and the socioeconomic effects, which will vary depending on whether the affected area is diverse, how developed the local economy is and to what degree it relies on the coal plant, which is undergoing closure (or repurposing). It's thus to be expected that regions which are dependent on a single industry (not diversified) will face challenges in adapting. On the bright side, moving away from coal can unearth new opportunities, and thus identifying and effectively managing these opportunities is key to laying the groundwork for sustainable development in the long run. Nonetheless, there are limited detailed case studies showcasing such successes (Bainton et al., 2018), while on the other hand, there is evidence suggesting that poorly managed transitions can have enduring negative effects (Beatty, 2019).

Transition to a greener and energy-efficient, renewables-based power sector in North Macedonia should be relatively easy as there are not many TPP capacities. However, the country notes decades of underinvestment in the energy sector and delays of coal phasing-out, thus resulting in a lagging energy sector. The fact remains that the energy transition faces obstacles, some of which are evidently the large coal dependence (thus energy (in) security) and, relatedly, the politics for energy in general.

A crucial aspect of achieving the energy transition to an environmentally sustainable—green transition, is a largely a matter of justice, as people which are either directly and indirectly affected in the particular territories are in vulnerable situations, and stand to suffer the most, therefore the way and approaches of how it is to be done matters vastly. Although these transitions are necessary and at this point only a matter of when and how, (not if) a top-down approach is not generating effective measures fast enough and to the desired levels (Rogelj et al., 2016; Schindler, 2019; UNEP, 2020 etc.) and the speeding up on the coal phase-outs will need further efforts of political action. Irrespective of the unique attributes of individual countries or regions, the conceptualization of energy transition, sufficiently addressing the imperatives of sustainable development encompassing decent work, regional development, and social equity, stands as a fundamental requirement for formulating local, regional, or national action plans (Sharpe & Martinez-Fernandez, 2021). The importance of local communities is central in energy just transition process, as it is more importantly of how the process of energy transition is being managed while not only when and if the set targets and objectives are achieved (for example see Hess, 2018; Van Veelen & der Horst, 2018).

In an effort to align with movements in the EU and reflect the EU's Just Transition Mechanisms (JTM) more closely, RNM developed a JT Roadmap along with an Investment plan for accelerated coal transition. Additionally, the country established a Coordination mechanism for governing JT. However, unlike the EU's JTM which requires preparation of Territorial Just Transition Plans on a regional level as a condition for utilizing EU resources, RNM's process lacks this place-based approach. The national RNM strategic documents aim to provide concrete support to all affected groups including employees, local businesses, and local/regional authorities while also aiming to attract investments that facilitate a smoother transition in line with NECPs as well as national and local strategies. Nevertheless, differences were observed in adopting a top-down approach versus the preferable bottom-up approach during initial stages of process and structure design structure.

The SWPR region in RNM lacks a regional or local perspective (place-based), as the policies are primarily set at a national level. Discussions[7] with local authorities and CSOs indicate that there is a limited involvement of local authorities and local stakeholders in the policy development and future planning. The strategic documents state that TPP Oslomej will be "closed", and that part of the energy generation will be substituted by PV RES investments (MoE, 2023a) while only nominally and hastily mentioning the significance of the local government's involvement in these processes (MoE, 2023b).

13.4 Territorial Socio-Economic Impact of Transition

13.4.1 Regional Context

The SWPR lags behind other regions in the country and is home to Oslomej, one of the oldest and outdated coal-fuelled thermal plants located in Kichevo, first in line to be closed. The region has one of the lowest average GDP per capita at 78% of the national average, contributes around 8% to the national GDP despite having 10% of the population, with no clear path towards convergence with the national average. Despite a decrease, unemployment remains high at 20.5%,[8] with an increasing inactive population, with rate of 48.6% in Y2022, leading to dependence on a limited number of working individuals and prone to significant outmigration from SWPR. Much of SWPR's employment is concentrated in trade sector (17%) and construction (10%), public administration (9%), with mining and energy production (4% and 1% respectively) and agriculture (6%). At the regional level, mining and quarrying and electricity and power generation[9] cumulatively represent total of 5% of available jobs, whereas the largest employment sector is in wholesale trade employing 17% and construction 10% (SSO, 2024).

The SWPR's Gross Value Added (GVA) is varied, with the majority stemming from real estate activities (12%), while mining and electricity generation contribute 7% to the region's GVA (SSO, 2024). The presence of TPP Oslomej in Kichevo area is particularly significant as it houses employees focused on lignite mining and power generation, accounting for 5% of employment. This amounts to over one thousand direct jobs, with additional indirect effects on employment across the regional economy through the supply chain estimated to be at least equally substantial.

[7] Discussions of authors with relevant stakeholders through series of interviews during 2023–2024.

[8] 6.1 pp above the national average of 14.4%, Y2022.

[9] NACE Rev. 2 sector B and Sector E.

High levels of unemployment, inactivity, and migration are likely to worsen due to the anticipated socio-economic effects of job loss and missed employment prospects. The local economy's reliance on the power plant decommissioning is expected to be significantly affected unless suitable social safety net measures are promptly implemented.

13.4.2 Contribution to the Employment and the Regional Economy

Given the link to the coal related energy production jobs it is estimated that the phasing-out with closure of the thermal plant in the region could lead to total of direct job losses of 1.3% of the region's employees (direct 965 FTEs) or 27% of the region's employees in mining, querying and electricity sectors. It is estimated that the TPP contributes with 7.65 mil EUR (which is 0.6% of the regional GVA) in the regional economy however furthermore the economic activity further generates additional contribution to the regional economy based on the interdependencies between different economic sectors and industries.

The input–output analysis method is used to assess the effects of economic shocks on an economy, such as those caused by TPP Oslomej. It is estimated that TPP Oslomej has a significant impact on indirect job creation and the regional value added, contributing to around 3% of regional employment and 2% of the regional economy. If the plant were to close without appropriate preparations for economic diversification and transition activities, it could severely affect the local economy. The manufacturing sector (secondary sector) would be most heavily affected, followed by the trade and agriculture sector (primary sector) to a lesser degree (see Table 13.1).

13.5 Balancing Costs and Justice Concerns

Various methods may be utilized to assess the costs and benefits of phasing-out coal and the economic effects of this transition. One commonly employed approach is cost–benefit analysis (CBA), which is frequently utilized to quantify the design and impact of policies, not just for decommissioning or repurposing coal plants but also in other energy sectors. Numerous studies highlight the importance of examining economic returns, community impacts, and the technical feasibility of solutions (see Table 13.2). These reveal that while methodologies like CBA and input–output analysis are effective for capturing economic implications, social aspects require further attention to mitigate adverse effects on affected communities. To better synthesize the findings of various studies addressing the socioeconomic impacts of coal plant decommissioning, we provide a short selected overview of key methodologies, focus areas, and implications via diverse approaches researchers have taken in assessing the transition and highlight the critical insights relevant to policy development.

Table 13.1 Estimation of employment and economic value added contribution (*Source* authors' own elaboration)

	#		*Mill. EUR*	Indirect effect distribution in other sectors in % (NACE)	
Direct jobs—FTE*	965	GVA*—Direct	7.65	B_E Mining, manufacturing, electricity, gas & water supply, etc	82%
Indirect jobs—FTE	1,233	GVA—Indirect	9.78	G_I Wholesale and retail trade	6%
Total	2,198	Total	17.4	A_ Agricul. forestry&fishing	4%
* FTE-full time equivalent; GVA-gross value added				F_Construction	2%
Contribution of potential losses in %				O_Q Public administration, etc	1%
Total Employment in SWPR (#)		75,292	2.9%	M_N Professional, scientific and technical activities, etc	1%
Total GVA—SWPR (mil. EUR)		848	2.1%	K Financial & insur	1%
**SWPR-Southwest planning region of North Macedonia				RSTU Arts, entertainment and recreation, etc	1%
Source authors calculation based on input–output analysis derived for SWPR; primary data based on SSO's: SIOT tables, labour statistics, regional statistics and information from ESM annual reports					
				L_Real estate	1%
				J_Information & comm	0%
				*NACE, Nomenclature of Economic Activities	

13.5.1 Several Possible Scenarios for the Case SWPR in North Macedonia—CBA

Following the commonly used approach of CBA, we as well employ the same approach in estimating the possibilities for the case in question. In this section we present the summary of the assessed costs and benefits of several policy scenario alternatives and compare these. We estimate the costs and benefits of the scenarios and convert into present value (NPV) to compare one policy alternative over the other. We particularly pinpoint the positive effect that can be accomplished by implementation of essential social safety net alternative to contribute to a more just process of coal phasing-out in the SWPR with the retirement of TPP Oslomej.

The study applies mixed-method by using direct quantitative and qualitative content analyses while assessing current planning policy efforts in relation

Table 13.2 Selected literature overview of methods used (*Source* authors' own elaboration)

Reference	*Method used*	*Focus area*	*Key findings & Implications*
Arrow et al. (1996)	CBA Review	Usefulness of CBA in policy design and impact incl. environment	CBA is crucial in policy debates regarding health and the environment. While not exhaustive for public policy formation, it facilitates improved processes and outcomes in energy sector evaluations
Tol (2001)	CBA, Climate policies	Climate policies and equity	Equity considerations can significantly alter CBA outcomes, affecting the political feasibility and social acceptance of climate policies
Dietz and Hepburn (2013)	CBA Framework Development	Environmental policies: Climate and Energy Projects	CBA has limitations in capturing long-term effects and uncertainties; Recommend dynamic evaluations for better climate policy insights via modified CBA with long-term environmental, economic, and social ramifications
Jindal and Shrimali (2022)	CBA, case study, comparative approach	Transition impacts	Highlights the balance between initial investments and long-term benefits of energy policies
Araújo et al. (2016)	CBA & Multi-Criteria Decision Analysis	Policy evaluation;	Evaluated multiple energy policy scenarios; identify optimal policy paths; importance of initial investment vs. Long-Term benefits and environmental benefits

Reference	*Method used*	*Focus area*	*Key findings & Implications*
Wang et al. (2015)	CBA, Multi-Factor Analysis, Case Study	Energy transition & energy-efficient policies	Emphasizes the need for supportive policies and stakeholder involvement to combat late transition costs
Sidhu et al. (2018)	Social CBA, case study,	Transition costs, project evaluation	Demonstrates positive economic and environmental impacts of grid-scale energy storage; ongoing evaluation is crucial.
Fleischman et al. (2013)	Economic Comparative Analysis	Economic returns of coal plants & transition to RES	Stresses the economic inefficiency of aging coal plants and advocates for decommissioning due to high external costs
Raimi (2017)	Economic Analysis	Decommissioning impacts, Role of public perception & community involvement	Identified economic ramifications of coal retirement; Emphasizes strategic planning for local economies; Highlights strategic planning and community involvement in coal plant retirements
Kefford et al. (2018)	Case Study Analysis	Social impact of plant retirement	Community engagement is critical for managing the transitions of fossil fuel power plants
Haggerty et al. (2017)	Policy Analysis	Local environmental policies	Advocates for cohesive transition policies to avoid adverse outcomes
Hamilton et al. (2017)	Policy Analysis	Support mechanisms post-transition	Aligning workforce development policies is essential for effective support

(continued)

Table 13.2 (continued)

Reference	*Method used*	*Focus area*	*Key findings & Implications*
Shrimali (2020)	Economic Analysis, Case Study, Policy Recommendations	Community perspectives in transition planning	Communities prefer aligned support initiatives; Suggests addressing local needs in transition planning; Employment and Social Considerations are underlined by addressing the social implications of plant retirements (retraining programs and economic adjustment strategies for local communities)
Heinrichs et al. (2017)	Input–Output Analysis (IOA), Integrated Assessment Model (IAM)	Job loss assessment, scenario analysis, stakeholder engagement and policy support	Quantified job losses and economic effects; Highlights importance of economic interdependencies, Evaluates the socio-economic consequences of a coal phase-out, including potential job losses in coal-dependent regions & strategies for JT
Feng et al. (2023)	Input–Output Models, labour market analysis, scenario development	Employment impacts, localized policy responses	Identifies significant job displacement in lignite sectors but notes opportunities in renewables. Stresses need for mitigation strategies; highlights significant potential job losses associated with coal phase-out, suggesting localized policy responses to address specific challenges faced by different communities

Reference	*Method used*	*Focus area*	*Key findings & Implications*
Pao-Yu (2019)	Economic Analysis	Social costs of coal phase-out	Identified significant social costs associated with job loss; Stresses importance of mitigating social impacts
Biehl et al. (2023)	Policy Analysis	Feasibility of transition strategies	Examined political and social obstacles to transitions; Suggests cohesive policy to overcome barriers; Need for Adaptive Governance to manage the uncertainties, advocating for holistic and adaptive policy measures
Jewell and Cherp (2020)	Conceptual Framework	Energy policy feasibility	Discussed political feasibility of energy transitions; Calls for multi-dimensional approaches to policy-making; Defines political feasibility in the context of climate change mitigation, emphasizing the importance of political will, public acceptance, and institutional capabilities

(continued)

Table 13.2 (continued)

Reference	*Method used*	*Focus area*	*Key findings & Implications*
Ribera (2020)	Comparative Analysis, conceptual framework	Transition pathways, JT is not merely an economic necessity but moral imperative	Evaluated various pathways for energy transitions; Recommends context-specific solutions, integrating climate and social goals, with both challenges and opportunities; Encourages stakeholders at all levels to take a proactive approach, emphasizing the leader roles in businesses, governments, and communities for JT
Rinscheid and Wüstenhagen (2019)	Case Study, Policy Recommendations	Challenge to transition; Public Opinion	Discrepancy between public desire for faster coal phase-out and government timelines; emphasizes the need for ambitious climate policies & public engagement

to the social component. In a first step, the focus is on estimation the direct and indirect interrelations of the sector with the other economic activities in the region. Accordingly, in order to derive more accurate estimated an input–output analysis (IOM) is applied for the impact estimates of the indirect impact in the region for the variables of regional economic value added and indirect labour. These are used as input in the next step for further scenario development. After estimating the indicators for the regional economy, for the scenarios, a CBA is conducted through constructing and estimating the costs and benefits, not only from purely financial aspect but incorporate the social and economic components though inclusion of selection of costs and benefits associated with the three different scenarios. Estimates on the monetization of the costs and benefits are derived on literature review and based on viable assumptions within array of strategic, policy and other documents specific to the country, region and the case in question.

Two alternative scenarios are being compared with a baseline scenario,[10] in which the plant is assumed to retire in five years without further activities (Table 13.3). Another less ambitious scenario involves decommissioning with labour substitution assistance programs, while a more ambitious one includes faster job replacement efforts and investment in renewable energy production as part of a just transition aspect.

The possibility of any of the scenario happening relies largely on the political decisions and timing. Each of the scenarios has an equal likelihood of being implemented. For example, the baseline scenario could happen because of the long-time prolonged closure as a policy option, while there is also an equal likelihood for the other two scenarios as they also align with national commitments for a just transition and recent policies, including boosting investments in energy transition and renewable energy. However, history also demonstrates instances where policy options were not put into action or implemented promptly, as well as cases of inconsistencies, delays, or even abandonment.

However, our objective is not to outline precise CBA assessed possibilities, due to limited resources for such an effort and the outcome being reliant on many variables such as speed, timeliness, as well as various policy and political determinants. Instead, our aim is to offer a comparison of alternatives that leans towards economically the most favourable direction. This also serves as another rationale for the urgent implementation of the "best" option based not only on financial considerations but also on its socio-economic benefits for the region, emphasizing the significance of timely incorporation of well-designed just transition measures which are ultimately most advantageous in the long term (Table 13.4).

[10] We are aware that baseline scenario usually entails "business as usual" however given the circumstances there is high likelihood that the operations of the TPP will cease.

Table 13.3 Scenario definition (*Source* authors' own elaboration)

Scenario 0	*Scenario 1*	*Scenario 2*
The baseline zero scenario assumes that the retirement of TPP Oslomej will take place by 2027 or up to five years. The plant closure will not entail any further activities and assumes retirement of the plant with ceasing of the production activities	The less optimistic or less ambitious scenario entails decommissioning with remediation, environmental abatement, demolition, and closure; just transition measures are planned to alleviate the economic, social, and environmental impact with partial labour substitution assistance program: part of the labour will be engaged in the process of decommissioning and the remainder will be gradually assisted to transition into other sectors through active labour measures and self-employment assistance programs comparable to current assistance programs and labour policies	S2: The more ambitious scenario that builds upon the previous and entails all the aspects of the decommissioning process, with remediation, environmental abatement, demolition, and closure while it assumes larger and faster job replacement efforts, as just transition aspect. The scenario also assumes carrying out of RES investment

13.5.2 Results of the Scenario Options

The future developments, options to be pursued, and the specific actions to take are subject to a high degree of uncertainty and cannot be precisely forecasted. As such, we consider the most likely options given the current context and policy directions based on our best-informed opinion. The transition of the regional economy in the SWP region towards a greener economy, focusing primarily on decarbonisation through ceasing operations of thermal plants, will inevitably impact the socio-economic landscape of the region as well as micro-locations. Supply chain interlinkages across economic sectors suggest potential direct and indirect job losses estimated to exceed two thousand full-time job equivalents. The mining/electricity production sectors will be directly affected by job losses that extend to related sectors, especially manufacturing, water supply & waste management, and remediation activities. These effects will also spill over into wholesale/retail trade, agriculture, and construction sectors. Therefore, implementing an appropriate just transition plan emphasizing reskilling and creating green jobs is vital for offsetting these job losses in order to mitigate any adverse effects from not taking action.

To mitigate regional economic losses and preserve the economic value added in the region, it is essential to strategically redistribute economic output to sectors that provide higher value added, particularly in the tertiary sectors

Table 13.4 Scenarios and selection of potential costs and benefits (*Source* authors' own elaboration)

Scenarios	*Costs*	*Benefits*
Scenario 0 —Closure of TPP—'as is' or baseline **Scenario 1** —Decommissioning with just transition process (labour assistance) **Scenario 2** —Decommissioning, just transition processes and (partial) RES/PV energy substitution	• **Environmental costs:** including degradation of the environment, land and air pollution, devastation, pollution of ground waters, deforestation, degradation of the ecosystem, etc., • **Social—workforce transition costs:** layoffs, retraining, or relocation and may include transition costs in severance packages, job retraining programs, and support services • **Economic impact:** reduction of the population due to internal and external migrations, reduction of the GDP entirely and per capita, local community effects loss of revenue, changes in property values, interlinked economic activity sector effects • **Health costs:** negative health effects in nearby communities • **Decommissioning costs:** decommissioning and dismantling costs for safe shutdown and removal of all equipment, structures, and facilities associated with the plant • **Cultural degradation:** eradication of traditional activities, aesthetic degradation of landscapes, destruction of the potential for development of alternative activities, one-sided development, etc	• **Environmental** benefits: decommissioning a thermal power plant can lead to environmental benefits in reduced air pollution, decreased greenhouse gas emissions, and improved water quality and ecosystem health. These environmental improvements can have positive impacts on public health, biodiversity, and the overall quality of the environment • Public **health** benefits: reduced emissions of harmful pollutants, which to respiratory problems, cardiovascular diseases, and other health issues, thus can contribute to improved public health outcomes and lower healthcare costs • **Climate change** mitigation benefits: as greenhouse gas emissions (carbon dioxide) reduce ecosystem disruptions • **Renewable** energy opportunities: create opportunities for the development of renewable energy sources, such as solar, wind, hydro, and geothermal power, which can help diversify the energy mix, increase energy security, and promote local economic development in the renewable energy sector • **Social benefits**: positive social impacts such as improved community wellbeing, reduced noise and visual pollution, and enhanced social cohesion • **Economic** benefits: besides the upfront costs, there are long-term economic benefits: job creation in the clean energy sector, increased energy efficiency, and reduced reliance on imported fossil fuels. Transitioning to cleaner energy sources can also lead to cost savings, innovation, and resilience in the face of changing energy markets

i.e. services and the secondary manufacturing & processing sectors, with partial and smallest emphasis on the primary agriculture sector. The last notion is relevant because although transitional, the region has strong agriculture-related activity. The primary sector has shrunk significantly, with a higher rate than the rest of the regions in the country, on one hand, but also because of the region's low agriculture productivity. The last is an assumption under the current context. Nevertheless, options for reviving the agricultural sector opportunity for absorption of the labour force will require additional and parallel efforts that may be deemed viable with other perspectives.

The cost–benefit analysis scenario assessments suggest that from a purely financial standpoint, closing the TPP Oslomej without any additional activities emerges as the most financially advantageous option (Table 13.5). This is because it does not require new investments while still generating revenues. However, there are indications of low efficiency due to the obsoleteness of the power plant, which could lead to a less favourable financial outcome; this possibility is not considered due to a lack of detailed information. On the other hand, decommissioning the plant with proper technological measures is a lengthy process involving substantial financial costs and, therefore, lacks profitability. Investing in renewable energy sources, on the other hand, not only covers decommissioning costs but also offers long-term profitability Fig. 13.1).

When considering socio-economic implications and weighing costs and benefits while calculating economic net present value (ENPV) with regard to

Table 13.5 CBA scenario assessment and comparison (*Source* authors' own elaboration)

	S0[11]	*S1*	*S2*
Financial—Cash flow	12.9	−38.7	8.7
Financial—NPV	12.0	−31.4	10.1
Costs			
Social costs (Dir. & Indir.)	−12.1	−5.0	−4.4
Labour reskilling costs		−6.9	−6.3
Health costs	−7.4	−3.0	−3.0
Environmental costs	−10.3	−8.3	−8.3
Reg. economy loss	−106.3	−30.9	−12.7
Benefits			
Environmental savings	28.4	28.4	28.4
Social savings	26.9	8.4	50.4
Health savings	16.4	19.3	19.3
Reg. economy gains	87.2	176.7	153.1
Economic evaluation—NPV	**41.9**	**84.9**	**143.8**
MIRR	9.7%	15.2%	**21.1%**
BCR	**1.3**	**3.4**	**5.6**

[11] Note: *S0—Scenario 0: Closure of TPP—'as is' or baseline, S1—Scenario 1: Decommissioning with just transition process (labour assistance), S2—Scenario 2: Decommissioning, just transition processes and (partial) RES/PV energy substitution.

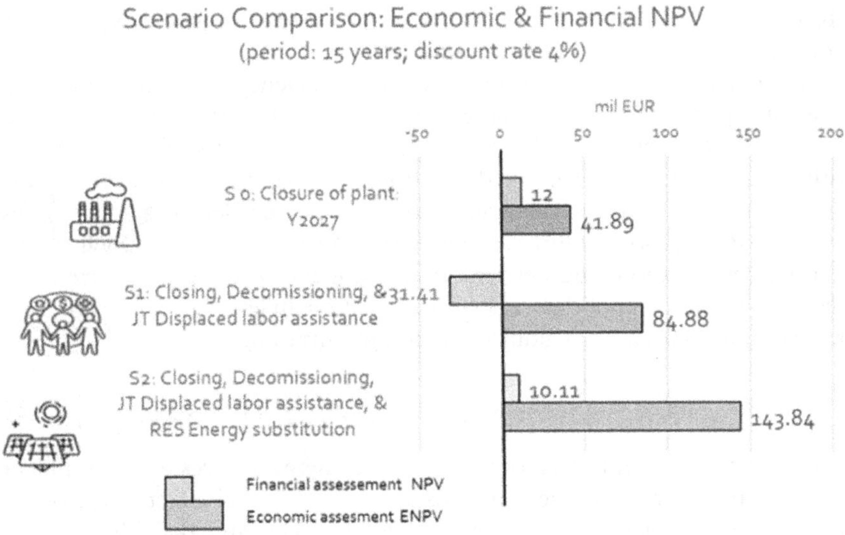

Fig. 13.1 Scenario comparison[12] (*Source* authors' own elaboration)

environmental, social, health, and labour reskilling factors across these options reveals that simply closing down the plant becomes least favourable in terms of gains by magnitude. Providing assistance for mitigating job losses and labour displacement during decommissioning becomes an economically more preferable choice when accompanied by just measures. Furthermore, combining this with investment in RES energy ultimately emerges as most beneficial choice (see Table 13.4).

The analysis summary discusses potential options for transitioning the Oslomej coal thermal plant, evaluating the economic and social aspects of several alternatives. These range from simply shutting down the plant to implementing social safety net measures for workers as part of a fair process, and additionally, integrating green energy substitution. One metric used to compare these scenarios is the benefit-to-cost ratio, which increases sequentially from 1.3 for a simple plant closure to 3.4 with decommissioning supported by labour assistance programs, and further to 5.6 with additional green energy substitution.

The main message and finding of the scenario assessment is that although purely from financial aspect the design and implementation of processes and activities aimed at providing assistance to the affected community (through the labour) may be costly and not the most preferred option, when the economic and social costs and benefits are incorporated in the outcome and these become the criteria for evaluation of the options the outcome is significantly different. In the case when appropriately designed and implemented option

[12] Note: period 15 years; discount rate 4%; in mill EUR.

which include assistance in just transition processes it becomes significantly more favourable both on mid and long term.

Moreover, the study uncovered a notable deficiency concerning local and regional expertise and substantial groundwork for the just transition process. Governance frameworks in RNM are newly formed and scarcely functional, with weak connections among the national government and local authorities, as well as the key stakeholders such as SMEs, academia, labour unions, and civil society organizations. Multi-stakeholder and multi-sectoral collaboration, coordination, and deliberation are imperative to address governance hurdles. Place-based policies can play a crucial role in leveraging the significant opportunities present in regions undergoing transition.

13.6 Discussion and Conclusions

The energy transition from coal to renewable energy sources presents a blend of a variety of opportunities and obstacles for the coal-affected regions in RNM, one of which is the southwest region. Achieving the full economic and social advantages demands a holistic and inclusive approach. Collaboration among the governmental entities, industries, and local communities is essential to ensure a smooth and fair transition to a sustainable local economy, leveraging territorial and place-based potentials, skills, and knowledge of the SWPR and the local economies within. Notably, government responsiveness to just transition concerns is proven to be heightened with comprehensive stakeholder representation in formal decision-making mechanisms. Tailoring place-based transition-supported policies to address the needs of the potentially affected by the thermal plant closure as a major step in the green transition of the region is crucial for fostering acceptance of the coal phase-out policies. Regarding the territorial approach and place-based design of specific localized policies and plans, they seem to be lacking in the SWPR. The nationally (centrally) set policies while aligning to the global, EU and regional trends and demands, in its onset is largely vertically set-up and centralized. The envisaged governing structure for JT is deemed as being imposed from the top or externally motivated, which poses potential for backlogs due to lack of acceptance, ownership and feasibility of the process not properly communicated and owned by stakeholders on a local level, including the local authorities. The absence of specific plans tailored for the SWPR presents shortcomings in both process and content. Apart from its centralized and top-down approach, the policy framework is characterized by insufficient funding, a focus on large-scale energy projects, and a lack of land rehabilitation and labour market assistance strategies. Consequently, additional resources are required, political consensus must be achieved, local communities in the micro-regions must play an active role in the transition, and they also need to be better informed and more engaged in consultation and governance procedures. While there is currently a lack of comprehensive consensus among all societal segments, particularly at lower government tiers, it is imperative for addressing this gap to implement a place-based just transition plan for phasing out coal. Effective and timely implementation of such a plan is essential in achieving its objectives. It is essential for the design and the implementation of the measures to be tailored to the specific requirements and challenges of a given territory considering that SWPR has high unemployment, long-term unemployment rates, and inactivity. These initiatives must be meticulously tailored with a profound comprehension of the distinctive socio-economic traits of the locale and its capacity of the society to foster sustainable green development. This nuanced approach not will acknowledge the diverse circumstances across the two regions but also will underscore the significance of aligning interventions with local conditions to optimize effectiveness and promote enduring positive outcomes. Furthermore, the design and operationalisation of labour market assistance programs and social safety nets programs designed for the microregion where TPP Oslomej is located needs to be targeted so that there are appropriate direct labour and indirectly affected labour should have access to the resources and support needed to successfully navigate the transition. Establishing and maintaining a balance between the aim of accelerating decarbonization and ensuring cost-effectiveness while upholding principles of just transition is vital. Prioritizing financial considerations alone may overlook the social implications of the transition, which are likely to come into play in the medium to long term as jobs and economic interests are at stake. The research findings underscore the necessity of taking a comprehensive approach to energy transition encompassing not only financial and technical considerations but also broader territorial, economic, and social implications of the transformation disruption. Although prioritizing financial aspects might initially suggest that designing and implementing processes and activities to support affected communities are costly and less preferable, a comprehensive evaluation incorporating economic and social dimensions reveals a significantly different outcome. When the economic and social costs and benefits are taken into account as primary evaluation criteria, the landscape shifts. In instances where well-designed and effectively implemented options include support for just transition processes, they emerge as markedly more favourable over both the medium and long term. This broader perspective not only accounts for immediate financial expenditures but also considers the broader societal and economic impacts, highlighting the importance of investing in sustainable and just transition measures. The CBA assessments of the designed possible scenarios show through several indicators such as the benefit-to-cost ratio (BCR) that the option of plant closure without any other actions is financially better option but socio-economically the worst option to be chosen (BCR, 1.3), while proper decommissioning with labour assistance programs while financially expensive, from socio-economic perspective presents substantially larger benefits (BCR, 3.4) and when enhanced with green energy substitution it is most favourable (BCR, 5.6).

References

Abhinav Jindal, A., & Shrimali, G. (2022). *Cost–benefit analysis of coal plant repurposing in developing countries: A case study of India, Energy Policy,* Vol. 164. https://doi.org/10.1016/j.enpol.2022.112911

Araújo, C., Almeida, M., Bragança, L., & Barbosa, J. A. (2016). Cost–benefit analysis method for building solutions. *Applied Energy, 173*, 124–133. https://doi.org/10.1016/j.apenergy.2016.04.005

Arrow, K. J., Cropper, M. L., Eads, G. C., Hahn, R. W., Lave, L. B., Nol, R. G., & Stavins, R. N. (1997). Is there a role for benefit-cost analysis in environmental, health, and safety regulation? *Environment and Development Economics, 2*(2), 195–221. https://doi.org/10.1017/s1355770x97220164

Bainton, N. A., & Holcombe, S. (2018). *The social aspects of mine closure: a global literature review.* Centre for Social Responsibility in Mining (CSRM), Sustainable Minerals Institute (SMI), The University of Queensland: Brisbane

Beatty, C., Fothergill, S., & Gor, T. (2019). *The state of the coalfields 2019.* https://www.coalfields-regen.org.uk/wp-content/uploads/2019/10/The-State-of-the-Coalfields-2019.pdf.

Berisha, E. (2025a). *Moving towards an inclusive green agenda in the Western Balkans.* Foundation for European Progressive Studies (FEPS). https://doi.org/10.13140/RG.2.2.12092.76162

Biehl, J., Missbach, L., Riedel, F., Stemmle, R., Jüchter, J., Weber, J., Kucknat, J., Odenweller, A., Nauck, C., Lukassen, L. J., Zech, M., & Grimm, M. (2023). Wicked facets of the German energy transition—examples from the electricity, heating, transport, and industry sectors. *International Journal of Sustainable Energy, 42*(1), 1128–1181. https://doi.org/10.1080/14786451.2023.2244602

Centre for Research on Energy and Clean Air (CREA) & CEE Bankwatch. (2021). *Briefing paper. Western Balkan coal power plants polluted twice as much as those in the EU in 2019.* https://energyandcleanair.org/wp/wp-content/uploads/2021/07/Western-Balkan-coal-plants-polluted-twice-those-in-the-EU-in-2019.pdf

Dietz, S., & Hepburn, C. (2013). Benefit–cost analysis of non-marginal climate and energy projects. *Energy Economics, 40*, 61–71. https://doi.org/10.1016/j.eneco.2013.05.023

EC. (2020). SWD (2020). *223 final. Guidelines for the implementation of the green agenda for the Western Balkans.* https://neighbourhood-enlargement.ec.europa.eu/system/files/2020-10/green_agenda_for_the_western_balkans_en.pdf

EC. SWD. (2021). *429 final. Commission Delegated Regulation supplementing EU 2021/1153, for methodologies for CBA assessments in renewable energy.* https://eur-lex.europa.eu/LexUriServ/LexUriServ.do?uri=SWD:2021:0429:FIN:EN:PDF#:~:text=Discounting%3A%20A%20social%20discount%20rate,to%20provide%20a%20societal%20benefit

EnC. (2006). *OJ L 198, 20.7.2006. The Energy Community Treaty.* 22006A0720(01) Treaty establishing the Energy Community. https://www.energy-community.org/legal/treaty.html

EnC. (2021). *Energy community. Disputes and settlements.* https://www.energy-community.org/legal/cases/

ERC. (2023). *Energy and water services regulatory commission of the Republic of North Macedonia.* Annual reports for ERC for 2023, 2022. https://www.erc.org.mk/page.aspx?id=342

ESM. (2022). *Annual reports and financial reports for different years*. https://www.esm.com.mk/?page_id=530

Feng, K., Song, K., Viteri, A., Liu, Y., & Vogt-Schilb, A. (2023). National and local labor impacts of coal phase-out scenarios in Chile. *Journal of Cleaner Production, 414*. https://doi.org/10.1016/j.jclepro.2023.137399

Fleischman, L., Cleetus, R., Deyette, J., Clemmer, S., & Frenkel, S. (2013). Ripe for retirement: An economic analysis of the U.S. Coal Fleet. *The Electricity Journal, 26*(10), 51–63. https://doi.org/10.1016/j.tej.2013.11.005

GoNM, MoE. (2020). *National Energy and Climate Plan (NECP)*. https://www.economy.gov.mk/content/Official%20NECP_EN.pdf

GoNM, MoE. (2023a). *Accelerating coal transition investment plan*. https://economy.gov.mk/content/documents/Pokani/%D0%9D%D0%B0%D1%86%D1%80%D1%82%20%D0%B8%D0%BD%D0%B2%D0%B5%D1%81%D1%82%D0%B8%D1%86%D0%B8%D1%81%D0%BA%D0%B8%20%D0%BF%D0%BB%D0%B0%D0%BD_06122023.pdf

GoNM, MoE. (2023b). *Just transition roadmap*. https://www.economy.gov.mk/en-GB/news/just-transition-roadmap.nspx

GoNM, MoE. (2019). *Strategy for energy development of the RNM up to 2040*. https://economy.gov.mk/Upload/Documents/Adopted%20Energy%20Development%20Strategy_EN.pdf

GoNM, MoEPP. (2023). *4th National climate change communication*. https://unfccc.int/documents/627667

GoNM, MoEPP. (2021a). *Enhanced nationally determined contribution*. https://unfccc.int/sites/default/files/NDC/2022-06/Macedonian%20enhanced%20NDC%20%28002%29.pdf

GoNM, MoEPP. (2021b). *NDC implementation roadmap for North Macedonia 2020–2030*. https://klimatskipromeni.mk/data/rest/file/download/c86929c13f43f00f201b38ef166822904cf3568a881e997bc608433de987eb8f.pdf

GoNM, MoEPP. (2021c). *Long-term strategy on climate action and action plan*. https://unfccc.int/sites/default/files/resource/MKD_ES_LTS_Nov2021.pdf

GoNM, MoEPP. (2022). *Revised national emission reduction plan (NERP)*. https://www.economy.gov.mk/content/Official%20NECP_EN.pdf

Haggerty, J. H., Haggerty, M. N., Roemer, K., & Rose, J. (2018). Planning for the local impacts of coal facility closure: Emerging strategies in the US West. *Resources Policy, 57*, 69–80.

Hamilton, L. A., Valova, R., & Rábago, K. R. (2017). *Transition support mechanisms for communities facing full or partial coal power plant retirement in New York*. Elizabeth Haub School of Law at Pace University.

Heinrichs, H. U., Schumann, D., Vögele, S., Biß, K. H., Shamon, H., Markewitz, P., Többen, J., Gillessen, B., Gotzens, F., & Ernst, A. (2017). Integrated assessment of a phase-out of coal-fired power plants in Germany. *Energy, 126*, 285–305. https://doi.org/10.1016/j.energy.2017.03.017

Hess, D. J. (2018). Energy democracy and social movements: A multi-coalition perspective on the politics of sustainability transitions. *Energy Research & Social Science, 40*, 177–189. https://doi.org/10.1016/j.erss.2018.01.003

Jewell, J., & Cherp, A. (2020). On the political feasibility of climate change mitigation pathways: Is it too late to keep warming below 1.5°C? *Wiley Interdisciplinary Reviews: Climate Change, 11*(1). https://doi.org/10.1002/wcc.621

Jindal, A., & Shrimali, G. (2022). Cost–benefit analysis of coal plant repurposing in developing countries: A case study of India. *Energy Policy, 164*. https://doi.org/10.1016/j.enpol.2022.112911

Kefford, B. M., Ballinger, B., Schmeda-Lopez, D. R., Greig, C., & Smart, S. (2018). The early retirement challenge for fossil fuel power plants in deep decarbonization scenarios. *Energy Policy, 119*, 294–306.

Noel, L., & McCormack, R. (2014). A cost benefit analysis of a V2G-capable electric school bus compared to a traditional diesel school bus. *Applied Energy, 126*, 246–255. https://doi.org/10.1016/j.apenergy.2014.04.009

Pao-Yu, O., Brauers, H., Herpich, P., Hirschhausen, C. V., Prahl, A., Wehnert, T., & Kemfert, C. (2019). *Phasing out coal in the German energy sector: Interdependencies, challenges and potential solutions.*

Raimi, D. (2017). *Decommissioning US power plants: Decisions, costs, and key issues.* Resources for the Future. https://www.rff.org/documents/578/RFF20Rpt20Decommissioning20Power20Plants.pdf

Ribera, T. (2020). *A just transition with climate and social ambition—Teresa Ribera. Social Europe.* https://www.socialeurope.eu/a-just-transition-with-climate-and-social-ambition

Rinscheid, A., & Wüstenhagen, R. (2019). Germany's decision to phase out coal by 2038 lags behind citizens' timing preferences. *Nature Energy*. https://doi.org/10.1038/s41560-019-0460-9

Rogelj, J., den Elzen, M., Höhne, N., Fransen, T., Fekete, H., Winkler, H., Schaeffer, R., Sha, F., Riahi, K., & Meinshausen, M. (2016). Paris Agreement climate proposals need a boost to keep warming well below 2°C. *Nature, 534*(7609), 631–639. https://doi.org/10.1038/nature1830

Schindler, H. (2019). *Managing the coal phase-out: A comparison of actions in G20 countries.* International Secretariat of Climate Transparency. https://www.climate-transparency.org/wpcontent/uploads/2019/05/Managing-the-phase-out-of-coal-DIGITAL.pdf

Shaker, Y., & Berisha, E. (2025). Just Green Transitions: Between terminological inexactitudes, conceptual fragmentation, and the exigency for a theoretical framework. European Spatial Research and Policy, 32(1), 121–138. https://doi.org/10.18778/1231-1952.32.1.05

Sharpe, S. A., & Martinez-Fernandez, C. M. (2021). The implications of green employment: making a just transition in ASEAN. *Sustainability*, *13*(13), 7389. https://doi.org/10.3390/su13137389

Shrimali, G. (2020). Making India's power system clean: Retirement of expensive coal plants. *Energy Policy, 139*, Article 111305.

Sidhu, A. S., Pollitt, M. G., & Anaya, K. L. (2018). A social cost benefit analysis of grid-scale electrical energy storage projects: A case study. *Applied Energy, 212*, 881–894. https://doi.org/10.1016/j.apenergy.2017.12.085

SSO RNM. (2024). *Energy, environmental, labour statistics, etc.* Available at database https://makstat.stat.gov.mk

Tol, R. S. (2001). Equitable cost-benefit analysis of climate change policies. *Ecological Economics, 36*(1), 71–85. https://doi.org/10.1016/s0921-8009(00)00204-4

UNEP. (2020). *United Nations environment programme. emissions gap report*. United Nations Environment Programme.

Van Veelen, B., & van der Horst, D. (2018). What is energy democracy? Connecting social science energy research and political theory. *Energy Research & Social Science, 46*, 19–28. https://doi.org/10.1016/j.erss.2018.06.010

Wang, X., Lu, M., Mao, W., Ouyang, J., Zhou, B., & Yang, Y. (2015). Improving benefit-cost analysis to overcome financing difficulties in promoting energy-efficient renovation of existing residential buildings in China. *Applied Energy, 141*, 119–130. https://doi.org/10.1016/j.apenergy.2014.12.001

CHAPTER 14

The Risks of Purposive Transitions in Albania—What to Expect by the Just Green Transition

Fiona Imami, Anila Bejko, and Besnik Aliaj

14.1 Introduction

The open-ended processes of power, political and societal transformation that initiated with the fall of the socialist regime/s around the beginning of 1990s, marked down the so-called post-socialism transition (Müller, 2019; Petrov, 2014) of the affected Central-East European (CEE) countries. The term "transition" in particular, was largely employed among scholars in political sciences, policy-makers and even lay people as a concept encompassing an historical period after the fall of communism (Petrov, 2014), to the extent that it would be used to label the CEE countries as "transition countries" (Carothers, 2002). This transition was signaling epochal change, and at the same time unfolding uncertain futures for the affected economies, which with time embarked on varying hybrid political models, regardless of the path designated by the Western agencies, relying on privatization, liberalization, and democratization (Müller, 2019). Within this context, Albania has moved from

F. Imami (✉) · A. Bejko · B. Aliaj
Co-PLAN, Institute for Habitat Development, Tirana, Albania
e-mail: fiona_imami@co-plan.org

A. Bejko
e-mail: anila_gjika@co-plan.org

B. Aliaj
e-mail: besnik_aliaj@universitetipolis.edu.al

B. Aliaj
Faculty of Planning, POLIS University, Tirana, Albania

E. Berisha et al. (eds.), *The Palgrave Handbook of Just Green Transitions in the Western Balkans and Beyond*,
https://doi.org/10.1007/978-3-031-95075-9_14

an almost feudal system in 1945, towards an extreme centralized economy. Then, in 1991, transformed "overnight" into another extreme: the liberal market economy.[1] After five decades of brutal repression, the post-communist vacuum of legislation, policies, and institutions has initiated a transition of people desperate for democracy and liberalism. The first decade of 90-s, ended up in a climate of increasing informality, characterized by problematic tenure recognition, illegal development, forced evictions, and rampant corruption. These matters have all been made worse by the failure of several political experiments which have contributed to making the extralegal economy the dominant reality in Albania, comprising more than half of the real estate and properties in the country. Although this does not seem to be a typical issue for EU countries, it remains the main technical and political struggle for Albania while it aims to join the EU. Hence, by the time transition began in mid-1992, Albania lacked a tradition of participatory governance. Given the past system's severe abuses, citizens perceived participatory governance not as an opportunity for engagement but as an extension of the forced labor that had been a cornerstone of their oppression. As with the rest of Eastern Europe, the radical changes of the early 1990s completely transformed Albania's political, social and economic realities. Albania's particular circumstances have made for a volatile transition, punctuated by chaotic and dramatic events, most notably the unprecedented economic collapse caused by the infamous pyramidal investment schemes[2] and the subsequent social unrest in 1997, which left the country temporarily without a government and required UN intervention and a multi-lateral peacekeeping force to restore order. It must be clear that Albania's transition has been uniquely complex and markedly different from that of most European countries. Unlike others, it emerged from the most rigid orthodox system of centralized planning in the former Eastern Bloc, making the shift to a market economy and democratic governance particularly challenging (Aliaj & Kacani, 2024). The 1999 Kosovo*[3] War further tested Albania's fragile social fabric, as the country absorbed a massive influx

[1] In line with the liberal policies, after massive expropriation of private properties by the collectivist communist state, the new pluralist authorities initiated a failed privatization process, trying to establish the rule of law but without a Cadastral Office. This has marked one of the most profound challenges, yet a pressing issue for the Albanian socio-economic context.

[2] The Albanian pyramid schemes of the mid-1990s were among the most devastating financial collapses in post-communist Eastern Europe. By 1996, an estimated two-thirds of Albanian households had invested in these fraudulent investment schemes, which promised exorbitant returns but operated as classic Ponzi schemes. The schemes thrived due to weak financial regulation, political complicity, and a lack of public financial literacy, fueled by Albania's rapid and unregulated transition to a market economy. By early 1997, the schemes collapsed, wiping out an estimated $1.2 billion—equivalent to nearly half of Albania's GDP at the time. The crisis led to widespread social unrest, mass protests, and the near-total collapse of state institutions, culminating in an intervention by a UN-sanctioned multinational peacekeeping force (Operation Alba).

[3] This designation is without prejudice to positions on status, and is in line with UNSCR 1244 and the ICJ Opinion on the Kosovo declaration of independence.

of refugees, compounding existing economic and social pressures. However, with the end of the Yugoslav Wars, regional stability improved, allowing Albania to enter a period of relative progress. The early 2000s saw a notable shift towards stability, economic growth, and social development, with GDP expanding at an average annual rate of 6 percent and significant advancements in socio-economic indicators. But, in 2010, problems began to resurface as the global recession, and the logic of stabilocracy[4] underlying financial and security aspects, slowing down both the economy and the democratization processes. However, since 2009, Albania has joined NATO as a member country, while it remains highly committed and progressing to the opening negotiations with the EU, in a world that becomes less and less stable and unpredictable. Yet, while still being in limbo with numerous political and economic reform, (and fairly to say still in transition for most of the started processes), by the end of the twentieth century, the GoA along with all the other WB governments, lined up as supporters and implementers of the European agenda for green transition. This is, of course, both a necessity to unite in a pan-European commitment to address climate challenges and ensure the planet's environmental sustainability, as it is a crucial requirement for integration. Hence, as (Geels, 2011) marks, the newly presented "green transition" is equally purposive for all Western Balkan countries, considering its embeddedness within the EU green agenda, the EU integration process, and the structural and technical weaknesses in the region which inhibit it from producing bottom-up niches of change. In this context, through this chapter we aim to examine the JGT within the broader context of Albania's post-socialist transformation, drawing parallels between these two pivotal periods of systemic change. First, we do so, by trying to decode the theoretical concept/ term of transition(s) in the following subchapter (14.2) below. Secondly, through a comparative approach, we try to examine Albania's post-1990 transformation, highlighting key governance, socio-economic, and spatial challenges, all while keeping notes on lessons relevant to the JGT and the process of exploring the multifaceted implications of JGT. Through this comparative lens we aim to then identify shared challenges and distinct features of the two transitions to draw actionable insights. The final section synthesizes the conclusions and a trajectory for further discussion while offering, at the same time, a few policy recommendations and underscoring the importance of inclusive, climate-sensitive governance for Albania's sustainable future.

[4] Stabilocracy is a term used to describe a political system where authoritarian or semi-authoritarian governance is tolerated—often by external actors like the EU or international organizations—as long as the regime provides regional stability. In these systems, democratic principles, rule of law, and human rights may be compromised or selectively applied, but the government remains in power with the tacit approval of external actors due to its role in maintaining geopolitical balance. The term is often used in the context of Western Balkans politics, where some governments exhibit democratic backsliding while still being supported by international stakeholders in exchange for stability and cooperation on key regional issues. For more information: https://www.cirsd.org/en/horizons/horizons-winter-2018-issue-no-10/the-rise-and-fall-of-balkan-stabilitocracies?utm_source=chatgpt.com.

14.2 Decoding Transition(s): A Theoretical Exploration of the Term

Transition studies initiate in biology and population dynamics, to continue with demographics, health, and most importantly [organizational] management and technological innovation scholarship (Rotmans, Kemp & van Asselt, 2001). Nowadays, transitions, broadly defined, are processes of profound and systemic change that reconfigure societal, economic, and environmental systems over time, presumably improved, to incorporate new features that are either desirable or contribute to solving the current problems (Loorbach et al., 2017). In the academic discourse, the term has been extensively analysed (Table 14.1) within the frameworks of socio-technical transitions, governance transitions, and socio-ecological systems (Geels, 2002); (Smith et al., 2005). The concept of socio-technical transitions, central to transition studies, focuses on the co-evolution of societal practices, technological advancements, and institutional frameworks. Sustaining this, (Geels, 2002) introduced the Multi-Level Perspective (MLP) as a foundational approach to understanding transitions, emphasizing three interconnected levels: niches (spaces of innovation), regimes (dominant systems), and landscapes (broader socio-political and environmental contexts). There are two particular features of transition that we want to dive deeper into in this chapter, which serves the purpose of discussing the purposive transition of JGT in Albania, and that is the concept of (i) sustainability transitions and (ii) justice in sustainability transitions.

In ***sustainability transitions****, the change is societal and multilevel, multi-institutional, and cross-sectorial, affecting multiple regimes* such as socio-technical, socio-ecological and political and governance regimes, as well as the set of values, ideas and perceptions about the world (Cedergren et al., 2022); (Hansen & Coenen, 2015); (Kemp et al., 2007); (Rotmans and Kemp, Managing Societal Transitions. Dilemmas and Uncertainties: The Dutch Energy Case Study, 2003), and ethics and morality of the society. These multiple transformations make sustainability transitions highly complex, and also hardly foreseeable, even in the best cases of planning for the transition (Hunsucker & Loos, 1989), due to the dynamic interplay between its dimensions of time, space, sectors and systems affected, actors, and purpose or targets. Sustainability transitions involve a number of special features such as spatiality, large scale of multiple transformations, focus on finite natural resources which are owned collectively, the legacy of the complexity that characterizes environmental sustainability, and the tensions that arise between growth sectors and private actors on one side and the public on the other regarding property rights and ethical implications on resources. The extent the stock of natural resources has to be maintained, can be adapted and can absorb the disturbance that causes the change, is a primary question of development for sustainability (Derissen et al., 2009) and it is responded by the resilience of the system. This means that while a system is dynamically learning to become

Table 14.1 Just transition timeline (*Source* UNRISD,[5] 2018)

Period	*Description*
1990s	In the United States, unionists and activists advocate for mechanisms to address job losses caused by new environmental regulations. The Just Transition Alliance is founded, uniting labor unions and environmental groups)
2000s	International trade union coalitions introduce the Just Transition concept into global climate discussions, including at COP15 (Copenhagen, 2009)
2011	At COP17 (Durban), South Africa initiates national discussions on a Just Transition framework
2015	The ILO Guidelines on Just Transition are adopted. The 2030 Agenda for Sustainable Development reinforces the concept, which is officially included in the Paris Agreement
2018	The Solidarity and Just Transition Silesia Declaration is signed by 50 countries at COP24 (Katowice, Poland), emphasizing the need for fair workforce transitions in climate policies
2019	At the UN Climate Action Summit, 46 nations commit to developing Just Transition strategies
2021	Just Transition gains prominence at COP26 (Glasgow): The Just Transition Declaration is signed by a group of developed countries to support fair transition conditions globally. The Global Accelerator on Jobs and Social Protection for Just Transitions is launched. The EU, US, UK, Germany, and France commit €8.5 billion to support South Africa's Just Transition
2022	The IPCC report, High-Level Political Forum (HLPF), and UN Secretary-General's priorities reinforce Just Transition as a key component of global climate action

robust and to increase its adaptive capacity (Perrings, 2006), it continuously self-modifies and reaches new equilibriums.

> *Sustainability* "meets the needs of the present generation without compromising the ability of the future generations to meet their own needs" and is therefore neither a fixed state of harmony, nor an identifiable state, coping with both the certainty and the uncertainty of change. (Kemp et al., 2007; United Nations, 1987, 24; Handmer & Dovers, 1996)

Sustainability, therefore, can be seen as the goal of the transition(s), but its exact meaning varies depending on the specific time and place. This is because sustainability evolves through these adjustments, with smaller, shifting objectives emerging to meet changing societal needs and demands. As such, sustainable development is not a fixed endpoint but a fluid and ongoing process of transformation, where sustainability transitions represent the continuous effort to balance ever-changing human demands with the biosphere's capacity to adapt and respond.

[5] report-jtrc-2018.pdf.

Due to the very nature of sustainability, its transitions impact and engage all societal layers and groups, with stakeholders taking on diverse roles—those implementing the transition are often not the same as those conceptualizing it, particularly in the case of complex purposive transitions. And it is in this context where we ought to bring the "justice" into the sustainability transitions.

A cry for ***"just transitions"*** popularized in the 1970s and 1980s by trade unions in the United States and Canada concerned about the consequences of environmental protection policies on jobs, and reflected the development of a justice-based political discourse in the nineteenth century, often linked to the distribution of income and wealth [(Abram et al., 2022); (Ciplet & Harrison, 2019); (Smith, 2017); (Knight, 1998)]. Since then, the concept of "just transition" is established in the International Labour Organisation (ILO) Guidelines[6] for a Just Transition Towards Environmentally Sustainable Economies and Societies for all. The Sustainable Development Goals (SDG) of the United Nations also represent the agenda for just transition since 2015, where the goals of decent work for all, clean energy for all, climate protection and poverty eradication have a particular attention on the matter (Smith, 2017).

With the SDG and furthermore the COP21 Paris Agreement, the need for change and shifting away from fossil fuel-based economies is not anymore simply an option and neither is questionable, but it rather is a designated overall societal and policy direction to be achieve through sustainability transitions and embracing multiple forms of justice (Loorbach et al., 2017); (Abram et al., 2022).

(Smith, 2017), emphasizes that the society has options in terms of how it manages the transition, so that not only the sustainability goals are achieved, but the just and fair dimensions of the transition are maintained as well. Choosing among the options or crafting the right ones, is a matter of transition governance, supported by transition policies, and most importantly involving transition stakeholders into transparent processes of communication and decision-making that aim to solve or address the transition tensions. These tensions are firstly rooted in the very core notion of sustainability, where a set of values such as utility, equality, fairness, welfare, poverty reduction, quality of life, and moral, ethical, cultural, aesthetic, and purely scientific reasons for

[6] The ILO (ILO) Guidelines for a Just Transition Towards Environmentally Sustainable Economies and Societies for all, endorsing seven policy principles: (i) strong social consensus on the goal and pathways towards sustainability; (ii) policies that respect principles and rights at work; (iii) policies and programs that take into account the gender dimension; (iv) coherent sectorial and cross-sectorial policies that provide a transition enabling environment for enterprises, workers, investors and consumers; (v) sectorial policies that provide a just transition framework for all; (vi) policies and programs that reflect the specific conditions of countries; and (vii) sustainable development strategies that foster international cooperation among countries. For more information read: https://www.ilo.org/publications/guidelines-just-transition-towards-environmentally-sustainable-economies.

nature conservation are at stake. Agreeing with (Abram et al., 2022),[7] the concept of just transition lacks a unified theoretical definition, partly due to the millennia-long debates surrounding the notion of justice and partly due to the ongoing theoretical exploration/evolution of theoretical perspectives on transitions. Hence for the purpose of this chapter, we will focus on the definition of "purposive transition", and investigate how the later was/is being manifested in both post-socialist transition and the just green transition.

14.2.1 Defining Purposive Transition

Stemming from (Petrov, 2014)'s definition of "transition" term and the extent of its use to mark the CEE countries, this type of transition was signaling epochal change, and at the same time unfolding uncertain futures for these affected economies. The post-socialism transition in the Western Balkans is what (Smith et al., 2005) call a "purposive transition", characterized by a deliberate goal pursued as of the outset of transformation, with an explicit set of societal expectations and drawing heavily on external resources, aid and political orientation. It is from this type of transition that the Transition Management approach takes its leap (Rotmans, Kemp & van Asselt, 2001), analyzing, among others, the role of external stakeholders and resources in carrying out (maybe even initiating) the transition. (Petrov, 2014), on the other hand, characterizes the post-socialist purposive transition as "teleological" for the CEE, with the regime change appearing as "purposeful, preordained, and therefore predictable, virtually unstoppable, and impelled by a future goal", which serves as the "defined end of the transition process" (p. 30). He also denotes further that the post-socialism transition of the CEE countries did not advance in the pathways or towards the goal defined by the Western agencies, but rather the uncertain futures affecting these economies, embarked on varying hybrid political models, regardless of the path designated by the Western agencies, lying on privatization, liberalization, and democratization (Müller, 2019). We would argue that this was, among others, because the transition itself, was not informed by its wider historical context, and because it was reduced to a unique joint model to follow, regardless of the heterogeneity and idiosyncrasies within the region, or Albania's context as a specific case. And it is these contextual, socio-cultural, and territorial factors, which should be accounted for in the respective of furthering the efforts towards transition. (Carothers, 2002) would call the latter "the underlying conditions ... the economic level, political history, institutional legacies, ethnic make-up, sociocultural traditions, or other 'structural' features", all of which

[7] To address this, they propose a comprehensive framework for justice in transitions that encompasses all facets of contemporary social justice, categorizing these dimensions into recognition, procedural, distributive, and restorative justice. Additionally, the spatial dimension of justice comes across the four dimensions mentioned above.

are equally relevant and important in defining how a possible new transition would unfold within the country. If utilizing the Multilevel Perspective (MLP) approach of socio-technical transitions, including that on sustainability for which we discuss earlier, these underlying factors correspond with the socio-technical landscape in which regimes are set and continuously change (transit), and count for the spatial structure of territories, economic growth and social factors, normative values and environmental challenges, as well as potential disasters and political dynamics. In Albania, the underlying conditions, coupled with the fatigue from a stagnating EU integration process, conflicting narratives stemming from the post-socialist era, dominant-power politics, institutional corruption, and deep territorial disparities, have created a sense of a "never-ending transition". However, Albania's history does not suggest resistance to change—on the contrary, the country has demonstrated openness to exploring different political, economic, and social trajectories. What has been missing, rather, is a cohesive governance and policy framework capable of structuring and guiding change in a systemic and strategic manner. This absence of a clear governance skeleton has historically hindered Albania's capacity to follow a structured transition path, and unfortunately, similar gaps persist in the case of the Just Green Transition as well.

14.3 Albania's Post Socialist Transition vis a vie the "New" Transition(s)

In the context of Albania, the post-1990 transition from a socialist to a market-driven economy exemplifies a regime shift precipitated by external pressures, including the collapse of socialist states in Eastern Europe and Albania's reintegration into global systems (Stark & Bruszt, 1998). Similarly, the current climate imperative under the JGT represents a landscape-level pressure requiring systemic changes to governance, energy systems, and societal behaviors. Through a comparative lens, the following sections offer a wide view panorama on both transitions.

14.3.1 A Retrospective on Albania's Post Socialist Transition

The collapse of Albania's communist regime marked one of the most dramatic transitions in Eastern Europe. This period of profound systemic change saw the country grapple with the challenges of democratization, economic liberalization, and the reorganization of its social and spatial structures. Albania's transition was characterized by political instability, rapid urbanization, economic hardships, and a significant shift in societal values, all of which left (and still has) a lasting impact on its development trajectory. On a systemic level, Albania's transition was uniquely complex and distinct from that of most European countries. Emerging from one of the most rigidly centralized and repressive communist regimes in the former Eastern Bloc, Albania's governance had been shaped by an extreme form of autarky, where decision-making was highly centralized, and citizens were systematically excluded from political processes. For 45 years, Albanians were denied fundamental freedoms of speech, religion, and movement, living under an isolationist regime often compared to present-day North Korea in terms of political and economic seclusion. The shift to a multi-party democracy was turbulent, marked by political instability, frequent government changes, and contested elections, which further eroded public trust in institutions (Biberaj, 1999). The adoption of a new constitution in 1998 was a critical milestone, laying the groundwork for Albania's gradual progress towards European integration. Simultaneously, efforts to decentralize governance sought to empower local authorities and promote regional development. However, these efforts were inconsistently implemented, with local governments lacking the necessary resources and institutional capacity to carry out their responsibilities effectively. Corruption and governance deficits further hindered the state's ability to implement coherent policies, fueling widespread public disillusionment with the transition process. In the same time, the spatial landscape of Albania underwent one of the most profound changes. Under communism, strict migration controls, and centralized planning had shaped the country's urban and rural dynamics. The removal of these controls in the 1990s led to rapid urbanization, with cities like Tirana and Durrës experiencing exponential growth as people migrated in search of better opportunities (Bejko, 2015). This population surge strained urban infrastructure and services, leading to the proliferation of informal settlements and unregulated construction. At the same time, rural areas were left behind, experiencing depopulation and neglect. The urban–rural divide widened, as rural regions struggled with inadequate access to infrastructure, education, and healthcare. The lack of coordinated spatial planning and governance further exacerbated these inequalities, leaving Albania with deeply entrenched spatial disparities. Economically, Albania's transition from a centralized to an open-market economy was abrupt and poorly managed. Privatization and liberalization were implemented rapidly but without proper regulatory frameworks, leading to widespread job losses and economic insecurity (Berisha et al., 2018, 2024). The 1990s saw severe financial instability, with hyperinflation, currency devaluation, and the collapse of pyramid schemes in 1997, which devastated household savings and deepened poverty. The private sector remained weak, while remittances became a crucial economic lifeline. Limited foreign investment, weak institutions, and unclear property rights further hindered sustainable growth, leaving Albania's economy fragile and highly dependent on external support. However, the 2000s saw modest recovery, driven by sectors such as agriculture and later, services and tourism. Remittances from Albanian migrant workers, particularly those in Greece and Italy, became a critical source of income for many households, highlighting the country's dependence on its diaspora for economic stability. The social fabric of Albania underwent profound changes during the transition. The shift from a collectivist, state-controlled system to an individualistic, market-driven society disrupted traditional social structures. Rising unemployment, poverty, and inequality created a sense of social fragmentation, while mass emigration transformed the country's demographics. Between 1991 and 2001, an estimated 600,000 Albanians emigrated, primarily to neighboring countries. This exodus included a significant portion of the skilled workforce, leading to a brain drain that hindered Albania's long-term development prospects. While remittances provided short-term economic relief, the loss of human capital presented a critical challenge for the country's future. Normative values also shifted as Albania embraced Western democratic ideals. However, the rapid modernization often clashed with entrenched patriarchal and rural traditions, creating tension between old and new societal norms. The process of adapting to these changes was slow and uneven, reflecting the complexities of transitioning from one ideological system to another. As a result all, the rapid urbanization of the 1990s further exacerbated environmental challenges (Berisha & Cotella, 2021, 2024). Under communism, the emphasis on heavy industry and large-scale agricultural production had left a legacy of environmental degradation. Deforestation, soil erosion, and pollution were widespread issues, particularly in industrial and mining areas. Informal settlements and unregulated construction contributed to waste accumulation, air pollution, and water contamination, particularly in urban centers like Tirana. Meanwhile, the lack of environmental awareness or policies during the early years of transition meant that these issues were largely ignored.

Only in the 2000s did Albania begin adopting environmental protection policies, driven in part by its aspirations for EU membership. These efforts included waste management reforms and the establishment of protected areas. However, implementation and enforcement remained weak, and significant gaps persisted in aligning environmental governance with sustainable development goals. Additionally, starting with the signing of the Stabilization and Association Agreement with EU- 2006/2007, following with the NATO integration (2009), and in 2014 gaining the EU candidate status, Albania was offered a trajectory for integration into the European Union, a commitment for which the country is still working on. This process, however, again consists of a prescribed list of structural and political reforms (Commission, 2018), that despite the financial and technical support provided, only recently has gained a new momentum, with the opening of negotiations for few[8] of the chapters in 2024.

To conclude, Albania's transition has unfolded (and still is) through three distinct yet interconnected phases: the political and ideological shift from a

[8] Albania applied for EU membership in April 2009 and was granted candidate status in June 2014. The first intergovernmental conference took place in July 2022. Negotiations officially began on October 15, 2024, during the second Accession Conference, focusing on Cluster 1: Fundamentals, including areas such as the functioning of democratic institutions, public administration reform, judiciary, justice, economic criteria, public procurement, statistics, and financial control. The third conference followed on December 17, 2024, opening Cluster 6 on external relations, including chapters on foreign, security, and defence policy.

rigid monist system to pluralism and democracy in the 1990s, the economic transformation from a centralized economy to a market-driven system in the 2000s, and the ongoing transition towards European and global integration over the past two decades.

At a global scale, Albania's transformation aligns with the major historical transition shifts of the post-war world: first, the Cold War era (1945–1989), which was dominated by the ideological and security struggle between the United States and the Soviet Union; second, the post-Berlin Wall period (1989–2006), which championed neoliberalism, human rights, and environmental concerns under a USA-led global economic order; and third, the post-2006 global recession era, marked by the decline of liberal interventionism and the rise of digital economies, geopolitical realignments, and increasing global competition between the Western alliance (USA/NATO) and emerging powers such as China and BRICS nations.

Albania's trajectory reflects both the opportunities and contradictions inherent in these transitions. While political pluralism, economic liberalization, and EU integration efforts have set Albania on a progressive path, deep-rooted governance weaknesses, institutional inefficiencies, and socio-economic disparities have hampered its ability to fully capitalize on the benefits of transition. Unlike many post-socialist countries in Central and Eastern Europe, Albania has not undergone a linear or structured transformation, but rather a fragmented and often reactive adaptation to shifting global trends bringing unstable economic development (Table 14.2). As the country now navigates the challenges of sustainable development and green transition, it faces yet another test of its institutional resilience and capacity for governance.

Table 14.2 Economic context of Albania: Indicators (*Source* IMF—World Economic Outlook Database)

Main Indicators	*2023 (E)*	*2024 (E)*	*2025 (E)*	*2026 (E)*	*2027 (E)*
GDP (billions USD)	22.82	26.13	27.99	29.92	31.85
GDP (Constant Prices, Annual % Change)	3.5	3.3	3.4	3.5	3.5
GDP per Capita (USD)	8,299	9,598	10,386	11,211	11,938
General Government Gross Debt (in % of GDP)	59.8	58.3	57.0	56.2	55.5
Inflation Rate (%)	4.8	2.2	2.4	2.6	3.0
Unemployment Rate (% of the Labor Force)	10.7	10.7	10.7	10.7	10.7
Current Account (billions USD)	−0.29	−0.21	−0.28	−0.36	−0.22
Current Account (in % of GDP)	−1.2	−0.8	−1.0	−1.2	−0.7

14.3.2 *Contextualizing the JGT in Albania*

While Just Green Transition concept and approaches are embraced at a global level, the process towards this transition in the WBs is progressing at a very slow pace and little is done in regard to societal preparedness for JGT. In Albania, contextualizing JGT (let alone conceptualizing it) is rather a complex task. Apart from requiring the involvement of several stakeholder and institutions from all countries in the region, both contextualization and conceptualization process is faced with the (i) theoretical gaps that subsist in the definition of basic concepts such as transition and justice, for which we try to navigate in the aforementioned chapter and (ii) lack of awareness in understanding the process it requires or the policy overview specter and the place based empirical evidences such transitions bring to the societal level. The JGT, both as a concept and as a policy framework, is being introduced in the Balkans via the Green Agenda for the Western Balkan. Formalized in 2020 with the Sofia Declaration, the GAWB is the basis for a new growth strategy for the region, embedded in the EU Economic and Investment Plan for the Western Balkans. The same year, during the Berlin Process Summit in Poznań, the WB Leaders (Albania's included) expressed their common willingness and readiness to meaningfully contribute to an ambitious Green Agenda. This entails the adoption of EU directives and transposition of regulation across policy areas and on a wide range of issues, from climate law, meeting energy efficiency goals, and air and water quality standards to the waste management framework, limiting pollution, increasing circularity, etc. This first step needs to be followed by implementation, leading to "real", tangible change in all industrial/economic sectors. Despite initial enthusiasm, the Western Balkan countries continue to lag behind the EU in their efforts to combat, mitigate, and adapt to climate change. This shortfall stems not only from inadequate progress in implementing climate actions but also from the absence of a comprehensive, integrated approach that transforms both national and regional policies into climate-sensitive frameworks capable of addressing the multifaceted challenges of environmental sustainability. Reaching the aforementioned goals cannot be accomplished overnight, as the implementation of such an ambitious agenda comes with multifaceted implications across social, environmental, economic, financial, technological, and even [geo]political dimensions. These complexities present significant challenges both regionally and nationally, primarily rooted in the lack of political commitment to improve or transform policies and insufficient financial resources to support basic services and economic activities, let alone the additional costs required to foster or nourish innovation.

14.4 Emerging Challenges and Opportunities

For countries such as Albania, embarking the just green transition does not come without challenges. Indeed, as continuously argued earlier, it is challenging for a country with yet unaccomplished transitions (towards democracy and unconsolidated governance, the EU integration or even the dream of a state with economic prosperity), to imagine re-entering a new transition, though of a different meaning and magnitude. To start with, there is a mentality barrier to overcome, which relates to a stigmatized "political transition": for both its temporal length and its outcome in terms of hybrid post-authoritarian regimes, diverged from the initially set objective of market capitalism and democracy (Toto et al., 2023). Overall, the promises of transitioning to a liberal democracy, a free-market economy, and eventual EU membership have sparked high expectations over the past few decades, only for many to become disillusioned by the inadequate performance of economic and social policies implemented and the uncertain future they have created. This situation, as we argue in one of our Policy Briefs (Imami et al., 2023), has produced a "transitions fatigue" or a general sense of skepticism, or even mistrust, towards policies prescribed by external entities, as well as the ability of the political class in countries like Albania to design comprehensive and place-sensitive policies. A meaningful challenge for just transitions is securing the funding and resources needed to support the transitions, and how are these allocated or distributed fairly and justly. Additionally, there are the impacts on workers, communities and consumers, coming out of the transformation of the involved sectors in terms of technologies, policies that guide the sectors, the employment and job opportunities or loses, and the affordability of green services. The efforts to securing funding and resources, and the impacts extend to the territories affecting quality of life from the personal to the ecological sphere, and from life enablers and maintenance to life flourishing dimensions (Sessa et al., 2020). In this context, without significant efforts to drive internal transformation, the Just Green Transition, put forward through the GAWB, risks becoming yet another unfulfilled transition. Moreover, the burden of action—or inaction—could disproportionately fall on the most vulnerable segments of society.

14.5 Conclusions

Definitions on JTG differ, but there exists a widespread recognition of the importance of context-specific analysis and strategies. Today just transition refers generally to strategies, policies or measures to ensure no one is left behind or pushed behind in the transition to low-carbon and environmentally sustainable economies and societies. The term has been also used in relation to other types of transitions, such as the transition towards digital and green economy (for example in the Global Accelerator on Jobs and Social Protection for Just Transitions launched by the UN Secretary-General in 2021). The origins of the concept lie in trade union activism, and the original emphasis was on addressing the loss of jobs resulting from the implementation of environmental regulations and policy. This remains a central element, but as more countries and a broader range of stakeholders have been engaged in understanding the challenges at hand and developing strategies for specific contexts, the definition of what a just transition consists of has expanded. When the ILO published its Guidelines in 2015, it defined just transition as "greening the economy in a way that is as fair and inclusive as possible to everyone concerned, creating decent work opportunities and leaving no one behind". The principles in its Guidelines refer, among others, to the need for social dialogue; respect, promotion and realization of rights at work; policies that address gender dimensions and promote the creation of decent jobs and provide social protection and skills development; and international cooperation. Approaches have varied in terms of how and to what extent they incorporate different dimensions of justice considered (distributive, procedural, and/or restorative. They also vary in terms of the degree to which a green transition would be transformative in terms of justice or equity: from a status quo approach whereby job losses are compensated for, to approaches that imply a deeper transformation of political and economic systems, or that consider the shift to a low carbon future an opportunity to correct historical inequities or injustices (The Just Transition Research Collaborative, 2018). Landing the narrative of a "new" transition in a country that has been subject to several transitions through history, and with a questionable level of success, remains a key concern. In these circumstances, the term of "Just Green Transition" is so far rather used as a buzzword, without fully generating the necessary awareness of the scale and transformative nature it implies (Imami et al., 2023).

The post-socialist transition, marked by abrupt shifts in governance, economy, and societal norms, left Albania grappling with systemic challenges, including governance deficits, social inequalities, and environmental degradation. These challenges have left a legacy that continues to shape the country's capacity to undertake new transitions. The JGT, like the transitions before it, is not merely a technical or economic transformation, it is a systemic, cross-sectoral, and deeply societal process. However, unlike past transitions, the JGT carries the added urgency of addressing climate change and the imperative of achieving sustainability in a just and equitable manner. At its core, the JGT is both an opportunity and a test of Albania's resilience and readiness to embrace transformative change. The lessons from the post-socialist era, including the importance of participatory governance, equitable policy frameworks, and robust institutional capacity, offer valuable insights for navigating this new transition. However, these lessons also highlight persistent risks, such as political inertia, lack of societal preparedness, and the uneven distribution of resources and opportunities.

For Albania, the success of the JGT will hinge on its ability to balance ambitious environmental goals with the socio-economic realities of its population.

This requires not only adopting the European Union's Green Agenda for the Western Balkans (GAWB) but also tailoring it to the country's unique context. The emphasis must be on fostering inclusive policymaking, strengthening institutions, and bridging the rural–urban divide to ensure that no community is left behind in the transition process. Moreover, the concept of "justice" in the JGT must be more than a rhetorical commitment; it must be operationalized through tangible actions that address existing inequalities and prevent new forms of exclusion. This includes providing targeted support to vulnerable groups, investing in education and awareness to build societal readiness, and creating mechanisms for meaningful stakeholder participation in decision-making processes. As this chapter has demonstrated, the JGT represents a critical juncture for Albania. It is an opportunity to not only align with global sustainability goals but also to redefine its development trajectory in a way that prioritizes environmental stewardship, social equity, and economic resilience. However, achieving this vision will require a concerted effort across all levels of governance, society, and international cooperation.

In conclusion, while Albania's journey towards a Just Green Transition is fraught with challenges, it also holds the promise of a more sustainable and equitable future. The JGT should be seen not only as a national process but also as part of the wider agenda of global climate justice, guided by the principle of common but differentiated responsibilities. It must recognize that, unlike high-income countries, many middle- and low-income nations—such as Albania—face persistent poverty, widespread informal labor, limited institutional capacity, and constrained fiscal space. These constraints make the balance between economic stability, social equity, and environmental commitments particularly complex. Hence, more is yet to be researched on JGT by looking in depth into more detailed pillars, such as: (i) partnerships for long-term sustainable urban development, cities and communities; (ii) expanding sustainable solutions via AI and digital technologies; (iii) equitable higher education for an inclusive future; (iv) Innovation towards carbon neutrality and clean energy; (v) Safeguarding biodiversity and improving natural resource management; (vi) Collaboration for universal health and wellbeing, etc. In this sense, the JGT should not be reduced to a rhetorical commitment; it must translate into concrete measures that build resilience, empower local economies, and ensure social inclusion. Learning from past transitions, Albania must focus on creating sustainable economic alternatives, investing in education and skills development, and establishing mechanisms for meaningful stakeholder engagement. Only through a comprehensive, context-sensitive, and globally informed approach can Albania navigate its Just Green Transition successfully, ensuring that it does not become yet another unfinished transformation in its complex transition history.

References

Abram, S., Atkins, E., Dietzel, A., Jenkins, K., Kiamba, L., Kirshner, J., Kreienkamp, J., Parkhill, K., Pegram, T., & Santos Ayllón, L. M. (2022). Just transition: A whole-systems approach to decarbonisation. *Climate Policy, 22*(8) (Informa UK Limited, trading as Taylor & Francis Group) 1033–1049. https://doi.org/10.1080/14693062.2022.2108365

Aliaj, B., & Kacani, A. (2024). How can Albania speed up the EU integration process? The role of reforms in the formalisation of the extralegal economy as an instrument for socioeconomic growth and for establishing the rule of law. In *Reconsidering 'transition': Albania 1990–2020 and the promise of democracy* (pp. 164–194). Konrad Adenauer Foundation (Konrad-Adenauer-Stiftung).

Bejko, A. G. (2015). Planning and financing Infrastructure in the city through land value capture instruments. The case of Tirana—Rinas Corridor. *The Scientific Journal of the Observatory of Mediterranean Basin.* https://doi.org/10.37199/o41001108

Berisha, E., Colic, N., Cotella, G., & Nedović-Budić, Z. (2018). Mind the gap: Spatial planning systems in the western Balkan region. *Transactions of the Association of European Schools of Planning, 2*(1), 47–62. https://doi.org/10.24306/TrAESOP.2018.01.004

Berisha, E., Colic, N., Cotella, G., & Nedović-Budić, Z. (2024). Potential future enlargement—Spatial planning in the Western Balkans. In V. Nadin, G. Cotella, & P. Schmitt (Eds.), *Spatial planning systems in Europe: Comparison and trajectories* (pp. 216–245). Edward Elgar. https://doi.org/10.4337/9781839106255.00022

Berisha, E., & Cotella, G. (2021). Territorial development and governance in the Western Balkans. In E. Berisha, G. Cotella, & A. Solly (Eds,), *Governing territorial development in the Western Balkans. Advances in spatial science.* Springer, Cham.

Berisha, E., & Cotella, G. (2024). Europeanisation of spatial planning. The Western Balkans between innovation and resistance. *Southeast European and Black Sea Studies*, 1–22. https://doi.org/10.1080/14683857.2024.2352890

Biberaj, E. (1999). *Albania in transition—The rocky road to democracy.* Routledge.

Carothers, T. (2002). The end of the transition paradigm. *Journal of Democracy, 13*(1).

Cedergren, E., Tapia, C., Sanchez Gassen, N., & Lundgren, A. (2022). *Just green transition—key concepts and implications in the Nordic Region.* Discussion Paper, NORDREGIO.

Ciplet, D., & Harrison, J. L. (2019). Transition tensions: mapping conflicts in movements for a just and sustainable transition. *Environmental Politics, 29*(3), 435–456. https://doi.org/10.1080/09644016.2019.1595883

European Commission. (2018). *A credible enlargement perspective for and enhanced EU engagement with the Western Balkans.* ec.europa.eu. Accessed December 13, 2024. https://ec.europa.eu/info/sites/default/files/communication-credible-enlargement-perspective-western-balkans_en.pdf

Derissen, S., Quaas, M., & Baumgärtner, S. (2009). *The relationship between resilience and sustainable.* Working Paper Series in Economics, No. 146, Leuphana Universität Lüneburg, Institut für Volkswirtschaftslehre, Lüneburg.

Geels, F. W. (2002). Technological transitions as evolutionary reconfiguration processes: A multi-level perspective and a case study. *Research Policy, 31*(8–9), 1257–1274. https://doi.org/10.1016/S0048-7333(02)00062-8

Geels, F. W. (2011). The multilevel perspective on sustainability transitions: Response to seven criticisms. *Environmental Innovation and Societal Transitions, 1*(1), 24–40. https://doi.org/10.1016/j.eist.2011.02.002

Handmer, J., & Dovers, S. R. (1996). A typology of resilience: Rethinking institutions for sustainable development. *Organization and Environment*, *9*(4), 482. https://doi.org/10.1177/108602669600900403; http://oae.sagepub.com/content/9/4/482

Hansen, T., & Coenen, L. (2015). The geography of sustainability transitions: Review, synthesis and reflections on an emergent research field. *Environmental Innovation and Societal Transitions, 17*, 92–109. https://doi.org/10.1016/j.eist.2014.11.001

Hunsucker, J. L., & Loos, D. (1989). Transition management—An analysis of strategic considerations for effective implementation. *Engineering Management International*, *5*(3), February: 167–178. https://doi.org/10.1016/S0167-5419(89)80014-6

Imami, F.,, Bejko, A., Giacometti, A., & Tapia, C. (2023). *Just green transition in the Western Balkans: Overcoming "transitions fatigue"*. Policy Brief, no 1, Albania: GreenFORCE HORIZON Project. https://greenforcetwinning.net/2023/10/09/just-green-transition-in-the-western-balkans-overcoming-transitions-fatigue-policy-brief/

Imami, F., Bejko, A., & Giacommetti, A. (2023). *Just green transition in the Western Balkans: Overcoming 'transitions fatigue'*. Policy Brief no.1, www.greenforcetwinning.net.

Kemp, R., Loorbach, D., & Rotmans, J. (2007). Transition management as a model for managing processes of co-evolution towards sustainable development. *International Journal of Sustainable Development & World Ecology*, *14*(1), 78–91.

Knight, J. (1998). Justice and fairness. *Annual Review of Political Science, 1*, 425–449.

Loorbach, D., Frantzeskaki, N., & Avelino, F. (2017). Sustainability transitions research: Transforming science and practice for societal change. *Annual Review of Environment and Resources,* 599–626. https://doi.org/10.1146/annurev-environ-102014-021340

Müller, M. (2019). Goodbye, postsocialism! *In Europe-Asia Studies, 71*(4), 533–550. https://doi.org/10.1080/09668136.2019.1578337

Perrings, C. (2006). Resilience and sustainable development. *Environment and Development Economics, 11*(4), 417–427. https://doi.org/10.1017/S1355770X06003020

Petrov, K. (2014). The concept of transition in transition. comparing the postcomunist use of the concept of transition with that found in soviet ideology. *Baltic Worlds, 1*(2014), 29–41.

Rotmans, J., & Kemp, R. (2003). *Managing societal transitions. dilemmas and uncertainties: The Dutch energy case study.* Report in a Workshop, OECD Workshop on the Benefits of Climate Policy: Improving Information for Policy Makers.

Rotmans, J., Kemp, R., & van Asselt, M. (2001). Transition management: A promising policy perspective. In M. Decker & F. Wütscher (Eds.), *Interdisiplinarity in technology assessment. implementation and its chances and limits* (pp. 164–197). Springer.

Sessa, C., Galvini, G., Biosca, O., DelCatillo, H., Ulied, A., Tödtling-Schönhofer, H., & Haider, J. et al. (2020). *ESPON QoL—Quality of life, measurements and methodology.* Project Final Report, ESPON.

Smith, A., Stirling, A., & Berkhout, F. (2005). The governance of sustainable socio-technical transitions. *Research Policy, 34*(10), 1491–1510. https://doi.org/10.1016/j.respol.2005.07.005

Smith, S. (2017). *Just transition, A report for the OECD.* https://www.ituc-csi.org/just-transition-centre

Stark, D., & Bruszt, L. (1998). *Postsocialist pathways: Transforming politics and property in East Central Europe.* Cambridge University Press.

The Just Transition Research Collaborative. (2018). *Mapping just transition(s) to a low-carbon world.* UNRISD.

Toto, R., Imami, F., Bejko, A., Poro, E., Garvanlieva, V., Gjoka, R., & Jeftic, M. (2023). *Report on Western Balkans just green transition conceptualisation.* Project Deliverable, GreenFORCE HORIZON Project. https://cordis.europa.eu/project/id/101059411

United Nations. (1987). *Our common future.* Report of the World Commission on Environment and Development. Oxford University Press.

CHAPTER 15

"Signs by the Roadside" On the Road to Just Green Transitions in Serbia

Zora Živanović, Vladimir Popović, Branko Protić, Aleksandar Djordjević, Velimir Šećerov, and Marija Jeftić

15.1 Just Green Transitions in Serbia

15.1.1 Global and EU Framework

All countries in the world have gone through several transition experiences in various sectors, driven by different motors. In the twenty-first century, there seem to be two main drivers for evolution and transition: digitalization and greening (Cavalheiro, 2020). Many countries are adopting ambitious strategies to reduce greenhouse gas (GHG) emissions and achieve climate neutrality, solving the problems of introducing eco-friendly technologies (including alternative energy), energy efficiency, the formation of a sustainable industry, and

Z. Živanović (✉) · V. Popović · B. Protić · A. Djordjević · V. Šećerov · M. Jeftić
Faculty of Geography, University of Belgrade, Belgrade, Serbia
e-mail: zora.zivanovic@gef.bg.ac.rs

V. Popović
e-mail: vladimir.popovic@gef.bg.ac.rs

B. Protić
e-mail: branko.protic@gef.bg.ac.rs

A. Djordjević
e-mail: aleksandar.djordjevic@gef.bg.ac.rs

V. Šećerov
e-mail: velimir.secerov@gef.bg.ac.rs

M. Jeftić
e-mail: marija.jeftic@gef.bg.ac.rs

E. Berisha et al. (eds.), *The Palgrave Handbook of Just Green Transitions in the Western Balkans and Beyond*,
https://doi.org/10.1007/978-3-031-95075-9_15

the transition to a circular economy (Zvezdanović Lobanova & Lobanov, 2023). Globally, significant frameworks guiding the green transition include the 2030 Agenda for Sustainable Development and the UN Framework Convention on Climate Change, along with its associated Paris Agreement. These frameworks set the stage for countries worldwide to commit to Sustainable Development Goals (SDGs) and take collective action against climate change.

Over the last 20 years, the EU has emerged as a prime mover of the global green transformation (World Bank, 2022). In 2019, the EU adopted the European Green Deal (EGD) as a new growth strategy to address the problems related to climate change and environmental degradation. This comprehensive, evolving, overarching strategy for a carbon–neutral Europe should enable GHG emissions reduction by over half by the end of the decade, climate neutrality achievement by 2050, boost the economy through green technology, create sustainable industry and transport, and cut pollution (European Commission, 2024). All member states are expected to implement policies and undergo profound economic, social, and environmental reforms to contribute to the EGD goals. Recently, the EU has been increasingly concerned about unfair competition from countries hosting potential "pollution heavens" and strives to coordinate shared climate policy and sustainable product standards and impose costs on countries that don't meet them (World Bank, 2022).

15.1.2 Serbia on Its JGT journey

The EGD (European Commission, 2019) particularly emphasizes supporting its immediate neighbours, underlining that the ecological transition for Europe can only be fully effective if the EU's immediate neighbourhood also takes effective action. As EU member states, candidate countries from Southeastern Europe, including Serbia, have to follow in the footsteps of the European climate policy and implement the developed plans to reduce the anthropogenic pressure on the environment (Zvezdanović Lobanova & Lobanov, 2023). Like many of its neighbours, several EU Member States and other countries around the world, Serbia is faced with a particular challenge related to the direct impacts of low carbon development in the coal sector and its workers, and in the coal regions and their communities (Cavalheiro, 2020). Serbia is one of the most energy-intensive economies in Europe, with a high carbon footprint and a high degree of vulnerability to energy price fluctuations. Its economy is four times more energy-intensive as compared to the EU27 average (World Bank, 2022). In other words, Serbia's businesses use much more energy to produce a unit of output or add much lower value to their energy use, than their European peers do, which is the primary cause of most environmental issues.

It is necessary to mention that the general public in Serbia has become increasingly concerned and vocal regarding environmental degradation, intensifying the pressure on the government to act (World Bank, 2022). A recent World Bank public opinion survey shows that Serbia's citizens consider a clean and livable environment as an essential factor in decisions related to where they want to live and bring up their children, with more than 75% of survey respondents stating that they felt that pollution, particularly air pollution, is worsening and with almost 85% stating that they felt that climate change was affecting their immediate environment. A high proportion of younger people (18–29 years) expressed similar sentiments, with a significant proportion even stating that they had considered immigrating as a result of these factors. This increased environmental awareness, particularly among the young, has resulted in increased green social activism, with intensified pressure on the government to take action.

15.1.2.1 JGT Development Framework and Progress in Serbia

Serbia is a signatory to the 2030 Agenda for Sustainable Development and is thus formally committed to the achievement of the Sustainable Development Goals (SDGs). It is also a signatory to the UN Framework Convention on Climate Change and its associated Paris Agreement. As a candidate for EU membership, Serbia is expected to implement certain actions in order to reduce the emission of harmful gases. Namely, in November 2020, Serbia signed the Sofia Declaration on the Green Agenda for the Western Balkans as a concrete plan for expanding the European Green Deal to Southeastern Europe, committing to work towards the 2050 target of a carbon–neutral continent together with the EU. In order to achieve long-term energy security, the Agenda implies the phasing out of coal-powered plants from the energy system and their replacement with renewable energy sources, as well as the allocation of investment aimed at increasing energy efficiency. In addition to the inevitable transformation of the power sector in Serbia, the Declaration also foresees the reduction and gradual abolition of subsidies for coal.

At the end of 2020, the Guidelines for the Implementation of the Green Agenda for the Western Balkans and the Investment Plan for the Western Balkans were adopted. A significant step forward was made with the adoption of the Law on the Use of Renewable Energy Sources and the new Law on Energy Efficiency and Rational Use of Energy, as well as amendments and additions to the Law on Energy and the Law on Mining and Geological Explorations. It is important to point out that the law provides for the formation of the Directorate for Financing and Encouraging Energy Efficiency within the Ministry of Mining and Energy and various subsidy programmes, with the main aims to increase energy efficiency, achieve energy security, strengthen the economy's competitiveness, and manage the adverse environmental side-effects of the energy sector. Following up on the obligations under the amendments to the Law on Energy, Serbia prepared an Integrated National Energy and Climate Plan covering the period from 2021 to 2030, with a vision until 2050. This plan was developed within the Instrument for Pre-Accession Assistance project "Further Development of Energy Planning Capacity", in order to define appropriate ways to increase the participation of renewable energy sources in the production of electricity.

15.1.2.2 Challenges of Transitioning to Greener Growth for Serbia

Despite the relatively developed policy framework, the progress of the green transition in Serbia is very limited: the dynamics of reducing greenhouse gas emissions is unsatisfactory, and the emissions intensity and energy intensity levels remain among the highest in the region (Zvezdanović Lobanova & Lobanov, 2023). In other words, while Serbia has ratified most major international environmental agreements, their implementation is mostly lagging behind (World Bank, 2022).

Serbia has committed to working to achieve the 2050 target of transforming the EU into a carbon-neutral zone. However, currently, its internal economic incentives and policies are not aligned with these commitments, and the progress towards achieving set goals is limited and inconsistent.[1] For example, a 2020 assessment of Serbia's progress towards the achievement of defined SDGs published by the Statistical Office of the Republic of Serbia (2020) indicated a number of adverse developments in key areas related to food security, water use, clean fuel use, energy intensity, and waste management. In 2021, the Energy Community Secretariat even launched proceedings against Serbia related to its failure to meet the National Emission Reduction Programme's targets for sulphur dioxide (SO_2), nitrogen oxide (NO_x), and dust emissions (Energy Community, 2024).

The most significant factors contributing to Serbia's poor environmental performance relate to the high energy and carbon intensity of its economy and a strong dependence on domestic lignite in energy production (World Bank, 2022). Despite significant improvements over the past two decades, Serbia's GHG level of emissions intensity is still more than 2.5 times higher than the EU average (IEA Serbia, 2024). The energy sector is by far the largest contributor to Serbia's GHG emissions, contributing to 76.1% of the total in 2018. Within the sector, the largest shares of GHG emissions originate from the production of electricity and heat, mainly from lower-quality coal (UNFCCC, 2024). The collapse of a number of coal power plants and resulting power outages in late 2021 resulted in a new sense of political urgency related to the need for the diversification of their energy systems and associated investments. However, there has been a lack of clarity and mixed signals from the government regarding whether or not these coal-driven facilities will be

[1] In this context, it is necessary to mention that there are even opinions that Serbian society is characterized by the existence of "hyperproduction" of regulations, strategic documents, etc. (Jovanović & Andonović, 2021; Rajčević, 2016), especially in the field of environmental protection (Todić 2017).

decommissioned and replaced with greener infrastructure, increasing the risk for potential investors (World Bank, 2022).

Overall, Serbia faces a significant number of challenges in implementing just green transition policies. Within their recent research, Zvezdanović Lobanova and Lobanov (2023) have summarized them in the following manner:

- Insufficient foreign and domestic investment in the green economy diverts resources to traditional sectors, impeding economic progress. Additionally, limited private investment in green initiatives persists due to businesses' reluctance to undertake risky, long-term projects.
- Serbian households face challenges in receiving compensation for adopting costly energy-efficient technologies in residential buildings. Similarly, obtaining loans for transition projects, like replacing internal combustion engine vehicles with electric cars, remains difficult.
- The outlook for the Concept of Open Strategic Autonomy is uncertain, raising doubts about the EU's capacity to achieve the goals of the EGD given its reliance on imports of critical raw materials and industrial intermediates.
- Replacing traditional primary energy sources with alternative ones in the context of limited subsidy opportunities may lead to an increase in the cost of electricity for households and producers, which will negatively affect the competitiveness of their products.
- There is insufficient motivation to set ambitious national targets, as there are many examples of reserved attitudes towards pan-European climate policy targets among the EU member states. In conditions when the largest emitters of greenhouse gases do not assume any obligations, Serbia may be inclined towards demonstrative image projects rather than effective ones (greenwashing).
- The COVID-19 crisis has had a dual impact on the prospects for achieving the EGD goals. On the one hand, the reduction in industrial activity and restrictions on the movement of people led to a decrease in greenhouse gas emissions, but on the other hand, a sharp increase in government spending on anti-crisis measures has slowed down the implementation of green transition projects, pushing them into the background.

One of the key challenges of promoting the EGD in Serbia is related to the destabilization of the global energy markets necessitating a reevaluation of plans to reduce the use of low-quality local brown coal in the energy sector due to energy security considerations. Thus, the phasing out of coal and the transition to imported natural gas and then to renewable energy sources are being postponed. Secondly, reducing dependence on Russian energy sources in a short time is costly due to higher prices from alternative suppliers and the financing of infrastructure projects (regasification terminals, pipeline networks, etc.). In general, two development scenarios are possible: a conservative approach with reliance on local energy sources and the absence of significant changes in the structure of electricity production or a progressive transition to renewables and the realization of energy efficiency projects driven by volatility in traditional energy prices (Zvezdanović Lobanova & Lobanov, 2023).

15.2 Research Methodology

Based on the conducted analysis of ratifications of international agreements and harmonization of the national legislative framework, the conclusion of Serbia's clear strategic commitment to the JGT goals is evident. On the other hand, from the presented quantitative data relating to the dynamics of reducing greenhouse gas emissions, emissions intensity, energy intensity, etc., as well as from the conclusions of recent scientific and expert research, it is clear that the results, i.e., the realization of defined goals and measures that would contribute to a JGT are lacking in the long term and to a large extent. This represents the primary problem addressed in our research.

With the aim of identifying the reasons for the discrepancy observed between the formal commitment to implementing green transition policies and the achieved results, the degree of implementation of green transition policies identified during the preparation of the National Mapping Report on Green Transition in Serbia is examined. The Report has been conducted in order to provide comparative research on just green transition processes through formulated policies, involving actors and implemented practices and initiatives, identified on the territory of the Republic of Serbia, which form a wider framework of support for the achievement of JGT goals (Živanović et al., 2022). The special aim of the report was to show which pillars of the Green Agenda for the Western Balkans were more strongly supported in the already existing framework, and which pillars had to be supported more intensively by the development of new instruments. As one of the outcomes of the Report, Djordjević et al. (2023) formed the Dataset for Green Transition Policies, Stakeholders, and Practices in Serbia, within which they identified JTG-relevant policies in Serbia, including policy documents; laws, strategies, recommendations, directives, regulations and others.

The Dataset directly facilitated the identification of "coverage" of the pillars of the Green Agenda for the Western Balkans by relevant policies and served as the primary resource for the subsequent detailed analysis, with the research objective focused on diagnosing the causes of insufficient concrete progress towards the policies' implementation. The subsequent steps were realized by thoroughly reviewing each document included in the Dataset. We've started from accessing how far Serbia has progressed from the legislative and strategic dimensions to concrete, actionable measures for achieving defined goals, which was determined by focusing on whether each policy has an adopted action plan and whether the prescribed measures contain clearly

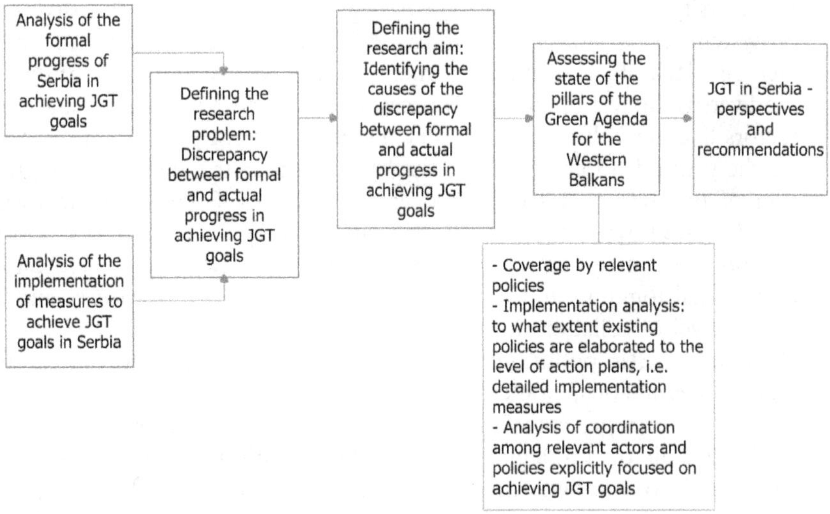

Fig. 15.1 Steps of the methodological procedure (*Source* authors' own elaboration)

defined elements (authority, implementation deadline, funding source, etc.) crucial for successful implementation. In the section dedicated to coordination, the focus is on determining the existence of defined measures for aligning relevant policies and fostering cooperation among stakeholders explicitly aimed at supporting the JGT process. Within this section, special attention is paid to the presence of "awareness" of each policy's contribution to the realisation of JGT goals, assessing the presence of this concept and its significance, as well as its representation within the goals and implementation measures of each policy. The criterion of the presence of the specified elements, determined through a detailed analysis of each document identified by the Dataset, forms the basis of the quantitative analysis presented in tabular and graphical formats in this study. The obtained results serve as a basis for formulating recommendations aimed at raising the level of real progress towards a just green transition and achieving formally adopted goals. Methodologically, the following algorithm can depict our research procedure (Fig. 15.1).

15.3 Green Transition Pillars in Serbia

Within the adopted Guidelines for the Implementation of the Green Agenda for the Western Balkans (European Commission, 2020), basic characteristics of the state of affairs in each of the defined areas, or pillars of the Green Agenda for the Western Balkans, have been recognized. In addition to existing potentials and achieved progress, challenges on the path of Agenda implementation have been identified, which generally apply to the Western Balkan countries, including Serbia:

- Biodiversity: In the Western Balkans region, protecting biodiversity and ecosystem services, aligning with EU legislation, is crucial. Key challenges include a lack of political commitment to improve the implementation of biodiversity policy, financial resources, and the environmental impact of sectors like agriculture and forestry. Despite existing policy frameworks, on-the-ground implementation lags, highlighting the need for integration of biodiversity into various sectors. Western Balkans partners are urged to align with the EU's Biodiversity Strategy for 2030 and support EU positions in global biodiversity negotiations. Addressing deforestation, illegal logging, and forest fires requires preventive measures and a higher level of collaboration with the EU.
- Depollution: Air pollution in the Western Balkans, largely caused by industrial emissions, domestic heating, and traffic, remains among the highest in Europe, impacting citizens' health significantly. Efforts to implement air quality legislation, including monitoring and assessment, are crucial to comply with EU directives. However, there's a gap in implementation, with exceedances of air quality standards, particularly for pollutants like particulate matter, nitrogen dioxide, and sulphur dioxide, requiring urgent action. Political commitment is essential to reduce emissions from key sectors like energy, transport, agriculture, and industry. In terms of water, the region faces challenges in protecting pristine rivers, managing wastewater, controlling pollutants like nitrates and pesticides, and ensuring water availability amid droughts and increasing hydropower capacity. Transboundary water cooperation is vital, but politically sensitive, especially regarding climate change impacts and water quality deterioration. Implementation of water-related legislation requires more resources for infrastructure investments and governance structures for sustainable water management. Soil erosion and land degradation pose significant problems in mountainous areas, requiring actions such as sustainable land use practices, ecosystem conservation, and restoration efforts, offering co-benefits for addressing other environmental issues like water pollution and biodiversity loss.
- Decarbonisation: Climate change presents a significant challenge for the Western Balkans, with projected temperature increases by the end of the century, contingent upon global efforts to reduce greenhouse gas emissions. The Paris Agreement commits Western Balkan signatories to limiting global warming below 2°C. The EU will reflect climate neutrality in bilateral relations and accession negotiations, urging Western Balkan countries to begin societal transformations accordingly. The European Commission is proposing a 55% reduction in greenhouse gas emissions by 2030 as part of the effort to achieve climate neutrality by 2050, necessitating significant changes to our current way of life. The EU can provide support in enhancing resilience and adapting to climate change impacts. Transitioning to clean energy requires political and technical assistance, including investment in renewable sources.

Smart and sustainable mobility is crucial, focusing on revitalizing rail networks, implementing intelligent transport systems, and promoting multimodal transport solutions. Innovative solutions, supported by the EU, are expected to contribute to emission reduction and environmental improvement in the Western Balkans.

- Circular economy: Resource extraction and processing contribute significantly to greenhouse gas emissions, emphasizing the necessity of transitioning to a circular economy to achieve climate neutrality by 2050. In order to achieve circularity, measures should address the entire product life cycle, emphasizing waste prevention, high-quality resource transformation, and stakeholder cooperation. Sustainable raw material supply, resource productivity enhancement, and waste minimization are crucial for a circular economy transition. Aligning with EU best practices and promoting innovation in production and consumption are essential for businesses in the Western Balkans. Waste generation in the region is increasing, necessitating investments in proper waste management infrastructure and recycling facilities. Implementing separate waste collection schemes and economic incentives is vital for quality recycling and waste reduction. Alignment with EU waste legislation and single-use plastic regulations is necessary. The situation is significantly worsened by inadequate solid waste management practices, insufficient infrastructure, and a general lack of public awareness regarding the consequences of their actions. Addressing these challenges requires concerted efforts to promote sustainable practices and infrastructure development in the Western Balkans.
- Sustainable food systems and rural areas: The agriculture sector in the Western Balkans, contributing significantly to GDP and employment, faces structural challenges such as small farm sizes and low productivity. Compliance with EU food safety standards varies, posing a challenge in aligning with initiatives like the Farm to Fork Strategy. Despite challenges, there's a growing demand for food, offering export potential and economic opportunities. Transforming the agri-food sector should prioritize sustainability to prevent rural unemployment and landscape deterioration while addressing climate change impacts. EU assistance, including IPARD, can aid in reforming the sector, enhancing competitiveness, and promoting resilience to climate change. Transitioning to a sustainable food system requires reforms in primary production, food processing, and consumption practices.

To ascertain the performance, progress, capacities, and obstacles in Serbia's journey towards overcoming these challenges, an analysis of the coverage of the pillars of the Green Agenda for the Western Balkans by relevant policies, the measures of their implementation, and the coordination of each policy with other relevant policies was conducted (Table 15.1):

Table 15.1 Analysis of policy coverage, implementation measures, and coordination in relation to the green agenda for the Western Balkans in Serbia (*Source* Djordjević et al. (2023); Authors' elaboration)

	Coverage	*Implementation*	*Coordination*
Biodiversity	Relatively well covered by relevant policies (10 identified adopted policies). They are thematically oriented towards the Protection and Restoration of Ecosystems	Half of the identified policies have adopted action plans with detailed implementation measures. The other half is still at the legislative or strategic level, or the implementation period of existing action plans has expired	There are no measures for coordinating relevant actors and policies explicitly directed towards the JGT process. The concept of JGT is not mentioned in any policy
Depollution	Only one identified policy, thematically focused on reducing air pollution. No identified policies directly related to soil pollution and water pollution	An action plan exists, and the implementation measures are thoroughly defined	The measures for coordinating relevant actors in the implementation of this policy are complex and detailed but are not explicitly aimed at supporting the JGT process. The concept of JGT is mentioned, albeit without detailed incorporation into the objectives and implementation and coordination measures

	Coverage	*Implementation*	*Coordination*
Decarbonisation	The pillar that is most extensively covered by relevant policies (currently 13 identified adopted policies). Thematically, the focus of these policies largely pertains to climate change (9 policies) and to a lesser extent to Smart and Sustainable Mobility (2 policies) and Clean Energy Transition (2 policies)	The vast majority (11 out of 13) of identified policies has adopted action plans with detailed implementation measures	Within the Integrated National Climate and Energy Plan for the period 2021 to 2030 with a vision until 2050 and the Climate Strategy and Action Plan, the importance of the JGT process is recognized, and certain measures for coordinating relevant actors in achieving JGT goals are prescribed. Half of the identified policies do not recognize the concept of JGT. Within the other half, it is mostly mentioned without detailed incorporation into the objectives and implementation and coordination measures

(continued)

Table 15.1 (continued)

	Coverage	*Implementation*	*Coordination*
Circular economy	Relatively well covered by relevant policies (currently 12 identified adopted policies). The thematic focus mainly revolves around Resources, Production, and Innovation (8 policies), while the rest are more directly oriented towards Waste and Plastic	The vast majority (9 out of 12) of identified policies has adopted action plans with detailed implementation measures. For the remaining policies, the implementation period of existing action plans has expired. A significant exception is the Law on the Use of Renewable Energy Sources, for which there is no unified strategy and action plan for implementation. Instead, it is envisaged at the level of each province and local self-government unit	In the vast majority of identified policies, the concept of JGT is not mentioned, nor are there measures for coordinating relevant actors in achieving JGT goals. In only two policies, this concept is mentioned and concisely described, without incorporation into the objectives and measures of their implementation
Sustainable food systems and rural areas	Only two identified policies	Still at the legislative level, without unified strategies and action plans	There are no measures for coordinating relevant actors and policies explicitly directed towards the JGT process. The concept of JGT is not mentioned in both policies

Generally speaking, it can be said that the number of policies of relevance on Serbia's path towards JGT is significant. Given the absence of an *optimal* number of these policies[2] (both collectively and for each pillar of the Green Agenda for the Western Balkans individually), we cannot make a comprehensive assessment of "coverage" either collectively or by pillars (for example, whether the 12 policies related to the Circular Economy pillar in Serbia are a "sufficient" number or not).[3] However, evident differences between pillars indicate the existence of issues to which greater attention is paid in Serbia (themes revolving around the Biodiversity, Decarbonization, and Circular Economy pillars), as well as issues that are in the initial stages of consideration (closer to the Depollution and Sustainable food systems and rural areas pillars). Further analysis was focused on the implementation of identified policies. Most of them have reached a level of elaboration that entails the adoption of action plans, i.e., defining detailed implementation measures. For policies where the absence of clearly defined implementation measures was identified, this, in addition to the absence of an action plan, implies the expiration of the existing action plan period, action plans in the drafting phase, prescribed elaboration by provincial and local plans (partially implemented), etc. However, regardless of the number of existing policies identified as significant on Serbia's path towards JGT, the fact remains that the vast majority of them do not recognize the concept itself, nor do they explicitly prescribe coordination and collaboration among relevant actors in achieving JGT goals.

15.4 Just Green Transitions in Serbia—Perspectives

15.4.1 Measures to Support JGT: Missing Linkages?

Among the identified territorial policies relevant to JGT, half are national laws adopted by the government. Slightly more than a quarter are strategies related to the national level. They are in the form of recommendations in our country, unlike laws that are binding in nature (Živanović et al., 2022). Almost all identified laws have initiated the adoption of strategies and action plans for their implementation.[4]

The measures within these action plans are thoroughly defined, containing necessary material and organizational elements. Certain guidelines for coordinating relevant actors in the implementation of defined measures are an integral part of the action plans, although they vary depending on the methodology of the plans' development, the topic, and the level (national, regional,

[2] The total number of identified policies in Western Balkan countries ranges from 34 in Albania to 62 in Bosnia and Herzegovina.

[3] The number of policies within this pillar in Western Balkan countries varies widely, ranging from only 2 policies in Bosnia and Herzegovina to 23 in Montenegro.

[4] The public policy document is the Strategy, which is further elaborated through an Action Plan. The Action Plan contains a set of measures (organizational, technical, financial, etc.) for achieving defined goals (Government of the Republic of Serbia, 2018).

or local) they relate to. In some cases, they are not significantly more detailed than listing the responsible actors for each defined measure. Additionally, the coordination of individual policies with other relevant and thematically related policies is mostly formulated indirectly, by assigning responsibilities for measures defined in action plans from multiple policies to the same institutions (e.g., the same sector/department within a particular ministry).

Overall, there is a large number of documents where certain analyses, goals, and measures are repeated, with the absence of a unified approach and explicitly defined responsibilities, obligations, and steps on Serbia's path towards achieving JGT goals. Additionally, there is a clear need for a stronger focus on recognizing the concept and importance of JGT and integrating this concept into public policy documents, objectives, and implementation measures. Only slightly less than a quarter of the identified policies recognize the concept of a just green transition, mostly minimally or not at all integrating it into objectives and implementation measures (Fig. 15.2).

The fact is that most of these policies contribute to some extent to the process of achieving JGT goals in Serbia. However, these goals are largely not explicitly and systematically incorporated into these policies. In other words, the implementers of these policies generally lack awareness of their involvement in the JGT process in Serbia and the necessity of acting towards achieving JGT goals. Overall, one of the main reasons for the lack of implementation, or for the gap between formal and substantial progress towards JGT, is the absence of a comprehensive JGT strategy explicitly focused on JGT or the lack of a systematic approach to addressing the issues, i.e., achieving goals

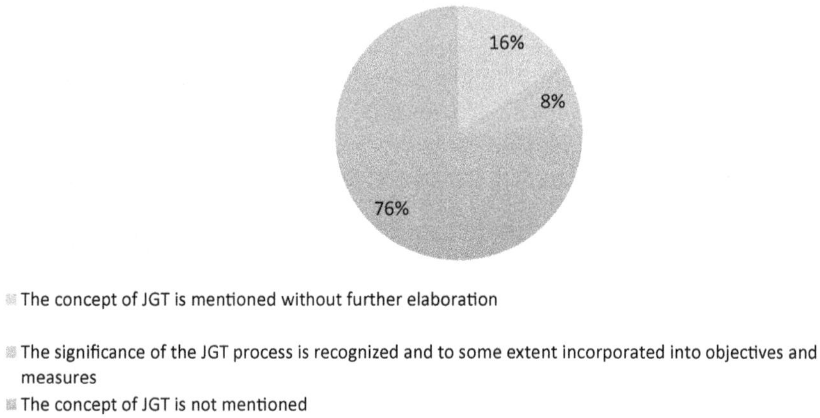

Fig. 15.2 The concept of JGT within the framework of identified policies (*Source* authors' elaboration)

declared as JGT goals.[5] Another important reason is the evident lack of coordination among relevant actors who would harmonize the implementation of adopted policies relevant to JGT. Consequently, effectiveness is lacking, goal and measure repetition, incomplete implementation, duplication of specific responsibilities (while some remain uncovered), irrational resource spending, etc., as well as the absence of systematic progress monitoring of Serbia on this path.

15.4.2 *Messages for Further Development*

By signing the Declaration of the Green Agenda for the Western Balkans, Serbia has committed to completely phasing out coal by 2050. Considering the pace of progress so far, and the evident gap between the formal commitment to implementing green transition policies and the actual results, there are even open doubts about Serbia's future on this path. For example, Zvezdanović Lobanova and Lobanov (2023), bearing in mind the current problems and imposed challenges, point out that Serbia is not in a position to simply diversify its energy resources, and, therefore, it is forced to postpone its plan to phase out coal-fired power plants in the coming period. Ćetković (2022) even argues that the EU energy and climate legislation is impossible to be transposed by Serbia since it is burdened with problems caused by weak institutional settings, outdated energy infrastructure, and the limited influence of grassroots environmental movements. Zvezdanović Lobanova and Lobanov (2023) also argue the Green Agenda is not yet given adequate priority in Serbia. In other words, a significant part of Serbian society is not ready for large-scale and costly reforms, and increasing their popularity in the near future is arguable. They point out that the initial stages of these reforms will most likely be accompanied by an increase in unemployment and societal polarization, which is not in the interest of political elites that usually lean towards measures with short-term benefits.

However, the transition towards low carbon development is ongoing and is expected to pick up speed all over the world, with particular intensity in the European Union. Serbia's transition to low carbon development will, like for its neighboring countries, become inevitable when the country joins the EU and it implements the EU acquis, although aspects of such development are also to some extent mandatory during the pre-accession period (further development of national strategies and action plans, further aligning the legislation and regulations with EU standards, preparing regular progress reports following EU standards, etc.) (Cavalheiro, 2020). Certain results exist, and

[5] It is important to note that, to maintain focus, this research emphasizes a specific, smaller number of reasons that systematically, "at the root," hinder the achievement of JGT goals. Other potential reasons are also significant and might include various issues, such as insufficient funding, lack of stakeholder engagement, inadequate technical prerequisites, etc.

their significance cannot be denied. In line with the commitments undertaken, Serbia is improving conditions in certain segments, both strategically and physically. However, these results are often insufficient and, due to the absence of a systematic approach, can largely be characterized as unconscious, spontaneous, accidental contributions to Serbia's progress on the JGT path. The overall progress of Serbia in achieving the goals of a just green transition can be symbolically illustrated by the contrast shown in Fig. 15.3.

Namely, a number of institutions in Serbia have certain, more or less direct competencies relevant to achieving the goals of a just green transition. The path from ratified international agreements to real changes in the environment falls under their jurisdiction, and the number of positive examples is certainly significant (Fig. 15.3a). On the other hand, the evident lack of coordination among relevant actors and adopted policies creates difficulties on this path, manifested by serious differences in effects in the ecological, socio-economic, and spatial dimensions (Fig. 15.3b).

The impression is that the road to a just green transition in Serbia cannot currently be seen as a single and clearly defined direction, but rather as a relatively incoherent set of narrower, sectoral paths of varying significance, support, and duration. The absence of a comprehensive JGT strategy and coordination among relevant actors explicitly focused on its implementation represents a systemic obstacle to significantly raising the level of real changes in

Fig. 15.3 Signs by the roadside (Municipality of Boljevac): (a) Bio Energy Point, a modern facility for energy production from renewable sources and (b) the municipal landfill, located just a few kilometers away from this facility[6] (*Source* authors' elaboration)

[6] This landfill is characterized by a range of environmental issues: the lack of prior waste disposal preparation, protective barriers, controlled waste disposal, waste compaction and daily covering with inert material, as well as the absence of channels for surface and procedural water drainage. This longstanding condition of the landfill leads to pollution of all environmental elements (Municipality of Boljevac, 2018).

the different dimensions that JGT entails (technical, economic, social, environmental, and territorial), and thereby achieving the goals to which our country is formally committed.

The fact is that a range of policies of varying importance exists on this path, and certain measures of actor and policy coordination are defined in most cases. However, these measures are not explicitly aimed at achieving JGT goals and can be characterized as partial and unsystematic. Such a situation prevents more serious progress, leading to some elements of JGT advancing, some stagnating, and some still in the initial phase. Without clear focus and coordination, the necessary reduction in Serbia's lag behind other countries in the region and Europe as a whole cannot be expected.

If the identified stagnation in Serbia continues, this lag may increase, making further progress more difficult and achieving desired results less certain. However, the growing number of policies, actors, and initiatives, from the most general to the most immediate level, is a definitive reflection of the increasing awareness of the importance of living in a quality environment. The general population, experts in various fields, and economic and political actors increasingly recognize the multiple benefits not only of reducing pollution in all elements of the environment but also of applying environmentally appropriate principles and means in their activities. This potential needs to be further mobilized, organized, and directed in the right way.

References

Cavalheiro, G. (2020). *Initiating the just transition in Serbia*. Belgrade: United Nations Development Programme.

Ćetković, S. (2022). Western Balkans: between clean energy sources and the dash for gas. *Agenda Pública*, June 30.

Djordjević, A., Živanović, Z., Protić, B., Jeftić, M., & Šećerov, V. (2023). Dataset for green transition policies, stakeholders and practices in Serbia (Final). *Zenodo*. https://doi.org/10.5281/zenodo.7785693

Energy Community (2024). *Case ECS-10/21: Serbia/environment*. https://www.energy-community.org/legal/cases/2021/case1021RS.html, accessed 12 February 2024.

European Commission (2019). *The European Green Deal*. COM(2019) 640 final. https://eur-lex.europa.eu/legal-content/EN/TXT/?uri=COM%3A2019%3A640%3AFIN, accessed 25 January 2024.

European Commission (2020). *Guidelines for the implementation of the green agenda for the Western Balkans*. https://neighbourhood-enlargement.ec.europa.eu/system/files/2020-10/green_agenda_for_the_western_balkans_en.pdf, accessed 25 February 2024.

European Commission (2024). *What we do: Green transition*. https://reform-support.ec.europa.eu/what-we-do/green-transition_en, accessed 14 February 2024.

Government of the Republic of Serbia (2018). *Law on Planning System*. Belgrade: Official Gazette of the Republic of Serbia (No. 30/2018).

IEA Serbia (2024). *Serbia: Energy system of Serbia*. https://www.iea.org/countries/serbia, accessed 12 February 2024.

Jovanović, Z., & Andonović, S. (2021). Publication of laws and other general acts: Hidden potentials of e-government. *Zbornik Radova Pravnog Fakulteta Nis, 60*(91), 163–176. https://doi.org/10.5937/zrpfn0-32514

Municipality of Boljevac (2018). *Local environmental action plan of the municipality of Boljevac.* Boljevac: Municipality of Boljevac.

Rajčević, T. (2016). Good Law. *Strani Pravni Zivot, 60*(2), 77–87.

Statistical Office of the Republic of Serbia. (2020). *Progress report on the implementation of sustainable development goals by 2030 in the Republic of Serbia.* Statistical Office of the Republic of Serbia.

Todić, D. (2017). Contemporary environmental legislation of the Republic of Serbia: From the "integral system" to the hyperproduction of regulations. *Zbornik Radova Pravnog Fakulteta, Nis, 56*(77), 1–16. https://doi.org/10.5937/zrpfni1777001T

UNFCCC (2024). *Nationally determined contributions registry.* https://unfccc.int/NDCREG, accessed 12 February 2024

World Bank (2022). *Supporting Serbia's transition to greener and more resilient growth: Policy and institutional reforms.* Washington, DC: World Bank. http://hdl.handle.net/10986/38271

Živanović, Z., Protić, B., Djordjević, A., Jeftić, M., & Šećerov, V. (2022). *National mapping report on Green Transition in Serbia.* Belgrade: University of Belgrade—Faculty of Geography.

Zvezdanović Lobanova, J., & Lobanov, M. (2023). Problems and perspectives of the European green deal with the focus on the SEE countries and Serbia. *The Review of International Affairs, 74*(1188), 51–77. https://doi.org/10.18485/iipe_ria.2023.74.1188.3

Social Impact of Green Transitions: Assessing the Distributional Implication of Coal Mines Closure in Bosnia and Herzegovina

Nermin Oruc, Amela Kurta, Ilma Kurtovic, and Fiona Imami

16.1 The End of Coal and the Promise of a "Just" Transition

The global transition towards climate neutrality necessitates profound shifts in energy production, industrial structures, and labor markets. Decarbonization policies outlined in the Paris Agreement (UNFCCC, 2015) and the European Green Deal (European Commission, 2019) emphasize the urgent need to phase-out fossil fuels, particularly coal, to meet international climate commitments. While this transition represents a fundamental step toward reducing carbon emissions and addressing climate change, it also brings significant socio-economic challenges, particularly for regions that remain heavily reliant on coal. In the Western Balkan context, Bosnia and Herzegovina (BiH) stands at the center of this complex transformation.

N. Oruc (✉) · A. Kurta · I. Kurtovic
Center for Development Evaluation and Social Science Research, Sarajevo, Bosnia-Herzegovina
e-mail: noruc@credi.ba

A. Kurta
e-mail: akurta@credi.ba

I. Kurtovic
e-mail: ikurtovic@credi.ba

F. Imami
Co-PLAN, Institute for Habitat Development, POLIS University, Tirana, Albania
e-mail: fiona_imami@co-plan.org

E. Berisha et al. (eds.), *The Palgrave Handbook of Just Green Transitions in the Western Balkans and Beyond*,
https://doi.org/10.1007/978-3-031-95075-9_16

Historical examples demonstrate that significant transitions in the coal extraction sector have often led to large, persistent, and painful social dislocation, as seen in the UK coal mining regions in the 1980s (Baeten et al., 1999; Caldecott, 2018; Del Rio, 2017; Fothergill, 2017; Gales & Hölsgens, 2017; Jonek Kowalska, 2015; Kok, 2017; Szpor, 2017).

While the national share of mining employment is typically low, its local concentration often makes it a dominant economic pillar, meaning that closures can have disproportionate and lasting effects on regional labor markets and community stability.

Coal has historically been the backbone of BiH's energy system, supplying much of its electricity while sustaining the economies of key mining regions like Tuzla, Zenica, and Banovići, where thousands of jobs depend directly and indirectly on coal extraction and power generation.[1] The anticipated phase-out of coal, driven by European decarbonization efforts and BiH's own commitments under its Nationally Determined Contributions (NDCs) strategy, poses a serious challenge for these regions. Although discussions on transitioning to greener energy sources have gained momentum, the absence of a clear and actionable roadmap for phasing out coal in BiH raises concerns about the profound socio-economic consequences of mine closures, particularly for workers, businesses, and local economies that rely heavily on coal and lack viable alternatives for income and employment.

The concept of Just Transition, originally advocated by labor unions in response to industrial restructuring and later institutionalized in global climate policies, serves as a critical framework for mitigating the socio-economic disruptions associated with decarbonization (McCauley, 2018; Morena et al., 2019; Smith, 2017), and it highly underscores that the shift toward a low-carbon economy must be socially inclusive and economically equitable, ensuring that workers and communities dependent on fossil fuels are not left behind in the process of structural change (European Commission, 2019; Newell & Mulvaney, 2013; UNFCCC, 2015). Following this, *The Economic and Investment Plan for the Western Balkans*[2] and the *Guidelines for the Implementation of the Green Agenda for the Western Balkans*,[3] both issued by the European Commission in 2020, emphasize that climate action must be inclusive, equitable, and participatory, ensuring that affected workers and

[1] This analysis focuses on the impact of coal mine closures, as they represent a direct and immediate challenge to employment and regional economies. However, the closure of thermal power plants, which are often located in the same regions and heavily reliant on coal supply, is also expected to have significant economic and social repercussions.

[2] **European Commission. (2020).** *An Economic and Investment Plan for the Western Balkans*. Brussels: European Commission. Available at: https://ec.europa.eu/neighbourhood-enlargement/news/economic-and-investment-plan-western-balkans-2020-10-06_en.

[3] **European Commission. (2020).** *Guidelines for the Implementation of the Green Agenda for the Western Balkans*. Brussels: European Commission. Available at: https://ec.europa.eu/neighbourhood-enlargement/news/green-agenda-western-balkans-2020-10-06_en.

communities are not left behind. Additionally, these documents recognize that coal-dependent regions require targeted support through reskilling programs, economic diversification initiatives, and strong social safety nets to mitigate the risks associated with the transition, making the issue particularly urgent in BiH, where the labor market is already fragile, and coal remains a significant employer in several municipalities.

The closure of coal mines is expected to disrupt local economies, leading to rising unemployment, loss of fiscal revenues, and social dislocation. While some sectors may benefit from decarbonization policies, such as renewable energy and energy efficiency industries, the spatial and sectoral distribution of benefits is unlikely to be even. While job losses and job creation may eventually balance out at the national level, the spatial distribution of economic risks and opportunities remains highly uneven, particularly in mono-industrial towns where coal dominates the local economy. In such regions, economic diversification is limited, making it significantly harder for displaced workers and businesses to transition into alternative industries (Baeten et al., 1999; Fothergill, 2017). The closure of coal mines, therefore, poses spatially asymmetric challenges, with rural and industrially specialized regions bearing a disproportionate burden compared to more economically diverse urban centres (Rodríguez-Pose, 2018).

This uneven distribution of economic risks underscores the deepening of regional disparities, reinforcing pre-existing inequalities between prosperous and struggling regions (Iammarino et al., 2019). In many cases, mono-industrial towns experience cumulative causation effects (Myrdal, 1957), where the decline of a dominant industry triggers a downward economic spiral—including outmigration, business closures, and reduced public revenues, further constraining local governments' ability to respond effectively (Tomaney et al., 2019). Without targeted policy interventions, such as place-based development strategies (Barca et al., 2012), affected regions risk long-term economic stagnation, exacerbating the social and territorial divides that often characterize post-industrial transitions.

Managing this transition requires a nuanced understanding of the labor market dynamics and fiscal implications of coal mine closures, thus we focused the research conducted under the GreenFORCE[4] project on precisely these aspects, offering an evidence-based analysis of how BiH can navigate the economic, social, and budgetary challenges associated with decarbonization. A key aspect of this research is identifying the expected employment losses, the budgetary constraints that municipalities will face due to declining tax revenues, and the potential policy measures that could mitigate these impacts.

[4] GreenFORCE—Foster Research Excellence for Green Transition in the Western Balkans. This EU Horizon 2020-funded project aims to enhance scientific research and innovation capacity in the Western Balkans' green transition. It focuses on strengthening the research profiles and management capacities of partner organizations in Albania, North Macedonia, and Serbia, contributing to convergence between Western Balkans and EU research capacities. More information is available at: https://greenforcetwinning.net/.

Nevertheless, beyond labor market implications, coal mine closures also pose serious challenges to local governance and public finance, especially those municipalities that rely on coal-related tax revenues (limiting their ability to invest in economic diversification or provide adequate social support). Declining fiscal resources often leave local governments struggling to sustain essential public services, while simultaneously, rising unemployment intensifies pressure on welfare systems, increasing the demand for unemployment benefits, retraining programs, and social assistance, thereby compounding the financial strain on public budgets (Rodrik, 2018; Tomaney et al., 2019). Adding to this, another major concern, especially for the Western Balkan context (and especially BiH), is the lack of comprehensive just transition planning at the national level. While strategies exist for coal phase-out, including mine closures and the shift to new energy sources, there are no comprehensive policies to address regional disparities and mitigate the localized socio-economic impacts of the transition.

While BiH has acknowledged the need for energy transition in policy discussions, concrete strategies for supporting affected workers and communities remain underdeveloped, and in many cases, decision-making processes lack transparency, and workers and local stakeholders are not sufficiently involved in shaping the transition agenda. A Just Transition cannot be achieved through top-down approaches alone, and we argue further in this chapter the necessity of bringing out strong policy interventions, as without them, these factors could trigger a vicious cycle of economic decline, depopulation, and social instability, particularly in regions where alternative employment opportunities remain scarce. Yet, to address these challenges an evidence-based approach is required, so it can assess the scale of the impact and identifies policy options for economic adaptation. In this regard, we build the argument in this chapter focusing on three key aspects (i) the expected employment losses; (ii) the fiscal and budgetary implications for affected municipalities, and (iii) potential mitigation strategies that could ease the transition process, which are subsequently presented below.

16.2 Coal Dependency in BiH, the Social and Economic Fabric of Mining Towns

Bosnia and Herzegovina (BiH) has long been a country deeply intertwined with coal, both as a critical energy resource and a key economic driver. The coal sector forms the backbone of BiH's energy system, with coal-fired power plants accounting for a significant share of the country's electricity generation. This dependency is not just a feature of the national economy but deeply ingrained in the socio-economic fabric of local communities, particularly in mono-industrial mining towns such as Banovići and Zenica. For decades, these towns have relied on coal mining to provide employment, support local businesses, and sustain public revenues. However, as the global push toward decarbonization accelerates, the future of these coal-dependent regions faces profound uncertainty.

The economic and social reliance on coal in BiH is starkly illustrated by its mining operations. Banovići, for instance, is home to one of the largest lignite ore reserves in Europe, with an estimated 208 million tons of coal and an annual production capacity of approximately 1.27 million tons. In contrast, Zenica, while still a major mining hub, has a smaller annual production capacity of 320,000 tons but remains economically tied to its mines. The difference in the scale of mining operations between these two towns highlights the varying levels of dependence on coal, with Banovići exhibiting a higher reliance on the sector. This discrepancy underscores the need for tailored transition strategies, as the socio-economic impacts of mine closures will vary significantly based on the level of dependence.

The country's historical reliance on coal is also reflected in its significant coal reserves, which are estimated at approximately 3.3 billion tons. In 2018, BiH produced 14.3 million tons of brown coal and lignite, which were primarily used for electricity generation in power plants located near mining sites. With such a high dependency, it is unsurprising that coal plays a dominant role not only in the energy sector but also in the livelihoods of many communities. Yet, this dependency comes with challenges, particularly as the country grapples with the need to meet its revised Nationally Determined Contributions (NDCs) under the Paris Agreement and align its energy policies with EU directives.

Adding to this, a major concern, especially for the Western Balkan context—and particularly for Bosnia and Herzegovina (BiH)—is the lack of comprehensive transition planning at the national level. This challenge is particularly acute in BiH due to several interconnected political, institutional, and economic factors that hinder strategic and coordinated policymaking.

First, BiH's complex and highly fragmented governance structure, resulting from the Dayton Peace Agreement[5] in 1995, creates significant barriers to national-level policy coordination. With a weak central government, energy and environmental policies are primarily managed at the entity level—divided between the Federation of Bosnia and Herzegovina (FBiH) and Republika Srpska (RS)—and further decentralized to cantonal and municipal authorities in FBiH. This governance fragmentation leads to overlapping responsibilities, inconsistent regulatory frameworks, and difficulties in implementing national strategies, which, as noted by (Keil & Perry, 2015) (Mujanović, 2018), exacerbate the risks associated with coal mine closures by enabling conflicting interests across jurisdictions and delaying much-needed reforms.

[5] ***The General Framework Agreement for Peace*** in Bosnia and Herzegovina, commonly known as the Dayton Peace Agreement, was signed in December 1995 in Dayton, Ohio, USA. It ended the Bosnian War (1992–1995) and established the constitutional framework of Bosnia and Herzegovina, creating a highly decentralized state with two entities: the Federation of Bosnia and Herzegovina (FBiH) and Republika Srpska (RS), alongside the Brčko District.

Second, economic dependence on coal remains deeply entrenched, making BiH more resistant to phasing out fossil fuels compared to other Western Balkan countries. The country generates over 60% of its electricity from coal (Energy Community, 2022), and its coal mining sector remains one of the largest state-subsidized industries. Unlike EU member states that have committed to structured phase-out plans, BiH has yet to define a comprehensive and just transition strategy. While a framework strategy for energy transition exists up to 2035, along with relevant strategies at lower administrative levels, these primarily focus on the technical aspects of transition rather than its social and economic impacts. Although there have been government-led initiatives to draft a just transition strategy, this process remains incomplete. Greater efforts are needed to develop and implement a holistic just transition strategy that provides certainty and alternative economic pathways for workers, local governments, and businesses (Frankhauser & Jotzo, 2018; Šćepanović & Artiles, 2020).

Third, the political economy of energy in BiH presents additional challenges, as the coal sector has historically been politicized, with state-owned mines and power plants functioning as key employment hubs for political patronage networks (Prelec, 2020). This dynamic makes coal transition not only an economic and environmental issue but also a political one, where vested interests in maintaining the status quo resist systemic change. Unlike cases such as Germany's Ruhr[6] region (for example), where transition policies were developed in consultation with trade unions and local stakeholders, BiH lacks a participatory decision-making process, further exacerbating uncertainty for coal workers and communities.

Finally, external pressures from the EU and international financial institutions (IFIs) are accelerating the urgency of transition, yet domestic capacity to respond remains weak. The Energy Community has repeatedly urged BiH to adopt decarbonization strategies aligned with EU standards, but delayed policy action risks not only economic stagnation but also potential exclusion from European funding mechanisms, such as the Just Transition Mechanism and the Western Balkans Green Agenda Investment Framework.

Given these structural weaknesses, the need for strong policy interventions becomes even more urgent, particularly as BiH's commitments under the Energy Community Treaty require the country to align its energy sector with EU standards. The Framework Energy Strategy of Bosnia and Herzegovina until 2035 outlines plans for the gradual decommissioning of inefficient

[6] The Ruhr region in Germany is often cited as a successful example of managing the transition from coal dependency to a diversified economy. This was achieved through proactive policies involving extensive consultation with trade unions, local stakeholders, and government authorities. The approach included targeted investments in infrastructure, education, and new industries, alongside robust social safety nets and retraining programs for affected workers. For more details, see: Hein, C. (2018). The transformation of the Ruhr region: Development strategies and lessons learned. European Planning Studies, 26(1), 38–57.

coal-fired thermal power plants, which will inevitably lead to mine closures. However, these closures will not only displace workers but also disrupt entire communities, particularly in areas where coal mining is a significant economic driver and in which, the absence of targeted interventions might lead these regions to fall into cycles of economic decline, depopulation, and social instability.

Box 1. Methodological Note

The research in which this book chapter is builds its arguments, employs a mixed-methods approach to assess the socio-economic impacts of coal dependency and mine closures in Bosnia and Herzegovina. The methodology integrates quantitative modeling and qualitative analysis to produce a comprehensive, policy-relevant analysis. Quantitative tools were used to estimate the labor market and fiscal effects of mine closures, focusing on employment losses, reduced tax revenues, and increased social welfare costs. The qualitative dimension involved case studies in two municipalities—Banovići and Zenica—selected for their varying levels of dependence on coal.

Additionally, the case study approach combined desk research, analysis of secondary data, and semi-structured interviews with stakeholders, including local government officials, representatives of coal companies, and affected workers. The research incorporated community consultations and socio-demographic profiling to understand the broader impacts on local economies and well-being. This triangulation of data ensures the robustness of findings, capturing the direct, indirect, and induced effects of mine closures.

For more information on the research, please visit: https://greenforcetwinning.net/wp-content/uploads/2024/09/D4.5_1st_research_study_report_Western-Balkans-5-cases_webpage.pdf.

16.3 The Great Shift: Expected Impacts of Coal Mine Closure

Job creation from green energy will not necessarily be concentrated in the same areas as the locations of mines, and for this reason, it is important to assess the local impact of such a policy, which should motivate appropriate policies in the most negatively affected areas.

The local economic impact will be examined through case studies of particularly vulnerable communities, with the research focusing not on critiquing the closure process but on illustrating planning strategies and practices adopted by governance and mines in BiH.

Regarding the location of the mine in BiH, it is crucial to consider both the region and the development level of the municipality, assessing the impact of potential mine closures on both more and less developed areas. For instance, the largest number of mines in Bosnia and Herzegovina is located in two cantons: Zenica-Doboj and Tuzla cantons. Banovići is a mining town located in the Tuzla Canton with an area of 185 km^2 and about 24,000 residents according to the latest population census (2013). Zenica, on the other hand is the administrative and economic centre of the Zenica-Doboj canton with an area of 558.5 km^2 and slightly more than 110,000 residents according to the latest population census (2013). It is also important to consider the size of the mine, as Banovići holds 208 million tons of lignite reserves, making it one of the largest in Europe and the world with an annual production capacity of 1.27 million tons of coal, while Zenica mine operates at a significantly lower capacity of 320,000 tons per year.

If we look at mines from the perspective of future closures, all mines in BiH are equally at risk of closure, given that BiH has taken decisive action by submitting revised NDCs with more ambitious GHG reduction goals in 2021. Nevertheless, it was recently announced that the FBiH Ministry of Energy, Mining and Industry, together with the World Bank, plans to implement a project that will close three brown coal mines, the first of which is the mine in Zenica.

Considering this, three main concerns rise from the research, and for which this chapter builds its arguments as following.

16.3.1 Unpacking the Labor Market Disruptions

The labor market in Bosnia and Herzegovina is characterized by low employment and high unemployment rates (see table below), especially for youth. In such a situation, even minor changes in the labor market are expected to have a significant impact on both the entire economy and the people living in the country (Table 16.1).

Statistics on poor productivity and the high share of employment in some sectors suggest that some industries have either not caught up with the technological advancements in the sectors or have not taken advantage of possibilities to improve skills of their employees, or both. Investing in human capital entails ongoing training and development of employees so they may pick up new skills, utilize cutting-edge technology, and eventually become more productive. In this context, the economic reliance on coal mining in BiH is stark. Findings from the case studies suggest that the short-term effect of mine closures will involve the dismissal of a substantial portion of the mining workforce. Estimates indicate that 19% of workers at the Zenica mine were dismissed in the previous year, and a similar trend is expected in Banovići.

The research scenarios developed during the research (Table 16.2), suggest a phased dismissal process, ranging from partial layoffs in the short term to full mine closures in the long term. In the absence of viable alternative industries, the displacement of coal workers is expected to result in heightened unemployment rates, particularly for low-skilled workers who face significant challenges in transitioning to other sectors.

The transferability of skills among mining workers is another critical challenge. While administrative workers may have greater mobility across sectors,

Table 16.1 Basic macroeconomic indicators relevant to the labor market in BiH, 2010–2020 (*Source* BiH Agency for Statistics)

Indicator	*2010*	*2011*	*2012*	*2013*	*2014*	*2015*	*2016*	*2017*	*2018*	*2019*	*2020*
LFP[7] rate 15–64 (annual %)	54.0	53.8	53.9	53.5	54.2	54.6	54.2	54.5	54.2	55.5	58.2
Employment rate 15–64 (annual %)	39.0	38.7	38.5	38.5	39.0	39.2	40.2	43.0	44.0	46.4	48.8
Unemployment rate 15–64 (annual %)	27.7	28.0	28.6	28.0	28.1	28.2	25.8	21.1	18.9	16.4	16.2

[7] LFP refers to Labor Force Participation. It refers to the percentage of the working-age population (typically ages 15–64) that is either employed or actively seeking employment. It is a key indicator of labor market activity and economic engagement.

Table 16.2 Scenario development (*Source* authors own elaboration)

	Scenario 1	*Scenario 2*	*Scenario 3*	*Scenario 4*
Description of the transition scenario	Short-term effect: Partial dismissal of workers without any transition to other sectors	Short-term effect: Partial dismissal of workers with transition into employment	Long-term effect: Full dismissal of workers with transition into employment	Long-term effect: Full dismissal of workers without transition into employment
Underlying assumptions	Part of the operations of mines are closed and workers are dismissed, there is no immediate transition into employment	Part of the operations of mines are closed and workers are dismissed, there is at least partial transition of workers into employment	Mine is permanently closing all operations, there is at least partial transition of workers into employment	Mine is permanently closing all operations, there is no immediate transition into employment
Adjustment of scenarios	Scenario adjusted based on qualitative data on the size of dismissal and transition to employment	Scenario adjusted based on qualitative data on the size of dismissal and transition to employment	Scenario adjusted based on qualitative data on the size of dismissal and transition to employment	Scenario adjusted based on qualitative data on the size of dismissal and transition to employment

technical mining skills are industry-specific and less adaptable to emerging labor market demands (Directorate-General for Employment Social Affairs and Inclusion, 2011). The potential for retraining programs to facilitate workforce mobility is severely constrained by the inadequate active labor market policies and inadequate vocational training systems in BiH. Without targeted retraining initiatives and investment in skill development, most displaced workers will face long-term unemployment or precarious employment in informal sectors.

In addition, findings from key informant interviews during the research case, suggested that emigration could become a significant coping strategy, particularly among younger workers who see no future in the domestic labor market. The impact of job losses varies between highly and less dependent municipalities; under the least severe scenario, a few hundred workers in highly coal-dependent communities will be dismissed, while in less dependent ones, the number could be up to 100. With 28% of displaced workers expected to transition into retirement, 2% to emigration, and 63% remaining unemployed, the long-term employment prospects in affected areas remain bleak. Moreover, the estimated employment loss was limited to direct jobs in coal mines, but interviews with local companies that rely heavily on contracts with mines suggest that the actual total job loss will be even greater.

16.3.2 *The Budgetary Puzzle and the Fiscal Consequences of the Coal Exit*

The fiscal implications of coal mine closures extend beyond job losses to the financial stability of municipalities that rely on coal-related tax revenues. The direct fiscal impact includes the loss of social security contributions, personal income tax, and value-added tax (VAT) revenue generated by coal sector workers. As evidenced by previous studies on coal transitions (Clark & Zhang, 2022), declining employment in the mining sector can lead to a substantial reduction in municipal revenues, undermining the capacity of local governments to finance public services and infrastructure projects.

In 2015, 170,619 households (16.5% of the total number of households) and 505,816 individuals (16.9% of the total number of individuals) in Bosnia and Herzegovina lived in relative poverty. There are geographical disparities in the distribution of poverty among different entities. The Republika Srpska has the highest rate of destitution among households, while Brcko District has the lowest.

Increased expenditures on unemployment benefits and pensions further compound the fiscal burden. With a large share of dismissed workers transitioning into social welfare programs, municipalities will face higher financial obligations while their revenue base declines. The estimated annual fiscal losses in Zenica and Banovići reveal a significant gap in municipal budgets, raising concerns about the long-term sustainability of public services.

Another key concern is the impact on local businesses and supply chains. The coal value chain extends beyond mining operations to include a network of contractors, service providers, and local enterprises that depend on mine-related economic activity. Research on coal-dependent regions in other countries has shown that the closure of mines often leads to a cascading effect of business closures, further weakening local economies.

In this regard, a preliminary conclusion might be drawn, related to the need of municipalities to develop alternative revenue streams to replace lost coal revenues. However, it should be kept in mind that the economic diversification in coal-dependent regions is a long-term process that requires substantial investment in infrastructure, education, and workforce development. In the absence of targeted national policies and financial support, many local governments will struggle to navigate this transition.

16.3.3 Community Resilience: Social and Spatial Implications

Lastly, the closure of coal mines is not merely an economic event—it is a social transformation that reshapes the identity and stability of entire communities. In BiH, where mining towns have historically been structured around coal employment, the transition threatens to erode social cohesion, exacerbate poverty, and accelerate depopulation trends. Findings from community-level indicators suggest that households in mining towns are more vulnerable to economic shocks than the national average. The loss of coal-related income is expected to increase household poverty rates, particularly for families that rely on a single breadwinner employed in the mines. Data from the Household Budget Survey (2015) indicate that poverty rates in BiH remain high, with 16.9% of individuals classified as poor. With 38.5% of disabled workers, 26.2% of unemployed individuals, and 18% of housewives living in poverty, the social implications of job losses will be profound. Another dimension of the transition is the geographic concentration of economic hardship. While some regions may experience economic renewal through green investments, others will be left behind, deepening spatial inequalities. Research from industrial transition studies suggests that mono-industrial towns often struggle to recover, as they lack the economic diversity needed to absorb displaced workers. Without targeted place-based policies, mining towns in BiH risk long-term economic stagnation and demographic decline. Finally, the psychological and cultural impact of mine closures should not be overlooked. Mining communities have strong generational ties to the industry, and the loss of coal employment represents not only economic displacement but also a rupture in collective identity. International experiences, such as the closure of mines in the UK and Germany, demonstrate that the loss of a dominant industry can trigger a sense of social dislocation and resentment toward policymakers, particularly when communities feel excluded from decision-making processes (McCauley & Heffron, 2018).

16.4 Conclusion: A Roadmap for a Fairer Future

The transition away from coal in Bosnia and Herzegovina (BiH) represents one of the country's most pressing socio-economic challenges. While the urgency of decarbonisation is undisputed, this research underscores the critical need for a Just Transition framework that safeguards workers, local economies, and social cohesion. The findings highlight that without targeted policy interventions, coal mine closures will exacerbate unemployment, deepen regional disparities, and strain municipal finances, leading to long-term socio-economic instability in the most affected regions. However, a fair and inclusive transition is possible, provided that proactive and strategic measures are taken. In this context, several recommendations can be formulated as a roadmap for a fairer future of green transition in these contexts:

A. ***Strengthening Policy and Institutional Coordination***

One of the greatest obstacles to a smooth transition in BiH is its fragmented governance structure, which has resulted in a lack of national coordination on coal phase-out policies. To address this, a unified national strategy for Just Transition should be developed, bringing together central, entity, and local governments. This strategy should: (i) Establish clear timelines and transition milestones aligned with BiH's energy commitments under the Energy Community Treaty and EU Green Agenda. (ii) Introduce regional transition task forces to ensure that local realities and economic specificities are factored into national decarbonisation efforts. (iii) Secure long-term funding commitments from domestic and international sources, including EU Just Transition Mechanisms and IFI funding.

B. ***Mitigating Labor Market Disruptions through Reskilling and Alternative Employment***

The research highlights that coal-dependent regions in BiH, particularly Banovići and Zenica, will face severe labor market disruptions as mines close, unless the following workforce-oriented measures are taken:

- First, national and local governments should prioritize reskilling and retraining programs by investing in vocational training that aligns with emerging sectors such as renewable energy, construction, and digital industries, ensuring that workers gain the skills needed to transition into sustainable employment opportunities.
- Second, national policies should support worker mobility by facilitating the transition of mining workers into alternative economic activities, providing subsidies for relocation and skills transfer programs to enhance employment opportunities in new industries.

- Third, entrepreneurship incentives should be strengthened by providing targeted financial support to small and medium enterprises (SMEs) in affected communities, fostering alternative employment opportunities and local economic resilience.
- Lastly, public sector employment programs should be expanded through temporary initiatives in infrastructure development and environmental rehabilitation projects, offering an immediate safety net for displaced workers while supporting community resilience and sustainable development.

C. ***Ensuring Fiscal Stability and Municipal Resilience***

Municipalities in coal-dependent regions face a dual economic shock: the loss of coal-related tax revenues and an increase in social welfare expenditures. In this regard, the transition roadmap should include: (i) Municipal Revenue Diversification: Financial support for local governments to develop alternative revenue sources, including property tax reform and investment in sustainable industries. (ii) Transition Funds for Local Budgets: The establishment of a Just Transition Fund, co-financed by international donors and national authorities, to compensate for revenue losses and fund economic diversification projects. (iii) Municipal Capacity-Building: Strengthening the ability of local governments to attract investment and implement economic restructuring programs through training and technical assistance.

D. ***Addressing Social and Spatial Inequalities***

The transition is expected to deepen existing regional inequalities, with mining towns facing economic stagnation and depopulation if not properly supported. To counter this, targeted place-based development strategies should focus on economic renewal in mining regions by fostering investment in local industries, tourism, and agri-business, creating diverse and sustainable employment opportunities. At the same time, strengthening social safety nets is crucial, requiring the expansion of unemployment benefits, pension adjustments, and targeted income support to protect vulnerable households from financial hardship. Ensuring equal access to green energy is also critical, as interviewees highlighted that projected capacities from renewable sources are currently well below existing energy consumption levels. Without adequate energy availability, disparities between regions may widen, limiting economic opportunities in affected areas. Equally important is community participation in transition planning, which can be facilitated through the establishment of local Just Transition councils composed of government representatives, businesses, labor unions, and civil society groups, ensuring that the voices and concerns of affected communities are meaningfully incorporated into decision-making processes.

E. ***Leveraging Green Investments and Innovation***

A successful Just Transition should go beyond mitigating losses and actively leverage new opportunities for growth by prioritizing investments in sustainable industries. At the national level, expanding wind, solar, and hydropower projects will be essential to replace coal as a key economic driver, ensuring a cleaner and more resilient energy sector. However, as these green projects are often located in different towns, effective regional planning is crucial to ensure a balanced distribution of economic opportunities. Additionally, reskilling and retraining coal mine workers for employment in other sectors will be essential to prevent widespread emigration. At the same time, sustainable industrial development should be promoted through investment in green technologies, circular economy initiatives, and sustainable infrastructure, fostering long-term economic stability. Equally important is green workforce development, which can be achieved by establishing strong partnerships between vocational training institutions and green industry leaders, enabling workers to access new job pathways in emerging sectors and ensuring that the transition creates inclusive and future-oriented employment opportunities.

To conclude, it is true that BiH stands at a crossroads, where the inevitable closure of coal mines can either lead to economic decline or serve as a catalyst for a more resilient and diversified future, and by committing to targeted investments, inclusive policies, and proactive planning, government, industry, and civil society can work together to ensure that no worker or community is left behind, transforming this transition into an opportunity for sustainable growth, innovation, and shared prosperity.

References

Baeten, G., Swyngedouw, E., & Albrechts, L. (1999). Politics, institutions, and regional restructuring processes: From managed growth to planned fragmentation in Belgium. *Regional Studies, 33*(3), 253–267.

Barca, F., McCann, P., & Rodríguez-Pose, A. (2012). The case for regional development intervention: Place-based versus place-neutral approaches. *Journal of Regional Science, 52*(1), 134–152.

Caldecott, B. (2018). *Stranded assets and the environment, risk, resilience and opportunity*. Routledge.

Clark, A., & Zhang, W. (2022). Estimating the employment and fiscal consequences of thermal coal phase-out in China. *Energies, 15*(3), 800. https://doi.org/10.3390/en15030800

Del Rio, P. (2017). *Coal transition in Spain: An historical case study*. Working Paper, IDDRI and Climate Strategies.

Directorate-General for Employment, Social Affairs and Inclusion. (2011). *Transferability of skills across economic sectors: Role and importance for employment at European level (KE-32-11-836-EN-C).* Publications Office of the European Union. https://pjp-eu.coe.int/bih-higher-education/images/eul14180_tos_110924_web_with%20erratum-3.pdf

Energy Community. (2022). Investments into the past—Analysis of direct subsidies to coal and lignite electricity production for the year 2020.

European Commission. (2019). *The European Green Deal.* EU Commission Report.

Fothergill, S. (2017). *Coal transition in the United Kingdom: An historical case study.* Working Paper, IDDRI and Climate Strategies.

Frankhauser, S., & Frank, J. (2018). *Economic growth and development with low-carbon energy.* WIREs Climate Change 9m no.1. https://doi.org/10.1002/wcc.495

Gales, B., & Hölsgens. R. (2017). *Coal transition in the Netherlands: An historical case study.* Working Paper, IDDRI and Climate Strategies.

Iammarino, S., Rodríguez-Pose, A., & Storper, M. (2019). Regional inequality in Europe: Evidence, theory and policy implications. *Journal of Economic Geography, 19*(2), 273–298.

Jonek Kowalska, I. (2015). Conditions of corporate restructuring in the mining sector in Poland in the context of socially responsible business. *Resources Policy, 44*, 28–36.

Keil, S., & Perry, V. (2015). *State-building and democratization in Bosnia And Herzegovina.* Routledge.

Kok, J. (2017). *The socio-economic consequences of coal transition: The case of Limburg.* Working Paper, IDDRI and Climate Strategies.

McCauley, D., & Heffron, R. (2018). Ust transition: Integrating climate, energy and environmental justice. *Energy Policy, 119*, 1–7.

Morena, E., Krause, D., & Stevis, D. (2019). *Just transitions: Social justice in the shift towards a low-carbon world.* Pluto Press.

Mujanović, J. (2018). *Hunger and fury: The crisis of democracy in the Balkans.* Hurst & Company.

Myrdal, G. (1957). *Economic theory and underdeveloped regions.* Duckworth.

Newell, P., & Mulvaney, P. (2013). The political economy of the 'just transition.' *The Geographical Journal, 179*(2), 132–140.

Prelec, T. (2020). *The persistence of elites and extractive practices in the energy sector in successor Yugoslav states (1980s–2010s).* Thesis, The persistence of elites and extractive practices in the energy sector in successor Yugoslav states (1980s–2010s). https://hdl.handle.net/10779/uos.23477996.v2

Rodríguez-Pose, A. (2018). The revenge of the places that don't matter (and what to do about it). *Cambridge Journal of Regions, Economy and Society, 11*(1), 189–209.

Rodrik, D. (2018). Populism and the economics of globalization. *Journal of International Business Policy, 1*(1–2), 12–33.

Šćepanović, V., & Martín Artiles, A. (2020). Editorial. *Transfer: European Review of Labour and Research 26*(1), 3–7.

Smith, S. (2017). *Just transition: A report for the OECD.* OECD Publishing.

Szpor, A. (2017). *The end of coal: Coal transition in Poland.* Working Paper, IDDRI and Climate Strategies.

Tomaney, J., Pike, A., & Rodríguez-Pose, A. (2019). *Local institutions and regional growth: Lessons for policy-making.* OECD Publishing.

UNFCCC. (2015). *Paris agreement.* https://unfccc.int/process-and-meetings/the-paris-agreement/the-paris-agreement

CHAPTER 17

The Montenegrin Waste Management Sector Transformation and Its Impacts Towards Just Green Transitions

Merita Toska, Rodion Gjoka, Ogerta Gjiknuri, and Azra Vukovic

17.1 Introduction

Waste is an urgent global challenge requiring the immediate attention of all stakeholders at any level. Population and economic growth, increased living standards, and urban density often drive increased municipal solid waste generation and volumes (Caputo & Pelagagge, 2002; Santa et al., 2023). In 2016, more than two billion tonnes of municipal solid waste were generated, averaging 0.74 kg per capita daily. Projections are for about 3.4 billion tonnes in 2050, according to (Kaza et al., 2018). Municipal solid waste [poor] management is also a complex problem globally, with adverse socio-economic, environmental, and health effects (Chatziaras & Constantinos, 2016; Kaza et al., 2018; Kahawalage et al., 2023; Oyedotun et al., 2021). According to Kaza et al. (2018), large quantities of waste remain dormant in landfills (about 40% globally), sometimes due to the lack of recycling practices

M. Toska (✉) · O. Gjiknuri
POLIS University, Tirana, Albania
e-mail: merita_toska@co-plan.org

O. Gjiknuri
e-mail: ogerta_gjiknuri@co-plan.org

R. Gjoka
Co-Plan Institute for Habitat Development, Tirana, Albania
e-mail: rodion_gjoka@co-plan.org

A. Vukovic
GreenHOME, Podgorica, Montenegro

E. Berisha et al. (eds.), *The Palgrave Handbook of Just Green Transitions in the Western Balkans and Beyond*,
https://doi.org/10.1007/978-3-031-95075-9_17

and the relative depositing fees applied (Santos et al., 2023). As these materials lose their physicochemical integrity, they become increasingly unfit for recovery as primary raw materials or viable energy sources. This deterioration renders them invaluable, emphasising the urgent need for improved waste management strategies focusing on timely recovery and recycling to prevent such irreversible losses. Landfills per se are of concern due to the associated potential risks of water and soil contamination, greenhouse gas (GHG) emissions, and land consumption. In addition, waste deposited in landfills locks a potential for materials and energy (Rezaei et al., 2020). In this regard, Kaza et al. (2018) point out that recyclables account for 16% of waste streams in low-income countries and 50% in high-income countries (plastic, cardboard, metal, glass, cardboard). Income levels, consumption patterns and preferences, climate and weather conditions are mentioned among the factors affecting waste composition diversification. Tapping into the waste potential through alternative fuel production via an energy-from-waste framework can simultaneously contribute to reducing waste in landfills (thus extending its lifetime), diversifying the energy mix (especially in industrial processes, reducing over-reliance on conventional fossil fuels) and fostering energy security (Gjoka & Toska, 2023). Additionally, shifting the focus from a disposal-oriented (linear) to a value-creation approach (circular) can incentivise investors (private and public) to reuse waste, prolonging the life cycle of resources, generating additional revenues, and creating jobs locally. Driven by cost-saving rationale, co-processing using waste streams as a substitute for fossil fuels is an opportunity and has the potential to tackle (at least partly) the solid waste challenge globally. Such an approach, using waste fractions to produce alternative fuels, has shown promising results, particularly in the cement industry (Chatziaras & Constantinos, 2016; Kahawalage et al., 2023). Therefore, co-processing waste streams in cement plants is not a novelty of our times, dating back to the 1980s following the oil crisis of 1970 (Daugherty, 1986). The World Business Council for Sustainable Development explains that "*selected waste and by-products with recoverable calorific value can be used as fuels in a cement kiln, replacing a portion of conventional fossil fuels, like coal, if they meet strict specifications. Sometimes, they can only be used after pre-processing to provide 'tailor-made' fuels for the cement process.*" (Broad Group, 2024). Separated in sorting facilities, combustible waste streams are processed into refuse-derived and solid-recovered fuels (RDF and SRF) as an alternative fuel for energy-intensive industries (like the cement industry). Switching, in part or totally, to alternative fuels is assessed to contribute to curbing CO_2 emissions, reducing fossil fuel dependency, reducing the quantity of waste deposited in landfills and contributing to lowering the incidence of energy costs in related industries' cost structure (Hajinezhad et al., 2016).

Waste management is of major concern and a primary source of environment-related issues in the Western Balkans, including air pollution, soil and water contamination, and the enormous loss of valuable materials by landfilling (Fig. 17.1). Waste generation per year reached about 537 kilograms per

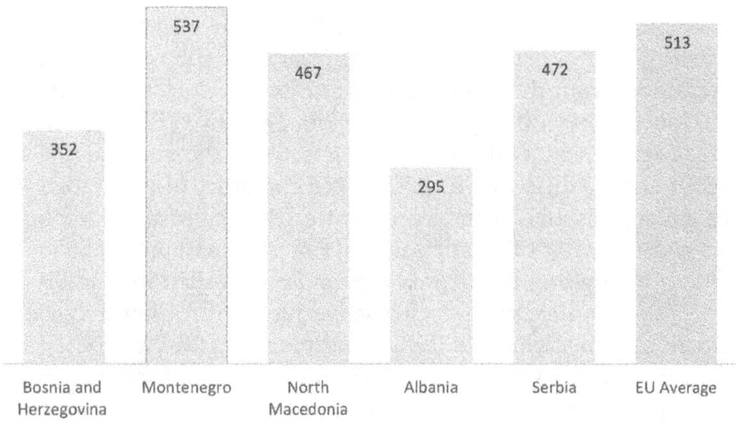

Fig. 17.1 Municipal waste generated, in kg per capita per year (*Source* Authors processing based on Eurostat data)

capita in Montenegro, about 24 kilograms per capita above the EU average and marked the lowest level in Albania (295 kg per inhabitant per year).

Waste collection coverage has increased in the last few years and ranges between 70% in Albania and 86% in Serbia and Montenegro, respectively (EEA, 2022). Nevertheless, there are significant differences between urban and rural areas, with the latter mainly lacking coverage of waste management services. In addition to financial constraints, inadequate infrastructure—including insufficient transportation fleets, underdeveloped road networks, and limited waste collection and treatment capacities—significantly impedes effective waste management services, particularly in rural areas. Separate waste collection remains largely unimplemented across the Western Balkan (WB) countries, with no country adopting a comprehensive systemic approach despite some isolated pilot projects. Consequently, recycling activities and associated infrastructure remain underdeveloped, contributing to marginal recycling rates in these countries, in stark contrast to the European Union (EU), where approximately 30% of total waste generated is recycled.[1] In Montenegro, the majority of collected waste is disposed of in sanitary landfills, further highlighting the region's limited progress in waste recovery and recycling efforts. The same patterns are observed in other WB countries, where alongside waste deposited in sanitary landfills, part of it still goes into dumpsites (EEA, 2022).

Incineration is also a common practice, e.g., in Albania, which completely bypasses the reduce-reuse-recycle and recovery principles. WB is moving towards EU integration; therefore, alignment with EU standards for an environmentally friendly, economically feasible, hygienically safe collection and

[1] Based on Eurostat data for 2022 on municipal waste treatment available at https://ec.europa.eu/eurostat/statistics-explained/index.php?title=Municipal_waste_statistics.

disposal of municipal waste is a priority. The waste management policy at the EU level aims to contribute to the circularity of the economy, extracting quality resources from the amounts of waste collected to the greatest extent possible. Concurrently, the Waste Framework Directive (EU legal framework for waste management and treatment) sets forth the foundations for EU waste management, the waste hierarchy, and the order of preference in waste management and disposal. The targets set for EU countries include increasing the reuse and recycling of municipal waste by a minimum of 65% by weight by 2035 and minimising the quantity of refuse deposited in landfills close to zero by 2050. The demand for alternative fuels with a lower carbon footprint is increasing following the EU commitment to global climate action and decarbonisation targets set forth (EU Green Deal—an economy with net-zero greenhouse gas emissions), particularly those linked with industrial activities subject to the EU Industrial Emissions Directive (IED).[2]

The research explores the potential and viability of RDF—Refused Derived and SRF-Solid Recovered Fuels from municipal waste that could be produced, marketed and co-processed in the case of Podgorica Region (targeted research area in Montenegro). The Montenegrin case study contributes to filling in a missing piece in the regional picture for assessing the viability of alternative fuel production as an opportunity to just green transition and SDGs.[3] The author's previous work has explored the potential transition from waste streams to alternative fuels in a regional project, "Waste to Energy for Western Balkans Cement Industry," for Albania, North Macedonia, and Serbia. Research in Albania, North Macedonia, and Serbia has shown promising results in the potential for the waste-to-energy setting up of facilities (RDF & SRF production). Such an approach, utilising waste streams with calorific embodied energy for alternative fuels production, might offer multiple benefits, including (i) reducing landfill waste, thereby enhancing environmental benefits; (ii) increasing the use of RDF/SRF in industry, decreasing dependence on fossil fuels, and consequently improving the competitiveness of local businesses in international markets through lower energy production costs; and (iii) reducing the GHG footprint through the co-processing of waste.

[2] The EU's Industrial Emissions Directive (IED) takes an integrated approach to controlling air, water and land pollution, and sets challenging industry standards for the most polluting industries. Source: https://www.era-comm.eu/IED/module_1/index.html.

[3] SDG 7 "ensure access to affordable, reliable, sustainable and modern energy for all" as the proposed research explores the use of alternative energy sources for industrial production; SDG 11 "make cities and human settlements inclusive, safe, resilient and sustainable" as the proposed research explores sustainable solutions for waste management at the local level; SDG 17, "global partnership for sustainable development," is the explored topic and envisages a multistakeholder collaboration at national, local, and private levels.

17.1.1 Stylised Facts on Waste Management in Montenegro

Montenegro is the smallest of WB countries, with an average population of 617,213 in 2022 (following a declining trend over the last three years)[4] and the lowest population density (45 inhabitants per square kilometre in 2022). The country is organised into 25 municipalities, including the capital, Podgorica and the Old Royal Capital, Cetinje. Podgorica is the largest municipality, with about 192,968 inhabitants in 2022, representing about 31% of the Montenegrin population. Waste management is a competence of municipalities based on the legal and regulatory framework in place (Figs. 17.2 and 17.3).

Despite the negative trend observed in population growth, municipal collected waste has been volatile and followed an upward trend during the last two years (+3.1 during 2022). In 2022, Montenegrin municipalities generated more than 335 thousand tons of waste, with a yearly average of 544 kg per capita (well above the EU average of 513 kg per capita in 2022). The increased tourism activity is assessed to have contributed to the increase in the indicators, alongside methodological improvement in data collection and reporting. According to Monstat data for 2022, 95% of the total municipal waste is disposed of in landfills. Only 0.3% is recycled (1.5 percentage points less than the data for 2021), constituting about 99% of the waste reported as treated (Fig. 17.4).

The waste management infrastructure in Montenegro includes two sanitary landfills with a carrying capacity of 672,849 m^3 operating in alignment

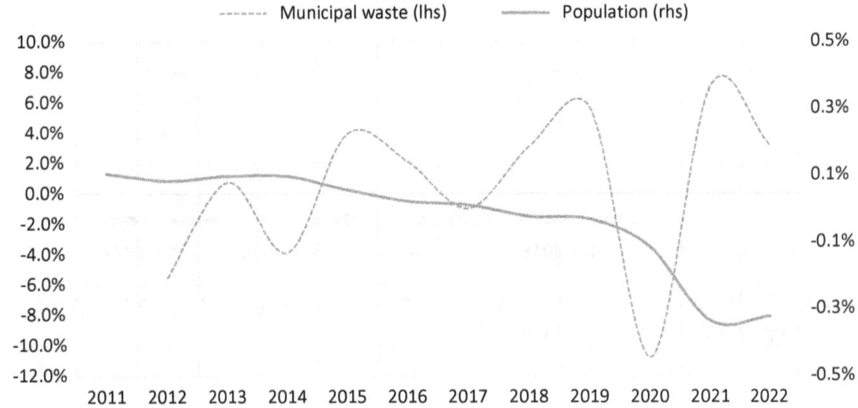

Fig. 17.2 Annual change in mid-year population and municipal waste (*Source* Authors processing based on MONSTAT data)

[4] Data from Monstat: Estimated mid-year number of population for 2022 available at: https://www.monstat.org/eng/page.php?id=234&pageid=48.

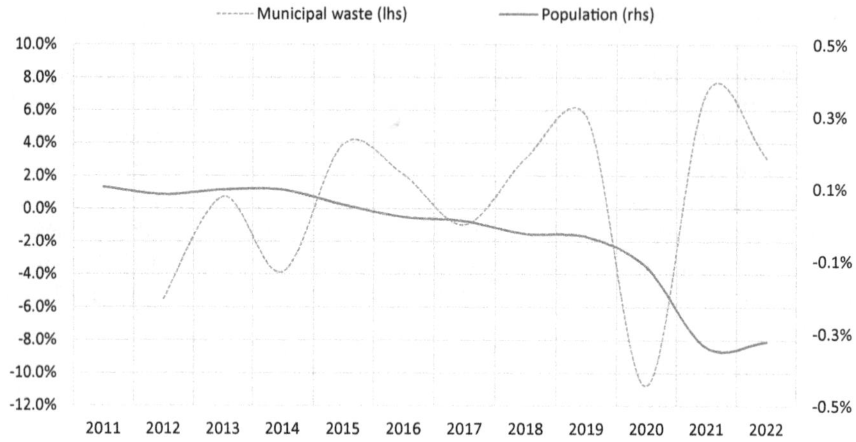

Fig. 17.3 Total generated municipal waste in kg per capita per year (*Source* Authors processing based on MONSTAT and Eurostat data)

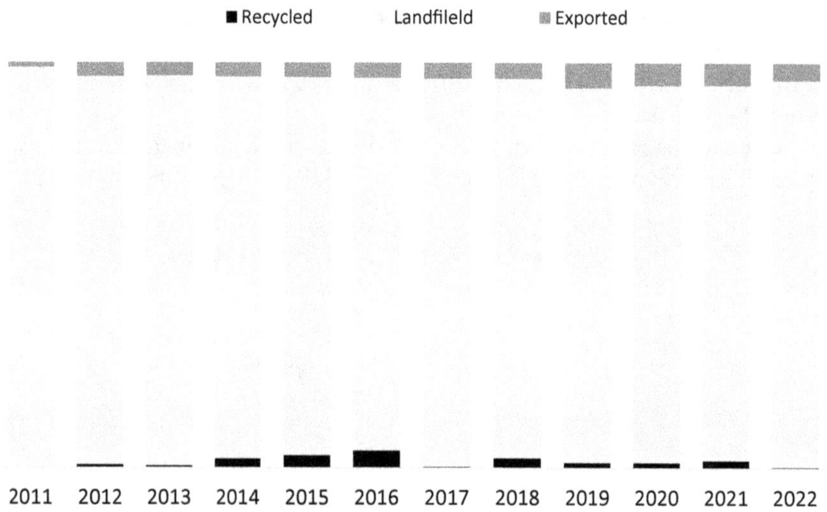

Fig. 17.4 Waste disposal: recycled, landfilled and exported (*Source* Authors processing based on MONSTAT and Eurostat data)

with the EU standard (located in Bar and Podgorica municipalities). In addition, there are also 17 controlled landfills used for temporary waste storage for a maximum of 2 years, which are to be converted into sanitary landfills or shut down (timing not defined). However, illegal waste depositing is present. Four material sorting facilities for dry recyclables are operative in Montenegro located in Podgorica, Žabljak, Kotor and Herceg Novi, with a total annual capacity of 108,250 tons. Recycling capacities are low and not fully used, and the recyclable materials derived from the sorting process are processed internally and, in part, exported (EEA, 2021). There are six recycling yards in Podgorica, one in Herceg Novi, one in Kotor, and one in Budva. Two transfer stations are located in Kotor and Herceg Novi, are dedicated medical waste processing facilities in Berane and Podgorica, electric and electronic waste processing facilities in Bar, and five vehicle waste processing plants in Podgorica (1), Berane (1) and Nikšić (3). Nevertheless, many unregulated and/or illegal landfills exist in Montenegro (more than 400).[5]

Waste management involves different actors at the central and local levels of governance. At the central level, the Ministry of Ecology, Spatial Planning and Urbanism is responsible for the strategic and legislative framework for waste management; the Nature and Environment Protection Agency (NEPA) is an executive agency in charge of licensing the activity of waste import, export and transit companies, waste data gathering and reporting, environmental monitoring and other.[6] At the local level, local governance units are responsible for administering the waste management service in the territory under administration. Waste management service is administered based on the local waste management plan (approved document by the local council) through PUCs or any other arrangement to guarantee the waste collection, transportation, and disposal services.

The strategic and legislative framework has been updated and includes ambitious targets related to waste management in Montenegro. Nevertheless, it lacks effective implementation. Law no 64 /2011, "On waste management," regulates the waste management sector in Montenegro.[7] A draft law on waste management was developed in 2021 to align with the EU legislative and regulatory framework. Stakeholder groups have broadly discussed and strongly contrasted the draft law adopted in April 2024. The new law's approval is critical and marks a turning point in waste management, as it sets the foundation for waste separation at the source, includes the extended producer's responsibility (an item in the previous law but never implemented), points to an increased recycling rate, and establishes an adequate waste management infrastructure. The National Waste Management Strategy of Montenegro 2015–2030,[8] adopted in 2015, sets the strategic vision for establishing an integrated waste management system aligned to EU standards in the field. The overarching goal is establishing a practical, functional and sustainable waste management system, accounting for countries' internal competencies

[5] Source: https://zerowastemontenegro.me/waste-management-status-montenegro/.

[6] Source: https://www.devex.com/organizations/nature-and-environment-protection-agency-of-montenegro-nepa-montenegro-126096.

[7] Published in the Official Gazette of Montenegro, No. 064/11 from 29.12.2011.

[8] Source: https://leap.unep.org/en/countries/me/national-legislation/montenegro-waste-management-strategy-until-2030 (title in original language Predlog strategije upravljanja otpadom u Crnoj Gori do 2030. godine).

and capacities. Among the objectives included in the strategy and relevant to the scope of the study:

- introduction and implementation of circular economy principles into the production and trade systems of Montenegro through the incentivisation of reuse practices of materials;
- develop and apply the primary and secondary waste selection system to maximise recyclables' collection and economic valorisation (through reuse and circular practices);
- preparation of the waste management system for the future use of energy-from_waste and/or thermal treatment of the same waste;
- minimisation of the quantity of waste permanently disposed in landfills;
- strengthening separation at the source of waste and alternative utilisation of biodegradable waste and sewage sludge treatment;
- increase separation, recycling, and disposal of construction (and demolition debris) waste.

The strategy paves the way for other essential state and local waste management planning documents (and other necessary documents for implementing the waste management policy). National Waste Management Plan (NWMP) 2015—2020 defined the system, objectives, measures, and activities to be implemented in the waste management field. A new draft of NWMP 2023–2028 has been prepared, with a particular goal of separation at source, developing recycling schemes, and promoting a circular economy. The plan also aims to design and implement a sustainable financing system based on cost recovery at the local level and strengthen governance authorities' capacities in implementing and enforcing NWMP measures. Local Waste Management Plans (LWMP) are the core documents at the local level that streamline efforts to administer the municipal waste service (implementation of the waste management policy and organisation of service delivery). Despite being mandatory, only a few municipalities complied to develop, adopt and implement waste management plans in Montenegro (the municipalities in the targeted study areas are equipped with LWMP—LWMPs for Capital Podgorica 2016—2020,[9] Danilovgrad Municipality 2016—2020,[10] Old Royal Capital Cetinje 2016—2020[11]). EEA (2021) and NEPA (2021) highlight among the reasons for the lack of financial and human resource capacities at the local level, the lack of penalties, and weak enforcement. Generally, municipalities implement waste management services through municipally owned public utility companies (PUCs).

[9] Local Waste Management Plan for Capital City of Podgorica for period 2016–2020.

[10] Local Waste Management Plan for Danilograd, available here: https://old.danilovgrad.me/files/xvii_sso_27122016/plan_upravljanja_otpadom_2016.pdf.

[11] Lokalni plan upravljanja komunalnim i neopasnim građevinskim otpadom Prijestonice Cetinje za

Table 17.1 Population and waste collected in the study area (*Source* Monstat[12]; Report on the implementation of NWMP for 2022[13])

Municipality	*Population (number)*	*Amount of waste collected (in tons)*
Podgorica	192,968	85,100
Cetinje	14,538	4,570
Danilovgrad	18,340	4,996
Total	225,846	94,666

Definition of the study area. The study area is defined by the units of Podgorica (the capital), Cetinje, and Danilovgrad, which are all part of the Montenegrin central region. These areas have a population of 225,846 inhabitants, and cumulatively, the amount of waste collected in 2022 is about 94,666 tons (Table 17.1).

There is a sanitary landfill, "Livade," in Podgorica, which, alongside the central region's areas, accepts waste from the municipalities of Nikšić, Plužine, and Šavnik, following primary and secondary separations.[14] While there has been a proposal to expand the Livade landfill, establishing three transfer stations in the respective municipalities is an effective solution alongside establishing the recovery facility in Livade (to produce alternative fuels from refused waste) (Fig. 17.5).

Podgorica counts on the most developed waste management infrastructure, including a sanitary landfill, a recycling centre (for a secondary selection of municipal waste), six recycling yards, and good coverage with service and containers (semi-underground and classic ones). There are eight unregulated landfills in Podgorica (all sanitised). Cetinje has 15 unregulated landfills (dumpsites of mixed municipal waste, construction non-hazardous waste and others). The waste from the three targeted municipalities is deposited in the Livade landfill. There are two unregulated landfills in Danilovgrad, Klikovače (two sites near the Spuž railway station) and Bogićevica.

17.2 Methodology

The study employs a mixed-method approach (Fig. 17.6) to assess the viability of alternative fuel production using pre-defined waste streams from the targeted area (three municipalities) to reduce carbon footprint and towards

[12] National Statistic Agency Monstat, available here: https://monstat.org/cg/page.php?id=273&pageid=48.

[13] Report on the implementation of the National Waste Management plan for 2021, available here: https://wapi.gov.me/download-preview/81b21759-e429-4599-b94c-8c9fb3bcf69f?version=1.0.

[14] The second landfill, "Možura" is located in Bar accepting waste from Ulcinj, Bar, Budva, Tivat and Kotor also. Other municipalities operate with dumps sites, causing a lot of environmental pollutions.

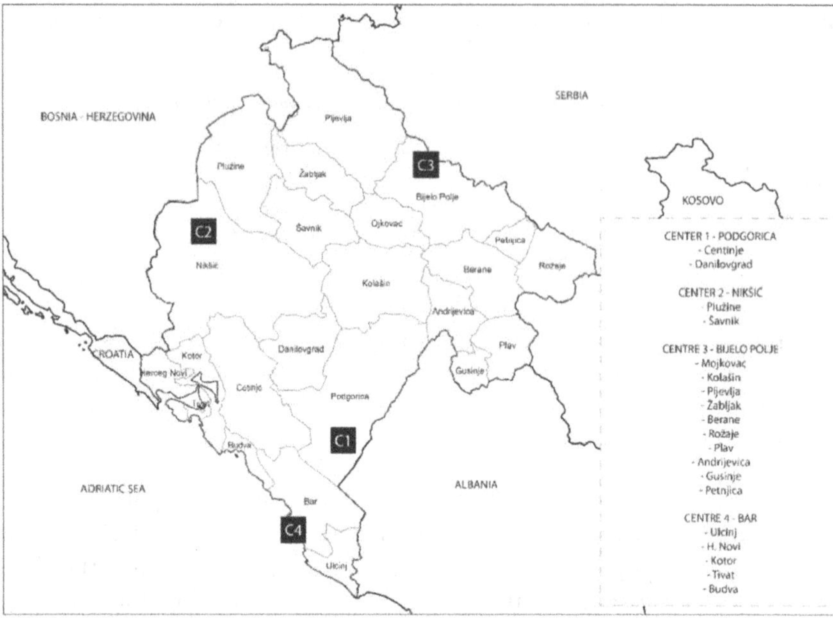

Fig. 17.5 Waste management areas (*Source* Decision on amendment of the NWMP[15])

implementing the Green Agenda for the WB. The viability of AF production is to be assessed from an economic and regulatory point of view. An exploratory sequential design combined qualitative and quantitative data collection and analysis at the local and national levels of waste management governance. The qualitative analysis (including a review of strategic, legal and regulatory acts, and semi-structured stakeholder interviews) provides a comprehensive and contextualised understanding of the waste management state of play and way ahead for the Montenegrin case.[16] In particular, interviews with stakeholders relevant to the research objective aimed at understanding the attitude to waste-to-energy initiatives, what the involved stakeholders foresee in the near future in their field of competence, and data collection on quantities and costs to be used in the cost–benefit analysis. Against this background, the quantitative analysis (including primary and secondary data on waste management) is used

[15] Decision on amendment of the National Waste Management Plan, available here: https://wapi.gov.me/download-preview/0e0eec95-a53b-4a98-b0f5-58e3c37ae887?version=1.0.

[16] Semi structured interviews were carried out with institutional stakeholders including: (i) Communal Utility Danilovgrad (Komunalno preduzeće Danilovgrad); (ii) Doo Deponija "Livade"; (iii) Ministry of Tourism, Ecology, Sustainable Development and Development of North; (iv) Eko Fund–Fund for Environmental Protection of Montenegro; (v) Društvo Mladih Ekologa Nikšić. More dettails included in (Vukovic et al., 2024).

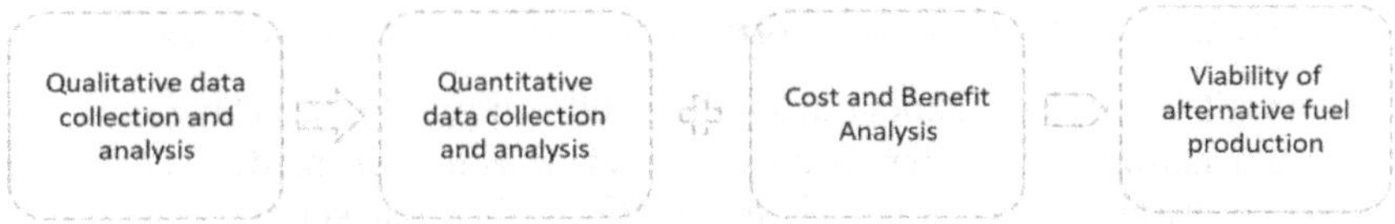

Fig. 17.6 Exploratory sequential design (*Source* Authors processing)

to assess the viability of alternative fuel production for the targeted areas. The viability of alternative fuel production is assessed using the environmental and social cost–benefit analysis instrument and scenario analysis.

17.2.1 Environmental and Social Cost–Benefit Analysis

The environmental and social cost–benefit analysis (ES-CBA) is a widely used instrument for assessing the economic feasibility assessment of any investment project, including economic, social and environmental impacts. The ES-CBA is a flexible instrument, allowing the estimation of the (net) costs and benefits stemming from an investment project. The information regarding costs and benefits enables decision-makers to rationally determine the viability of an investment project and compare competing investment projects in terms of value added. The main steps followed in the ES-CBA preparation process included:

- Identification of costs and benefits deriving from project implementation, combining information from desk review, interviews with stakeholders and information provided by the municipality of Podgorica;
- Quantifying and monetising gross outcomes regarding costs and benefits from project implementation (when possible, conditional on data availability and/or benchmarks).
- Assessment of net impacts, as the difference between the "with the project" versus "without the project" scenarios (the counterfactual). The counterfactual (without the project scenario) considers the amount of change that might have happened, regardless of the intervention. This approach captures the "net change" that can be explicitly attributed to the proposed intervention.
- Assessment of cost and benefits behaviour over the reference period—assess how costs and benefits will change in time (using average growth rates, trends when available or introducing assumptions);
- Discounting of costs and benefits to obtain present values—convert future costs and benefits into present value using a social discount rate;
- Calculating economic performance indicators: the net present value (NPV), the internal rate of return (IRR), and the benefits-to-costs ratio (BCR).

17.2.2 Data, General Assumptions, and Economic Indicators

The data. The cost and benefit analysis is built on publicly available information from national and international sources, information made available by Podgorica municipalities, and information from direct stakeholder interviews (Table 17.2).

General assumptions introduced for the ES-CBA include the following:

- The EU guidelines for waste management infrastructure suggest a reference period of 25–30 years (European Commission, 2015). However, the reference period in the ES-CBA is 16 years, starting in 2024 and ending in 2040. Choosing a shorter reference period is considered more appropriate given the expected structural changes in waste management

Table 17.2 Data used in the SE-CBA (*Source* Authors' calculations)

Indicator	*Value*	*Unit*	*Source*
Starting year	2024		Expert assumption
End year	2040		Expert assumption
Exchange rate USD/Euro Year 2023	0.924	Euro per unit of USD	https://www.exchangerates.org.uk/USD-EUR-spot-exchange-rates-history-2023.html
Social Discount Rate (UE Guidelines)	3.0%		Source: DG CLIMA (2021) / Economic Appraisal Vademecum 2021–2027
Production capacity	50,000	tons per year	Expert assumption, based on stakeholder interviews and available data
RDF & SRF Average selling price, minimum	27	Euro per ton	Assumption based on information from Holcem operations
RDF& SRF Average selling price, maximum	60	Euro per ton	Assumption based on information from Holcem operations
RDF& SRF Average selling price, average	44	Euro per ton	Expert calculation (average of min and maximum prices)
Landfill depositing fee in Montenegro	35.0	Euro per ton	Expert assessment and municipality of Podgorica
CO_2 and GHG savings from engaging in RDF&SRF production	17,890	Tons per year	Expert assessment

(policy and practice) and the speed of technological changes (production line).

- The EU guidelines suggest including a residual value of the investment at the end of the reference period. In a conservative approach, the residual value at the end of the reference period is assumed to be 10% of the initial investment (excluding spare parts).
- Lacking a national reference, the EU social discount rate (SDR) of 3% was employed to calculate the net present value of future cash flows.
- Cost and benefits from all sources are presented net of VAT as suggested in the EU Guidelines.
- General maintenance costs are estimated at 5% of the investment costs.
- Operating costs are presented at a lump sum based on technical suggestions (including personnel and other unidentifiable costs) and plant electricity consumption.
- An incremental approach was adopted to account for project implementation's economic, social and environmental costs and benefits. The "without the project scenario" was built considering the continuation of the situation with a business-as-usual approach (when possible). The counterfactual baseline scenario is defined by European Commission (2015) as "*what would happen in the absence of the project*" (assuming business as usual or a do minimum approach).
- All values in other currencies are converted into Euros using the average exchange rates for 2023.

17.2.3 Economic Indicators

The CBA for the proposed investment project employs three widely used economic performance indicators calculated on incremental cash flows.

- The *economic net present value indicator (NPV)* is a standard indicator calculated as the difference between the discounted total benefits and costs arising from the project.

$NPV = \left(\left[\sum \frac{B_i}{(1+sdr)^i}\right]\right) - \left(\left[\sum \frac{C_i}{(1+sdr)^i}\right]\right)$ summed over 0 to n years.
B_i = project benefits in year i, for i {0,1…n};
C_i = project costs in year i, for i {0,1…n};
n = total number of years for project duration—the time horizon;
SDR = social discount rate.

Projects with positive NPV are considered viable, while those with negative net present value are not viable.

- The *internal economic rate of return (IRR)* is the ratio that produces a zero value for the economic NPV. As a rule of thumb, the IRR should be higher than the discount rate used in the analysis (it creates value added). On the contrary, the project destroys value if the IRR is lower than the discount rate.

$$\left(\left[\sum \frac{B_i}{(1+irr)^i}\right]\right) - \left(\left[\sum \frac{C_i}{(1+irr)^i}\right]\right) = 0$$

- The *benefits-to-cost ratio (BCR)* is calculated as the ratio between the sum of discounted economic benefits and costs over the time horizon considered in the analysis. The BCR informs us of how many euros are generated by the intervention for each euro 1 invested over the time horizon of the project. The following formula summarises the information included in the BCR:

$$\text{BCR} = \frac{\left[\sum B_i/(1+sdr)^i\right]}{\left[\sum C_i/(1+sdr)^i\right]}$$

B_i = project benefits in year i, for i {0,1…n};
C_i = project costs in year i, for i {0,1…n};
N = total number of years for project duration—the time horizon;
SDR = social discount rate.
Based on the results obtained:

1. BCR < 1.0 means that the project's costs exceed its benefits in economic terms. In other words, we are destroying the value of each 1€ invested.
2. BCR = 1.0 means that the project's costs equal its benefits in economic terms. Therefore, while it can proceed with the project, its viability is uncertain. In other words, no added value is created for each one Euro invested.
3. BCR > 1.0 means that the project's benefits exceed its costs in economic terms. In other words, we are creating value for each one Euro invested.

17.3 Results

The economic feasibility of the waste-to-energy approach using the CBA starts with identifying costs and benefits related to the intervention. Once identified, costs and benefits will be monetised (when possible) and planned over the selected period.

Capital expenditure. The estimated initial investment costs for the W2E plant are USD 2.8 million, as detailed in Table 17.3. A detailed technical description of the proposed W2E plant and its costs is presented in Vukovic et al. (2024). Investment is planned to be completed within two years, 30% in the first and 70% in the second year. In addition to the initial investment costs, the methodology foresees estimating the potential replacement costs (costs for replacing short-life equipment). Replacement costs are expected to

Table 17.3 Capital expenditure (*Source* Authors' calculations)

Investment project	*Cost (in USD)*
I1 Initial investment	2,835,622
I2 Replacement costs	339,900
Total capital expenditure	3,022,265
RV Residual value	850,687

occur starting in the third year and are equally distributed during the project's duration.

Current expenditures. The current expenditure categories include the cost for the facility's regular maintenance and operationalisation (personnel, energy, and other costs) and energy consumed from the production facility (as input for the production process), as presented in Table 17.4.

C1. General maintenance. General maintenance costs are presented in Table 17.5 and were calculated as a fixed percentage of initial investment costs (excluding replacement costs. From the alternatives presented, based on experience and the European Commission (2015).

Maintenance costs were assumed to be about 5% of the initial investment, or about Euro 131 thousand. Maintenance costs change over time and are scaled using a confidence indicator of 75% during Y1-Y3, 95% in Y4-Y8, and 100% for the rest of the reference period.

- *C2. Operative Costs.* The operationalisation of the W2E production plant is expected to account for about Euro 1 million (Vukovic, 2024),

Table 17.4 Current expenditures (*Source* Authors' calculations)

Current costs	*Quantified*	*Monetised*	*Included CBA*
C1. General maintenance	Yes	Yes	Yes
C2. Operating costs	Yes	Yes	Yes
C3. Energy	Yes	Yes	Yes

Table 17.5 General maintenance costs (*Source* Authors' calculations)

Percentage	*Value*
1% of initial investment	Euro 26,204
2% of initial investment	Euro 52,408
3% of initial investment	Euro 78,612
5% of initial investment	Euro 131,020
7% of initial investment	Euro 183,428
10% of initial investment	Euro 262,040

including personnel, administrative, and other unidentifiable costs. Operative costs are scaled down by the same confidence indicator as general maintenance costs.

- *C3. Energy Consumption.* The W2E production plant has an overall installed production capacity of RDF 250 tons daily and a daily consumption of 650–900 kWh daily (Vukovic et al., 2024). Electricity costs were estimated using Monstat data on electricity prices for final non-household customers in Montenegro (including an annual 3% increase in energy prices).

The assessment identified three main categories of benefits derived from the investment project in RDF&SRF production. All three benefit sources are included in the cost–benefit analysis, as presented in Table 17.6.

- *B1. Revenue from Selling Activity of Alternative Fuels.* Revenues from selling activity of alternative fuels are the main source of income in the CBA, among the others. The data from the market research indicate an average minimum selling price of 27 euros per ton and a maximum average selling price of 60 Euros per ton. The analysis considered an average price of 44 euros per ton (average of minimum and maximum average prices) to estimate potential revenue over the project duration. Based on the increasing trend in energy prices over the last five years, it was assumed there would be an annual price increase of 5% for alternative fuels. As per the counterfactual scenario, from the overall waste quantity usable for alternative fuel production during one year, at least 75% of it would have been landfilled. Using an average depositing fee of about 35 euros per ton, it was assessed the amount of money spent in the business as usual (or counterfactual scenario). The net effect from engaging in alternative fuel production was derived as a difference between the revenues from alternative fuel selling and the counterfactual scenario costs for waste landfilling.
- *B2. Increased Landfill Life.* While engaging in alternative fuel production requires a large quantity of land fillable waste to be diverted to the production plant, less waste would be deposited annually in the landfill. Therefore, engaging in alternative fuel production extends landfill life, meaning there are fewer costs for investments in the short-medium term.

Table 17.6 The benefits sources (Authors' calculations)

Benefit item	*Quantified*	*Monetised*	*Included in CBA*
B1. Revenue from alternative fuels selling	Yes	Yes	Yes
B2. Increased life of the landfill	Yes	Yes	Yes
B3. Saving on CO_2 emissions	Yes	Yes	Yes

The landfill's extended life was quantified and monetised by applying the same landfilling fee over the reference period.

B3. Saving in CO_2 and GHG Missions. In the assessment, engagement in alternative fuel production is expected to save about 17,890 CO_2 and GHG emissions per year (Vukovic et al., 2024). These benefits have been quantified, monetised and included in calculating economic indicators in the CBA, using shadow prices of CO2eq from DG Regio (2014) and DG Regio (2021).

Economic Indicators Results. The identified cost and benefit sources have been monetised and assessed over the reference period (16 years). Using the social discount rate suggested in DG Regio (2014) and DG Regio (2021), four economic indicators were calculated to assess the viability of alternative fuel production for the case of Podgorica (see Table 17.7). Results for net present value indicate that the project's benefits exceed the costs associated with its implementation. This finding is corroborated by the internal rate of return (IRR), which significantly surpasses the social discount rate, indicating the proposed investment's profitability. The benefits-to-costs ratio (BCR) also signifies that each Euro invested in the project will generate societal value. In addition, the investment is expected to yield a simple payback period within the seventh year of operations (Table 17.8).

Table 17.7 Economic indicators (Authors' calculations)

Indicator	*Unit*	*Value*	*Comment*
Capital Investment	Euro	2,620,398	Estimated capital investment (including initial investment and replacement costs)
Net Present Value (NPV)	Euro	15,618,312	The NPV suggests the project is viable; thus, the benefits exceed its costs
Internal Rate of Return (IRR)	%	18%	The internal rate of return (the one that sets NPV = 0) is considerably higher than the discount rate, suggesting a profitable investment
BCR	Euro	1.7	The BCR > 1 indicates that the project's benefits exceed its costs. In other words, for each 1 Euro invested, we create 1.7 Euro of value
Simple payback period	7th	Year	The investment presents a simple payback period in the seventh year of operations

Table 17.8 Summary of CBA results (Authors' calculations)

Project	*GreenFORCE—WASTE TO ENERGY MONTENEGRO*							
Method	*Cost Benefit Anlysis (CBA)*							
Scenario	2							
Key Assumptions								
Social discount rate	3.0%							
Appraisal period (years)	2040							
Currency	Euro							
		0	**1**	**2**	**3**	**4**	**5**	**6**
Year		**2024**	**2025**	**2026**	**2027**	**2028**	**2029**	**2030**
Discount factor		1.000	0.971	0.943	0.915	0.888	0.863	0.837
CAPEX								
I1	Investment 1	7,86,119	18,34,279	–	–	–	–	–
I2	Spare parts	–	–	–	22,436	22,436	22,436	22,436
RV	Residual value	–	–	–	–	–	–	–
Total CAPEX		**7,86,119**	**18,34,279**	–	**22,436**	**22,436**	**22,436**	**22,436**
Present value of CAPEX		7,86,119	17,80,853	–	20,532	19,934	19,353	18,790
Cumulative PV of CAPEX		26,42,567						
Benefit sources								
B1	Revenues from selling of RDF & SRF	(14,43,687)	(14,43,687)	7,31,313	8,40,063	9,54,251	10,74,147	12,00,039

Project	*GreenFORCE—WASTE TO ENERGY MONTENEGRO*						
Method	*Cost Benefit Anlysis (CBA)*						
Scenario	*2*						
B2	Increased life of the landfill	–	–	15,46,808	15,46,808	15,46,808	15,46,808	15,46,808
B3	Saving on CO2eq emmissions	–	–	1,95,359	2,13,607	2,31,854	5,00,204	5,36,700
Total Benefits		**(14,43,687)**	**(14,43,687)**	**24,73,479**	**26,00,477**	**27,32,912**	**31,21,159**	**32,83,547**
Present Value of Benefits		(14,43,687)	(14,01,638)	23,31,491	23,79,805	24,28,157	26,92,339	27,49,919
Cumulative PV Benefits		4,69,82,743						
Cost sources								
C1	General maintenace costs	–	–	98,265	98,265	98,265	98,265	1,24,469
C2	Operating costs (personnel and other administrative)	–	–	7,50,000	7,50,000	7,72,500	7,95,675	8,19,545
C3	Energy bill for the plant operation	–	–	1,95,359	2,13,607	2,31,854	5,00,204	5,36,700
Total Costs		-	-	**10,43,624**	**10,61,872**	**11,02,619**	**13,94,144**	**14,80,714**
Present Value of Costs		–	–	9,83,715	9,71,763	9,79,663	12,02,601	12,40,075
Cumulative PV of Costs		2,50,18,075						
Undiscounted cash flows								
CAPEX		(7,86,119)	(18,34,279)	–	(22,436)	(22,436)	(22,436)	(22,436)
Total costs		–	–	(10,43,624)	(10,61,872)	(11,02,619)	(13,94,144)	(14,80,714)
Total benefits		(14,43,687)	(14,43,687)	24,73,479	26,00,477	27,32,912	31,21,159	32,83,547

(continued)

Table 17.8 (continued)

Project	*GreenFORCE—WASTE TO ENERGY MONTENEGRO*							
Method	*Cost Benefit Anlysis (CBA)*							
Scenario	*2*							
Net Cash Flow	(26,72,460)	(22,29,806)	(32,77,966)	14,29,856	15,16,170	16,07,857	17,04,579	17,80,397
Discounted cash flows (SDR1)								
CAPEX		(7,86,119)	(17,80,853)	–	(20,532)	(19,934)	(19,353)	(18,790)
Benefits		(14,43,687)	(14,01,638)	23,31,491	23,79,805	24,28,157	26,92,339	27,49,919
Costs	(26,42,567)	–	–	(9,83,715)	(9,71,763)	(9,79,663)	(12,02,601)	(12,40,075)
Net Cash Flow	(26,42,567)	(22,29,806)	(31,82,491)	13,47,776	13,87,510	14,28,560	14,70,385	14,91,054
Cummulative cash flow	(26,42,567)	(48,72,373)	(80,54,864)	(67,07,088)	(53,19,578)	(38,91,018)	(24,20,633)	(9,29,579)
ENPV	1,56,18,312							
EIRR	18%							
B/C ratio	1.7							
cumulative discounted benefits	*4,69,82,743*							
cumulative dicounted costs	*2,76,60,642*							

Project	*GreenFORCE—WASTE TO ENERGY MONTENEGRO*									
Method	*Cost Benefit Anlysis (CBA)*									
Scenario	*2*									
Key Assumptions										
Social discount rate										
Appraisal period (years)										
Currency										
	7	8	9	10	11	12	13	14	15	16
Year	**2031**	**2032**	**2033**	**2034**	**2035**	**2036**	**2037**	**2038**	**2039**	**2040**
Discount factor	0.813	0.789	0.766	0.744	0.722	0.701	0.681	0.661	0.642	0.623
CAPEX										
I1	–	–	–	–	–	–	–	–	–	–
I2	22,436	22,436	22,436	22,436	22,436	22,436	22,436	22,436	22,436	22,436
RV	–	–	–	–	–	–	–	–	–	(2,62,040)
Total CAPEX	**22,436**	**22,436**	**22,436**	**22,436**	**22,436**	**22,436**	**22,436**	**22,436**	**22,436**	**(2,39,604)**
Present value of CAPEX	18,242	17,711	17,195	16,694	16,208	15,736	15,278	14,833	14,401	(1,49,313)
Cumulative PV of CAPEX										
Benefit sources										
B1	13,32,225	14,71,021	16,16,756	17,69,779	19,30,452	20,99,159	22,76,301	24,62,301	26,57,600	28,62,664

(continued)

Table 17.8 (continued)

Project	*GreenFORCE—WASTE TO ENERGY MONTENEGRO*									
Method	*Cost Benefit Anlysis (CBA)*									
Scenario	2									
B2	15,46,808	15,46,808	15,46,808	15,46,808	15,46,808	15,46,808	15,46,808	15,46,808	15,46,808	15,46,808
B3	5,96,810	9,85,381	10,75,547	11,65,712	16,74,504	17,90,431	19,06,358	25,27,857	26,72,766	28,17,675
Total Benefits	**34,75,843**	**40,03,210**	**42,39,111**	**44,82,298**	**51,51,763**	**54,36,398**	**57,29,467**	**65,36,965**	**68,77,173**	**72,27,147**
Present Value of Benefits	28,26,179	31,60,171	32,48,925	33,35,251	37,21,743	38,12,980	39,01,488	43,21,704	44,14,196	45,03,719
Cumulative PV Benefits										
Cost sources										
C1	1,24,469	1,24,469	1,24,469	1,24,469	1,31,020	1,31,020	1,31,020	1,31,020	1,31,020	1,31,020
C2	8,44,132	8,69,456	8,95,539	9,22,405	9,50,078	9,78,580	10,07,937	10,38,175	10,69,321	11,01,400
C3	5,96,810	9,85,381	10,75,547	11,65,712	16,74,504	17,90,431	19,06,358	25,27,857	26,72,766	28,17,675
Total Costs	**15,65,411**	**19,79,306**	**20,95,555**	**22,12,587**	**27,55,601**	**29,00,031**	**30,45,316**	**36,97,052**	**38,73,107**	**40,50,095**
Present Value of Costs	12,72,822	15,62,482	16,06,068	16,46,372	19,90,705	20,34,023	20,73,712	24,44,187	24,86,000	25,23,885
Cumulative PV of Costs										
Undiscounted cash flows										
CAPEX	(22,436)	(22,436)	(22,436)	(22,436)	(22,436)	(22,436)	(22,436)	(22,436)	(22,436)	2,39,604

Project	*GreenFORCE—WASTE TO ENERGY MONTENEGRO*									
Method	*Cost Benefit Anlysis (CBA)*									
Scenario	2									
Total costs	(15,65,411)	(19,79,306)	(20,95,555)	(22,12,587)	(27,55,601)	(29,00,031)	(30,45,316)	(36,97,052)	(38,73,107)	(40,50,095)
Total benefits	34,75,843	40,03,210	42,39,111	44,82,298	51,51,763	54,36,398	57,29,467	65,36,965	68,77,173	72,27,147
Net Cash Flow	18,87,997	20,01,468	21,21,120	22,47,276	23,73,726	25,13,931	26,61,716	28,17,477	29,81,631	34,16,656
Discounted cash flows (SDR1)										
CAPEX	(18,242)	(17,711)	(17,195)	(16,694)	(16,208)	(15,736)	(15,278)	(14,833)	(14,401)	1,49,313
Benefits	28,26,179	31,60,171	32,48,925	33,35,251	37,21,743	38,12,980	39,01,488	43,21,704	44,14,196	45,03,719
Costs	(12,72,822)	(15,62,482)	(16,06,068)	(16,46,372)	(19,90,705)	(20,34,023)	(20,73,712)	(24,44,187)	(24,86,000)	(25,23,885)
Net Cash Flow	15,35,114	15,79,977	16,25,662	16,72,184	17,14,830	17,63,220	18,12,499	18,62,684	19,13,795	21,29,147
Cummulative cash flow	6,05,535	21,85,513	38,11,175	54,83,359	71,98,189	89,61,410	1,07,73,908	1,26,36,592	1,45,50,388	1,66,79,535
ENPV										
EIRR										
B/C ratio										
cumulative discounted benefits										
cumulative dicounted costs										

17.4 Conclusions

Waste management is stagnating in the WB countries. While great efforts have been channelled into aligning the regulatory and strategic planning framework to EU requirements, implementation remains weak and lags behind European peers. Integrated waste sector governance still presents serious problems, posting challenges to today's and future generations regarding environmental, economic, and social dimensions.

Waste management sector transformation is essential for Montenegro to align with the EU directives and advance the country's green agenda. The study reveals that the current system has produced poor outcomes in waste management, with low recycling rates, heavy reliance on landfilling, and inappropriate waste management infrastructure with widening disparities among urban and rural areas. Despite recent developments, such as the approval of the new waste management law, its implementation remains far from what was projected and expected, contributing to increased environmental risks and inefficient use of resources. Our findings emphasise the importance of a substantial switch and accelerating the implementation of circular economy principles and earmark investments into waste management infrastructure improvement.

The potential for transforming waste streams into alternative fuels, such as Refuse Derived Fuel (RDF) and Solid Recovered Fuel (SRF), presents a viable solution for reducing waste (which builds up year after year in landfills) and promoting energy security. Exploring co-processing in the industrial sector could potentially lead to decreasing reliance on fossil fuels (thought at a progressive substitution rate), reducing landfill capacity consumption and contributing to reducing GHG emissions, supporting, at the same time, economic and environmental goals. Establishing a production facility for alternative fuel production within the Livadei Landfill premises entails an initial investment of about Euro 2.6 million. All the economic indicators employed in the analysis suggest the financial and economic viability of the solution, which is in line with the findings and authors' previous work on the case of Albania, North Macedonia, and Serbia (Gjoka et al., 2023). The Montenegrin case study fills in a missing piece in the regional picture for assessing alternative fuel production's viability as an opportunity to foster just green transition and SDGs, among others. Nevertheless, the results are limited to the targeted area (and data availability), and further investigation is needed to assess alternative fuel production at a broader level.

References

Broad Group (2024). *Broad group*. [Online]. https://www.broadgroup.com/news/rdf-refuse-derived-fuel-explained#:~:text=The%20World%20Business%20Council%20for,if%20they%20meet%20strict%20specifications

Caputo, A. C., & Pelagagge, P. M. (2002). RDF production plants: II economics and profitability. *Applied Thermal Engineering, 22*(4), 439–448.

Chatziaras, N., & Constantinos, C. (2016). Use of waste derived fuels in cement industry: A review. *Management of Environmental Quality: An International Journal, 27*(2), 178–193.

Daugherty, K. E. (1986). *An identification of potential binding agents for densified fuel preparation form municipal solid waste Phase 1, Final Report.* US: s.n.

DG Regio (2014). *Guide to cost-benefit analysis of investment projects—Economic appraisal tool for cohesion policy 2014–2020,* s.l.: European Union.

DG Regio (2021). *Economic appraisal Vademecum 2021–2027 General principles and sector applications,* European Union: European Union.

EEA (2021). *Municipal waste management in Western Balkan Countries—Country Profile Montenegro,* s.l.: European Environment Agency–European Topic Centre on Waste and Materials in a Green Economy.

EEA (2022). *Briefing no. 06/2022.* s.l.:s.n.

European Commission. (2015). *Guide to cost-benefit analysis of investment projects: Economic appraisal tool for Cohesion Policy 2014–2020.* Publications Office of the European Union. https://doi.org/10.2776/97516

Gjoka, R., & Toska, M. (2023). *Waste-to-energy: Unlocking the potential of waste as an alternative source of energy in the Western Balkans.* Tirana: GreenFORCE.

Gjoka, R. et al. (2023). *Waste to energy for Western Balkans cement industry–Feasibility study for the production of alternative fuel at Bushati Landfill Albania,* Tirana: Co-PLAN, Institute for Habitat Development.

Hajinezhad, A., Halimehjani, E. Z., & Tahani, M. (2016). Utilisation of refuse-derived fuel (RDF) from urban waste as an alternative fuel for cement factory: a case study. *International Journal of Renewable Energy Research, 6*(2).

Kahawalage, A. C., Melaaen, M. C., & Tokheim, L.-A. (2023). Opportunities and challenges of using SRF as an alternative fuel in the cement industry. *Cleaner Waste Systems,* p. 1000072.

Kaza, S., Yao, L., Tata, P. B., & Woerden, F. V. (2018). *What a waste 2.0—A global snapshot of solid waste management to 2050.* Washington: International Bank for Reconstruction and Development / The World Bank.

NEPA (2021). *Questionnaire to Western Balkan countries for providing information regarding municipal waste management and waste prevention—Montenegro—and comments provided during the review of the draft country profile,* s.l.: Nature and Environment Protection Agency of Montenegro.

Oyedotun, T. D. T., et al. (2021). Evaluation of waste dynamics at the local level: The search for a new paradigm in national waste management. *Environmental Challenges, 4,* Article 100130.

Rezaei, H., Panah, F. Y., Lim, C. J., & Sokhansanj, S. (2020). Pelletisation of refuse-derived fuel with varying compositions of plastic, paper, organic and wood. *Sustainability, 12,* 4645.

Santos, S. M., Nobre, K., Brito, P., & Gonçalves, M. (2023). Brief overview of refuse-derived fuel production and energetic valorisation: Applied technology and main challenges. *Sustainability, 15*(13), 10342.

Vukovic, A., Gjoka, R., & Toska, M. (2024). *Waste management sector transformation and its impact towards JGT in Montenegro,* s.l.: s.n.

CHAPTER 18

Advancing Energy Efficiency and Waste Management In Kosovo: Policy Implementation, Challenges, and Pathways to Sustainability

Elvis Ndreka and Fiona Imami

18.1 Introduction

Across Europe, decarbonization has become a central policy objective, driven by the *EU Green Deal*,[1] the *Fit for 55 packages*,[2] and commitments to achieving climate neutrality by 2050 (Shaker & Berisha, 2025). The energy sector, responsible for the largest share of greenhouse gas emissions, is undergoing rapid transformation through coal phase-outs, renewable energy expansion, and investments in energy efficiency. While Western European

[1] *The European Green Deal* is the European Union's strategic plan to achieve climate neutrality by 2050. It encompasses a broad range of policies aimed at reducing greenhouse gas emissions, promoting energy efficiency, fostering clean energy investments, and ensuring a just transition for all regions and sectors. The initiative aligns with the EU's commitments under the Paris Agreement and includes legislative measures to accelerate the transition to a sustainable economy.

[2] *The Fit for 55* package is a set of legislative proposals introduced by the European Commission in 2021 to align EU policies with the goal of reducing greenhouse gas emissions by at least 55% by 2030 (compared to 1990 levels). It includes revisions to key policies such as the EU Emissions Trading System (ETS), the Renewable Energy Directive, and the Energy Efficiency Directive, aiming to accelerate the decarbonization of various sectors, including energy, industry, transport, and buildings.

E. Ndreka (✉) · F. Imami
Co-PLAN Institute of Habitat Development, Polis University, Tirana, Albania
e-mail: elvis_ndreka@co-plan.org

F. Imami
e-mail: fiona_imami@co-plan.org

E. Berisha et al. (eds.), *The Palgrave Handbook of Just Green Transitions in the Western Balkans and Beyond*,
https://doi.org/10.1007/978-3-031-95075-9_18

countries have advanced in transitioning to wind, solar, and hydrogen-based systems, Southeastern Europe and the Western Balkans face structural challenges in reducing carbon dependency (Berisha, 2025). Many countries in the region, including Kosovo, Serbia, Bosnia and Herzegovina, and North Macedonia, remain heavily reliant on lignite, the most carbon-intensive fossil fuel, making decarbonization both an economic and political challenge. Despite commitments under the Energy Community Treaty to align with EU climate policies, progress is hindered by outdated infrastructure, financial constraints, and socio-economic dependence on coal-related industries. Kosovo's energy sector is at a critical juncture, shaped by a historical reliance on lignite, outdated infrastructure, and growing demands for sustainable development. Approximately 95% of Kosovo's electricity is generated from lignite-fired power plants, making it one of Europe's most carbon-intensive economies (Couture, 2021). This reliance on lignite contributes to significant greenhouse gas emissions and environmental degradation, while on the other hand, energy infrastructure, much of which dates back decades, is inefficient and unable to meet the demands of a growing economy, leading to substantial energy losses and elevated emissions. The country's strategic focus, as outlined in its National Energy and Climate Plan (NECP) and Energy Strategy (2022–2031), provides a roadmap to address these challenges through several targeted measures, including (i) Phasing Out Lignite Dependency; (ii) Enhancing Energy Efficiency; (iii) Leveraging Renewable Energy; (iv) Modernizing Infrastructure; and more importantly (v) Undertaking Policy and Market Reforms. Building on this roadmap, Kosovo aims to achieve a 32% reduction in primary energy consumption by 2028, focusing on public buildings, industrial sectors, and energy-intensive industries, aligning with broader EU directives and demonstrating Kosovo's commitment to the Just Green Transition (Kosovo Ministry of Economic Development, 2019). A crucial pillar of Kosovo's energy transition is improving energy efficiency, particularly in public buildings, some of the country's largest energy consumers. Schools, hospitals, and government offices, often housed in outdated and inefficient structures, are being retrofitted with high-efficiency insulation, LED lighting, and advanced HVAC systems. These upgrades have already shown tangible results, with some facilities achieving up to 30% energy savings, reducing operational costs and improving comfort levels for users. Beyond public buildings, Kosovo's industrial sector, including energy-intensive industries such as steel and cement production, is also undergoing modernization. The adoption of energy management systems and process optimization has led to energy savings of 15–20%, proving that efficiency measures are not just environmentally beneficial but also economically viable. On the next frontier, while energy efficiency measures help reduce demand, the long-term goal remains diversifying Kosovo's energy mix and cutting its dependence on lignite. To this end, renewable energy integration has emerged as a key priority. Pilot programs exploring solar and wind energy potential are being discussed, testing the feasibility of large-scale deployment, with the aim of marking the beginning of a long-awaited shift towards a more sustainable and decentralized energy system. However, despite the progress, Kosovo faces persistent challenges—aging infrastructure, regulatory gaps, and the need for stronger market incentives continue to slow down the transition. To this end, this chapter critically assesses the implementation and impact of Kosovo's NEEP, examining how its initiatives contribute to energy efficiency, renewable energy integration, and waste management practices as an integrated approach to addressing the decarbonization commitments of the country.

18.2 Energy Efficiency in Kosovo—From Policy to Monitoring

The NEEP is designed in alignment with the EU Energy Efficiency Directive (2012/27/EU), which emphasizes the need for clear and achievable targets to enhance energy use across sectors, reduce energy wastage, and improve energy performance. It specifically sets ambitious targets for Kosovo (showcasing at the same time serious commitment), to reduce primary energy consumption by 168 GWh per year by 2028, supporting the broader EU objective to reduce energy consumption by 32% in 2030. At the regulatory level, the legislative foundation for implementing energy efficiency measures in Kosovo is the *Law on Energy Efficiency (06/L-079)*, enforcing several critical policies, such as mandatory energy audits for large enterprises, the introduction of building energy performance certificates, and energy-saving obligations for utilities. These regulations are designed to create an enabling environment for energy efficiency improvements across various sectors, including industry, transport, and, particularly, the building sector. However, despite their potential benefits, this framework remains largely under-implemented, hindered by institutional inertia, enforcement challenges, and limited financial support, slowing down the transition towards more efficient energy use. Accounting for over 40% of Kosovo's total energy consumption (Kosovo Ministry of Economy), the building sector is the country's largest energy consumer, making it a key target for energy-saving measures, including retrofitting public and private buildings, improving insulation, upgrading heating systems, and promoting energy-efficient technologies such as advanced lighting, heating, and cooling systems, as emphasized in the NEEP. However, while improvements in energy efficiency within the building sector play a crucial role in reducing demand, they alone cannot offset Kosovo's overwhelming dependence on lignite, making a broader transition towards cleaner energy alternatives an urgent priority. As Kosovo advances its energy transition, it faces the critical challenge of reducing its heavy dependence on lignite, which currently generates over 90% of the country's electricity (Shllaku, 2023). This reliance on one of the most carbon-intensive fossil fuels significantly contributes to Kosovo's greenhouse gas emissions, accounting for 86% of total emissions—approximately 8.7 million tons of CO_2 annually. The environmental and economic urgency of phasing out lignite underscores the need to accelerate energy efficiency efforts and integrate renewable energy sources, ensuring a more sustainable and diversified energy mix that aligns with EU climate targets. Beyond the legislative framework, the progress of Kosovo's energy efficiency reforms (tackling one and foremost the abovementioned concerns), heavily relies on the institutional oversight and regulatory enforcement, primarily led by two key entities: the Kosovo Energy Efficiency Agency (KEEA) and the Energy Regulatory Office (ERO). These institutions play a pivotal role in translating national policies into tangible actions that drive progress toward Kosovo's decarbonization and energy transition goals. The KEEA is the main institution responsible for implementing the NEEP and overseeing energy efficiency projects across different sectors. Its primary focus is on retrofitting public buildings, modernizing heating and cooling systems, and integrating renewable energy technologies. A significant aspect of KEEA's work is its collaboration with municipalities to develop and implement local energy efficiency plans, particularly targeting public infrastructure such as schools and administrative buildings. Complementing these efforts, the Energy Regulatory Office (ERO) plays a critical regulatory and enforcement role in Kosovo's energy efficiency landscape. Tasked with ensuring compliance with energy efficiency regulations, ERO also designs energy tariffs that incentivize lower energy consumption, encouraging a more efficient and responsible use of energy resources. Moreover, ERO enforces energy-saving obligations for utilities, ensuring that energy suppliers meet mandatory efficiency targets, which in turn fosters a market dynamic that promotes both supply-side and demand-side efficiency improvements (European Energy Community, 2021). Underscoring the need for stronger institutional coordination, increased investment, and more effective policy implementation to fully unlock the potential of Kosovo's energy efficiency strategy, the following chapter sets the stage for a deeper examination of the practical implementation measures, financing mechanisms, monitoring frameworks, and stakeholder engagement strategies that shape the country's transition. As the discussion unfolds, the focus will shift to how these elements interact, the challenges faced in execution, and the opportunities for enhancing Kosovo's alignment with EU climate goals and broader sustainability objectives.

18.2.1 Implementation of Energy Measures

Kosovo's energy efficiency agenda has gained significant momentum in recent years, supported by international financing, national policies, and growing private sector engagement. While legislative frameworks and institutional oversight have played a crucial role in shaping reforms, the implementation of concrete projects—ranging from public building retrofits to industrial efficiency improvements—demonstrates the tangible progress made towards reducing energy demand and enhancing sustainability.

The sector has undergone notable advancements, particularly in the public sector, where government-led initiatives have driven improvements in building performance and energy use optimization. A major focus has been on the retrofitting of public buildings, which account for a significant share of the country's overall energy consumption. As of 2024, over 50 public buildings in Pristina and other major cities have been upgraded with improved insulation, LED lighting, and advanced heating, ventilation, and air conditioning (HVAC) systems, leading to measurable reductions in energy demand and operating costs.

In parallel, the residential sector has seen a steady rise in energy efficiency initiatives, spearheaded by the Government of Kosovo in collaboration with international donors. The "Home Energy Efficiency Program", a government-backed initiative, provides subsidies for homeowners to invest in solar panels, energy-efficient heating systems, and insulation improvements. To date, approximately 10,000 families have benefited from this program, with participating households reporting an average 20% reduction in energy consumption (Ministry of Economy of Kosovo, 2024). This energy transformation, has been largely enabled by international funding and financial mechanisms that support both public and private sector initiatives. The European Union, through its Instrument for Pre-Accession Assistance (IPA), has allocated over €25 million since 2019 to finance energy efficiency projects. Additionally, the World Bank has played a key role in providing financial resources for Kosovo's energy efficiency sector, offering low-interest loans and direct grants to support public and private sector investments in energy-saving technologies. A recent World Bank-funded project involved the installation of energy-efficient lighting and heating systems in 15 municipal buildings, leading to estimated annual energy savings of 5 GWh. On the other hand, while international and public sector financing has driven much of Kosovo's energy efficiency progress, private sector involvement is also expanding. Energy Service Companies (ESCOs) have gained traction, offering performance-based contracts that allow businesses and public institutions to invest in energy efficiency upgrades without upfront capital costs. A notable ESCO-financed project in Pristina included the energy-efficient retrofit of a large commercial complex, with HVAC system upgrades and advanced lighting installations, expected to cut energy consumption by 30% and generate annual savings of €150,000.

18.2.2 Key Stakeholders in Energy Efficiency Implementation

The successful implementation of energy efficiency measures in Kosovo hinges on the involvement of a diverse network of stakeholders, ranging from government institutions and regulatory bodies to private sector actors, financial institutions, and end-users.

First to support this momentum, the Kosovo Fund for Energy Efficiency (KFEE), a central financial institution supporting energy efficiency initiatives in public buildings, industrial facilities, and residential properties, established in 2015, has enabled a wide range of projects through its financing mechanisms. This includes:

- *Public building retrofitting*, with over 40 schools, health centers, and municipal offices have undergone energy-efficient renovations, resulting in energy savings of up to 35% in some cases
- *Industrial Efficiency Improvements* in which KFEE has funded efficiency upgrades in key industries, including the cement and steel sectors, where process optimizations and machinery upgrades have led to a 12% reduction in energy intensity.
- *The Residential Energy Efficiency Programs* for which grants and subsidies provided by KFEE have helped households adopt energy-efficient windows, insulation, and solar water heaters, contributing to an average 15% reduction in household energy use.

Beyond direct project financing, the KFEE plays a crucial role in technical assistance and capacity-building, equipping local stakeholders with the knowledge and tools to maximize energy savings and improve efficiency standards. Through energy audits, specialized training for municipal energy managers, and workshops on best practices, KFEE fosters a more skilled and informed approach to energy efficiency implementation. The impact of these efforts is already evident, with KFEE-supported initiatives leading to an annual reduction of approximately 70,000 tons of CO_2 emissions, reinforcing Kosovo's commitment to decarbonization and sustainability. Looking ahead, the Fund aims to broaden its impact by integrating advanced energy storage systems and expanding efficiency measures into the transportation sector. At the government level, the Ministry of Economy provides strategic oversight and is responsible for developing long-term energy efficiency policies, while the Ministry of Environment and Spatial Planning ensures that these policies are aligned with local regulatory frameworks and environmental sustainability goals. Municipal governments, in turn, play a pivotal role in the implementation process, overseeing local energy efficiency projects and ensuring compliance with building codes and energy performance standards. The Energy Regulatory Office (ERO), is another central institution acting as the key enforcement body, monitoring energy efficiency regulations, overseeing compliance, and setting tariffs that promote energy savings. The Kosovo Energy Corporation (KEK) and Distribution System Operators (KEDS/KESCO) play a direct role in the country's energy efficiency landscape by aligning energy production and distribution with efficiency targets.

While the abovementioned entities have demonstrated commitment to improving energy distribution practices, challenges remain in reducing transmission losses and ensuring that efficiency gains translate into lower consumption at the household and industrial levels. In particular, grid modernization efforts have been slow, and the integration of renewable energy into the distribution network remains limited, highlighting the need for stronger financial and technical investments in energy system upgrades. On the other hand, despite the technical, financing or even advisory support, challenges in accessing funding persist, particularly for local governments and small businesses. Many of these stakeholders lack the technical capacity to prepare funding proposals, navigate complex application procedures, or meet the eligibility criteria for international financing programs. This disparity in access often results in larger, well-resourced municipalities and companies benefiting disproportionately, while smaller, resource-limited stakeholders struggle to implement energy efficiency measures (Purnell, 2017).

The private sector, mentioned above as well, has become an increasingly important player in energy efficiency investments, particularly through Energy Service Companies (ESCOs), construction and engineering firms, and suppliers of energy-efficient technologies. ESCOs, in particular, have introduced performance-based contracts, allowing businesses and institutions to implement energy efficiency upgrades without upfront costs, with repayment structured around energy savings achieved over time. However, the ESCO market in Kosovo remains underdeveloped, primarily due to regulatory uncertainty, lack of standard contracts, and limited awareness among businesses and municipalities. In many cases, public institutions hesitate to engage in performance-based contracting, citing uncertainties about long-term cost savings and financial risks (Greenstone, 2012). This institutional hesitancy, coupled with a lack of incentives for private sector involvement, hampers the scalability of energy efficiency investments in Kosovo. Lastly, while policy frameworks, financial incentives, and private sector investments are essential for energy efficiency progress, public engagement remains one of the most critical yet overlooked elements of Kosovo's energy transition. Environmental NGOs, consumer advocacy groups, and local civil society organizations have actively worked to raise awareness about energy efficiency, encouraging adoption of energy-saving technologies and behavioral changes among consumers. However, public understanding of energy efficiency benefits remains limited, with studies highlighting low participation rates in energy-saving programs due to lack of awareness and misinformation (Jensen, 2016). A major barrier in Kosovo's stakeholder engagement process is the lack of systemic information-sharing and co-designing practices in energy efficiency planning. In many cases, energy efficiency projects are designed and implemented without direct input from local communities, businesses, or end-users, leading to low uptake rates and public resistance to efficiency measures. Research suggests that public participation in energy policy design enhances acceptance, accelerates adoption, and improves the effectiveness of efficiency programs (Stern, 2020). However, Kosovo's policy environment still lacks structured mechanisms for inclusive dialogue, participatory decision-making, and ongoing consumer education on energy efficiency benefits.

18.2.3 Monitoring and Evaluating Progress

Kosovo's approach to EE monitoring is largely guided by its Kosovo Energy Strategy Implementation Program (KESIP) (2022–2025), which lays out five key strategic objectives: improving system resilience, decarbonization, energy efficiency, regional cooperation, and consumer protection. The government reports indicate that EE measures have been successfully implemented in public and residential buildings, with projected energy savings of up to 1.67 ktoe/year in 2023, an increase from 0.78 ktoe/year in 2022 (mainly linked to the subsidies aimed at vulnerable consumers and financial incentives to encourage energy efficiency investments).

However, a closer look at these reports reveals a fragmented monitoring process. While the Ministry of Economy (ME) is responsible for overseeing the implementation, much of the reported progress relies on self-reported data from municipalities and implementing agencies, with a significant gap in data and without a rigorous independent verification mechanism.

While documents report savings figures in ktoe, there is limited transparency on how these figures are calculated, whether independent audits have been conducted, and the extent to which local beneficiaries actually experience reductions in energy costs. Furthermore, progress rates are declining, with 2023 seeing only a 51.8% achievement of its EE objectives, down from 73.7% in 2022. The reasons cited for this decline in these reports include: (i) Delays in project financing and ratification of agreements by the Kosovo Assembly; (ii) Bureaucratic bottlenecks in permitting processes, particularly for energy efficiency renovations in buildings and (iii) Market fluctuations in construction costs, making initial cost estimates unreliable.

In this context, a fundamental question must be raised: Is Kosovo's EE progress merely a policy on paper, or is it translating into real impact?

And we argue that it is the disconnection between Policy and local realities, that should translate the real question with regards to implementation of EE measures and furthering green transition efforts for the country. In this regard two distinct layers should be observed:

- *Public perception and stakeholder engagement in monitoring*—While government reports present EE progress as a technical and financial matter, the human dimension of energy efficiency is often overlooked. Interviews and reports from local municipalities indicate that many residents remain unaware of available EE programs or face significant barriers in accessing them. For instance, while subsidies for vulnerable consumers are highlighted in official documents, how many of these consumers actually benefit from lower energy costs in winter months? Are these measures truly lifting households out of energy poverty, or are they merely symbolic gestures? A critical analysis of Kosovo's monitoring systems suggests that there is a lack of direct engagement with affected communities, making it difficult to assess the real socio-economic impact of EE measures. Moreover, local businesses that could potentially benefit from EE programs often find themselves excluded from financing schemes due to complex administrative procedures or a lack of clarity on eligibility criteria.

- *Political and Institutional Barriers to effective Monitoring*—Energy efficiency in Kosovo is not just a technical challenge—it is also a political and governance issue. The EU Commission's Kosovo 2024 Report points to persistent institutional weaknesses, including limited government capacity for enforcement and monitoring. Additionally, Kosovo's legal and regulatory framework for EE, while aligned with EU Directives, remains weakly implemented and enforced. A deeper look into EE project implementation reveals significant reliance on donor-funded initiatives. This raises concerns about sustainability—will Kosovo's government be able to maintain and monitor EE programs once external funding decreases? Or is the current approach merely dependent on short-term donor-driven projects?

On a first reflection, a shift to outcome-based monitoring is needed, with independent audits, consumer feedback, and longitudinal assessments to ensure real energy savings and socio-economic benefits. Strengthening local implementation capacities, including municipal EE monitoring units and public–private partnerships, would improve accountability. Additionally, EE policies must be better integrated with social justice goals, ensuring that vulnerable populations truly benefit from subsidies and winter heating support. Without stronger verification mechanisms and a human-centered approach, Kosovo's EE progress risks remaining a symbolic rather than transformative effort.

18.3 Waste Management vis a vie EE in Kosovo: Overview and Challenges

Waste management and energy efficiency (EE) are fundamental components of Kosovo's transition towards a sustainable economy. However, these two sectors are often treated in isolation, despite their deep interconnections. As Kosovo aligns itself with the Just Green Transition (JGT), ensuring that social equity, economic stability, and environmental sustainability progress hand-in-hand, integrating waste management and energy efficiency is critical. Before analyzing their interconnectedness, lets review shortly the situation with regards to waste management sector in Kosovo.

18.3.1 Current State of WM in Kosovo, a Crisis of Landfills and Fragmented Policies

Kosovo's waste management is governed by the Law on Waste and the National Waste Management Strategy (NWMS) 2021–2030. These documents outline an ambitious framework aimed at reducing landfill dependence, improving recycling rates, and enhancing hazardous waste management. The strategy sets a target of 50% recycling for municipal solid waste by 2022, yet current recycling rates stand at merely 5–7%, far below the EU's 55% goal for 2025 (EEA, 2021). Institutional responsibility for waste management is divided between the Ministry of Environment and Spatial Planning (MESP) and municipal-level authorities, but weak enforcement, lack of funding, and inefficient coordination have hampered progress, mainly because of the disconnect between policy ambition and actual implementation, underscoring the broader governance challenges Kosovo faces in environmental management, as it was the case with the EE.

Kosovo generates approximately 450,000 tons of waste annually, with 80–90% ending up in landfills (Country Factsheet, EEA, 2021). However, landfill management is far from EU standards, with most sites lacking proper waste separation, pollution controls, and monitoring mechanisms, making it the default yet the most unsustainable option. In this regard, Kosovo's seven major landfill sites, of which only Mirash landfill operates under relatively acceptable conditions, are reaching capacity. Moreover, illegal dumping in rural areas and riverbanks poses severe environmental and health risks.

On the other hand, recycling (a nascent sector yet with structural barrier), remains largely informal, with waste pickers playing a crucial yet unrecognized role in recovering valuable materials. However, the lack of formalized waste-sorting infrastructure and incentives for recycling businesses prevents the sector from scaling up.

18.3.2 Intersections Between WM and EE, a Missed Opportunity?

One of the most underexplored intersections between waste management and energy efficiency in Kosovo is Waste-to-Energy (WtE) solutions. With landfills overburdened and 95% of Kosovo's electricity generated from coal, WtE can simultaneously reduce landfill dependency and enhance energy security.

Waste-to-Energy (WtE) is a practice that has evolved as part of broader efforts to manage waste sustainably while simultaneously addressing energy needs. The concept of converting waste into energy dates back to the late nineteenth century, when the first incineration plant was built in Nottingham, England, in 1874. The modern development of WtE systems gained momentum in the mid-twentieth century, as industrialized nations faced increasing waste management challenges due to rapid urbanization and economic growth. Technological advancements in combustion, gasification, and anaerobic digestion have since enabled WtE to become a viable alternative to traditional waste disposal methods, such as landfilling, while also contributing to energy security and reducing greenhouse gas emissions (Arena, 2012).

Today, WtE is widely implemented in several countries, particularly in Europe and Asia, where land availability for waste disposal is limited, and environmental regulations are stringent. Denmark, Sweden, and the Netherlands have led the way in WtE adoption, integrating it into their broader waste management and energy policies. Sweden, for example, has developed an extensive WtE network, where nearly 50% of its municipal solid waste is incinerated to generate electricity and district heating (Avfall Sverige, 2021). Similarly, Japan and South Korea have embraced WtE as part of their national strategies to address urban waste management, with Japan operating over 1,000 WtE facilities as of 2020 (Themelis, 2003).

The benefits of WtE are multi-faceted, making it an attractive solution for cities struggling with high waste generation and energy deficits. First, it reduces landfill dependency by diverting significant portions of municipal solid waste (MSW) into energy recovery processes. This minimizes land use, mitigates methane emissions from decomposing waste, and extends the lifespan of existing landfill sites. Second, WtE contributes to energy security by converting non-recyclable waste into electricity or heat, reducing reliance on fossil fuels and enhancing the stability of national energy grids (Panepinto et al. n.d.). Additionally, WtE can promote circular economy principles by integrating with material recovery systems, ensuring that valuable resources are extracted before incineration or thermal treatment (European Environment Agency, 2019). Neverthless, despite its advantages, WtE implementation faces several challenges that must be addressed for effective deployment. One of the main concerns is the environmental impact of emissions from incineration processes, including carbon dioxide (CO_2), dioxins, and particulate matter. Advanced emission control technologies, such as flue gas treatment and carbon capture systems, have mitigated these risks in countries with stringent environmental regulations (Astrup et al., 2015). Another challenge is the high capital investment required to establish WtE facilities, with costs often exceeding those of traditional landfills (Scarlat et al., 2015).

Moreover, public opposition to incineration projects due to perceived health risks and environmental concerns has led to delays and project cancellations in some regions, highlighting the need for transparent communication and community engagement in WtE planning (Zaman & Lehmann, 2013).

Neverthless, for economies like Kosovo, dependent on lignite, exploring these opportunities, could offer a different trajectory to enhancing energy efficiency, promoting circular economy, improving waste management service and eventually contributing to the efforts towards JGT.

With approximately 441,000 tons of waste generated annually, and 80% of this waste being landfilled, WtE technologies present an opportunity to both reduce landfill dependency and recover valuable energy. Urban areas like Pristina experience the highest waste generation, making WtE a particularly viable option for reducing the waste burden while contributing to the energy supply, where waste incineration has the potential to notedly contribute to Pristina's district heating needs. This would reduce reliance on coal and diesel, both of which dominate Kosovo's energy mix, with 95% of electricity currently generated from coal.

Although promising, several challenges must be addressed for successful WtE implementation. Kosovo's existing infrastructure is insufficient, and significant investments are needed to build modern WtE facilities, with costs estimated between €60–€100 million depending on the scale and technology. Additionally, effective waste separation is crucial to prevent recyclable materials from being incinerated, but Kosovo's recycling rate is currently less than 5%. Scaling up WtE solutions requires improvements in recycling, waste collection, and separation practices.

Beyond WtE, adopting circular economy principles can further enhance the linkage between energy efficiency and waste management. Circular economy models prioritize reducing waste, reusing materials, and recovering energy, creating a closed-loop system that minimizes environmental impact and conserves resources. For instance, recycling 1 ton of aluminum saves up to 95% of the energy needed to produce new aluminum (The Aluminium Association, 2023), and recycling plastics can cut energy use by 88%.

Another area of potential is improving the energy efficiency of waste management facilities. Current landfilling and waste treatment processes in Kosovo are energy-intensive and inefficient. By upgrading facilities with modern, energy-efficient technologies, such as solar-powered treatment plants, Kosovo can lower operational costs and reduce carbon emissions. Optimized transport logistics and energy-efficient sorting systems could further enhance the efficiency of waste management operations.

To this end, the combination of WtE technologies and circular economy practices has the potential to yield substantial economic and environmental benefits for Kosovo. Reducing landfill waste through energy recovery would not only mitigate greenhouse gas emissions like methane, which is 25 times more potent than CO_2, but also contribute to energy savings. Achieving EU-level recycling rates could result in up to 2,000 GWh of energy savings annually, while also enhancing public health and reducing environmental harm.

18.4 Conclusions

The examination of Kosovo's energy efficiency (EE) and waste management (WM) systems within the broader context of decarbonization and the Just Green Transition (JGT) reveals a stark contrast between policy ambition and practical implementation. As outlined in the abstract, this chapter sought to critically assess the state of EE and WM, explore their intersections, and examine the institutional, financial, and governance constraints that hinder progress. While substantial commitments have been made through national strategies such as the NEEP (2022) and NWMS (2021–2030), the structural and operational shortcomings identified throughout this discussion raise fundamental concerns about the effectiveness and sustainability of Kosovo's transition efforts. Energy efficiency remains a cornerstone of Kosovo's climate and energy policies, yet progress has been fragmented and uneven. While official reports indicate some level of achievement—such as the retrofitting of public buildings and the implementation of demand-side management

(DSM) strategies—this progress has not translated into systemic improvements. The persistent disconnect between output-oriented monitoring and outcome-based impact assessments prevents a clear understanding of how these measures contribute to long-term energy savings and social equity. The government's reliance on donor-driven programs further raises concerns about sustainability, as financial and technical support is external rather than locally embedded.

Moreover, the absence of independent verification mechanisms undermines the credibility of reported gains. Many EE initiatives rely on self-reported municipal data, which lacks rigorous oversight. This issue, coupled with bureaucratic delays in project implementation and the underfunding of regulatory bodies, suggests that Kosovo's approach to EE remains largely reactive rather than proactive. To ensure meaningful impact, the following recommendations emerge:

- *Shift towards outcome-based monitoring*: Energy efficiency programs should incorporate third-party audits and longitudinal assessments to measure actual energy savings and socio-economic benefits.
- *Strengthen local institutional capacities*: The role of municipalities in implementing EE measures must be reinforced through capacity-building initiatives and dedicated EE monitoring units.
- *Integrate EE with social justice policies*: Targeted subsidies and incentives should prioritize vulnerable populations, ensuring that energy savings contribute to reducing energy poverty.

Kosovo's waste management system is emblematic of the broader governance and infrastructure deficits that hinder environmental progress. The overwhelming reliance on landfilling, coupled with insufficient recycling infrastructure and poor regulatory enforcement, has created a waste crisis that is both environmental and social. Despite the adoption of the NWMS, which aims to reduce landfill dependency and improve waste separation, practical implementation has been slow. The contradiction in reported landfill rates (80% vs. 90%) further highlights the inconsistencies in data collection and transparency.

The informal sector plays a crucial role in waste recovery, yet its contributions remain unrecognized and unsupported. The exclusion of informal waste pickers from policy frameworks represents a missed opportunity to integrate circular economy principles into waste governance. Moreover, hazardous waste remains largely unregulated, posing long-term risks to environmental and public health. Addressing these challenges requires:

- *Formalizing waste separation and recycling systems* where investments in modern waste-sorting infrastructure and the adoption of Extended Producer Responsibility (EPR) schemes can improve recycling rates.

- *Strengthening enforcement and compliance with r*egulatory bodies empowered with better monitoring tools and financial resources to ensure adherence to waste management laws.
- *Recognizing the informal sector* and waste pickers as integral to the formal waste management frameworks, receiving support through training programs and financial incentives.

Lastly, we argue that the interconnection between energy efficiency and waste management remains an underexplored yet crucial avenue for Kosovo's sustainability transition. However, scaling up WtE solutions requires overcoming significant financial and operational barriers. The high capital costs (estimated between €60–€100 million) and Kosovo's limited waste separation practices present major constraints. Additionally, public resistance to incineration-based WtE solutions due to environmental concerns must be addressed through transparent stakeholder engagement and stringent emissions controls. Ultimately, achieving a sustainable and just transition requires a governance model that moves beyond short-term fixes towards structural transformations that address the root causes of inefficiency and waste proliferation. Kosovo has the potential to redefine its energy and waste policies into an integrated, forward-looking framework, but doing so will require a fundamental rethink of how policies are designed, financed, and monitored. The time for incremental changes is over—Kosovo's path forward must be one of decisive, coordinated, and accountable action.

References

Arena, U. (2012). Process and technological aspects of municipal solid waste gasification. A review. *Waste Management, 32*(4), 625–639.

Allcott, H., & Greenstone, M. (2012). *Is there an energy efficiency gap?* Vol. 26. American Economic Association. https://doi.org/10.1257/jep.26.1.3

Astrup, T., Moller, J., & Fruergaard, T. (2015). Incineration and co-combustion of waste: Accounting of greenhouse gases and global warming contributions. *Waste Management & Research, 33*(8), 705–715.

Avfall Sverige. (2021). *Swedish waste management 2021.* Swedish Waste Management Association Report, Malmo.

Berisha, E. (2025). *Moving towards an inclusive green agenda in the Western Balkans.* Foundation for European Progressive Studies (FEPS). https://doi.org/10.13140/RG.2.2.12092.76162

Couture, T. D. (2021). *Accelerating power sector transformation in the republic of Kosovo: de-risking analysis and policy options.* E3 Analytics. https://www.e3analytics.eu/project/accelerating-power-sector-transformation-in-kosovo/?utm_source=chatgpt.com.

David, Paul A. (1985). *Clio and the Economics of QWERTY.* Vol. 75. 2 vols. JSTOR. https://www.jstor.org/stable/1805621.

European Environment Agency (EEA). (2021). *Municipal waste management—Country fact sheet Kosovo*. European Environment Agency (EEA). https://www.eea.europa.eu/themes/waste/waste-management/municipal-waste-management-country/kosovo-municipal-waste-factsheet-2021

European Energy Community. (2021). *Energy efficiency policies and measures in the Western Balkans*. Vienna, Vienna: European Energy Community.

European Environment Agency. (2019). *The role of waste-to-energy in a circular economy*. EEA Report.

Foundation Ellen MacArthur. (2013). *Towards the circular economy Vol. 1: An economic and business rationale for an accelerated transition*. Ellen MacArthur Foundation. https://www.ellenmacarthurfoundation.org/towards-the-circular-economy-vol-1-an-economic-and-business-rationale-for-an

Geels, F. W., Sovacool, B. K., Schwanen, T., & Sorrell, S. (2017). *Sociotechnical transitions for deep decarbonization*. Vol. 357. Science. https://doi.org/10.1126/science.aao376

Gellings, C. W. (2005). The concept of demand-side management for electric utilities. *Institute of Electrical and Electronics Engineers*. https://doi.org/10.1109/PROC.1985.13318

Jensen, M. B., Møller, B., & Madsen, H. (2016). *Local strategies to promote energy retrofitting of single-family houses: A case study from Denmark*. Vol. 9. Springer. https://doi.org/10.1007/s12053-018-9653-5

Ministry of Economy of Kosovo. (2024). *Home energy efficiency program*.

Panepinto, D., Tedesco, V., Brizio, E., & Genon, G. (n.d.). Environmental performances and energy efficiency for MSW gasification treatment. *Waste and Biomass Valorization, 6*, 409–422. https://doi.org/10.1007/s12649-014-9322-7

Scarlat, N., Motola, V., Dallemand, J. F., Monforti-Ferrario, F., & Mofor, L. (2015). Evaluation of energy potential of municipal solid waste from African urban areas. *Renewable and Sustainable Energy Reviews, 50*(0), 1269–1286.

Shaker, Y., & Berisha, E. (2025). Just Green Transitions: Between terminological inexactitudes, conceptual fragmentation, and the exigency for a theoretical framework. *European Spatial Research and Policy, 32*(1), 121–138. https://doi.org/10.18778/1231-1952.32.1.05

Shllaku, L. (2023). *What's the right energy mix? How to plant the green agenda in the Kingdom of coal?* RIIVEST.

Stern, P. C. (2020). Public acceptance of energy efficiency policies: The role of consumer engagement. *Energy Research & Social Science, 60*, 101401. https://doi.org/10.1016/j.erss.2020.101506

The Aluminium Association. (2023). *Infinitely recyclable*. https://www.aluminum.org/Recycling?utm_source=chatgpt.com

Themelis, N. J. (2003). An overview of the global wwaste to energy industry. *Waste Management World 2003–2004 Review Issue* 40–47.

United Nations Framework Convention on Climate Change (UNFCCC). (2015). *Paris agreement*. Paris: UNFCCC Secretariat. https://unfccc.int/sites/default/files/resource/parisagreement_publication.pdf.

Unruh, G. C. (2000). *Understanding carbon lock-in*. ELSEVIER.

Velenturf, A. P., & Purnell, P. (2017). Resource recovery from waste: Restoring the balance between resource scarcity and waste overload. *Sustainability, 9*(9). Leeds: MDPI. https://doi.org/10.3390/su9091603

Zaman, A., & Lehmann, S. (2013). The zero waste index: A performance measurement tool for waste management systems in a 'zero waste city.' *Journal of Cleaner Production, 50*, 123–132.

CHAPTER 19

Climate Change Adaptation Green Technologies for Coastal Risk Prevention: Case Studies Summary from Coastal Countries

Irma Kraja Temali

19.1 Introduction

Anthropogenic factors have a rising impact on climate change, and according to scientists since the middle of the twentieth century, human activities have been the main reason for global warming (IPCC, 2022b). The sensitivity of the coastline to climate change has led to the development of new European directives and guidelines to help manage risk. But it is also evident that coastline cities are already under assault due to unplanned coastal development, lack of coordination between land use planning and development strategy, overpopulation, overfishing, pollution, habitat loss, and other damaging activities. A significant decline in biodiversity is predicted and is likely to involve local loss of pollinators, which, along with other threats, will jeopardize food production (IPCC, 2022b). Coastal communities are mostly affected by hazards like flooding and erosion. Ecosystem disruption, agriculture, and livestock loss, prompt governments to seek alternative ways to cope with these threats. By the 2080s, sea level rise could cause coastal flooding to increase tenfold, impacting over 100 million people per year (IPCC, 2022b). According to the United Nations more than 40%, of the world population lives by 100 km in of the coast, and often economic activities are located in these areas (United Nations, 2007). In the context of climate change, it can be argued that threats depend

Present Address:
I. K. Temali (✉)
Planning and Territorial Development Department, Municipality of Shkodra, Shkodër, Albania
e-mail: irmatem@gmail.com

E. Berisha et al. (eds.), *The Palgrave Handbook of Just Green Transitions in the Western Balkans and Beyond*,
https://doi.org/10.1007/978-3-031-95075-9_19

on how the environmental system is affected. From the data taken from the Intergovernmental Panel on Climate Change, if the global temperature rises between 1.5 °C and 2 °C in 2100 ecosystem will suffer from an incursion of seawater into groundwater related to the increase in sea levels (IPCC, 2024). Prolonged drought periods will affect the availability of drinking water, rising acidity and decline oxygen levels in marine ecosystem, with the projection of 70–90% loss of coral reefs (IPCC, 2022c). Studies point out that alterations in weather patterns influence the distribution and equilibrium of ecosystems. As a consequence of global warming, the alteration of geographical ranges by bacteria and viruses facilitates the dissemination of diseases (USAID, 2009). In light of this background, how has the implementation of these adaptive technologies advanced locally? It is feasible to identify and evaluate their effect in the affected area making them replicable models for other sites? Despite the sensitivity of these efforts, the peer-reviewed literature is lacking in critical evaluation of many climate change responses (Abbass et al., 2022).

In recent years there is a great need to identify new references to protect the territories affected by the risks from climate change. Referring to the report of the European Commission (2013), transition from traditional forms to green interventions is crucial for urban areas. Defined by the European Environment Agency, green infrastructure is characterized as organized network of natural and semi-natural areas with features that are planned and controlled to provide various ecosystem benefits (European Environment Agency, 2019). Requalification intervention on existing building stock action will trigger a path to improve urban welfare and social growth of communities. At the same time, the realization of ecological intervention will affect positively reducing land consumption, encouraging regeneration, soil renaturation, and water conservation (Geneletti, 2020). Green infrastructure, based on adaptation strategy, makes urban systems more resilient and less susceptible to damage, enhancing environmental quality while safeguarding biodiversity (Janiszek & Krzysztofik, 2023).

The concept of 'Just transition' is a new added topic to the ongoing process of climate change. In order to achieve climate neutrality, preserve biodiversity, and guarantee social justice, a more sustainable future is required, and this can only be achieved through a comprehensive socio-technical shift known as a 'just green transition' (Cedergren et al., 2022). An important step toward reducing environmental risks would be the adoption of ecological technology solutions to the complex problem of climate adaptation. The effects of climate change are unequally distributed and action to adapt to risk should be based in social equity. It is to be expected that the most vulnerable have less to benefit from adaptation action; for example, informal settlements destroyed by flooding or smallholder farmers by heavy rainfalls. Climate change intensifies pre-existing inequities, encompassing gender, income, age, and ethnicity (IPCC, 2022a). Nature-based solutions serve as an inspiration for European city policies on climate change. Furthermore, the possibility of finding effective measures that could serve as a template for ecological adaptation actions is the foundation of this study. Selected practices are comprehensively described, first referring to efficiency perspectives which take into account the urban environment within the ecological and infrastructure territorial systems, and then discussing the strategic adaptation alternatives of retreat, co-existence, and defence. In order to find a framework of climate-adaptive green measures, this study divides specific practices into individual strategies and actions that may be applied in comparable territorial conditions. There are limitations because these models are not fully exportable to other contexts without a detailed analysis of the regulatory planning instruments. This research focused on theoretical and methodological references that aim to classify some specific technologies and actions in a general context to benchmark for other sites. The purpose of this contribution is to highlight site-specific action to be analysed and discussed in the future for the benefit of the territorial systems and be integrated into policy strategy and urban planning.

19.2 Methods

Determining reference points for relevant green coastal protection technologies that could be effective and exportable is the main focus of this study. By raising awareness and increasing information for adaptation, it will help make adaptation solutions exportable, providing real opportunities to achieve transitions toward an ecological approach.

Green technology for coastal risk prevention was first identified through a series of selected studies utilizing web-based documents, with a focus on adaptation technology in coastal countries within their urban contexts. The study used three adaptation approaches, 'defence', 'co-exist', and retreat as a proxy for identifying and characterizing green adaptation technologies (Zhu et al., 2010). Adaptation technology was identified through the selection of relevant literature on climate change studies, primarily in English. A Google search was conducted using specific keywords, such as 'coastal area', flooding', 'sea level rise', and phrases related to green adaptation technologies, climate change adaptation, green infrastructure, adaptation and mitigation. The system involved in intervention refers to 'Environmental-Ecological' and 'Infrastructure-Service' contexts.

The case study selection process takes into consideration the following elements: the relevant phenomenology linked to the three adaptation approaches involving nature-based solutions, the territorial system, and a critical analysis of specific actions that provide replicable models. The data collection was carefully analysed based on reports from platforms such as Climate Adapt (2024a), Oppla (2024), Naturvation (2024) or Ecoshape (2024), materials published on institutional websites, scientific articles, policy documents, funding initiatives, and meetings with municipalities' staff involved in climate projects. The application of green technology options led to the creation of a matrix that connects the systems and strategies employed in the adaptation intervention, indicating possible courses of action. These documents offer valuable insights into various adaptation approaches that facilitate the Just Green Transition process. The prescribed adaptation options are categorized based on their focus and response to address the challenge of disaster risk reduction in coastal areas. The limitations in assessing long-term disaster risk reduction, including its impact on social equity and the insufficient evaluation of costs associated with green technologies, pose a significant constraint in this study. Despite these limitations, these experiences can serve as a potential theoretical and practical reference for innovative planning in different territorial contexts.

19.3 Identify Coastal Protecting Schemes

Identifying the technologies most appropriate for a specific country's climate change circumstances is the foundation for effective climate action (Sara, 2022). This paper focuses on coastal climate change mitigation in relation to flood risks. This study has resulted in the formulation of two concurrent strategies. The conceptualisation of the adaptation schemes in coastal areas can be classified by three approaches: 1. Defend; 2. Co-exist/accommodate; and 3. Retreat/relocation, or as a. Hardware, which means 'hard technologies', b. Software-'soft technologies', c. Orgware- organisational technologies (Christiansen et al., 2011).

Climate Technology Centre and Network [CTCN] introduces 102 water technologies by first identifying the water adaptation challenge, second exploring adaptation options and third learning applicable adaptation technologies (Bertule et al., 2018).

The process of decision makers selecting one or a set of technologies to address water challenges requires a strong ability to identify criteria for assessment and performance of these technologies.

- Defend is seen as the most common response that involve measures including hard and soft technologies. Beach nourishment, salt marshes, dune restoration and creation, saltwater intrusion barriers, floating agricultural systems, seagrasses and mangrove are employed as green technologies in coastal defence. Sea walls, barriers and gates against storm and flood, land claims, sea walls, dikes, are predominant response used as hard structural option (ibid). They can be movable or fixed used in the developed world like Maeslantkering Barrier Rotterdam. The disadvantages of movable barriers are high capital and maintenance cost, while closure dams and fixed barriers obstruct an estuary or river. In both instances, those technologies serve as physical barriers that inhibit storm surge events, which can significantly alter the chemical, physical, and biological characteristics of the estuarine system by modifying the inflow and outflow of water (Orton et al., 2023). Modification to water salinity, temperature, suspended nutrients and particulates could impact local organism communities (ibid.). Maintenance costs for certain types of hard infrastructure defence are poorly detailed resulting in greater uncertainty (Harvey & Nicholls, 2008).

- Accommodate or co-exist option requires communities to accept climate risk and periodically react to hazards. This implies expensive cost and continuously exposure to stress and crisis event that may include loss of life. Various technology are applied like aquaculture, saline resistant crops, floating agriculture improve drainage, improve building code, planning and technology, evacuation and early-warning systems, financial and technical assistance to farmers, secure equipment above forecasted flooding etc.
- Retreat options involve preventing development in endangered coastal areas or through withdrawal of communities from areas affected by climate change to areas less vulnerable. Relocation, managed realignment, and coastal setbacks are the most used approach. The cost of land of reallocation and psychological and socio-cultural implication make this solution a radical approach.

Among these three approaches, the globe has adopted "Nature-based Solution" [NbS] as a response to reduce environmental challenges to climate change in a sustainable way. Wetlands habitats represent important habitats that significantly contribute to coastal flood and erosion management Salt marshes are coastal wetlands which are flooded and drained by tides. They are marshy because the soil may be composed of deep mud and peat. Preserving and restoring wetland impacts the flooding areas due to their function of floodwater storage. Flood risk management strategy by implementing green infrastructure could facilitate the rehabilitation of degraded biodiversity of wetlands (Yamanaka et al., 2019). Dunes serves as natural coast defence and many coastal countries rely upon these natural defences. Restoring dunes represents a natural protective measure while increasing existing or new ones can have an ecological effect, but those are mostly temporary effects (Dean, 2002). Beach nourishment is not a permanent solution because of the disadvantages it entails. This method requires regular re-investment and usually is combined with other hard structure measures that have led to cost increases. Depositing sediments onto beaches cause direct burial of organisms residing there caused by agitation that affect the types of animals or rare species (ibid.). The deposition of fill material on the beach can interfere with marine habitats, including bird and sea turtle breeding, particularly if the grain size or composition of the sand does not correspond with that of the indigenous beach (Grain et al., 1995). In the future there may be a higher demand for nourishment project consequently. But also, the demand for sand has damaged coastal dunes of other countries like India and Morocco, where the sand robbery and its illegal trade has reached alarming values (Pilkey et al., 2022). Deltas are as well ecosystem largely impacted by fluctuations in the sea level and alteration in freshwater runoff from terrestrial sources

(IOC, 2009). Deltas are vulnerable to flooding, sediment shortage, inadequate flood protection schemes or receptor of pollutants from upstream. But delta ecosystems possess a notable adaptive and resilient ability to regenerate, exemplified by the restoration of estuaries through their reconnection with rivers and the sea (Deltares, 2008). Currently, there is insufficient evidence regarding the circumstances in which human intervention is necessary for ecosystem recovery. Hard measures protection offers the most reliable safety against coastal hazards, having been extensively implemented along developed European coastlines, although with a negative impact on the landscape (Vousdoukas et al., 2020). Integrated coastal management zones (ICMZs) can also be a powerful tool for coordinating sustainable development and conservation efforts along the coast. Adopting ICMZ strategies that combine an ecosystem approach with sustainable development goals can enhance coastal resilience. Some adaptation options to protect the coastline are used widely in several countries (Table 19.1).

19.4 Critical Description of Case Studies Selected

A specific solution related to flooding phenomena could be implemented referring to territorial dimension [Environmental system, infrastructure and service system] to provide an exportable and replicable model to cope with climate change.

19.4.1 The Rotterdam Climate Initiative [RCI]

The illustration presented here is part of 'The Rotterdam Climate Initiative (RCI) 2007–2025' where green factors and CO_2 emission goals were the key assets for ensuring the city protection against the impact of climate change. Waterplain in Benthemplein was design between 2011 and 2013 in the city of Rotterdam and aim to manage the heavy rains flooding thanks to the construction of rainwater catchment (Fig. 19.1). Once an empty square, now holds three large rainwater collection ponds, which, in good weather, can be used as amphitheatres for sports and greenery. Three catchment spaces at different heights have been located on the surface, with less cost to the water storage system, which has a defensive approach. The total area requalified 9000 m^2, cost 4,500,000 euro, and the construction period lasted one year. The benefit goes beyond and is not limited to one function. With square becoming ponds, that brings dynamism in the urban space (De Urbanisten, 1970).

19.4.2 Room for the River, 2006–2015

The Dutch government adopted a nationwide program in response to flooding due to the Rhine River. The objective was to increase current river velocity flows by ensuring more space to the river. The concept emerged from the conviction that integrating resilient measures is more advantageous than

Table 19.1 Option for responding to flooding and erosion (*Source* author's own elaboration)

Strategy	*Type*	*Technologies*	*Countries*	*Comments*
Defend	Hard technologies	Sea Walls Storm Surge barriers Closure dams Dikes Land Claim [LC]	Maldives, Fiji island, Nigeria, Maeslantkering Barrier Rotterdam Thames Barrier London, St Petersburg, Mose Project in Venice Feni closure in Bangladesh Dikes in Vietnam, East Asia Netherlands LC in Singapore, Dubai and Hong Kong,	Movable barriers are high capital and maintenance cost By affecting water inflow and outflow, closure dams or surge barriers can modify the chemical, physical, and biological characteristic of estuarine system Zhu X (2010)
Retreat	Soft technologies	Beach nourishment Salt marshes Dune restoration and creation Saltwater intrusion barriers Use of agricultural floating Seagrasses, Mangrove, Wetland rehabilitation	Beach extension United Kingdom Netherlands, Brazil, Malaysia, Korea, Ghana ect.] Uk, USA, Netherlands, Bangladesh Burma Vietnam	Depositing sediments onto beaches can affect birds or sea turtle nesting Grain A (1995) Demand for sand damaged other coastal beaches
Relocation Managed realignment Coastal setbacks			In Ihe Isle de Jean Charles, Louisiana Native Americans resettle with a cost about $48 ml Zhu X (2010). Reallocation has psychological and socio-cultural implication which make this solution a radical approach	
Co-exist	Various technology is applied like aquaculture, saline resistant crops, floating agriculture, mangrove improve drainage, improve building code, planning and technology, evacuation and early-warning systems, financial and technical assistance to farmers, secure equipment above forecasted flooding, flood hazard mapping etc. Accommodate or co-exist lead to expensive cost and continuously exposure to stress			

Fig. 19.1 Photo taken by the author November 2019, Waterplain in Benthemplein, Rotterdam

resisting natural phenomena. In summary, a study conducted by Stowa indicates the project is a transition from vertical flood protection like sea walls or embankments to horizontal widening of rivers and from top-down control to more equitable forms of multi-actor governance (Deltares, 2024). Different defensive and co-exist measures like removing obstacles, lowering part of the floodplain, relocating a dike inland, deepening, creating water channels bypass, were implemented in 30 locations. More than 2 billion euros was also determined to be a financially feasible amount to spent on enhancing flood protection in Dutch rivers. As compared to the grey strategy, which provides the same level of service but costs more in the long run, the Rotterdam case study found that green methods were somewhat more cost-effective over 50 years (ibid.).

19.4.3 Creating Green Corridor Along the Lower Danube

Four countries. Romania, Bulgaria, Moldova, and Ukraine, in 2000, initiated an agreement concerning the Lower part of the Danube River, with the aim to create a green corridor along its length. Along a 1000 km stretch, the initiative preserved 935,000 hectares, strengthened protection for 775,000 hectares and added protection for an additional 160,000 ha, all with the goal of restoring wetlands and re-establishing the river's connection to natural floodplains and boosting local businesses. Positive outcomes include economic diversity, biodiversity, and enhanced flood resilience. A potent instrument for the transformation in the Danube River Basin has been the international agreement for improved river and water management. The adaptive approach of the project is to restore foodplants and preserve unprotected areas to foster biodiversity (WWF, 2024).

19.4.4 Build Toolkit for Sava River Ecosystem

The main river basin of South-Eastern Europe, is home to 8.1 million people, covers approximately 97,200 km^2, and expands in Croatia, Slovenia, Montenegro, Serbia, Bosna and Herzegovina and a small section of northern Albania. The cooperation that starts on June 2001, contains a vast network of alluvial wetlands, Ramsar sites where several undamaged foodplants that copes flooding and support biodiversity (Sava River Basin, 2024). Forecasts indicate a rise in precipitation, with flood peaks anticipated to be 13% higher during the period of 2011–2040 compared to 1990–2013, and an increase of 23% in peaks projected for 2071–2100 (De Roo et al., 2016). To enable a coordinated response to extreme risks, a toolbox was created that includes a real-time knowledge sharing platform and a GIS model of the river basin (Sava River Basin, 2015). The function of the toolbox was to enhance resilience and safeguard individuals and the ecosystem from flooding and pollution. The system was successfully tested in Slovenia during severe flooding that occurred in 2023, showing its effectiveness in responding to and preparing for flood risks (Climate Adapt, 2024b). The toolbox is expected to remain operational and undergo maintenance for at least 10 years.

19.4.5 Protecting the Marche Region's Coastline from Erosion

The project seeks to restore the coastal area of the Marche region, which has retreated due to erosion, and protect against the risk of storms, flooding and landslides. Anthropic activities, for example, the modification of rivers in the region, have decreased the amount of sediments that are carried to the sea. The action in this case refers to two municipalities, Sirolo and Numana, with the main objectives to protect the beach and settlements of the coastal area and strengthen the recreational opportunities provided by the shoreline. Under the ICZM Plan 9 2005–2016, over 176 km of coast protection measures were implemented. The intervention consists of removing breakwaters, re-naturalization, and beach nourishment with materials taken from inland quarries and river deposits, stabilizing the cliff base, and integrating adaptation strategies in coastal management plans. The intervention along the two coastline cities was implemented during a 2-year period with the necessity to carry out regular beach replenishment and sand movement interventions (Climate Adapt, 2020). Natural coastal habitats have diminished to narrow strips of dune vegetation as a result of protection measures that have artificially reinforced coastal systems (Bisci et al., 2022).

19.5 Albania in Front of Green Technologies

Albania has one of the richest systems of coastal lagoons and water basins. The country's watersheds face a variety of hazards, particularly those related to flooding. Albania has seven water basins, of which four out of seven are management draft plans in process (AMBU, 2024). In Albania, climate change is predicted to make temperatures and rainfall more unpredictable, which with flood prevention infrastructure being limited (UNDP, 2022), creating endangered areas, especially in rural localities where the poor inhabitants will be more susceptible. As a result of climate change, excessive amounts of water over a period of time will cause significant damage, especially in the coastal region, which includes a high concentration of population and agriculture (ibid). The budget-driven implementation of new infrastructure fails to evaluate the risks associated with climate change, and the technical aspects of adaptation. Strategic investment project in the coast line brings up serious concern regarding the preserving of existing ecosystems. Interventions adjacent to ecosystems seem to lean out to hard defence technologies. This is due to the fact that the cost is easier to calculate and the impact effects are short-lived. One example is the construction of Shkodra By-pass (act as dam and roadway) in Albania as a result of the 2010–2011 flooding, which was built with a total initial project value of 17.5 ml of euro[i]. Supposed to function like a dike to prevent floods, it was open for vehicles in 2022. It has not been fully completed, which makes it difficult to assess the structure's impact on flood protection. However, some NbS interventions have been made in Kune-Vain Lagoon by rehabilitation dunes at a length of 2000 m a new tidal inlet channel, 7 hectares of riparian forest (Kune Vain, 2018), or the project 'Recommendation for an Natural Based Solution intervention in Elbasan Municipality, Albania, through the NbS Assessment of the Shkumbini River (IUCN, 2023). An ongoing process is unfolding from April 2021 named "Buna Delta Masterplan-Integrating Green–Blue Connection for Enhanced Flood Risk Management' by Deltares with technical assistance to the municipality of Shkodra based on NbS solution with reference and benchmarks to 'Room for the River Program'. Detailed Plans for Areas of National Importance have been developed with the aim of controlling planning and preserving the ecosystem in lagoons or protected areas, but adaptation strategies and the consequences they involve are not expressed in technical measures. The lack of determined local experts in this field, trained in designing with technical measures that affect the ecosystem, has a negative impact. Addressing green technologies in Albania is somewhat difficult due to chaotic legislation, property concerns, the evaluation of financial resources, separate initiatives, and technical issues regarding these interventions.

19.6 Conclusion

Develop a just transition strategy that includes green solutions to work with ecosystems in the mitigation and adaptation strategies imply actions to protect, conserve, and restore natural or alternated coastal ecosystem, promoting resilience human health, services, and ecological benefits (Durschinger, 2024). By increasing local resilience, cities can decrease their impact on greenhouse gas emissions and protect themselves from climate-related losses and damages. The duration of floods can change the dynamics of CO2 emissions, which in turn affects the relationship between grassland ecosystems and hydrologic regimes (Amanuel et al., 2017). Coastal cities are incubation areas where new approaches to adapting to climate change are carried out. Here, the worldwide community has stressed the significance of action taken at the municipality level in reaction to the threat of climate change. Research on how cities have used technology to adapt to climate change has shown that adaptation through green technology can be less complicated and expensive than hard adaptation ones. Therefore, considering local circumstances, such as the limited financial capacities of local governments, it is possible to apply some of them also in Albania. The cases analyzed above could be helpful for Western Balkan territories when appropriate adaptation technologies are implemented as listed in Table 19.2.

Considering or not the worst scenario, adaptation through the green technology identified provide a potential strategy to be considered in local and national level. The main challenge is finding the best technology option

Table 19.2 Green technology application options on flooding and erosion phenomena (*Source* author's own elaboration)

Challenge	*System*	*Strategies and action*		
		Defence	*Co-exist*	*Retreat*
Heavy rain and river flood	Environmental/ ecological	Planting native trees and riparian forest	Protection of foodplant	Relocate agricultural production
		Create green channels to convey flood waters	Modify agricultural practice	No development in exposed area
	Infrastructure and service	Urban parks with wetlands		
		Smart tools for monitoring	Create catch basin as storage in public space	
		Minimise paved surface	Integrated planning system	
		Integrated gutters		
Sea level rise and erosion	Environmental/ ecological	Dune preservation and restoration by building fences, plant vegetation	Switch to aquaculture	
		Saltmarshes restoration	Planting desalination plants	
		Smart tools for monitoring	Balance between preservation and development	Coastal setbacks
	Infrastructure and service	Vegetation corridors	Modify land use	Change Land Use planning in threatened area
		Wetland protection and restoration		
		Suspended walkways and cycling route		

with a positive cost–benefit and reduction of negative environmental impacts to respond to menaces in a specific region. Usually, these studies require time and field specialists, the right tools and long-term forecasts, becoming more complicated than short-term and invasive interventions. Issues tracking successful adaptive technology followed by poor economic evaluation, limited capacity to evaluate and report environmental impact over time, have made it challenging to set up reference models that are relevant, particularly in emerging countries. Achieving resilience requires in-depth study of technologies, as coastal areas may present several not replicable characteristics making it difficult to assess the environmental benefits and consequences of full implementation. But it becomes essential to assure that climate risks are addressed and managed with justice, supporting vulnerable communities rather than redistributing the burden of risk to those with the lowest capacity to respond. Coastal cities must come together to address local climate challenges constructively and long-term. The most crucial step in the process involves guaranteeing the right adaptation technologies by developing local planning tools, assessing the overall efficiency, and analysing the cost–benefit of the technology being employed. National and local governments must be heavily involved, cooperative, and as part of this, established a well-defined plans for predicting their requirements and addressing the effects of climate change in their communities. Measures reported in this research provide a preliminary result that will help to modify urban planning tools by integrating exportable models in order to implement adaptation strategies through site-specific activities that consider the risks associated with climate phenomena. Long-term planning solution adopted by developed countries through the use of eco-sustainable technologies is the most suitable alternative that must be taken into consideration in developing countries to avoid financial shortcomings.

References

Abbass, K., Qasim, M. Z., Song, H., Murshed, M., Mahmood, H., & Younis, I. (2022). A review of the global climate change impacts, adaptation, and sustainable mitigation measures. *Environmental Science and Pollution Research*. https://doi.org/10.1007/s11356-022-19718-6

Amanuel, W., Osborne, G. B., & Orr, P. (2017). Flooding-related increases in CO2 and NO2 emissions from a temperate coastal grassland ecosystem. In: *Biogeosciences*. https://doi.org/10.5194/bg-2016-522. Accessed 20 February 2024.

AMBU. (2024). Agjencia E Menaxhimit Te Burimeve Ujore. In *AMBU*. http://www.ambu.gov.al/. Accessed 20 July 2024.

Benthemplein, W., DE—URBANISTEN D. (1970). Watersquare Benthemplein, Rotterdam. DE URBANISTEN. In: DE URBANISTEN. https://www.urbanisten.nl/work/benthemplein. Accessed 15 June 2024.

Bertule, M., Appelquist, L. R., Spensley, J., Trærup, S. L. M., Naswa, P. (2018). *Climate change adaptation technologies for water: A practitioner's guide to adaptation technologies for increased water sector resilience*. UNEP DTU Partnership.

Bisci, C., Cantalamessa, G., Casavecchia, S., Tramontana, M., & Spagnoli, F. (2022). *Coastal dunes along the Marche littoral (Adriatic side of central Italy)*. Firenze University press eBooks. https://www.academia.edu/110866387/Coastal_dunes_along_the_Marche_littoral_Adriatic_side_of_central_Italy_. Accessed 24 June 2024.

Cedergren, E., Tapia, C., Gassen, N. S., Lundgren, A. (2022). *Just green transition* (Nordregio discussion paper). https://doi.org/10.6027/WP2022:2.1403-2511. Accessed 8 February 2024.

Christiansen, l., Olhoff, A., & Trærup, S. (2011). *Technologies for adaptation: Perspectives and practical experiences*. UNEP Risø Centre.

Climate Adapt. (2020). *Addressing coastal erosion in Marche region, Italy*. Climate Adapt. https://climate-adapt.eea.europa.eu/en/metadata/case-studies/addressing-coastal-erosion-in-marche-region-italy. Accessed 5 June 2024.

Climate Adapt. (2024a). *Discover the key services, thematic features and tools of climate-ADAPT*. Climate Adapt. https://climate-adapt.eea.europa.eu/en. Accessed 13 November 2023.

Climate Adapt. (2024b). *Toolbox for transboundary water contingency management in the Sava River Basin*. https://climate-adapt.eea.europa.eu/en/metadata/case-studies/toolbox-for-transboundary-water-contingency-management-in-the-sava-river-basin. Accessed on 17 June 2024.

De Roo, A., Bisselink, B., Beck, H., Bernahard, J., Burek, P., Reynaud, A., Pastori, M., Lavalle, C., Jacobs, C., Baranzelli, C., Zajac, Z., & Dosio, A. (2016). *Modelling water demand and availability scenarios for current and future land use and climate in the Sava River Basin*. Publications Office of the European Union.

Dean, R. G. (2002). *Beach nourishment*. World Scientific.

Deltares. (2008). *Towards sustainable development of deltas, estuaries and coastal zones*. Deltares.

Deltares. (2024). *Room for the Rhine branches in The Netherlands*. Stowa. https://www.stowa.nl/sites/default/files/assets/DELTAFACTS/Deltafacts%20E%20PDF%20nieuw%20format/Room%20for%20the%20river%20EN.pdf. Accessed 20 June 2024.

Durschinger, L. (2024). *Delivering a just transition through nature-based solutions*. Terra Global Capital, LLC.

EcoShape. (2024). *Cases archive*. EcoShape—EN. https://www.ecoshape.org/en/cases/. Accessed 25 September 2023.

European Commission. (2013). EUR-Lex—52013DC0249—EN—EUR-Lex. EUR-Lex. https://eur-lex.europa.eu/legal-content/EN/TXT/?uri=celex%3A52013DC0249. Accessed 10 December 2023.

European Environment Agency. (2019). *Green infrastructure*. Environment. https://environment.ec.europa.eu/topics/nature-and-biodiversity/green-infrastructure_en. Accessed 10 December 2023.

Geneletti, D. (2020). *Planning for ecosystem services in cities*. Springer.

Grain, A., Bolten, A. B., & Bjorndal, K. A. (1995). *Effects of beach nourishment on sea turtles*. Wiley https://doi.org/10.1111/j.1526-100X.1995.tb00082.x. Accessed 19 December 2023.

Harvey, N., & Nicholls, R. (2008). *Global sea-level rise and coastal vulnerability*. Spring Nature. https://doi.org/10.1007/s11625-008-0049-x Accessed 16 December 2023.

Intergovernmental Oceanographic Commission. (2009). *Hazard awareness and risk mitigation in integrated coastal area management*. UNESCO. https://unesdoc.unesco.org/ark:/48223/pf0000183253.locale=en. Accessed 19 December 2023.

Intergovernmental Panel on Climate Change (IPCC). (2022a). *Climate Change 2022—Impacts, adaptation and vulnerability Cambridge core*. https://doi.org/10.1017/9781009325844. Accessed 30 January 2024.

Intergovernmental Panel on Climate Change (IPCC). (2022b). *Global Warming of 1.5°C*. Cambridge Core. https://doi.org/10.1017/9781009157940. Accessed 15 February 2023.

Intergovernmental Panel on Climate Change (IPCC). (2022c). *Sea level rise and implications for low-lying islands, coasts and communities (Chapter 4)—The ocean and cryosphere in a changing climate*. Cambridge Core. https://doi.org/10.1017/9781009157964.006. Accessed 30 February 2024.

IUCN. (2023). *Recommendations for an NbS Pilot intervention In Elbasan municipality, Albania, defined through the NbS baseline assessment of the Shkumbini River Basin*. Oppla. https://oppla.eu/casestudy/29281. Accessed 30 June 2024.

Janiszek, M., & Krzysztofik, R. (2023). *Green infrastructure as an effective tool for urban adaptation—Solutions from a big city in a postindustrial region*. MDPI. https://doi.org/10.3390/su15118928. Accessed 11 January 2024.

Kune Vain. (2018). *Kune Vain—Island in the Ardiatic Sea*. Kune Vain. https://www.kunevain.com/. Accessed 15 May 2023.

NATURVATION. (2024). *Cities*. NATURVATION. https://naturvation.eu/cities.html. Accessed 8 September 2023.

Oppla. (2024). *Case studies*. Oppla. https://oppla.eu/case-study-finder. Accessed 13 September 2023.

Orton, P., Ralston, D., Van Prooijen, B., Secor, D., Ganju, N., & Chen, Z. (2023). *Increased utilization of storm surge barriers: A research agenda on estuary impacts*. Earth's Future. https://doi.org/10.1029/2022EF002991. Accessed 2 February 2024.

Pilkey, O. H., Longo, N. J., Neal, W. J., Rangel-Buitrago, N. G., Pilkey, K. C., & Hayes, H. L. (2022). *Vanishing sands: Losing beaches to mining*. Duke University Press.

Sara, T. (2022). *The role of climate technologies in green transition pathways*. Open Edition Journals. https://journals.openedition.org/factsreports/6885 Accessed 25 September 2023.

Sava River Basin. International Sava River Basin Commission. https://www.savacommission.org/en/sava-river-basin/7. Accessed 18 July 2024.

The international Sava River Basin Commission. (2015). Sava Gis Geoportal. https://savagis.org/map;jsessionid=7D62B54087C78DF0F0E614FD31D21618. Accessed 17 June 2024.

UNDP. (2022). *The Fourth National Communication of Albania On Climate Change*. UNDP. https://www.undp.org/albania/publications/fourth-national-communication-albania-climate-change. Accessed 20 July 2024.

United Nation. (2007). *Percentage of total population living in coastal areas*.

United states agency for international development (Usaid). (2009). *Adapting to coastal climate change. A guidebook for development planners*.

Vousdoukas, M. I., Mentaschi, L., & Hinkel, J. (2020). *Economic motivation for raising coastal flood defenses in Europe—Nature Communications*. Nature. https://doi.org/10.1038/s41467-020-15665-3. Accessed 17 April 2024.

WWF International Danube-Carpathian Programme Lower Danube Green Corridor: Floodplain Restoration for Flood Protection. Climate Adapt. https://climate-adapt.eea.europa.eu/en/metadata/case-studies/lower-danube-green-corridor-floodplain-restoration-for-flood-protection. Accessed 18 July 2024.

Yamanaka, S., Ishiyama, N., Senzaki, M., Morimoto, J., Kitazawa, M., Fuke, N., & Nakamura, F. (2019). *Role of flood-control basins as summer habitat for wetland species—A multiple-taxon approach*. Elsevier. https://doi.org/10.1016/j.ecoleng.2019.105617. Accessed 20 December 2023.

Zhu, X., Linham, M. M., & Nicholls, R. J. (2010). *Technologies for climate change adaptation—Coastal erosion and flooding*. Danmarks Tekniske Universitet.

CHAPTER 20

Green EU Integration: The Role of the Green Agenda in Accelerating Just Green Transitions in the Western Balkans

Erblin Berisha

20.1 Introduction

The Western Balkan (WB) countries are experiencing the effects of multiple crises, which require a profound redefinition of their economic, social and institutional systems (Berisha, 2025). Indeed, climate change (Knez et al., 2022), technological revolution (Popović et al., 2020), and energy crises (Morina, 2022) are straining the regional existing governance structures (Berisha, 2025). Even though not explicitly addressed to the WB, according to the Just Transition Mechanism introduced by the European Union in 2019, these simultaneous transition challenges could be addressed more equally, fairly and justly (Shaker & Berisha, 2025). Since then, the European Union (EU) has actively promoted the European Green Deal as a multi-annual political framework to achieve climate neutrality by 2050. Through this initiative, the EU has defined ambitious targets and supports countries and territories in managing the transition they face. The European effort toward a green transition also extends to candidate countries, such as the WB, to accelerate their convergence with the EU (Berisha, 2023) by increasing cooperation in various fields (Pinnavaia & Berisha, 2021). The transition fatigue in the Western Balkans is particularly evident when it comes to restructuring the existing economic and social models (Imami et al., 2023). These structural challenges must be contextualised into the EU Integration process that sees

E. Berisha (✉)
Interuniversity Department of Regional and Urban Studies and Planning (DIST), Politecnico di Torino, Turin, Italy
e-mail: erblin.berisha@polito.it

E. Berisha et al. (eds.), *The Palgrave Handbook of Just Green Transitions in the Western Balkans and Beyond*,
https://doi.org/10.1007/978-3-031-95075-9_20

the WB countries involved (Berisha & Cotella, 2021; Berisha et al., 2021). The six WB countries are, to some extent, involved in the integration process recently renovated by the EU with the introduction of the new methodology, "Enhancing the accession process—A credible EU perspective for the Western Balkans," launched in 2020. The EU Integration process, which is seen more as an institutional and technocratic approach, is accompanied by the introduction of several regional focus initiatives like the introduction of the Green Agenda for the Western Balkans passed in 2020 (referred to as the Agenda).

This chapter has the ambition to present the state of the art of the EU Integration process and the just green transition in the WB. It offers an overview of what is happening in the region in relation to the different initiatives activated with a particular focus on the adoption, implementation and monitoring of the Agenda. More specifically, after this brief introduction, the following section offers an overview of the EU integration path. It discusses where the countries stand in relation to the integration process and what they are doing to cope with. This section is then followed by a detailed exploration of where the region stands in relation to the just green transition issue, presenting some good practices and identifying structural pitfalls. This is followed by a detailed presentation of the Green Agenda for the Western Balkans as part of the EU Green Integration process, which discusses the content, objectives, and trajectories. The section on areas of improvement offers a detailed selection of regional priorities where regional actors are expected to intervene to ensure a smooth and fast EU integration process. Finally, the chapter concludes by reflecting on some takeaways and future developments.

20.2 EU Integration Path

The WB are crossing two transformative challenges: the transition toward democratic and free markets linked to EU integration since the early 90s and, more recently, the transition to a sustainable and green economy (Berisha, 2025). Although apparently unrelated, recent developments show these processes are deeply interconnected, reflecting a shared vision of regional stability, economic modernisation, and environmental responsibility. The dual challenges of aligning with European norms and addressing pressing environmental issues offer opportunities and hurdles as the region seeks to redefine its role in a rapidly changing global context. According to Aliaj and Sulmina (2022), indeed, EU integration is more than a political aspiration for the WB; it represents a comprehensive framework for reforming institutions, modernising economies, and fostering regional cooperation. The integration process seeks to align the WBs' political, legal, and economic systems with the *EU acquis communautaire* (Berisha et al., 2021). Progress often reveals entrenched issues such as weak governance structures, pervasive corruption, and inefficient public administration, all of which complicate the reform process (Qorraj et al., 2024). The topic of green transition, meanwhile, is emerging as a parallel and equally vital transformation that requires deep institutional and social reconfiguration (Djatkov, 2024). Currently, the WB's countries are dealing with the EU integration path. Since the introduction of the new methodology of WB integration in 2020,[1] countries are facing the opening of the so-called "clusters" (Table 20.1). According to the current state of EU Integration, Serbia, Montenegro, Albania, and, in part, North Macedonia are the countries that are performing better: Serbia has 22 of 35 chapters opened; Montenegro has 33 chapters opened, while Albania is focusing on Cluster 1 and Cluster 6, finally, North Macedonia has opened Cluster 1—fundamental.[2] When it comes to Bosnia and Herzegovina and Kosovo*,[3] instead, the process of EU Integration is relatively slow. Indeed, while the former has gained its candidate status in 2022, the latter is still a potential candidate country—in the initial phase of the process.

In parallel with the formal EU Integration process, the European Green Deal has extended its reach to include the WB through the Green Agenda for the Western Balkans (Berisha, 2025). This initiative emphasises the region's role in achieving Europe's climate goals while addressing local environmental crises that have long plagued it. According to the Agenda, the green transition, however, is not merely an environmental imperative; it has to be considered a pathway to economic resilience, energy independence, and social cohesion. Both processes (i.e., EU Integration and green transition) demand structural changes that are deeply interdependent and mutually influenced (Berisha, 2025). Indeed, aligning environmental policies with EU standards is a critical component of accession negotiations, requiring countries to adopt measures to reduce greenhouse gas emissions, invest in renewable energy, and protect biodiversity (Imami et al., 2023). For a region, as the WB, traditionally heavily reliant on coal and other fossil fuels, this shift represents a profound economic and social challenge (Nikolov et al., 2024). The transition away from coal is both a matter of compliance but also a necessity for addressing the region's disproportionate contribution to air pollution and carbon emissions (Knez, Štrbac & Podbregar, 2022). To cope with these simultaneous transitions, the EU integration is more institutional-based, and the green transition is

[1] The EU Commission communication has adopted in 2020 the "*Enhancing the accession process—A credible EU perspective for the Western Balkans*'" of 5 aiming to reinvigorate the accession process by making it more predictable, more credible, more dynamic and subject to stronger political steering. To make the process more credible, the revised EU integration process gathers the 35 chapters into 6 clusters: 1—Fundamentals, 2—Internal Market, 3—Competitiveness and inclusive growth, 4—Green agenda and sustainable connectivity, 5—Resources, agriculture and cohesion and 6—External relations. For more info—Enhancing the accession process - A credible EU perspective for the Western Balkans - European Commission.

[2] The EU Commission presented to the Council further screening reports on both countries (Albania and North Macedonia): in May 2024 for cluster 2—internal market, and in October 2024 for cluster 6—external relations and cluster 3—competitiveness and inclusive growth.

[3] This designation is without prejudice to positions on status, and is in line with UNSCR 1244 and the ICJ Opinion on the Kosovo declaration of independence.

Table 20.1 EU Integration stages and clusters opened by Western Balkan countries (*Source* European Commission, 2024a)

Country	*Integration stage*	*Clusters opened/under negotiation*
Albania	Candidate (Accession Negotiations Started 2022)	Cluster 1: Fundamentals (e.g., judiciary, economic criteria) Cluster 6: External Relations (e.g., foreign policy)
Montenegro	Candidate (Negotiations Started 2012)	All 33 chapters within 6 clusters opened Working on provisional closure of key chapters
Serbia	Candidate (Negotiations Started 2014)	22 of 35 chapters opened Focus on reforms in the rule of law and relations with Kosovo
North Macedonia	Candidate (Negotiations Started 2022)	Screening process ongoing Assessing alignment with acquis in clusters such as Fundamentals
Bosnia and Herzegovina	Candidate (Status granted 2022)	Screening process started Preparing to open specific clusters
Kosovo	Potential Candidate	Early stages of integration Working to meet criteria for candidate status

more economic and social-based, the EU has framed a series of economic and financial packages (Qorraj et al., 2024). Indeed, the EU provides financial support through instruments such as the Instrument for Pre-Accession Assistance (IPA) and the Western Balkans Investment Framework (WBIF). These funds aim to catalyse projects that promote energy efficiency, renewable energy, and environmental protection. In 2023, the EU adopted the Growth Plan for the Western Balkans, which incentivises enlargement partners' preparations for EU membership by bringing forward some of its benefits ahead of full integration into the EU. The plan aims to integrate the Western Balkan partners into the EU's single market, advance regional economic cooperation, deepen EU-related reforms and increase pre-accession funding because of accelerating socio-economic convergence of the Western Balkans to the EU. This joint effort to accelerate the EU integration process not only from an institutional perspective, has been concretised by the adoption of several agendas and promoted by several regional initiatives. The following sections present the state of the art alongside the just green transition process with a particular focus on the Green Agenda for the Western Balkans.

20.3 Just Green Transition: Where the Region Stands

As mentioned, the WB faces a complex and multifaceted transition process as it works toward a Just Green Transition aligned with global sustainability goals and EU standards (Berisha, 2025a). Across the region, progress varies significantly, with most countries having started conceptualising Just Green Transition policies, mainly at the strategic level, but struggling with operationalisation, institutional coordination, and implementation (Imami et al., 2023). From December 2020 to December 2023, the EU implemented its first Initiative for coal regions in transition in the Western Balkans and Ukraine[4]—an initiative due to be followed up in 2024. During this first period, the differences in the Western Balkans countries' just transition processes increased. In Albania, for instance, the Just Green Transition seems to be a cornerstone of its broader strategy to align with EU climate goals, while mitigating the socio-economic impacts of industrial transformation (UNECE, 2024). The country emphasises decarbonisation, sustainable industrial practices, and the creation of green jobs: to these purposes the country in 2019 has introduced the Climate Change Strategy and Action Plans in 2019. However, its efforts remain at the strategic level, with no Territorial Just Transition Plans (TJTPs) having been adopted yet. The transition focuses primarily on regions with high concentrations of energy-intensive industries, such as mining and brownfield sites, as identified in a 2024 UNECE study. While there is recognition of the need for workforce upskilling and regional economic recovery, especially in mining areas, tangible policies to address these challenges remain underdeveloped. To tackle with issue, Bosnia and Herzegovina has adopted a central strategy for climate adaptation and low-emission development, covering sectors like construction, waste management, agriculture, and forestry. Key measures include improving building energy efficiency and transitioning to greenhouse gas (GHG)-efficient heating systems. However, the strategy lacks detailed institutional roles and implementation mechanisms, leaving progress largely stagnant and not well implemented. Importantly, the fragmented governance structure in BiH further complicates implementation, as the central government lacks enforcement instruments, and strategic and regulatory frameworks are yet to be developed by the Federation of BiH and Republika Srpska entities' institutions. Kosovo's transition framework, instead, is more advanced at the strategic level, with clear objectives outlined in its Strategy for Industrial and Business Development Support 2030.[5] Specific sub-objectives include promoting recycling, green products, and technological advancements for energy efficiency. However, concrete measures to achieve these goals remain undefined. The strategic groundwork is strong, but without operationalisation and actionable policies, Kosovo's transition

[4] 2024_04_Local-authority-planning-for-A-just-transition.pdf.

[5] 8CE08B27-3741-4544-809F-BF643B97233F.pdf.

efforts risk losing momentum (Battali, 2023). In Montenegro, the Just Green Transition seems to be at an early stage, with TJTPs and coordination mechanisms still to be conceptualised. In this respect, the central government has established a Coordination Body for Industrial Policy to align the country's industrial framework with EU standards (UNDP, 2023). This body operates on both managerial and operational levels, ensuring intra-ministerial and inter-ministerial cooperation. Despite this organisational structure, the absence of concrete implementation plans and performance metrics limits Montenegro's progress in transitioning to a greener economy. Regarding North Macedonia, the country has made significant strides in integrating Just Transition into its climate and energy policies (CEE Bankwatch Network, 2024). Its Long-Term Strategy on Climate Action, adopted by the Ministry of Environmental and Physical Planning, defines Just Transition as a framework to ensure workers' rights and livelihoods during the shift to sustainable production. The country's roadmap focuses on phasing out coal, promoting renewable energy, and fostering socio-economic regeneration in coal-dependent regions like Pelagonia and Southwest North Macedonia (Ministry of Economy, 2023). However, despite announcing an ambitious plan to phase-out coal by 2027, the country extended its timeline to 2030 in 2023. North Macedonia's integrated approach to energy transition and employment alternatives is a promising model, though gaps remain in funding and operationalisation, given its non-eligibility for the EU Just Transition Fund. Finally, Serbia prioritises the Just Green Transition through comprehensive measures to reduce GHG emissions across various sectors, including energy, industry, and transport. The government plays a central role in planning and implementing these policies, while local self-governments monitor and report climate action. However, adopting local climate action plans remains sporadic, and concrete measures for Just Transition are rare. The centralisation of responsibilities and limited local capacity hampers the broader adoption and implementation of green policies.

Besides that, countries have launched several initiatives and projects to cope with just green transition. An example is the Durrës Brownfield Redevelopment Project, an integrated approach to redeveloping former industrial sites in Durrës (World Bank Group, 2022). This project combines environmental restoration with economic revitalisation strategies. It has successfully transformed these areas into hubs for green industries, supported by local SMEs and funded through the Just Transition Fund and other national instruments (Ministria e Financave dhe Ekonomisë, 2021). The project also strengthens local value chains by promoting the use of locally sourced materials and labour. Successful experiences are implemented also in North Macedonia, like, for instance, the revitalization of the Kichevo region, the Climate Coalition launched by CSOs and the Establishing Aarhus Centre. More in detail, in Kichevo region, where the mine feeding the Oslomej power plant has been depleted for years, regional re-development is already taking shape—however, without a clear participatory process to guide it. Most of the employment there is generated by the trade, processing industry and agricultural sectors, so the domination of the coal sector in employment figures has already been eroded. Coupled with the conversion of old mining fields into solar PV plants, this is a prime example of how the need for transition motivated joint actions by the community, business sector, and decision-makers.[6] Notably, the Climate Coalition launched by CSOs in 2020 has allowed the CSOs active in education and awareness raising on environmental issues in North Macedonia to launch the Climate Coalition to increase citizen participation in decision-making processes and monitor the implementation of energy, climate and environmental legislation. Climate Coalition fosters local synchronised action, strong social influence, and transformation by building climate culture. This initiative can be a good example of how non-public actors can exert bottom-up pressure and get involved in decision-making (OCED, 2022). Moreover, the country has been engaged in establishing the Aarhus Centre, which was set up in 2019 in support of the Aarhus Convention, which establishes specific rights for individuals and CSOs regarding the environment. This has provided a platform to engage citizens, governments and the private sector in dialogue on environmental challenges, support public participation in energy, climate and environment, facilitate access to environmental information and assist governments in implementing the Aarhus Convention. In Bosnia and Herzegovina, instead, the Strategy for Climate Change Adaptation and Low Emission Development (2020–2030) is a starting point for tackling the issue of Just Green Transition. The strategy defines clear objectives and integrates international cooperation to facilitate progress. A notable example is the "Scaling-up Investment in Low-Carbon Public Buildings" initiative, which successfully mobilises significant co-financing for sustainable retrofitting projects. However, challenges remain in translating these strategic plans into actionable policies at the entity and cantonal levels, where regulatory harmonisation is still lacking. Montenegro stands out for its comprehensive approach, driven by the Law on Protection from Negative Effects of Climate Change. This legal framework not only supports broader climate goals but also underscores the importance of a just green industrial transition. Montenegro's Industrial Policy (2024–2028) prioritises green industrial practices and incorporates detailed key performance indicators and monitoring mechanisms to ensure accountability and progress. Despite these strengths, the implementation of industrial greening policies often faces delays due to funding constraints and limited private-sector engagement. Finally, Serbia demonstrates a robust regulatory foundation, with strong interlinkages between climate policies and sectoral strategies. The country has prioritised sustainable energy transitions, focusing on renewable energy and energy efficiency initiatives. However, the private sector plays a limited role in these efforts, as high operational costs and market uncertainties deter the widespread adoption of green technologies. Serbia's industrial sector, while

[6] North Macedonia pioneering energy transition in the Western Balkans: Just Transition.

leading in certain green investments, is still slow to adopt comprehensive decarbonization measures.

Besides these relatively positive examples, Just Green Transition in the WB remains a regional challenge, far from being solved Djatkov (2024). In Albania, the absence of a cohesive strategy for decarbonization sets the stage for stagnation. Despite the introduction of the Green Agenda for the Western Balkans, political will remains lukewarm, and communities remain disengaged. The abandonment of former industrial strongholds, like mining regions, underscores a broader issue: the lack of investment and innovation that could rejuvenate these areas. Compounding this, a "coordination gap" hampers the alignment of national strategies like the National Energy Strategy with regional and local efforts. Small and medium-sized enterprises (SMEs) struggle to adapt to new green standards due to inadequate resources and support, while public awareness of the transition's benefits remains minimal, particularly in rural communities. In Bosnia and Herzegovina, a strategic framework exists, but without the regulatory backbone required to implement sustainable plans at the entity and cantonal levels, progress is stymied. Similarly, Kosovo has yet to operationalize its JTPs, leaving roles and responsibilities across governance levels unclear and unaddressed. Montenegro, too, grapples with inertia, with its strategic and regulatory frameworks for green transition still awaiting adoption. North Macedonia presents a paradox. It boasts ambitious goals, such as the Western Balkans' most aggressive coal phase-out plan, but execution falters. Delayed adoption of key policies, such as the Law on Climate Action and the National Energy Efficiency Action Plan, reflect a broader trend of slow legislative progress. Challenges in attracting private investment for renewable energy projects—particularly in coal-reliant regions like Bitola—impede momentum. Workers in traditional industries face uncertainty, with retraining programmes lagging behind. Meanwhile, plans for new fossil gas units risk undermining the country's renewable energy aspirations, swapping one dependency for another. Serbia, while facing a slower pace of industrial greening, contends with economic vulnerabilities tied to global trends. Only the largest enterprises have significantly invested in sustainable practices, leaving SMEs and smaller industries struggling to keep pace. The high costs and productivity losses associated with green transitions further dissuade broader adoption. Across the region, the common barriers to progress across the region are clear: fragmented governance, regulatory delays, and insufficient stakeholder engagement (Djatkov, 2024). Besides countries' specificities, it can be said that the region faces some shared challenges and opportunities (Table 20.2). Among them, it is essential to mention the gap in public awareness and buy-in, with rural communities often sceptical of the benefits of green practices. At the same time, industries remain hesitant, citing financial and operational constraints.

Table 20.2 Comparison of WB countries Just Green Transition challenges and opportunities (*Source* author's own elaboration)

Country	*Strengths*	*Key challenges*	*Opportunities*
Albania	National Energy Strategy in place	Lack of comprehensive decarbonisation plans Weak coordination among governance levels	Improve coordination between national, regional, and local levels
		Public awareness gap Neglect of former industrial regions	Engage SMEs with better-targeted support mechanisms
Bosnia and Herzegovina	Strategic framework present	Regulatory framework not yet adopted Need for action at entity and cantonal levels	Develop and implement regulatory frameworks to operationalise strategies
Kosovo	JTP framework exists	Lack of operationalisation Undefined governance roles and responsibilities	Clarify governance structure and implement existing JTPs
Montenegro	Growing acknowledgement of the need for JGT	Strategic and regulatory frameworks pending adoption	Finalise and adopt frameworks to enable transition
North Macedonia	Ambitious coal phase-out plan Just energy transition investment platform launched (2023)	Delays in adopting laws (e.g., Law on Climate Action, NEAP)	Leverage investment platform to attract private capital
		Non-compliance with emission reduction plans	
		Slow private investment	Accelerate implementation of climate and energy laws

(continued)

Table 20.2 (continued)

Country	*Strengths*	*Key challenges*	*Opportunities*
		Worker retraining program,es delayed	
Serbia	Large enterprises leading green investments	Industry hesitant due to high costs and productivity losses Economy vulnerable to global trends Limited SME participation	Introduce incentives to encourage SME participation Broaden investment efforts beyond large enterprises

20.4 The Green Agenda for the Western Balkans

To facilitate the transition of the WB from a carbon-based to a carbon-free economy, the EU introduced the Green Agenda for the Western Balkans in 2020, also known as the "Sofia Declaration" (European Commission, 2022a). This initiative was further strengthened with the Action Plan presented during the EU-Western Balkans Summit held in Brdo pri Kranju, Slovenia, in 2021. The Agenda is structured around five main pillars: (1) climate action, encompassing decarbonisation, energy, and mobility; (2) promoting a circular economy by addressing waste management, recycling, sustainable production, and resource efficiency; (3) biodiversity conservation, aimed at safeguarding and restoring the region's natural ecosystems; (4) combating pollution across air, water, and soil; and (5) fostering sustainable food systems and rural development. Additionally, digitalisation serves as a cross-cutting enabler, supporting the dual transitions to green and digital economies. The Action Plan provides a structured approach to facilitate, support, and monitor the Agenda's implementation. It outlines 58 specific actions divided into seven components, with provisions for annual evaluation and monitoring to ensure effective execution. Scheduled updates are planned for 2024 and 2027. Financial support for the Agenda's implementation is sourced from various international funding agencies (see for instance: GIZ; SIDA; SDC; USAID and EBRD activities in the region) (Berisha, 2018), EU funds, including the Instrument for Pre-Accession Assistance, the Western Balkans Investment Framework, and tailored funding initiatives. A key financial mechanism supporting this transition is the Economic and Investment Plan for the Western Balkans (2021–2027), which allocates up to €9 billion in EU funds and aims to leverage up to €20 billion in investments through the Western Balkans Guarantee Facility. Since 2021, the European Commission has committed €1.25 billion to the Green Agenda, focusing on technical assistance, energy efficiency, renewable energy, phasing out coal, and environmental management investments (European Commission, 2023). European Commission President Ursula von der Leyen reaffirmed this commitment to mitigating transition fatigue, who stated, "Our €30 billion Investment Plan for the Western Balkans is delivering, with this new €1.2 billion package. It strengthens regional connections with quality infrastructure and brings us closer together, which is our ultimate goal."[7] These initiatives underline the EU's acknowledgement of the significant environmental and economic challenges faced by the Western Balkans. The region remains dependent on coal, struggles with severe pollution, and suffers from ageing infrastructure. Nevertheless, Western Balkan countries are striving to align with the EU by meeting environmental standards, as outlined in the European Green Deal.

[7] European Commission announces additional €1.2 billion investment package for infrastructure and support to entrepreneurship in the Western Balkans—European Commission.

These efforts come when the EU is undergoing major economic and social transformations, further intensifying the region's push toward integration.

Under the EU's revised enlargement methodology adopted by the European Commission (2020b), Cluster 4, titled "Green Agenda and Sustainable Connectivity," encompasses Chapter 27, "Environment and Climate Change." This chapter emphasises helping candidate countries align with EU standards across various environmental areas, such as air and water quality, waste management, nature conservation, industrial pollution control, chemicals, noise, climate change, and civil protection. The alignment process involves adapting to over 200 EU legal acts, making Chapter 27 one of the most resource-intensive and institutionally demanding in the accession process (European Commission, 2024b). It also necessitates significant shifts in societal and industrial practices. Consequently, this endeavour should not be seen merely as compliance with legislative frameworks but as part of a strategic approach that integrates alignment with a transparent regional strategy. The Green Agenda provides a politically oriented framework for candidate countries to outline their strategies within a shared EU-endorsed vision. This Agenda is intrinsically linked to EU accession aspirations, as environmental reforms are central to the *acquis communautaire*. Adhering to EU environmental standards is a critical milestone in progressing toward membership. Recognising its dual role, the European Commission acknowledges that the Agenda helps Western Balkan (WB) countries meet EU environmental and climate goals and is a pivotal element of the broader enlargement process (European Commission, 2020c). The transition to a green economy is widely seen as a way to strengthen regional integration into the EU by showcasing candidates' commitment to sustainable development and economic modernisation. The Agenda's first Implementation Report (2022), prepared by the Regional Cooperation Council (2023), shows that countries are gradually implementing their priorities and action plans. While progress monitoring is still limited due to the short implementation timeframe, the report indicates a general commitment across countries to adhere to the Agenda's goals. The implementation process has been acknowledged as gradual and incremental, with its societal and economic impacts not yet fully evident.[8] Nonetheless, measurable progress is reported in several areas. For example, five of the

[8] Civil organisations have largely criticised the process of monitoring and assessment due to the lack of transparency and involvement of key stakeholders. More here: This year's Green Agenda Action Plan update must be participatory - Bankwatch.

six WB economies have enacted or are in the process of enacting climate-related legislation,[9] and all have set energy and climate targets in Nationally Determined Contributions or government strategies.

Specific achievements include North Macedonia and Albania advancing renewable energy, particularly in hydropower and solar energy (International Energy Agency, 2021). Despite environmental concerns about hydropower, Albania generates most of its electricity from renewable sources and is exploring new renewable technologies.[10] North Macedonia has implemented laws to encourage solar energy investments, aiding its greenhouse gas emission reduction goals. All WB nations are committed to cutting emissions under the Paris Agreement. However, the region's energy market remains relatively detached from the EU, experiencing price volatility exacerbated by the Ukraine conflict. For instance, North Macedonia declared an energy crisis (Nikolov et al., 2024). Coal-fired power plants still dominate energy production in the region, with lignite accounting for around 70% of domestic electricity generation, except in Albania, where nearly 100% comes from renewables like hydropower and solar.[11] This disparity in energy resources contributes to widespread energy poverty, with up to 40% of the population affected, compared to just 10% in the EU (Robić, 2016). High energy costs have sparked political tensions and protests, as households often spend over 10% of their income on energy (Esser et al., 2018). To address these challenges, countries are devising integrated energy and climate plans (see Fig. 20.1). Additionally, they are adopting climate adaptation strategies, promoting nature-based solutions, and supporting initiatives for sustainability. Although legislative and regulatory frameworks are evolving steadily, implementing them remains a significant hurdle. Differences in economic structures and stages of EU integration mean each country follows its unique transition path, making the Agenda's implementation uneven and complex compared to neighbouring EU member states.

[9] Albania adopted the Law on Climate Change, Bosnia and Herzegovina still has not started working on the Climate Law. Kosovo* and North Macedonia are currently drafting the Law on Climate Change with the Energy Community Secretariat" active support. Montenegro is preparing a new Law on Protection against Negative Impacts of Climate Change with the aim to transpose elements of EU Climate Law. Serbia adopted its Law on Climate Change in March 2021.

[10] The implementation of hydropower plants has significant impacts on the environment by impacting on water quality and life as have been underlined by several external observatory and NGOs—Hydropower in Albania: Vanishing rivers strangle rural life.

[11] According to the data of International Energy Agency (IEA), Albania is ranked as the first (together with Iceland) in Europe in renewables share of electricity generation in 2022.

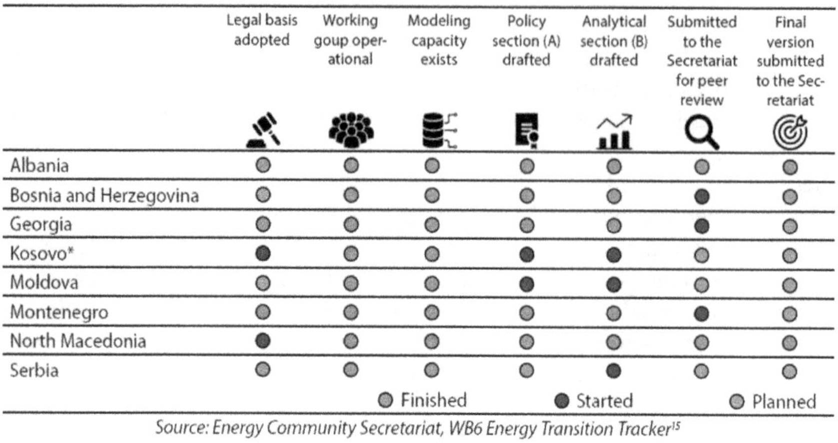

Fig. 20.1 State of National Energy and Climate Plans preparation (*Source* Regional Cooperation Council[12])

20.5 Remaining Gaps

Although progress has been uneven across the region, significant challenges remain. For example, the region's water and waste management systems are often inadequate or poorly maintained (European Environmental Agency, 2022). Untreated wastewater is frequently discharged into rivers, and illegal waste dumping is widespread (World Bank, 2018). The River Drina, heavily polluted by industrial waste, exemplifies these issues, while waste recycling rates in the WB fall far short of EU standards. A lack of modern waste management infrastructure presents a critical problem requiring substantial investment. Enhanced efforts are needed to upgrade wastewater treatment facilities and adopt sustainable waste management practices, such as recycling and waste-to-energy projects.

In addition to water and waste management issues, air pollution poses one of the most urgent problems, especially in major metropolitan areas like Sarajevo, Skopje, and Belgrade (World Bank, 2021). These cities regularly record air quality levels exceeding EU safety limits due to coal-based energy and heating systems, outdated industrial processes, and vehicular emissions (World Health Organization, 2020). Coal-fired power plants, particularly prevalent in Serbia, Bosnia and Herzegovina, and Kosovo, are significant contributors to poor air quality. This pollution results in high rates of respiratory diseases and premature deaths. Notably, Serbia leads European countries in pollution-related mortality rates and ranks ninth globally, surpassing even India, according to the World Bank. A 2020 study by the European Commission's Joint Research Centre emphasised the need for the WB to prioritise

[12] Regional Cooperation Council|Green Agenda for the Western Balkans Action Plan—Implementation Report 2022.

reductions in air pollutants and greenhouse gas emissions, though alignment with the EU's *acquis* on environment and climate remains at an early stage.[13]

Energy poverty is another pressing issue in WB. Programs aimed at addressing this challenge, such as financing schemes for household renovations and ensuring basic living standards, are still in their infancy. Despite the Tirana Declaration's[14] focus on the EU's €1 billion Energy Support Package, which aims to support vulnerable families, SMEs, and energy transitions, many households allocate a significant portion of their income to energy expenses. The transition to renewable energy must be managed carefully to avoid exacerbating social inequality (UNDP, 2015). Affordable energy prices are critical during this shift to ensure equity. As highlighted by the energy crises following the Russian invasion of Ukraine in 2022, energy provision has become a geopolitical issue. Some WB countries, like Serbia, have hesitated to align their energy systems with EU standards. Serbia's reliance on Russian gas remains high, averaging 74% dependency between 2009 and 2019. While Serbia has opened Chapter 15—Energy in its EU integration process to modernise its energy sector, it continues to benefit from discounted Russian natural gas delivered via the Black Sea, Turkey, and Bulgaria, bypassing Ukraine (Brkić, 2024).

The region's vulnerability to climate change is also increasing, with more frequent and severe weather events, such as floods and droughts. The devastating floods of 2014 in Serbia and Bosnia and Herzegovina underscored the WB's lack of resilience to climate-related disasters. These events are becoming more common and intense, as demonstrated by the floods and landslides in Bosnia and Herzegovina in October 2024, which caused extensive infrastructure damage and human casualties. Such disasters have already cost the region billions of euros and displaced thousands of people. Agriculture, a cornerstone of many WB economies, faces significant risks due to shifting weather patterns and reduced water availability. Without improved water resource management and agricultural risk mitigation, the region is likely to experience a further 20% decrease in precipitation and an increase in dry days, exacerbating vulnerabilities (Županić et al., 2021).

20.6 Conclusions

The WB is making significant strides toward adopting a sustainable model to transition from a carbon-dependent to a carbon–neutral society. These efforts are largely guided by the Green Agenda for the Western Balkans under the EU's supervision and economic support (Berisha, 2025). However,

[13] JRC Publications Repository - Status of air pollutants and greenhouse gases in the Western Balkans (europa.eu).

[14] The so-called Tirana declaration was adopted during the EU-Western Balkans Summit of 2022 by the European Union (EU) and its Member States in consultation with Western Balkans leaders and in the presence of regional and international stakeholders. More info here: tirana-declaration-en.pdf.

the path is fraught with challenges, particularly in reducing coal reliance, improving waste and water management, and addressing energy poverty. In parallel, the EU integration process demands extensive structural reforms, requiring additional efforts to meet EU standards. These dual challenges—transitioning to sustainability and advancing EU integration—could become mutually reinforcing if approached with unwavering political commitment. This commitment was emphasised in the Hamburg Declaration on the Green Agenda for the Western Balkans,[15] issued in Germany on October 8, 2024. WB6 affirmed their "commitment to the full and timely implementation of the Green Agenda for the Western Balkans and its Action Plan 2021–2030," underscoring the importance of accelerating efforts, fostering cross-sectoral collaboration, and embracing multi-level governance.

A Regional Adaptation Strategy, expected by 2026, aims to serve as a cornerstone for this transition. However, while progress is evident, the Agenda and its Action Plan could benefit from further refinement. Implementing and monitoring the Agenda has often been criticised as overly insular and technocratic. The Regional Cooperation Council's role in overseeing progress has also faced scrutiny. Outsourcing the monitoring phase to regional research centres, universities, and NGOs could enhance transparency and enable meaningful contributions from civil society, local communities, businesses, and youth. Additionally, translating the collaborative principles of the Hamburg Declaration into actionable reforms will be vital. Strengthening regional governance frameworks and institutional capacities will be key to addressing persistent challenges. Moreover, to accompany these dual challenges of EU integration and transition, it seems crucial to focus on improving coordination between different governance levels in a multi-level perspective (including cooperation with the EU). Notably, the involvement of various stakeholder categories seems paramount, as well as increasing transparency of the monitoring and evaluation processes. It can be summarised that the WB faces a critical juncture where the dual challenge of green transition and EU integration can either synergies or result in fragmented efforts (Berisha, 2025). The question remains whether the WB approach to "green integration" can serve as a model for broader EU integration processes or if it will persist as an isolated, inadequately coordinated initiative. As Mitrovic (2022) stated, the agenda can be a dead letter of genuine opportunity, and it's up to regional actors to take the momentum toward green-based EU integration.

[15] hamburg-declaration-on-the-green-agenda-for-the-western-balkans_1729083604.pdf.

References

Aliaj, J., & Sulmina, V. (2022). European Union integration—Perspectives and challenges of Western Balkans. *Interdisciplinary Journal of Research and Development, 9*(4.S1), 85. https://doi.org/10.56345/ijrdv9n4s111

Battali, L. (2023). Framing the just transition in Kosovo. *Kosovo Foundation for Open Society.* Retrieved from 65084be18b210585324254.pdf

Berisha, E. (2018). *The evolution of spatial planning systems in the Western Balkan Region: Between international influences and domestic actors* (Doctoral dissertation). Politecnico di Torino. Retrieved from https://hdl.handle.net/11583/2707105

Berisha, E. (2023). The Adriatic Ionian region: Challenges and opportunities in the time of geopolitical tensions and the rule of EUSAIR. *Economia Marche-Journal of Applied Economics, 42*(1). https://doi.org/10.57638/1120-9593AIREGION

Berisha, E. (2025). *Moving towards an inclusive Green Agenda in the Western Balkans.* Foundation for European Progressive Studies (FEPS). https://doi.org/10.13140/RG.2.2.12092.76162

Berisha, E., & Cotella, G. (2021). Territorial development and governance in the Western Balkans. In E. Berisha, G. Cotella, & A. Solly (Eds.), *Governing territorial development in the Western Balkans.* Advances in Spatial Science. Springer. https://doi.org/10.1007/978-3-030-72124-4_2

Berisha, E., Cotella, G., & Solly, A. (2021). Governing territorial development in the Western Balkans: Conclusive remarks and future research perspectives. In E. Berisha, G. Cotella, & A. Solly (Eds.), *Governing territorial development in the Western Balkans* (pp. 357–381). Springer. https://doi.org/10.1007/978-3-030-72124-4_17

Brkić, D. (2024). Serbian energy sector in a gap between East and West. *Energy Exploration & Exploitation, 42*(1), 330–340. https://doi.org/10.1177/01445987231215445

CEE Bankwatch Network. (2024). *Local authority planning for a just transition: Why Western Balkans coal regions should create their own redevelopment plans and how it can be done.* Retrieved from 2024_04_Local-authority-planning-for-a-just-transition.pdf

Djatkov, D. (2024). *Policy report on the green transition in the Western Balkans.* Retrieved from Report_Green-Transition_final.pdf

Esser, S., Schulz, S., Amon, A., Littlecott, C., & Reitzenstein, A. (2018). Energy sector trends in the Western Balkans. In *High carbon lock-in vs. low carbon opportunity in the Western Balkans: Critical investments and the EU accession process* (pp. 13–23). E3G. http://www.jstor.org/stable/resrep21858.7

European Commission. (2020a). *Green agenda for the Western Balkans.* Retrieved from https://ec.europa.eu/commission/presscorner/detail/en/ip_20_1812

European Commission. (2020b). *A more credible, dynamic, predictable, and political EU accession process.* Retrieved from https://ec.europa.eu/commission/presscorner/detail/en/ip_20_181

European Commission. (2020c). *Guidelines for the implementation of the Green Agenda for the Western Balkans.* Retrieved from https://neighbourhood-enlargement.ec.europa.eu/system/files/2020-10/green_agenda_for_the_western_balkans_en.pdf

European Commission. (2024a). *Communication on EU enlargement policy.* Retrieved from 2024 Communication on EU Enlargement Policy - European Commission

European Commission. (2024b). *Chapter 27*. Retrieved from https://neighbourhood-enlargement.ec.europa.eu/enlargement-policy/conditions-membership/chapters-acquis_en

European Environment Agency. (2022). *Municipal waste management in the Western Balkan countries*. Retrieved from https://www.eea.europa.eu/publications/municipal-waste-management-in-western

Imami, F., Bejko Gjika, A., Giacometti, A., & Tapia, C. (2023). *Just green transition in the Western Balkans: Overcoming transition fatigue*. Retrieved from https://greenforcetwinning.net/wp-content/uploads/2024/02/POLICY-BRIEF-OVERCOMING-TRANSITION-FATIGUE.pdf

International Energy Agency. (2021). *Energy transition in the Western Balkans: Prospects and challenges*. Retrieved from https://www.iea.org/reports/western-balkans-energy-transition

Knez, S., Štrbac, S., & Podbregar, I. (2022). Climate change in the Western Balkans and EU Green Deal: Status, mitigation, and challenges. *Energy, Sustainability and Society, 12*(1). https://doi.org/10.1186/s13705-021-00328-y

Ministria e Financave dhe Ekonomisë. (2021). *Strategjia Kombëtare për Zhvillim dhe Integrim 2021–2027* [National Strategy for Development and Integration 2021–2027]. Ministria e Financave dhe Ekonomisë. Retrieved from https://www.financa.gov.al

Ministry of Economy. (2023). *Just transition roadmap*. Ministry of Economy of the Republic of North Macedonia. Retrieved from economy.gov.mk

Mitrovic, S. (2022). The Green Agenda for the Western Balkans—A dead letter or a genuine opportunity? *European Policy Center, CEP Insight*. Retrieved from The-Green-Agenda-for-the-Western-Balkans.pdf

Morina, E. (2022). *Power of need: Energy security in the Western Balkans*. Retrieved from https://ecfr.eu/article/power-of-need-energy-security-in-the-western-balkans/

Nikolov, M., Garvanlieva Andonova, V., Gotovska, E., Giacometti, A., & Moodie, J. (2024). *Building up governance pathways for just energy transition: The case of North Macedonia*. Retrieved from POLICY-BRIEF-NO-3-FINAL-DESIGNED.pdf

OECD. (2022). *Multi-dimensional review of the Western Balkans: A green recovery in North Macedonia*. Retrieved from https://www.oecd-ilibrary.org/docserver/4223758a-en.pdf

Pinnavaia, L., & Berisha, E. (2021). The role of cross-border territorial development: Evidence from Albania. In E. Berisha, G. Cotella, & A. Solly (Eds.), *Governing territorial development in the Western Balkans: Challenges and prospects of regional cooperation* (pp. 309–332). Springer. https://doi.org/10.1007/978-3-030-72124-4_15

Popović, G., Erić, O., & Stanić, S. (2020). Effects of technological changes and trade liberalisation on industrial development in the Western Balkan countries. *Economic Research-Ekonomska Istraživanja, 34*(1), 2873–2889. https://doi.org/10.1080/1331677X.2020.1845967

Qorraj, G., Hajrullahu, A., & Qehaja, D. (2024). Absorption capacity and the European integration of the Western Balkans. *Regional Science Policy & Practice, 16*(8), Article 100043. https://doi.org/10.1016/j.rspp.2024.100043

Regional Cooperation Council. (2023). *Green agenda for the Western Balkans action plan—Implementation report 2022*. Retrieved from https://www.rcc.int/pubs/162/green-agenda-for-the-western-balkans-action-plan--implementation-report-2022

Robić, S. (2016). *Energy poverty in South East Europe: Surviving the cold*. Retrieved from https://www.door.hr/wp-content/uploads/2016/10/Energy-Poverty-in-South-East-Europe_Surviving-the-Cold.compressed.pdf

Shaker, Y., & Berisha, E. (2025). Just green transitions: Between terminological inexactitudes, conceptual fragmentation, and the exigency for a theoretical framework. *European Spatial Research and Policy, 32*(1), 121–138. https://doi.org/10.18778/1231-1952.32.1.05

UNDP. (2015). Climate change adaptation in the Western Balkans. *United Nations Development Programme*. Retrieved from https://www.europe.undp.org/

UNDP. (2023). Just transition leading to a more prosperous society. *Conference "Dialogue for Just Transition in Montenegro"*. Retrieved from https://www.undp.org

UNECE. (2024). *Just transition assessment of Albania: Sector decarbonization and the just transition framework applied to coal mining*. Retrieved from https://unece.org/sites/default/files/2024-04/UNECE%20ALBANIA%20Just%20Transition%20And%20Decarbonization%20Report%20FINAL.pdf

World Bank Group. (2022). *Multilateral development banks' collaboration: Infrastructure investment project briefs*. Port of Durrës. Retrieved from https://ppp.worldbank.org/public-private-partnership/sites/ppp.worldbank.org/files/2022-03/MDB_Brief_AlbaniaPortDurress.pdf

World Bank. (2018). *Western Balkans wastewater and solid waste management report*. Retrieved from https://www.worldbank.org/en/region/eca/brief/western-balkans-environment

World Bank. (2021). *Going green means cleaner air, healthier living in the Western Balkans*. Retrieved from https://www.worldbank.org/en/news/feature/2021/05/17/going-green-means-cleaner-air-healthier-living-in-the-western-balkans

World Health Organization. (2020). *Air pollution in Europe: Health impacts and policy responses*. Retrieved from https://www.who.int

Županić, F. Ž, Radić, D., & Podbregar, I. (2021). Climate change and agriculture management: Western Balkan region analysis. *Energy, Sustainability and Society, 11*, 51. https://doi.org/10.1186/s13705-021-00327-z

PART V

Just Green Transitions Practice in Europe and Beyond

CHAPTER 21

Just Green Transitions in the EU: Comparing the Cases of Italy, Sweden, Poland, and the Netherlands

Erblin Berisha, Giancarlo Cotella, Erika Puntillo, and Yahya Shaker

21.1 Introduction

The urgency of addressing climate change has placed sustainability at the forefront of policy agendas worldwide (van Bommel & Höffken, 2023). Within this global movement, the European Union (EU) has led in promoting the European Green Deal (2019), aiming to achieve climate neutrality by 2050. According to the European Commission (2019), the European Green Deal is a new growth strategy that seeks to transform the EU into a just, fair and prosperous society (Berisha, 2025a, 2025b; Shaker & Berisha, 2025) with a modern, resource-efficient and competitive economy where there are no net emissions of greenhouse gases in 2050 and where economic growth is decoupled from resource use (Cotella et al., 2016). However, to pursue the transition effectively remains a complex socio-technical problem (Valkenburg & Cotella, 2016). As the EU pursues its ambitious decarbonisation goals,

E. Berisha (✉) · G. Cotella · E. Puntillo · Y. Shaker
Interuniversity Department of Regional and Urban Studies and Planning (DIST), Politecnico di Torino, Turin, Italy
e-mail: erblin.berisha@polito.it

G. Cotella
e-mail: giancarlo.cotella@polito.it

E. Puntillo
e-mail: erikapuntillo1@gmail.com

Y. Shaker
e-mail: yahya.shaker@polito.it

E. Berisha et al. (eds.), *The Palgrave Handbook of Just Green Transitions in the Western Balkans and Beyond*,
https://doi.org/10.1007/978-3-031-95075-9_21

there is a pressing need to ensure this transition remains socially equitable and spatially balanced (Berisha, 2025a, 2025b). Therefore, the just transition concept has emerged as a cornerstone of the EU's approach, reflecting a commitment to achieving environmental targets without leaving any region or community behind (García-Mira et al., 2024). Accordingly, the EU just green transitions (Shaker & Persico, 2024) seeks to balance economic, social, and environmental priorities, focusing on regions heavily reliant on fossil fuels and carbon-intensive industries (Koasidis et al., 2023). For these regions, the shift to a green economy represents both an opportunity and a challenge, as it involves significant structural changes and the potential for job displacement and economic restructuring (Theodoropoulou et al., 2024). To address these complexities, the EU has developed a range of supportive policies and financial mechanisms, including the Just Transition Mechanism (JTM) and the Just Transition Fund (JTF), which aim to support affected regions in diversifying their economies, investing in green industries, and fostering workforce adaptability (Berisha, 2025a, 2025b). This chapter provides a comparative analysis of the Just Green Transitions (JGT) frameworks in four diverse EU geographically located countries—Italy (South), Poland (East), Sweden (North), and the Netherlands (West). Each of these countries has developed its "transition" strategies shaped by distinct economic, ecological, and social contexts. Italy and Poland, having a strong industrial base in high-carbon sectors, have focused on regional strategies to manage socio-economic impacts towards transition. Sweden's approach centres on ecological connectivity and biodiversity, while the Netherlands emphasises regional adaptation, innovation, and stakeholder engagement to foster a sustainable industrial transition.

The chapter is structured as follows. After this brief introduction, Sect. 21.2 describes the JGT within EU Policies: It elaborates on the EU's foundational JGT policies, including the JTM and Territorial Just Transition Plans (TJTP), and their role in guiding Member States. Section 21.3 gives a detailed analysis of the JGT frameworks adopted in Italy, Poland, Sweden, and the Netherlands, complemented by comparative tables highlighting differences in policy focus, regulatory frameworks, and key actors. After that, Sect. 21.4 compares and discusses the identified commonalities and differences across the four cases, evaluating the effectiveness of each approach in balancing environmental goals with socio-economic needs. Finally, the conclusion section provides a synthesis of the insights from the comparative analysis, discusses the importance of tailored approaches within the EU's overarching JGT policies framework, and presents recommendations for future JGT policy efforts.

21.2 Just Transition Within EU Policies

The EU's Just Transition framework is part of the European Green Deal, aiming to support sustainable development across Member States by facilitating a climate-neutral economy while minimising socio-economic disruption (Berisha et al., 2025; Moesker & Pesch, 2022). Recognising that the transition from fossil fuels and carbon-intensive industries can lead to economic shifts that disproportionately affect certain regions (Shaker & Berisha, 2024), the EU established the JTM to provide targeted support (European Commission, 2019). This JTM is designed to help Member States address the structural, economic, and social challenges associated with shifting to a low-carbon economy, ensuring no region or community is left behind. The JTM operates through three main pillars: the JTF, the Invest EU Scheme, and the Public Sector Loan Facility (PSLF) (García-Mira et al., 2024). Each component of the JTM has a specific technical function in promoting transition (Fig. 21.1).

The JTF is the foundational element of the JTM, with a substantial allocation of €17.5 billion for 2021–2027 (European Commission, 2019). This fund supports regions most vulnerable to economic disruptions from the green transition, especially those with industries that rely heavily on fossil fuels, such as coal, peat, and oil shale. The JTF provides these regions with financial support to diversify their local economies, fund workforce reskilling programmes, and improve social cohesion by investing in projects that retain community engagement and support local identity in the face of industrial shifts (Bujdosó et al., 2022). Unlike other EU funds, the JTF is "spatial dependent", meaning that the funds can be used only in the regions/areas with a Territorial Just Transition Plan approved underlined as a priority by the Member States and the European Commission (Moesker & Pesch, 2022). As for the sectors concerned, the JTF is often directed towards projects that promote alternative industries, such as renewable energy, sustainable agriculture, and eco-tourism, which offer new economic opportunities for regions undergoing significant transformation. By funding job training programmes, the JTF also helps workers transition from carbon-heavy sectors to roles in green industries, such as solar and wind energy or sustainable construction, providing long-term employment stability in sectors aligned with the EU's

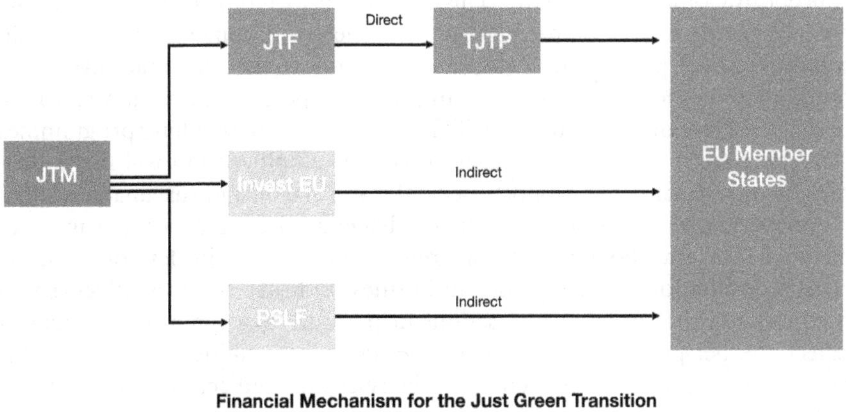

Fig. 21.1 Financial mechanism for Just Green Transitions (*Source* authors' own elaboration)

climate goals. The Invest EU Scheme, a complementary pillar of the JTM, broadens the scope of support by leveraging private investments in sustainable projects across the EU. Where the JTF provides direct funding, the Invest EU Scheme mobilises private finance to develop renewable energy infrastructure, digital technologies, social development projects, and more. The difference is that the funds guaranteed by the Invest EU package can also be used in territories not targeted by any Territorial Just Transition Plan. By using financial instruments such as guarantees and loans, Invest EU helps reduce investment risk, making sustainable projects more attractive to private investors. This risk-sharing model ensures that regions can access sufficient funding for large-scale projects essential for a robust JGT.

Finally, the Public Sector Loan Facility is the third pillar of the JTM and specifically addresses projects that fall outside the scope of private investment but are vital for a comprehensive green transition. This scheme provides preferential loans to public sector bodies, with the EU budget and the European Investment Bank (EIB) co-financing the facility. This allows municipalities, regions, and other public entities to undertake essential projects, such as upgrading infrastructure for energy efficiency, retrofitting social housing with green technologies, and developing sustainable public transportation networks. By financing public initiatives that directly benefit citizens, the Public Sector Loan Facility enhances community resilience and supports a more inclusive transition. Theoretically, these three pillars are designed to work in tandem, creating a coordinated support system that allows Member States to tailor their JGT strategies to specific regional needs. To access JTM funding, each Member State must prepare and submit TJTPs, which must be place-based and region-specific blueprints for managing the transition across the different EU countries. The development of the TJTPs is supposed to be a comprehensive process that begins with assessing each region's socio-economic context and identifying sectors most affected by the transition (European Commission, 2023). This assessment includes analyses of employment dependency on high-emission industries, the vulnerability of certain communities to job displacement, and the capacity for economic diversification. Based on these assessments, Plans prioritise specific actions tailored to the needs of each region. For instance, TJTPs may prioritise reskilling programmes if a region has a high concentration of workers employed in fossil fuel industries or provide targeted support for SMEs and startups in sustainable sectors if economic diversification is a primary challenge. In the paper, the involvement of local stakeholders is considered a crucial factor in developing these plans; consultations with regional authorities, industry representatives, social partners, and civil society organisations help ensure that the proposed actions reflect the perspectives and needs of the local communities. This collaborative process is supposed to strengthen the relevance and acceptance of TJTPs, fostering a sense of ownership and long-term commitment to the green transition among local stakeholders. However, the level of public engagement has varied across the different countries' experiences (García-Mira et al., 2024).

Once the Plans are completed, they undergo a review to ensure alignment with EU-wide climate and social goals. Approved TJTPs receive funding from the JTF, often combined with national resources and additional EU funds, maximising the impact of the financial support. Progress is closely monitored, and performance indicators are set to track the effectiveness of each TJTP in meeting the specified goals. This continuous monitoring allows for adjustments as regional needs evolve, ensuring that the JGT remains responsive to sub-regional dynamics and emerging challenges.

21.3 Presentation and Comparison of the Four Case

This section compares the experiences of four different EU Member States in conceptualising and operationalising their TJTPs, showing a high heterogeneity across them (Fig. 21.2). Indeed, when analysing these four countries implementing their TJTPs, each conceptualises the JGT to reflect its industrial, social, and environmental contexts. Italy and Poland, for instance, focus more on reducing dependency on coal while protecting jobs in fossil-fuel-dependent regions, ensuring economic stability and social support. Sweden, instead, combines a strong interest in green innovation with efforts to reduce social inequalities, aiming for climate neutrality by 2045. Last but not least, the Netherlands pursues a balanced strategy targeting diverse geographical needs, emphasising economic resilience and environmental sustainability.

In Italy, the JGT is seen as essential for achieving sustainable industrial development in line with the European Union's climate objectives (Niewitała-Rej, 2024). According to the data presented by the International Energy Agency (2024), the country is the third largest greenhouse gas emitter in the EU, and its energy sector is the main contributor to total greenhouse gas emissions in 2022.[1] That being said, Italy's approach prioritises decarbonising energy-intensive industries, advancing the circular economy, and fostering a socially equitable transition. Key objectives include job protection, workforce upskilling, and ensuring that no region is overlooked in the transition. The TJTPs focus on regions heavily affected by the move away from fossil fuels, specifically those with economic reliance on high-carbon industries. Italy's strategy identifies the urban area of Taranto in Apulia and Sulcis Iglesiente in southwestern Sardinia as the hardest-hit territories. More specifically, Taranto, home to one of Europe's largest steel mills and one of Italy's largest coal-fired power plants, is a significant contributor to industrial pollution and employs around 10,000 people directly in steel production, with another 10,000 working in related industries. In Sulcis Iglesiente, Italy's last coal mine, Monte Sinni, is set to cease coal production by 2025, impacting 350 jobs in a region already challenged by high youth unemployment, low income, and an ageing population. The area is already characterised by a high

[1] For more info please visit: Europe – Countries & Regions – IEA.

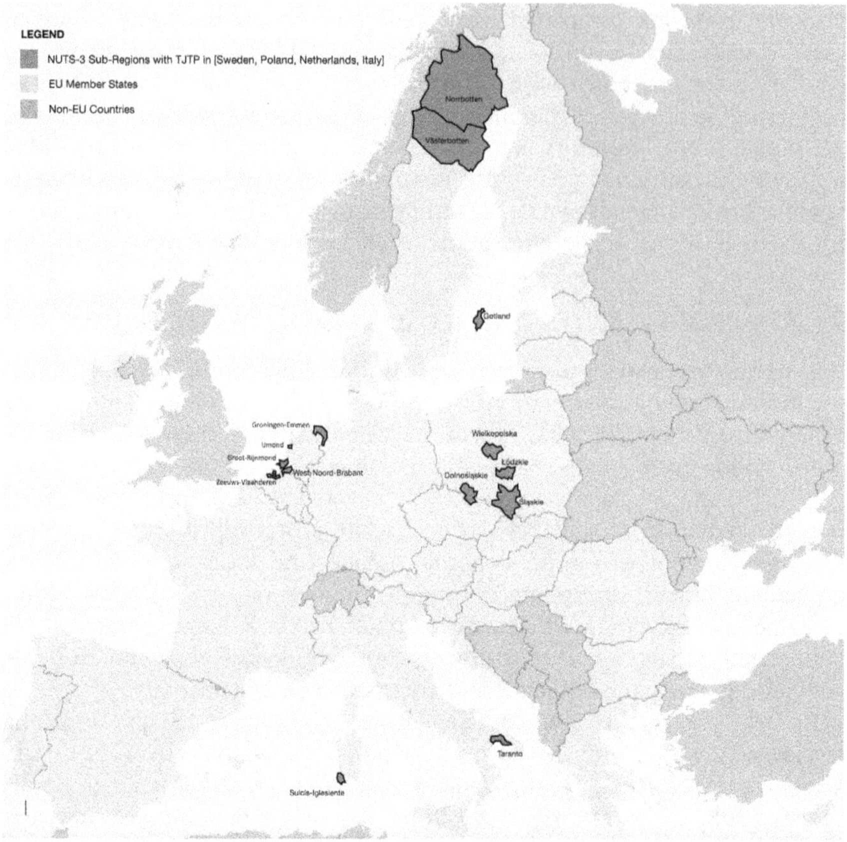

Fig. 21.2 TJTPs in Italy, Poland, Sweden, and the Netherlands (*Source* authors' own elaboration)

percentage of elderly inhabitants, few young graduates, high youth unemployment (35.7%), low per capita income and an overall low quality of life. For these reasons, Italy's National JTF Programme concentrates investment in these two areas, aligning the Territorial Plans with the National Integrated Plan for Energy and Climate (PNIEC). The Italian TJTPs are coordinated with regional programmes financed by the European Regional Development Fund (ERDF) and the European Social Fund Plus (ESF+), as well as other local initiatives, like the Sulcis Plan and Institutional Development Agreement of Taranto.[2] The TJTPs in Italy are closely aligned with the broader national and regional strategies for industrial transition. Accordingly, for each identified area, the TJTPs are designed and implemented consistently with the

[2] Signed on 30 December 2015, this agreement aims to redevelop and develop the area that affects the municipalities of Taranto, Statte, Massafra, Crispiano and Montemesola.

Integrated National Energy and Climate Plan, which lays down Italy's guidelines to decarbonise its economy and achieve climate neutrality by 2050. The National Integrated Energy and Climate Plan (PNIEC) serves as a key policy instrument that complements the TJTPs. The PNIEC sets out Italy's 2030 energy and climate goals, focusing on renewable energy development and energy efficiency in industrial processes. Furthermore, the National Recovery and Resilience Plan (NRRP) inevitably plays a central role in this coordination, as it outlines Italy's green transition goals in line with the EU's broader climate objectives. The PNRR is linked to the TJTPs, particularly in its sections on energy transition, industrial decarbonisation, and workforce reskilling, ensuring that regional efforts align with national targets for reducing carbon emissions and promoting sustainable industries. In terms of funding, the TJTPs are designed to ensure that the JTF is used effectively in conjunction with other national and regional financial instruments, such as regional operational programmes (POR) under EU structural funds. This multi-level coordination between the TJTPs, NRRP, PNIEC, and JTF ensures that both national and regional objectives are met while addressing local socio-economic and environmental challenges.

In the Polish context, the term JGT does not exist, primarily, it is both called and focused on the Just Transition. The Polish expected multiple transitions (Tarasova, 2024) are coordinated at the National level between four ministries (The Ministry of Development Funds and Regional Policy, the Ministry of Climate and Environment, the Ministry of Industry, and the Ministry of National Assets), managed, conceptualised, and operationalised at the regional level and implemented at a sub-regional level (Ślimko et al., 2021). The Polish transitions are targeting economic diversification by supporting new green sectors such as sustainable manufacturing and promoting social equity through extensive reskilling and upskilling initiatives, but still in need of a cohesive place-based territorial policy (Nowakowska et al., 2021). Poland's Territorial Just Transition Plans have been developed at the sub-regional level of five regions with high reliance on coal and fossil fuel industries, selected for their socio-economic vulnerability and need for significant support (Ślimko et al., 2021). Poland's transition is distributed at the NUTS-3 level in five mining regions (*Voivodeship*) of Śląskie, Małopolska, Wielkopolska, Dolnośląskie, and Łódzkie, and precisely in only four subregions reliant on carbon-intensive industries like hard coal and lignite mining (European Commission, 2023). Śląskie is the largest coal-producing area in the EU, which is undergoing economic diversification, especially in the IT and healthcare sectors. Dolnośląskie is facing several legal and juristically court cases at a cross-border scale with its neighbouring transitioning regions in the Czech Republic and Germany. Wielkopolskie is expected to be the first to shut down the mine in Konin by the next four years. Łódzkie hosts the biggest coal mine in Poland and maybe the EU in Bełchatów, which has been operating since the 1970s and is expected to be shut down by 2035 and has various alternative transitioning scenarios (Ruszkowski, 2021); for instance, to be transformed into a massive artificial lake and maybe hosting the second nuclear power plant to be finished by 2035. Finally, Małopolskie has not been able to allocate the JTF, but it is in the shadow of the transition of Śląskie meaning that the effects of shutting down the mines in Śląskie will affect its sub-regions even if most of the mines are already closed; thus, the TJTP of Śląskie and Małopolskie aims at cleaner production methods and green industry growth around Kraków region. Additionally, the Lubelskie and Pomorskie Regions were initially included in the first discussions on the transition process, but the European Commission ultimately excluded them from the final list due to their lesser socio-economic dependence on coal compared to the other five regions. In the programming period 2021–2027, 76 billion euros from the cohesion funds and €3.85 from the JTF are supporting the Polish regions in promoting economic, social and territorial cohesion and embracing a Green and Digital transition. The Green Transition's main objectives include investing in nuclear power plants and offshore Baltic windmills farm expected to generate up to 51% of electricity from renewables by 2040 in Pomorskie Region (which is not a JTF region nor does it have a TJTP) and new ideas on the horizon to invest in a second power plant in the Bełchatów sub-region in Łódzkie. Other sectoral investments are in Transport decarbonisation and the development of the Trans-European Transport networks. The green dimension falls into the systematic decarbonisation of the major coal-mining sub-regions that were able to allocate the JTF to close most of the mines by 2049 and towards carbon emission reductions (European Commission, 2023). The majority of the JTF is directed to the biggest mining companies, such as PGE while trying to support the miners financially, especially those who might lose their jobs by 2050. The just transition in Poland is claimed to be investing in reskilling and retraining of the miners and more into infrastructure and new green and digital jobs in the IT sector.

In Sweden, the term JGT is not used officially; instead, the discourse is pretty much split between the Clean Industrial Transition, Green and Digital Transition, and the Social Just Transition, a multidimensional effort that addresses technical, social, and territorial aspects to move towards a low-carbon economy with fairness and inclusivity (Moodie et al., 2021). The Swedish expected transitions are coordinated, managed, and conceptualised at the National level (Tillväxtverket—Swedish Agency for Economic and Regional Growth), operationalised at the Regional level and implemented at the Local level, which is subject to a great level of competitiveness (Gärdebo, 2023). Sweden Just Transition is distributed at the NUTS-3 level in three decarbonisation regions (*Region*).

Sweden's TJTPs are centred on the Nordic regions: the steel industry in Norrbotten, the metal industry in Västerbotten, and the cement industry and high-emission sectors and fossil fuel energy production in Gotland Island in the South of Sweden. These three regions were chosen based on their economic dependence on difficult-to-decarbonize industries, and the Swedish TJTPs aim to support these industries in transitioning by promoting green technology investments, reskilling workers, and diversifying local economies. Each TJTP is tailored to address both the technical aspects of transitioning away from high-emission production methods and the social challenges associated with these changes, particularly in regions where jobs are closely tied to carbon-intensive industries. Sweden's plans aim to leverage regional strengths and align with national climate goals, creating a transition that not only reduces emissions but also fosters long-term regional development and socio-economic stability. The Green dimension of the transition seeks carbon neutrality by 2045. The Green Transition in the Nordic Region is being promoted as a digital innovation in various sectors, including green hydrogen energy in northern Sweden (Nordic Council of Ministers, 2021). This transformative endeavour is being realised through close collaboration with both major industrial players as Luossavaara-Kiirunavaara Aktiebolag (LKAB),[3] Svenskt Stål AB (SSAB),[4] Heidelberg Materials Cement Sverige (former Cementa)[5] and small to medium-sized enterprises within the region, ensuring a dynamic partnership that drives sustainable innovation and regional growth. Sweden is investing in Hydrogen Energy in what is called H2-Labs, testing the possibilities for climate-neutral hydrogen production (RISE Research Institutes of Sweden, 2022). Additionally, Sweden is seeking to bring around 100,000 highly skilled workers and their families to the Nordic Regions as the industrial transition focuses on an opening up approach and investing in Green Steel mining, besides the reskilling and retraining programme in the so-called green jobs and reducing social and spatial inequalities, which is contrary to the impact of the so-called green mining in the Saami lands (Larsen et al., 2022). In the programming period 2021—2027, 2.2 billion euros from the cohesion funds and 156 million from the JTF are supporting the Swedish regions in the industrial transformation. Planned initiatives in these regions include RandI, the adoption of clean energy technologies and infrastructure. Key focus areas include exploring technologies for large-scale renewable hydrogen production and carbon capture and storage. These efforts are projected to significantly reduce greenhouse gas emissions, with supported enterprises expected to cut emissions by approximately 42% in CO2 equivalent annually. Additionally, a portion of the fund will be allocated to enhancing Gotland's electricity grid capacity (European Commission, 2023).

The Netherlands conceptualises its JGT as a balanced approach towards sustainability, ensuring social equity, economic inclusivity, and environmental resilience. The country's main objectives are achieving social equity and

[3] LKAB is a state-owned Swedish mining company active in Kiruna and Gällivare mines, in northern Sweden. For more info, please visit: LKAB.

[4] SSAB is a global steel company operating in high-strength steels and related services. For more info, please visit: SSAB.

[5] Heidelberg Materials Cement Sverige (formerly Cementa) is one of Sweden's largest building materials companies, which manufacture and sell cement and offer knowledge on the use of Cement products. For more info, please visit: Heidelberg Materials Cement Sweden.

inclusion, bolstering economic resilience and job creation, and fostering environmental sustainability through community engagement. Accordingly, the Commission has adopted 2022 the Dutch TJTP, making €623 million available under the JTF to support a fair transition to a climate-neutral economy. The TJTPs are developed based on regional economic structures, environmental risks, and social demographics, each tailored to address the specific needs of the area. The Fund supports six Dutch regions—Groningen and Emmen, IJmond, Groot-Rijnmond, Zeeuws-Vlaanderen, West-Noord-Brabant and Zuid-Limburg—helping them move away from an economy centred on fossil-fuel extraction or carbon-intensive industries. For instance, in industrial regions like Rotterdam-Rijnmond, the focus is on economic diversification and emissions reduction, given its dependence on high-carbon industries. In urban centres such as Amsterdam, Rotterdam, and Utrecht, the TJTPs prioritise urban sustainability initiatives, promoting green infrastructure and ensuring equitable access to green jobs. Rural and agricultural areas, like Gelderland and Overijssel, focus on sustainable farming practices, renewable energy adoption, and economic diversification to reduce dependency on traditional industries. Coastal and flood-prone areas, including Zeeland and North Holland, concentrate on enhancing climate resilience and sustainable water management. In addition to economic dependency and environmental vulnerability, the Netherlands also considers social needs when determining which regions to prioritise, supporting areas with higher social vulnerabilities. The TJTPs in the Netherlands are closely coordinated with other industrial transition policies and funding instruments to ensure a cohesive approach to achieving a green economy. Netherlands' Climate Agreement and the Subsidy for Stimulation of Sustainable Energy Production and Climate Transition complement the TJTP by reducing and transitioning from carbon-intensive industries and providing subsidies for renewable energy and CO2 reduction technologies, respectively.

21.3.1 Main National Policy Documents and Strategies

Table 21.1 provides an overview of how different countries are addressing the concept of a JGT through national policy documents and strategies. Each of the four countries takes a different approach, shaped by their specific socio-economic and environmental contexts, while also grappling with some shared challenges. In Italy, the focus is on decarbonisation and ensuring social equity in the transition process. In this respect, the country has identified the implementation of the Integrated Plan for Energy and Climate, the National Strategy for Sustainable Development and the National Strategy for Circular Economy as key in addressing the transition process. Together, they are expected to accompany the country in reducing regional disparities, also tackling challenges such as energy poverty, ensuring affordable access to clean technologies, and, in the meantime, securing investments to finance green infrastructure. Poland, on the other hand, is heavily focused on transitioning away from coal and modernising its industrial sector. In this respect, the country has introduced the National Energy and Climate Plan for 2021–2030, the Responsible Development Strategy and Energy Policy of Poland until 2040. These documents are intended to support the country in transitioning from a carbon-depend to a carbon-free society, considering that it is still deeply reliant on coal for energy and employment. Yet, the process is complicated by the socio-economic impact on coal-dependent regions, such as Silesia, and the challenge of integrating small and medium enterprises (SMEs) into the transition framework, balancing industrial modernisation with decarbonisation efforts. With the introduction of the Swedish Industrial Strategy and the National Energy and Climate Plan, the country emphasises competitiveness, energy efficiency, and further decarbonisation, demonstrating its position as a leader in sustainability efforts. Nevertheless, the country is trying to face risks like deindustrialisation and regional imbalances and ensure a smooth energy transition without exacerbating social or territorial inequalities. Lastly, the Netherlands is focused on industrial decarbonisation and regional adaptation to climate goals. To do so, the country has adopted the National Climate Agreement, which focuses on collaborative efforts across sectors, energy, industry, transport, agriculture, and built environment, the Regional Energy Strategy, which focuses on creating a sustainable energy system benefiting all regional stakeholders, and the National Programme for Sustainable Industry, which aims to lead the country towards sustainability through innovation, technological advancement and collaboration. However, while the policies are forward-looking, challenges such as job displacement, land-use conflicts, and public opposition to certain green initiatives remain significant obstacles.

21.3.2 Organisational Framework and Regulatory Environment

Table 21.2 examines the regulatory and legal settings that each country has outlined to make the transition more feasible. The table showcases distinct national priorities, responsible institutions, and strategies for adapting to emerging trends within the scope of sustainable development and climate goals. For instance, in Italy, Decree-Law No. 80 of 2021, overseen by the then Ministry for the Ecological Transition, focuses on strengthening public administrative capacity to implement JGT policies effectively. The framework is closely aligned with the National Recovery and Resilience Plan, reflecting Italy's emphasis on leveraging EU recovery funds to ensure a transition that supports both climate and economic recovery goals. Poland partially addresses its transition through the Act on Electromobility and Alternative Fuels (2018), for instance. The law is primarily centred on promoting the use of alternative fuels and the development of electromobility, aiming at reducing emissions in the transport sector and advancing the demand for new technologies and

Table 21.1 Main National Policy documents and strategies on JGT (*Source* authors' elaboration)

Country	*Key documents*	*Policy focus*	*Challenges*
Italy	Integrated Plan for Energy and Climate, National Strategy for Sustainable Development, National Strategy for Circular Economy	Decarbonization, social equity	Regional disparities, energy poverty, securing investments
Poland	National Energy and Climate Plan for 2021–2030, Responsible Development Strategy, Energy Policy of Poland until 2040	Transition from coal, industrial modernisation	SME integration, managing socio-economic impacts in coal regions
Sweden	Swedish Industrial Strategy Basis for the government's upcoming climate action plan and climate report National Energy and Climate Plan (NECP) for Sweden	Competitiveness, decarbonisation, greenhouse gas reduction, energy transition, energy efficiency	Avoid de-industrial, reduce territorial unbalance and ensure energy transition
Netherlands	National Climate Agreement, Regional Energy Strategy, National Programme for Sustainable Industry	Industrial decarbonization, regional adaptation	Job displacement, land-use conflicts, public opposition

infrastructure. Sweden takes a leadership role with its Climate Policy Framework (2017), managed by the Swedish Government in collaboration with the Climate Policy Council and the Swedish Environmental Protection Agency. The framework targets net-zero greenhouse gas emissions by 2045 and a full transition to renewable energy by 2040; to achieve these goals, the framework integrates climate goals into all sectors and requires the government to adopt a climate action plan every four years. The Netherlands, under the Climate Act (2019) and the Ministry of Climate and Green Growth, aims for climate neutrality by 2050, with an interim target of 49% less greenhouse gas emissions by 2030. The law includes innovative measures such as the introduction of a CO_2 tax for large emitters and expanded support for the circular economy (such as subsidies for CO_2 reducing technologies), illustrating a forward-thinking approach to integrating economic and environmental sustainability into its transition strategy.

Table 21.2 National Regulatory/Legal Frameworks on JGT (*Source* authors' elaboration)

Country	*Law name*	*Responsible body*	*Main focus*	*Adaptation to emerging trends*
Italy	Decree-Law No. 80 of 2021 "Urgent Measures for Strengthening the Administrative Capacity of Public Administrations for the Implementation of the National Recovery and Resilience Plan (PNRR) and for the Efficiency of Justice", 2021	Ministry for the Ecological Transition	Strengthening public capacity for implementing JGT policies	New framework specifically aligned with the PNRR
Poland	Act on Electromobility and Alternative Fuels, 2018	Ministry of Climate and Environment	Supporting alternative fuels and clean transport	Updated to promote zero-emission transport infrastructure
Sweden	Sweden's Climate Policy Framework, 2017	Swedish Government, Climate Policy Council, Swedish Environmental Protection Agency	Achieve net-zero greenhouse gas emissions by 2045	Update the climate action plan
Netherlands	Climate Act, 2019	Ministry of Climate and Green Growth	Achieving climate neutrality by 2050	Introduces CO_2 tax and expanded support for circular economy

21.3.3 *Instruments and Levels of Just Green Transition*

Table 21.3 outlines a comparative analysis of the main instruments introduced by each country. It distinguishes between central and regional levels and details the specific instruments, roles, and competencies involved in the implementation process. Generally, national strategies and funding instruments set overarching frameworks at the central level, ensuring alignment with regional and local plans. These include measures such as Italy's National JTF Programme and the Netherlands' Subsidy for Stimulation of Sustainable Energy Production and Climate Transition, emphasising financial support and regulatory incentives. At the regional level, localised strategies focus on implementing transition goals in vulnerable or industrial areas, such as Poland's

TJTPs and Sweden's Regional Climate and Energy Strategies. More in detail, these plans prioritise economic diversification, coal phase-out, and regional development through close collaboration with national policies and local authorities. Importantly, the table also highlights the critical linkages between central and regional efforts, showcasing coordinated approaches to achieving sustainable development and addressing region-specific challenges.

Table 21.3 Instruments and Levels of JGT (*Source* authors' elaboration)

Country	*Level*	*Instrument name*	*Role and competences*	*Linkages*
Italy	Central	National Just Transition Fund (JTF) Program, National Recovery and Resilience Plan (PNRR)	National strategy, coordinating across regional plans	Aligns with regional operational programmes
	Regional	Sustainable Development Regional Strategies	Local implementation, with focus on vulnerable regions	Coordinates with central strategic plans
Poland	Central	Industrial Policy, Responsible Development Strategy	National strategy, supports coal-reliant regions	Integrates regional and local plans
	Regional	Territorial Just Transition Plans (TJTPs)	Regional economic diversification and coal phase-out	Links with local governments to support JGT
Sweden	Central	Sweden's Framework, Just Transition Fund	The regulatory framework, Financial support	Linked with regional and local authorities through County Administrative Boards (CABs) and national agencies
	Regional	Regional Climate and Energy Strategies, Territorial Just Transition Plans (TJTPs)	Regional plans with a focus on the most exposed transition regions	County Administrative Boards coordinate with regional and local au-thorites,
Netherlands	Central	Subsidy for Stimulation of Sustainable Energy Pro-duction and Climate Transition (SDE++), National C02 Tax	National incentives for renewable energy and CO_2 reduction	Collaborates with regional green industry strategies
	Regional	Regional Energy Strategy, Industrial Plans	Regional adaptation for sustainable industry goals	Links local initiatives with national JGT targets

21.3.4 Key Actors in TJTP Development and Implementation

Table 21.4 identifies the key governmental and administrative actors involved in Territorial Just Transition Plans' development and implementation phases across the four countries under investigation. The table distinguishes between central and regional levels, highlighting the roles played by national ministries, agencies, and regional authorities in each phase. In the development phase, central-level actors such as Italy's Ministry of Environment and Energy Security, the Department for Cohesion Policies and the South, and Sweden's Swedish Agency for Economic and Regional Growth are responsible for designing overarching strategies and frameworks to guide the Just Transition process. These institutions set priorities, funding mechanisms, and policy directions. During the implementation phase, regional and provincial authorities coordinate, ensuring that TJTPs are executed effectively and tailored to local needs. For example, Marshal Offices in Poland and Regional Governments in Italy (e.g., Sardinia and Puglia) or in the Netherlands (e.g., Groningen) play a crucial role in adapting central strategies to regional contexts. The table underscores the collaborative approach required between central and regional actors to achieve a successful transition. While central actors provide strategic direction and funding, regional entities oversee practical implementation, adapting policies to address region-specific socio-economic and environmental challenges.

Furthermore, Table 21.5 identifies the key sectoral and non-governmental actors involved in Territorial Just Transition Plans' development and implementation phases across the four countries. The table distinguishes between the Energy, Mining and Industry sectors and non-governmental sectors such as NGOs and local communities, highlighting the key actors in each phase. From the analysis, it emerges that key actors from the energy and mining sectors are involved in both the development and implementation phases in all countries except for the Netherlands, where, instead, the main sectoral actors involved in the shift to green technologies are from the industry and business sectors. On the contrary, in Poland, the industry sector is not relevant, while key actors from energy (i.e., *PGE Polska Grupa Energetyczna, Tauron Polska Energia, Enea*) and mining (i.e., *Polska Grupa Górnicza (PGG)* play an important role in both the development and the implementation phases. Regarding non-governmental actors, environmental NGOs are involved in all the countries but with differences in the phases; for instance, in the Netherlands, they participated in consultations to ensure that the plans considered both environmental and social impacts, and similarly, in Sweden, they were involved in the development phase. On the other hand, in Poland, environmental NGOs are involved in the implementation phase of the TJTP only, while in Italy, their presence can be traced both in the development and implementation phases. Finally, for what concerns local communities, they are similarly involved in the Netherlands and Sweden for the development phase where they are consulted,

Table 21.4 Key governmental and administrative actors in TJTP development and implementation (*Source* authors' elaboration)

Country	*Level*	*Phases*	*Key actors*
Italy	Central	Development	Ministry of Environment, Department for Cohesion Policies and the South
	Regional and Provincial	Coordination/ Implementation	Regional Governments (e.g., Sardinia, Puglia) and Provincial authorities (e.g., Sulcis Iglesiente and Taranto)
Poland	Central	Development	Ministry of Development Funds, Ministry of Climate and Environment
	Regional	Implementation	Marshal Offices, Regional Development Agencies
Sweden	Central	Development	Swedish Agency for Economic and Regional Growth, Swedish Energy Agency, Swedish Environmental Protection Agency, Swedish Public Employment Service
	Regional	Implementation	Regional Authorities responsible for adapting TJTPs over-seeing execution in collaboration with sectoral
Netherlands	Central	Development	Ministry of Economic Affairs, Ministry of Climate Policy
	Regional and Provincial	Implementation	Regional and Provincial Authorities (e.g., Groningen)

especially in Sweden, to ensure that the transition reflects local needs and mitigates adverse socio-economic impacts. Conversely, in Poland, they are involved in the implementation phase, while in Italy, they have a further coordination responsibility through the Local Action Groups.

Table 21.5 Key sectoral and non-governmental actors in TJTP Development and Implementation (*Source* authors' elaboration)

Country	*Sector*	*Phases*	*Key actors*
Italy	Energy	Development/ Implementation	National energy companies (i.e., ENI, Enel), Regional energy agencies (i.e., Agenzia Regionale per l'Energia, ASSET—Agenzia regionale Strategica per lo Sviluppo Ecosostenibile del Territorio; ASSET—Agenzia regionale Strategica per lo Sviluppo Ecosostenibile del Territorio
	Mining	Development/ Implementation	Mining companies (i.e., Carbosulcis)
	Industry	Development/ Implementation	Industrial associations (i.e., Confindustria)
	Others	Development/ Implementation	Research institutions (i.e., Consiglio Nazionale delle Ricerche (CNR)
	NGOs	Development/ Implementation	Environmental NGOs, Labour Unions (i.e., Legambiente Sulcis, CGIL, CISL)
	Local communities	Coordination/ Implementation	Local Action Groups (i.e., LAG Sulcis Iglesiente Capoterra e Campidano di Cagliari), third-sector associations, civic associations
Poland	Energy	Development/ Implementation	PGE Polska Grupa Energetyczna, Tauron Polska Energia, Enea and other state-owned companies involved in the energy and fuel industry
	Mining	Development/ Implementation	Polska Grupa Górnicza (PGG) and other state-owned companies involved in the energy and fuel industry
	Others	Implementation	Trade unions

(continued)

Table 21.5 (continued)

Country	*Sector*	*Phases*	*Key actors*
	NGOs	Implementation	WWF Poland, Greenpeace Poland, Polish Green Network (Polska Zielona Sieć)
	Local communities	Implementation	Local society organisations, local business associations
Sweden	Energy	Development/ Implementation	Vattenfall, Svenska Kraftnät
	Mining	Development	LKAB, SSAB, Cementa AB, SMA Mineral AB
	Industry	Development/ Implementation	Cementa AB (Gotland), Preem AB (Västra Götaland)
	NGOs	Development	Environmental and civil society organisations, academic institutions
	Local communities	Development	Local communities and citizens
Netherlands	Industry	Development/ Implementation	Businesses and industries
	NGOs	Development	Social Partners and Civil Society: Unions, business associations, and environmental NGOs

21.4 Discussion

The comparative analysis of the cases of Italy, Poland, Sweden, and the Netherlands reveals distinct approaches to the JGT, reflecting on the one hand each country's different social, economic, and environmental contexts and, on the other hand, that different administrative and spatial governance traditions that characterise them. In particular, the respective levels of centralisation/ decentralisation that characterise each country's approach to spatial governance[6] seem to play a role in providing the institutional anchoring to the domestic implementation of the EU JGT framework. While all four countries adhere to the EU's JGT framework, their strategies diverge in the policy focus, regulatory framework, and stakeholder engagement.

Policy Focus and Regional Targeting

[6] For a thorough, comparative overview of spatial governance in Europe see: Nadin et al. (2024).

Italy and Poland have prioritised decarbonising high-carbon regions through targeted investments in coal-reliant areas, but their strategies reflect different historical and economic dependencies. Italy, with its industrial regions of Puglia and Sardinia, has crafted TJTPs to balance decarbonisation with social equity by addressing job displacement and workforce reskilling. Poland's plans for Silesia and other coal-heavy regions underscore the importance of economic diversification, focusing on retraining programmes and promoting green industries as a replacement for coal-dependent employment. In contrast, Sweden's focus on ecological connectivity and biodiversity aligns closely with its national values of environmental stewardship and industrial transition (Steel in Norbotten e Västerbotten and Cement in Gotland). In this respect, the Netherlands' emphasis on regional customisation and stakeholder engagement represents a middle ground. The Netherlands combines industrial decarbonisation and ecological sustainability elements by developing adaptable TJTPs through extensive consultation with regional stakeholders. This approach fosters a high degree of local ownership, especially in regions like Groningen and the Port of Rotterdam, where industrial transformation is critical.

Regulatory Frameworks and Funding Coordination

Regulatory frameworks across the four countries reflect varying levels of centralisation and regionalisation. To some extent, all countries have experienced a hybrid approach to designing and implementing TJTPs. In general terms, while the central level has been leading in coordinating national interests when dealing with EU institutions, the regional (and provincial level in Italy and in the Netherlands, too) has been coordinating contextual needs and priorities. In this respect, the role of the Voivodship in Poland is a good case when it comes to discussing the regionalisation process of TJTPs. The Netherlands stands out for its integrated funding approach, combining national programmes like the SDE++ subsidies with EU funding sources like the JTF. This structure simplifies access to funding and aligns financial incentives with local priorities, allowing for targeted support in industrial and urban areas. However, the complexity of navigating these funding streams may disadvantage smaller businesses and local communities needing more resources to engage with the available support mechanisms fully.

Stakeholder Engagement and Inclusivity

When it comes to stakeholder engagement, the four countries under investigation present a complex interplay of central and regional actors, as well as sectoral and non-governmental stakeholders. Key actors at the central level provide strategic direction and funding, while regional authorities ensure context-sensitive applications. For instance, central ministries and agencies in Italy and Sweden are responsible for the frameworks to guide the transition, but local adaptation is pivotal, as seen in Poland through the Marshal Offices and in Italy with the role of provincial authorities, tailoring strategies to unique socio-economic and environmental conditions. Sectoral actors from energy, mining, and industry further influence these processes, with distinct national patterns reflecting the different economic landscapes. Poland's reliance on energy and mining sectors contrasts with the Netherlands' focus on industrial and business entities driving green technology adoption. Non-governmental stakeholders such as environmental NGOs and local communities underscore the need for inclusivity. NGOs, while variably engaged across countries, provide critical oversight to balance environmental and social considerations, particularly in Sweden and the Netherlands' development phases. Local communities, similarly, ensure grounded, equitable transitions, their involvement ranging from consultative roles in Sweden to coordination responsibilities in Italy. These dynamics highlight that achieving a just transition is not merely a technical or economic challenge but a deeply collaborative endeavour. By weaving together central strategy, regional implementation, and diverse stakeholder input, TJTPs navigate the delicate balance of addressing immediate economic impacts while steering towards sustainable, long-term transformation.

21.5 Conclusion

The JGT are a key component of the European Green Deal, intended to ensure that the shift towards a climate-neutral economy is both environmentally sustainable, socially equitable and territorial balanced. This chapter has shown that Italy, Poland, Sweden, and the Netherlands have adopted varied approaches to the JGT, reflecting these Member States' distinct economic, ecological, social and institutional contexts. Italy and Poland have focused their TJTPs on tackling coal and carbon-intensive industries (the case of Taranto in Italy is emblematic) and have developed transition strategies that focus heavily on socio-economic resilience and industrial decarbonisation. Italy's TJTPs are committed to addressing regional disparities, focusing on workforce reskilling and promoting investment in vulnerable post-industrial areas. Poland's strategy similarly focuses on coal-dependent regions like Silesia, where economic diversification and retraining programmes are prioritised to mitigate the social impact of coal phase-out. Sweden's approach emphasises the transition of coal-depended regions: the steel industry in Norrbotten, the metal industry in Västerbotten, and the cement industry in Gotland. The Netherlands presents an adaptable model of regional engagement and innovation, driven by a combination of stakeholder consultation and extensive use of funding mechanisms like SDE++ subsidies and the JTF. This stakeholder-driven approach allows for regionally tailored TJTPs that address industrial decarbonisation and social equity. Including local governments, industry representatives, and environmental organisations fosters a collaborative environment for JGT, although the complexity of the funding system may pose challenges for smaller entities. The four cases do not represent the heterogeneity of approaches adopted by all EU Member States. However, they offer a wide spectrum of possible approaches. In conclusion, the varied approaches of Italy, Poland, Sweden, and the Netherlands underscore the importance of tailoring JGT strategies to the specific challenges and opportunities within each Member State. Effective JGT implementation requires alignment with EU policies and a flexible framework for local adaptation, stakeholder engagement, and cross-sectoral integration. This study represents a preliminary comparison of the current situation without considering the implementation consequences and the impact of the TJTPs in each country. However, it offers a good reference for Western Balkans countries that have not yet started to think about transition plans.

References

Berisha, E. (2025a). *Moving towards an inclusive Green Agenda in the Western Balkans*. Foundation for European Progressive Studies (FEPS). https://doi.org/10.13140/RG.2.2.12092.76162

Berisha, E. (2025b, in this volume). Green EU integration—The role of the green agenda in accelerating Just Green Transitions in the Western Balkans. In E. Berisha, J. Moodie, L. Allkja, & M. Jeftić (Eds.), *The Palgrave handbook of just green transitions in the Western Balkans and beyond*. Palgrave Macmillan.

Berisha, E., Moodie, J., Allkja, L., & Jeftić, M. (2025, in this volume). Just Green Transitions—An introduction. In E. Berisha, J. Moodie, L. Allkja, & M. Jeftić (Eds.), *The Palgrave handbook of just green transitions in the Western Balkans and beyond*. Palgrave Macmillan.

Bujdosó, Z., Bordás, A., Hegyi, B., Piskóti-Kovács, Z., Vaszkó, C., Nagy, R., & Kovács, G. (2022). The involvement of stakeholders in the decarbonisation process in the coal region of Northern Hungary. *Europa XXI, 42*, 81–93. https://doi.org/10.7163/Eu21.2022.42.1

Cotella, G., Crivello, S., & Karatayev, M. (2016). Chapter 2—European Union Energy Policy evolutionary patterns. In P. Lombardi & M. Gruenig (Eds.), *Low-carbon energy security from a European perspective* (pp. 13–42). Academic Press. ISBN 9780128029701. https://doi.org/10.1016/B978-0-12-802970-1.00002-4

European Commission. (2019). The European Green Deal. Retrieved at EUR-Lex - 52019DC0640 - EN - EUR-Lex. Accessed 14 December 2024.

European Commission. (2023). *Cohesion 2021–2027: Forging an ever stronger union report on the outcome of 2021–2027 Cohesion policy programming*. Commission Staff Working Document. SWD (2023) 134 Final. Brussels. https://ec.europa.eu/regional_policy/sources/reports/2021-2027-programming-outcome/report-outcome-2021-2027-cohesion-policy-programming-part1.pdf

García-Mira, R., Garha, N. S., Michas, S., Mey, F., Basu, S., & Süsser, D. (2024). Leave no one behind: Engaging communities in the just transition process towards climate neutrality. In E. Galende Sánchez, A. H. Sorman, V. Cabello, S. Heidenreich, & C. A. Klöckner (Eds.), *Strengthening European climate policy*. Palgrave Macmillan. https://doi.org/10.1007/978-3-031-72055-0_8

Gärdebo, J. (2023). Transitioning unions: What constitutes a Just Transition for Swedish Trade Unions? *Oxford Open Climate Change, 3*(1). https://doi.org/10.1093/oxfclm/kgac006

International Energy Agency. (2024). *Total CO2 emissions*. Retrieved at Europe – Countries and Regions - IEA

Koasidis, K., Nikolaev, A., Gaydarova, E., Karamaneas, A., Todorov, T., Georgiev, G., Irimie, S., Valmaseda, C., Gonzalez, I., Frankovic, A., Mazur, A., Ogrodniczuk, J., & Süsser, D. (2023). *Report on the current status of coal-dependent regions in the EU—JUSTEM "D2.1: Current Status of the Regions" and supplementary material*. Zenodo.

Larsen, R. K., Boström, M., Muonio Reindeer Herding District, Vilhelmina Södra Reindeer Herding District, Voernese Reindeer Herding District, & Wik-Karlsson, J. (2022). The impacts of mining on Sámi lands: A knowledge synthesis from three reindeer herding districts. *The Extractive Industries and Society, 9*(101051). ISSN 2214-790X. https://doi.org/10.1016/j.exis.2022.101051

Moesker, K., & Pesch, U. (2022). The just transition fund—Did the European Union learn from Europe's past transition experiences? *Energy Research and Social Science, 91*, Article 102750. https://doi.org/10.1016/j.erss.2022.102750

Moodie, J., Tapia, C., Löfving, L., Gassen, N. S., & Cedergren, E. (2021). Towards a territorially just climate transition—Assessing the Swedish EU territorial just transition plan development process. *Sustainability, 13*(13), 7505. https://doi.org/10.3390/su13137505

Nadin, V., Cotella, G., & Schmitt, P. (Eds.). (2024). *Spatial planning systems in Europe*. Edward Elgar Publishing. https://doi.org/10.4337/9781839106255

Niewitała-Rej, M. (2024). Aspetti territoriali e distributivi della Transizione Giusta nella bozza aggiornata del Piano Nazionale Integrato per l'Energia e il Clima Italiano, Reform Institute, Varsavia. Retrieved at IT_NECP_Assessment_Report_IT.pdf

Nordic Council of Ministers. (2021). *Nordic co-operation programme on energy policy 2022–24*. PolitikNord 2021:731. ISBN 978-92-893-7155-1. https://doi.org/10.6027/politiknord2021-731

Nowakowska, A., Rzeńca, A., & Sobol, A. (2021). Place-based policy in the "just transition" process: The case of Polish coal regions. *Land, 10*(10), 1072. https://doi.org/10.3390/land10101072

RISE Research Institutes of Sweden. (2022). Prestudy H2ESIN: Hydrogen, energy system and infrastructure in Northern Scandinavia and Finland. Luleå University of Technology. In M. Vendt & C. Wallmark (Eds.), *Vinnova (Sweden's innovation agency)*. RISE rapport 2022:120. ISBN (RISE): 978-91-89757-00-4. https://www.diva-portal.org/smash/get/diva2:1719894/FULLTEXT01.pdf

Ruszkowski, P. (2021). Bełchatów 2030: Alternative transformation scenarios and development. *Warsaw Forum of Economic Sociology, 12*(23), 45–55. https://econjournals.sgh.waw.pl/wfes/article/view/3159

Shaker, Y., & Berisha, E. (2025). Just Green Transitions: Between terminological inexactitudes, conceptual fragmentation, and the exigency for a theoretical framework. European Spatial Research and Policy, 32(1), 121–138. https://doi.org/10.18778/1231-1952.32.1.05

Shaker, Y., & Berisha, E. (2024). *Transizioni verdi giuste: Tra imprecisioni semantiche e necessità di nuovi modelli di governance*, (pp. 193–203). Società Italiana degli Urbanisti: PLANUM. http://www.planum.net/planum-magazine/planum-publisher-publication/atti-della-xxv-conferenza-nazionale-siu-volume-nove-formato-vingelli

Shaker, Y., & Persico, S. (2024). Mapping the just green transitions in the socio-political virtual space. *European Journal of Spatial Development, 21*(1), 61–82. https://doi.org/10.5281/zenodo.13817780

Ślimko, E., Bartecka, M., & Pogoda, A. (2021). *Territorial just transition plans for polish coal regions*. Polish Green Network. Briefing October 2021. CEE Bankwatch Network. https://bankwatch.org/wp-content/uploads/2021/10/2021-10-20_TJTP_Poland.pdf

Tarasova, E. (2024). Multiple just transitions? A study of how a just transition is envisioned in Poland. *Energy Policy, 194*(114322). ISSN 0301-4215. https://doi.org/10.1016/j.enpol.2024.114322

Theodoropoulou, S., Sabato, S., & Akgüç, M. (2024). *National eco-social policies in the framework of EU just transition: The cases of Greece, Italy, Portugal and Spain*. Global Social Policy. https://doi.org/10.1177/14680181241291122

Valkenburg, G., & Cotella, G. (2016). Governance of energy transitions: About inclusion and closure in complex sociotechnical problems. *Energy, Sustainability and Society, 6*, 20. https://doi.org/10.1186/s13705-016-0086-8

van Bommel, N., & Höffken, J. I. (2023). The urgency of climate action and the aim for justice in energy transitions—Dynamics and complexity. *Environmental Innovation and Societal Transitions, 48*, Article 100763. https://doi.org/10.1016/j.eist.2023.100763

CHAPTER 22

Understanding Territorial Governance Dynamics in the Swedish EU Territorial Just Transition Planning Processes: Lessons for the Western Balkans

Aida Shaneh, John Moodie, Carlos Tapia, Alberto Giacometti, and Elin Cedergren

22.1 Introduction

In recent years, the European Commission has emphasized the need for greater territorial governance (TG) and place-based policymaking processes, established on the principle of "active subsidiarity," which supports an increased role for regions and municipalities in European Union (EU) policymaking (Moodie et al., 2022). There is a strong TG logic within the EU's

A. Shaneh (✉)
Interuniversity Department of Territorial Sciences, Planning and Policies (DIST), Politecnico di Torino, Torino, Italy
e-mail: aida.shaneh@polito.it

J. Moodie · C. Tapia · A. Giacometti · E. Cedergren
Nordregio Research Institute for Regional Development and Planning, Stockholm, Sweden
e-mail: john.moodie@nordregio.org

C. Tapia
e-mail: carlos.tapia@nordregio.org

A. Giacometti
e-mail: alberto.giacometti@nordregio.org

E. Cedergren
e-mail: elin.cedergren@nordregio.org

A. Giacometti
School of Business and Law, University of Stavanger, Stavanger, Norway

E. Berisha et al. (eds.), *The Palgrave Handbook of Just Green Transitions in the Western Balkans and Beyond*,
https://doi.org/10.1007/978-3-031-95075-9_22

just transition policies that lay at the heart of the EU Green Deal (Moodie et al., 2021a). Firstly, the EU Just Transition Mechanism (JTM) is committed to identifying and supporting EU regions facing major technical, social, and territorial impacts of transitioning to a low carbon economy, focusing on those regions highly dependent on fossil fuels and energy-intensive processes. Secondly, the core instrument of the JTM is the Just Transition Fund (JTF), which aims to facilitate socio-economic diversification and reconversion in EU regions most affected by climate transitions. Thirdly, EU member state access to the JTF, implemented within the remit of EU Regional Policy, is conditional on the development of Territorial Just Transition Plans (TJTPs). TJTPs are designed to be territorially specific, outlining targeted socio-economic and environmental actions and governance mechanisms needed to help meet the regional challenges and opportunities posed by transitions (European Commission, 2021). TJTPs are also meant to be formulated through open and inclusive processes, based on local knowledge and expertise, to ensure that just transition policies meet the needs of citizens and leave no one region or person behind (European Commission, 2019).

This chapter employs a TG lens to investigate the development of EU TJTPs in Sweden. The decentralized nature of governance structures in Sweden provides regional and municipal authorities with a significant role in delivering climate and energy policies due to their proximity to citizens, responsibility for spatial planning, and close relationships with key regional transition stakeholders. Within this context, Sweden should be a good testing ground for place-based just transition policymaking processes based on TG approaches. The chapter first assesses whether the processes adopted in the preparation of EU TJTPs in Sweden reflect a TG approach, driven by regional specificities, local knowledge, and sub-national level institutions and actors. The final section reflects on the importance of developing effective place-based just transition policies, and what other regions in Europe and the Western Balkans can learn from the Swedish experience of developing EU TJTPs.

22.2 Establishing a Territorial Governance Framework

Territorial governance (TG) has been advocated by the Organization for Economic Co-operation and Development (OECD) and the European Union (EU) as an important concept for fostering place-based policymaking (European Commission, 2008; OECD, 2020). TG examines the relationship between governance and territory, with a particular emphasis on exploring how territorial specificities and place-based knowledge are recognized, understood, and included in policymaking processes (Van Well & Schmitt, 2015). Initially developed in the field of European Planning Studies (Stead, 2014), TG has been referenced as an approach for coordinating regional planning processes (Smas & Lidmo, 2018) and enhancing regional resilience (Van Well et al., 2018). More recently, TG has gained prominence in political science discussions as a tool for empowering sub-national level institutions and actors in EU Cohesion Policy (Cotella, 2018; Moodie et al., 2021b).

There are no specific references to the TG concept in relation to the EU Green Agenda and Just Green Transition policies; however, TG has been officially recognized by the OECD as an approach for assisting in the delivery of United Nations Sustainable Development Goals (OECD, 2020). While TG is not explicitly referenced, the spatial and territorial dimensions of JGTs are embedded within the EU Green Deal. For example, the European Commission highlights the important role of regions and cities in guiding just transition processes: "Citizens, depending on their social and geographic circumstances, will be affected in different ways. Not all Member States, regions and cities start the transition from the same point or have the same capacity to respond. These challenges require a strong policy response at all levels." (European Commission, 2019) Indeed, EU just transition policies are explicitly territorial in focus, with the EU JTM and JTF targeting regions most vulnerable to the impacts of shifting to carbon neutral societies and economies. TJTPs are meant to address territorial just transition challenges and opportunities that meet the needs of local citizens. TJTP processes, therefore, require a significant role for regional and local institutions, sectors, and other expert stakeholders (i.e., universities, businesses, and civil society groups) with a knowledge and understanding of the region in which they reside (European Commission, 2021).

Within this context, the TG concept can be a useful tool for building, implementing and evaluating EU TJTP processes, as it: (1) emphasizes bottom-up processes that devolve power to the lowest governance levels so policymaking is driven by regional and local knowledge and expertise (Schmitt & Van Well, 2016); (2) focuses on bringing together local stakeholders in an open, transparent, deliberative, and consensus-orientated process of policy co-creation (Davoudi et al., 2008); and (3) fosters the development of concrete regional strategies built around specific place-based knowledge and expertise (Böhme et al., 2015). The academic literature on TG identifies five key dimensions of the approach, as outlined in Table 22.1.

The five dimensions highlighted in Table 1 can be used as a framework for assessing whether TJTP processes conducted across EU regions are exemplified by a TG approach. A TJTP process based on a TG approach would: (1) reflect a multi-level governance process primarily driven by sub-national level institutions and actors; (2) create synergies between different public and private actors at the sub-national level; (3) represent an open and inclusive process including a range of relevant and interested public and private stakeholders, and possibly involving direct citizen engagement; (4) be a reflexive process adapting to changing territorial contexts and new information; and (5) lead to the development of a place-based strategy that addresses regional challenges and maximizes opportunities. The following section uses this TG framework to provide an empirical analysis of the TJTP process conducted in Sweden and to what extent they reflected a TG approach. The paper focuses

Table 22.1 Key Dimensions of TG (Authors' adaptation from Davoudi & Cowie, 2016; Davoudi et al., 2008; Schmitt & Van Well, 2016; Van Well & Schmitt, 2015; Van Well et al., 2018)

TG dimension	*TG dimension key features*
Coordinating actors and institutions	TG strengthens vertical links between EU, national and sub-national authorities, but with policymaking agency devolved to the sub-national level
Integrating policy sectors	TG enhances horizontal synergies between public and private actors at the local level (e.g., sectors, businesses, universities, and civil society groups), enabling the development of new networked ecosystems that add to territorial cohesion
Mobilizing stakeholder engagement	TG emphasizes open and inclusive policymaking, providing a voice to all relevant and interested actors, allowing them to shape the policymaking process and provide enhanced regional autonomy and accountability
Adaptive to changing contexts	TG is reflexive and adaptive to changes in territorial context, with actors able to review and revise territorial policies considering new information and learning
Place-based/territorial specificities	TG is a process driven by local knowledge that focuses on identifying territorial challenges and opportunities and the development of place-based visions and strategies

on evaluating the planning process of TJTPs within eligible Swedish regions, as part of the JTF program.

22.3 The EU TJTP Process in Sweden

EU TJTPs were developed in four Swedish regions, focusing on industrial sectors that contribute to the largest share of carbon emissions in Sweden—the steel industry in Norrbotten, the cement industry on Gotland, refineries and the petro-chemical industry in Västra Götaland, and the metal industry in Västerbotten. The TJTP process has involved the same governance and administrative levels within all four Swedish regions, as outlined in Table 22.2. The Swedish Agency for Economic and Regional Growth (Tillväxtverket) was tasked by the government to manage the EU JTM and JTF processes. The government also identified the Swedish Public Employment Service, the Swedish Energy Agency, and the Swedish Environmental Protection Agency as important governmental agencies to be involved in the preparations. These government agencies have worked in close collaboration with the County Administrative Boards and regional and municipal authorities in the preparation of EU TJTPs (Moodie et al., 2021c).

Table 22.2 Overview of the different governance and administrative levels involved in the TJTP process in Swedish regions (table adapted from Moodie et al., 2021a)

<table>
<tr><th>Level</th><th colspan="3">Role</th></tr>
<tr><td>European</td><td colspan="3">European Commission Support for the development of the TJTPs is provided by DG REFORM</td></tr>
<tr><td>National</td><td colspan="3">National Authorities and Agencies
The national authorities and agencies are responsible for implementing the decisions made by the Parliament and the Government
Role in TJTP process:
An authority group has been established to lead and coordinate the process, align priorities as well as to assist with expertise and data. These consist of:
• Swedish Agency for Economic and Regional Growth (managing authority)
• Swedish Public Employment Service
• Swedish Energy Agency
• Swedish Environmental Protection Agency
County Administrative Boards
National authorities operating at county level, responsible for the state administration in the county (in those areas where no other authority is responsible for special administrative tasks). The County Administrative Boards shall work to ensure that national goals have an impact in the county, while also taking account of regional conditions
Role in TJTP process: The county administrative board has the mission to promote, coordinate and lead regional work in the implementation of the government's policy regarding energy conversion. Together with the regions, they have important roles in the implementation of the TJTPs. County Administrative Boards have participated in meetings at regional level and provided input during the development of the TJTPs</td></tr>
<tr><td></td><td>County Administrative board of Norrbotten</td><td>County Administrative board of Västra Götaland</td><td>County Administrative board of Gotland</td></tr>
<tr><td>Regional/ county</td><td colspan="3">Regional Authorities
Led by political assemblies with responsibilities for e.g., healthcare and care, public transportation and regional development. Counties and regions cover the same geographical area
Role in TJTP process: The regions have the responsibility for regional development and a key role in the implementation of the TJTPs. The region has an active role in moving the regional economy from fossil dependence to a sustainable society and an especially important role in linking sectors as well as regional and national level actors in triple-helix innovation collaborations</td></tr>
<tr><td></td><td>Region Norrbotten</td><td>Region Västra Götaland</td><td>Region Gotland</td></tr>
</table>

(continued)

Table 22.2 (continued)

Level	*Role*
Local	**Local authorities** *Local government responsible for e.g., primary and secondary school, preschool activities, elderly care, physical and comprehensive planning, roads, water and sewage issues and energy issues. Also issue different types of permits, such as building permits* **Role in TJTP process:** Municipalities play an important role in Sweden's climate work due to their proximity to citizens and their role in spatial planning. Municipalities drive local development in collaboration with companies, organizations, residents and will therefore play a significant role in the implementation of TJTPs. *The municipalities participated in the drafting of the TJTPs through a written consultation process on the plans and related outputs* • Gällivare municipality • Öxelösund municipality • Luleå municipality • Lysekil municipality • Stenugnssund municipality • Gotland Municipality

TJTPs were also drafted in dialogue with key industrial sectors in each region, such as Cementa AB and the Heidelberg Cement Group representing the cement industry on Gotland; SSAB, Lulekraft AB, and LKAB Kiruna from the HYBRIT initiative in the Norrbotten county; as well as the chemical and refining industries, including Preem AB, Borealis AB, and ST1 in the Västra Götaland county. Other key sectoral agencies, stakeholders, and actors have also played an important part in formulating the plans. These include Vattenfall and Svenska Kraftnät in the energy sector, higher education institutions, academia, and other interested civil society and labor organizations. Nordregio Research Institute was also involved in the process conducting socio-economic impacts assessment of transitions in each Swedish region (Moodie et al., 2021c).

The TJTPs for each transition region have the same structure, including the following seven core elements; one, an overview of the national climate policy framework, highlighting key national climate targets; two, a national level assessment of the territorial impacts of climate change, identifying which regions within the country contribute the most to gas and carbon emissions; three, an assessment of the potential economic, social, and environmental impacts of the transition on the selected region; four, an analysis of the development needs required in each region to meet climate targets; five, an outline of the main type of actions planned to deliver the regional transition; six, performance-related output and results indicators are highlighted for measuring the impact of the proposed actions; and seven, the governance structured and key stakeholders needed to develop and implement the plans (Moodie et al., 2021c).

The following sub-section analyze this process through the lens of the five core key TG dimensions. The methodology involved a case study approach, combining document and policy analysis with semi-structured interviews of stakeholders at various levels. Data collection included a literature review and detailed interpretation of TJTPs of Sweden, highlighting strategies, challenges, and policy measures. The study employed a framework encompassing the five dimensions of TG to structure research questions (open-ended) and themes for analysis. Thematic analysis of interviews followed a deductive approach, with codes derived from interview content and aligned with the TG dimensions. The interview sample consisted of ten online interviews in 2022 and one interview in 2024 with stakeholders, selected using non-probability and snowball techniques, starting with primary JTF program coordinators and expanding to include other relevant stakeholders (e.g., Regional County administrative board, Environmental Protection Agency (EPA), etc.) recommended by initial stakeholders and experts.

22.3.1 Coordinating Actors and Institutions

Tillväxtverket, the national level Swedish Agency for Economic Growth and Regional Development, were given responsibility for managing the formulation of the TJTPs by the national government. In line with this TG dimension, Tillväxtverket coordinated a range of actors and institutions from across governance levels in the development of the different regional plans. According to one interviewee, the JTF program was a well-coordinated process by Tillväxtverket:

> The Just Transition Fund program proceeded with vertical coordination across institutional actors and dialogue between non-state stakeholders. To begin the process, Tillväxtverket started at the state level, county council, to consider all the relevant stakeholders to the program which was a long and complicated process. The national agency was successful in meeting the goal set by the Swedish government to include all necessary stakeholders in the program. (Interview, 2022)

While the process was professionally managed and coordinated by Tillväxtverket, the process did not meet the full scope of this TG dimension, as policymaking agency and responsibility was not devolved to the sub-national level (Shaneh, 2023). A full TG approach would have seen regional and local authorities taking the management and coordinating role for developing the plans. The decision of the national government to grant management responsibilities to Tillväxtverket restricted the role of sub-national institutions and actors, which limited the potential for bottom-up TJTP processes. This was confirmed by one interviewee at the Environmental Protection Agency who noted:

> The decisions on which regions to include were mainly driven by politics and national considerations. The goal was to make the just transition fund as focused as possible, to make a big contribution in fewer sectors and regions, but there were also other political considerations that influenced the final decision. (Interview, 2022)

The limited role for sub-national actors was further exacerbated by the national government decision to focus the plans on individual sectors in each region, which gave these sectors a primary role in the development of the plans. Consequently, while the European Commission advocated a bottom-up approach for formulating the plans, the process in Sweden was predominantly top-down and sector-driven (Shaneh, 2023). As one interviewee working in a County Administrative Board noted:

> The process of green transition is at full swing for the moment, and it is quite early to make definite conclusions regarding the coordination process. At this stage, the process is improving. Initially, it was heavily dominated by top – down approaches, but now the civil society and other actors and institutions are being included and having greater influence which is crucial for society. (Interview, 2022)

This last comment indicates that the TJTP preparation phase did not meet the TG criteria in relation to the first dimension, as the process was predominantly top-down, with significant power asymmetries between sub-national actors, national level policymakers and agencies, and sectoral representatives. However, the comment also suggests that the process of implementing the TJTPs is reflecting a more TG approach, where sub-national level actors are having a more significant role, including civil society organizations who represent citizens at the local community level.

22.3.2 Integrating Policy Sectors

The TJTP process in Sweden did not meet the criteria for the TG dimension on integrating policy sectors. The decision to focus the plans on the largest polluting sectors in each region meant that the final plans, and planning process, were sector specific; consequently, the TJTPs in Sweden did not consider horizontal synergies between public and private actors at the regional and local level or enable the development of new networked ecosystems that add to territorial cohesion (Shaneh, 2023).

Socio-economic impact assessments conducted in each region pointed to the potential for negative knock-on effects on businesses reliant on supply chain materials from the largest regional sectors that were the primary focus of the plans (Moodie et al., 2021a). Despite this information, the planning process did not include a stage in which different sectors were brought together in a dialogue to discuss the potential repercussions of individual sector transformations for other businesses in the regions. Of particular significance was the lack of dialogue and co-operation between sectors and regional and local electricity grid-owners regarding conditions for energy supply which is a prerequisite for large-scale electrification of industries. One interviewee noted that the short timeline granted by the European Commission for preparing the plans left limited opportunities for cross-sectoral discussions. Creating a dialogue between different public and private actors was further challenged by the Covid-19 pandemic that reduced opportunities to get sectoral actors together in the same room (Shaneh, 2023).

Across all Swedish EU just transition eligible regions, high levels of co-operation and social capital have been built up between public and private actors through various EU, national and regional projects focused on just green transitions. One interviewee at Luleå University of Technology expressed disappointment that the EU TJTPs did not build on these existing collaborations and promote further integration:

> There is a synergy and integration among universities as institutions, regions, municipalities, and ministers across policy sectors to achieve the just and green development in the region. There is need for stakeholders' engagement at a regional level, and this is where the responsibility should lie according to the structure of the Just Transition Fund. (Interview, 2022)

The TJTP process did not promote cross-sector dialogue and integration; furthermore, the final actions within the plans did not address or facilitate public–private collaboration, therefore, this TG dimension was not adequately reflected in the process overall (Shaneh, 2023). The governance section of the final plans in each region highlighted a range of different regional and local level actors that need to work together in the delivery of the plans. This suggests that the implementation stage of the TJTP process might promote cross-sector collaboration in line with this TG dimension.

22.3.3 Mobilizing Stakeholder Engagement

Tillväxtverket collaborated with regional and municipal authorities to identify relevant local stakeholders and groups most impacted by transitions and provide frameworks for them to participate in the TJTP process. Tillväxtverket raised awareness and transparency through various communication techniques, such as media coverage (Shaneh, 2023). They also enhanced openness and inclusivity by offering stakeholders the opportunity to attend workshops, bilateral meetings, and produce written consultations on draft plans. This TG dimension was, therefore, partially reflected in the TJTP process as the opportunity to participate was offered to a wide range of stakeholders. In practice, however, the plans were developed by a small number of key actors and then made available to wider stakeholders for consultation. The timing of stakeholder involvement also limited the impact that other actors could make on the content of the plans. Furthermore, certain sectors were less engaged in the process; for example, SMEs and schools were less involved as they had insufficient regional representation or limited capacity and resources to participate (Interview, 2022).

The European Commission places a strong emphasis on the social justice element of the just transition within the European Green Deal (European Commission, 2019). The Commission highlights the need for an open and inclusive transition process including communities and citizens most affected by transitions to promote trust and acceptance of policies. They continue that the transition can be successful only if policies are designed with the involvement of citizens and accepted by them. More specifically, an "active social dialogue" is recognized as an essential element to ensure that the transition is successful and accepted by workers and companies. In the development of the Swedish TJTPs, societal groups, NGOs, and citizens were largely excluded from this process. As one interviewee noted, the national government decision to focus EU TJTPs on sectoral processes made it difficult to engage citizens and least engaged communities as the plans became very technical. They continued that citizen engagement would have been easier if the plans were developed on the basis of a territorial focus rather than sector-based approach. The notable exception with regards the involvement of societal groups was the direct role of the Sámi Parliament in the development of TJTP in Norrbotten. The role of the Sami Parliament was essential as minor conflicts arose regarding the impact of transitions on Sámi culture and land use. Nonetheless, Sámi representatives have argued that policies and legislation to support the climate transition have so far not adequately reflected their interests and that their traditional rights have not been respected (Moodie et al., 2021a).

In Sweden, greater consultation of societal groups and citizens in the TJTP process might have acted as a counterweight to the political and technical priorities of the Swedish government and sectors, creating a balance between the technical and social actions outlined in the TJTPs. If this TG dimension is going to be fully implemented, it is vital that societal groups and local citizens are consulted in the development, implementation, and evaluation of future transition policies that impact on the regional and local levels (Moodie et al., 2021a). It is essential that transition processes are open and transparent to facilitate participatory governance throughout the policy cycle, which involves labor unions, civil society organizations, and citizens, particularly societal groups most vulnerable to the impacts of transitions.

22.3.4 *Adaptive to Changing Contexts*

The TG dimension adaptive to changing contexts was only partially reflected in the TJTP process in Sweden as there was very little scope for actors to review and revise the drafts of the plans considering new information and learning (Shaneh, 2023). The TJTPs became narrowly defined by political decisions during the planning process. Consequently, the final plans have a quite explicit technical focus and direction but lack flexibility to be adapted to a new context unless exceptional circumstances arise, such as new policy regulations from the European Commission or national government. The point was supported by one interviewee at Tillväxtverket who argued that:

> The JTPs changed during the process. Initially, the JTPs were broad and allowed for a wider range of applications, but the goal to directly address high CO2 emissions in ETS installations through JTPs led to negotiations with the European Commission.... As a result, the plans are now more specific and focused than they were at the beginning of the process. (Interview, 2022)

Changes in the TJTPs resulted mainly from internal discussions between the national government and regional and municipal authorities. Furthermore, the socio-economic impact assessments developed by Nordregio research institute were also influential in persuading national level actors to further reflect on the potential social and economic impacts of transitions on citizens in the different regions (Moodie et al., 2021a). Consequently, a degree of flexibility and adaptability was built into the process in response to the socio-economic impact reports, as the original draft plans were redeveloped to include new targeted actions promoting citizen reskilling and retraining in the regions, which helped balance out the other targeted actions focused on technical process development. Further adaptability and flexibility could have been built into the process had other key actors, including universities, civil society organizations, and schools been granted a more significant role in discussions.

A range of different actors need to be included in TJTP process if this TG dimension is to be fully implemented, as different viewpoints generate new ideas and information that cultivates adaptivity and flexibility in the consideration of different policy options. There are several valuable lessons to be drawn from this experience that can be applied in other regions developing TJTPs. Firstly, the early involvement of interested and relevant stakeholders is crucial to ensure all voices and different perspectives are heard. Secondly, it is important to avoid rushing the process and to practice patience and tolerance, considering the viewpoints of all parties involved. Discussions on place-based solutions require extra time, and stakeholders should be engaged on an equal basis to help merge different needs and propose territorial solutions that are supported by all actors and citizens. This requires open and inclusive multi-actor processes with the inbuilt flexibility and adaptability at the core of the TG approach (Shaneh, 2023).

22.3.5 Place-Based/Territorial Specificities

The TJTPs developed in eligible Swedish regions are grounded in place-based territorial specificities and local knowledge. Place-based knowledge and the territorial regional and local specificities were integrated in the process in a number of different ways: (1) the Swedish regions and municipalities are elected, therefore, regional territorial needs were voiced by elected regional and municipal authority representatives; (2) the regional perspective was also represented through the involvement of other key regional based actors, including the County Administrative Boards and local universities; (3) EU requirements stipulated that regional smart specialization strategies (S3) were in line with the plans and that the plans also reflected goals and objectives of regional climate and energy plans; (4) sector roadmaps might take a national perspective but as the sectors are based in a territory, the roadmaps address local specificities relevant to citizens, such as the need for more skilled workers.

One interviewee at Tillväxtverket argued that the plans were place-based and focus on different regional territorial specificities:

> The JTF program and TJGTPs are place-based, hence there are several territorial or place-based specificities that are most relevant to this context: "Gotland is an island with specific challenges in access to electricity and a lack of competences in certain areas.... Northern Sweden is dominated by the big steel and iron industry, leading to a cultural connection to the industry.... There are also territorial issues with the indigenous population in northern Sweden, causing conflicts with the opening of new mines and clearing of interests. (Interview, 2022)

Certainly, these different place-based inputs and specificities are well documented in the contextual parts of the plans, which are grounded in local knowledge, data and information pertaining to different social, environmental, economic and technical transition challenges and opportunities in each region. For example, in the Norrbotten plan, issues such as a shortage of skilled labor, cultural barriers in the steel industry are highlighted, and challenges in actions like opening new mines and working with ETS installations pose significant threat to the transition process. Furthermore, the Norrbotten plan notes there are difficulties for transitions, such as attracting more people to the northern counties and focusing on reskilling and upskilling are deemed essential. Whereas, the Gotland plans highlight low education levels, particularly in technical fields, and challenges obtaining planning permissions for mining in the cement industry (Shaneh, 2023).

While the plans are contextualized in place-based specifics regarding the environmental and socio-economic impacts of transitions in the region, the extent to which the final actions outlined in the plans deal with these specificities is open to question as the targeted actions predominantly focus on transforming internal sectoral energy processes. As one interviewee pointed out, if the plans adopted a territorial rather than a sectoral focus, the plans and final targeted action might have been more balanced and focused on place-based territorial social, economic and environmental issues (Interview, 2024). Overall, it is important that a TG place-based approach is adopted in the preparation of TJTPs, as each region has their own unique territorial specificities, challenges and opportunities. These need to be carefully considered and addressed in the plans if the targeted actions developed are going to meet the needs of local citizens.

22.4 Discussion and Conclusions

The empirical analysis indicates that Sweden did not follow a full TG logic in the preparation of their EU TJTPs. Even in Sweden, where responsibilities for climate and the environment policies are shared across administrative levels, the role of regional and local institutions and actors was largely curtailed in a TJTP process dominated by the national government and sectoral groups (Moodie et al., 2021c). As one interviewee noted, the decision of the national government to focus the TJTPs on specific regional sectors limited the scope of the plans to the technical and economic dimensions of shifting to carbon neutral processes within industries, rather than fully addressing the territorial social, and environmental challenges and opportunities presented by transitions (Interview, 2024). Although the plans highlight different territorial specificities in each region, contextualizing the plans in place-based knowledge and information, the technical focus of the targeted actions indicates that these other important aspects of the transition have been neglected in the planning process.

From the outset, the single sector focus of the plans effectively restricted the role of other regional and local actors. The regional perspective was represented and voiced but within the context of the sectoral framing. One interviewee noted that restrictions on sub-national level actors were also exacerbated by the short time frame allocated for developing the plans, which left little room for widespread consultation and collaboration with other regional and local stakeholders and citizens (Interview, 2024). In the end, the targeted actions outlined in the TJTPs in Sweden prioritized internal transition processes within individual sectors, which meant there was little consideration given to the potential for cross-sectoral integration. Furthermore, there was relatively little adaptability and flexibility shown during the planning process. Initial drafts of the plans were only amended to include societal driven actions when the results of regional socio-economic impact assessments revealed the potential for high unemployment if retraining and reskilling measures were not introduced. In the final analysis, the European Commission's efforts to promote place-based policymaking concepts like TG and active subsidiarity is to be commended, but they are not enough to empower regional and local institutions and actors in EU TJTP processes. If sub-national level actors are going to be truly empowered to develop TJTPs that address regional challenges and opportunities, then the planning process needs to be built on a TG logic at the very outset.

22.4.1 Message for the Western Balkans

Western Balkans countries can take a TG approach as they embark on their just green transition journeys. For Western Balkan countries embarking on just green transition processes, there is much that can be learned from examining the EU TJTP process in Sweden through a TG lens; (1) national governments need to delegate responsibility for the planning process to regional and municipal authorities who possess the prerequisite local knowledge and expertise and can mobilize existing regional stakeholder networks; (2) the largest polluting industries in the region have a central role to play in the process, but a cross-sector dialogue is needed to address the knock-on effects of single sector transitions on other industries and businesses, and the potential for cross-sectoral diversification; (3) adequate time needs to be allocated to the planning process so relevant regional stakeholders can be consulted in the development of the first draft of plans, particularly societal groups with an understanding of the potential socio-economic impacts of transitions for citizens; (4) explore the potential for directly involving citizens through the use of digital and non-digital citizen engagement tools to enhance awareness and support for transitions; (5) ensure that transition plans reflect a balance between regional Environmental Impact Assessment and Socio-economic Impact Assessment; (6) enhance the institutional capacity of regions and municipalities to facilitate these processes as the limited resources of smaller municipalities can affect their ability to provide essential guiding infrastructures. A JGT in the Western Balkans built on a TG approach can help produce JGT policies that address regional challenges and opportunities and meet the needs of citizens.

References

Böhme, K., Zillmer, S., Toptsidou, M., & Holstein, F. (2015). *Territorial governance and Cohesion Policy*. European Parliament: Directorate General for Internal Policies. IPOL_STU(2015)563382_EN.pdf

Cotella, G. (2018). Editorial. EU cohesion policy and domestic territorial governance. What chances for cross-fertilization? *Europa XXI, 35*, 5–20. https://doi.org/10.7163/eu21.2018.35.1

Davoudi, S., & Cowie, P. (2016). Guiding principles of 'good' territorial governance. In Schmidtt & Van Well (Eds.), *Territorial governance across Europe: Pathways, practices and prospects* (pp. 48–59). Routledge.

Davoudi, S., Evans, N., Governa, F., & Santangelo, M. (2008). Territorial governance in the making: Approaches, methodologies, practices. *Boletín de la Asociación de Geógrafos Españoles, 46*, 33–52. Territorial governance in the making. Approaches, methodologies, practices - Dialnet

European Commission. (2008). *The green paper on territorial cohesion—The way ahead, Inforegio Panorama* (pp. 4–7). Inforegio - Panorama 28 - Green Paper on Territorial Cohesion, the way ahead

European Commission. (2019). *Communication from the Commission—The European Green deal, COM, 2019, 640 Final*. https://eur-lex.europa.eu/legal-content/EN/TXT/?uri=COM%3A2019%3A640%3AFIN.

European Commission, (2021). *The Just Transition mechanism: Making sure no one is left behind*. https://ec.europa.eu/info/strategy/priorities-2019-2024/european-green-deal/actions-being-taken-eu/just-transition-mechanism_en

Interview. (2022). *10 Interviews conducted with TJTP stakeholders as part of Master's Thesis*.

Interview. (2024). *1 Interview conducted with Tillväxtverket Agency for Economic and Regional Growth as part of the Horizon Europe DUST project*.

Moodie, J., Tapia, C., Löfving, L., Gassen, N. S., & Cedergren, E. (2021a). Towards a territorially just climate transition—Assessing the Swedish EU territorial just transition plan development process. *Sustainability, 13*(13), 7505. https://doi.org/10.3390/su13137505

Moodie, J. R., Wøien Meijer, M., Salenius, V., & Kull, M. (2021b). Territorial governance and smart specialisation: Empowering the sub-national level in EU regional policy. *Territory, Politics, Governance, 11*(7), 1392–1412. https://doi.org/10.1080/21622671.2021.1905542

Moodie, J. R., Salenius, V., & Wøien Meijer, M. (2021c). Why territory matters for implementing active subsidiarity in EU regional policy. *Regional Studies, 56*(5), 866–878. https://doi.org/10.1080/00343404.2021.1966404-

Moodie, J. R., Salenius, V., & Kull, M. (2022). From impact assessments towards proactive citizen engagement in EU cohesion policy. *Regional Science Policy & Practice, 14*(5), 1113–1133.

Organisation for Economic Co-operation and Development (OECD). (2020). *A territorial approach to sustainable development goals: Synthesis report, OECD urban policy reviews*. Development Co-operation Report 2020|OECD

Schmitt, P., & van Well, L. (2016). *Territorial governance across Europe*. Routledge. https://doi.org/10.4324/9781315716220

Shaneh, A. (2023). *Towards a just and green transition. The making and implementation of Just Transition Plans in Sweden and what can we learn from it*. https://webthesis.biblio.polito.it/25898/1/tesi.pdf

Smas, L., & Lidmo, J. (2018). Organising regions: Spatial planning and territorial governance practices in two Swedish regions. *Europa XXI, 35*, 21–36. https://doi.org/10.7163/Eu21.2018.35.2

Stead, D. (2014). The rise of territorial governance in European policy. *European Planning Studies, 22*(7), 1368–1383. https://doi.org/10.1080/09654313.2013.786684

Van Well, L., & Schmitt, P. (2015). Understanding territorial governance: Conceptual and practical implications. *Europa Regional, 21*(4), 209–221. https://nbn-resolving.org/urn:nbn:de:0168-ssoar-457127

Van Well, L., van der Keur, P., Harjanne, A., Pagneux, E., Perrels, A., & Henriksen, H. J. (2018). Resilience to natural hazards: An analysis of territorial governance in the Nordic countries. *International Journal of Disaster Risk Reduction, 31*, 1283–1294. https://doi.org/10.1016/j.ijdrr.2018.01.005

CHAPTER 23

Synergy Effects of Just Green Transitions: From Overall Challenges to Innovative Responses in the Case of a Slovakia 2030

Maros Finka

23.1 Introduction: Transition, Visioning, and Planning

After three decades since 1990 in which the societal and academic debates on transition, especially in Europe, have focused on economic transition and transition economies changing from the centrally planned economies to market economies undergoing a set of structural transformations intended to liberalise economics, developing market-based institutions, privatising state-owned enterprises and resources, transforming financial sector to facilitate macroeconomic stabilisation, creating environment for free movement of private capital, the end of the second decade of new millennium brought significant shift in the content of these debates. As a crucial turning point the publishing of the European Green Deal (COM, 2019) by the European Commission can be perceived, announcing a new start towards meeting the goals of the Agenda 2030 (UN, 2019), commitments in the field of climate change and environmental challenges, together with the transformation of the EU into a modern, resource-efficient and competitive economy. Discussing pressing problems and challenges of these transition processes the awareness about lesson learned on complexity of societal contexts, interrelations and effects from the history of economic transitions was not adequately present (see e.g. Lavigne, 1995, Filipov & Dobritz, 2003, Raskin, 2016). The macroeconomic dimension of

M. Finka (✉)
Faculty of Management, SPECTRA Centre of Excellence of the EU, Comenius University, Bratislava, Slovakia
e-mail: maros.finka@stuba.sk

E. Berisha et al. (eds.), *The Palgrave Handbook of Just Green Transitions in the Western Balkans and Beyond*,
https://doi.org/10.1007/978-3-031-95075-9_23

the transition has dominated the balance between all 3 dimensions (economic, social and environmental) of the sustainability of the transition processes and effects, although the history of crucial transitions in mankind's development clearly shows how fatal such imbalances can be. As explained in the book Epochal Transitions (Smil, 2021), the modern world has been created by five relatively rapid and interconnected transitions. All of them are predominantly of economic nature (increase of agricultural productivity, transition from traditional biofuels to fossil fuels, extensive industrialisation of goods production, robotization, and digitalisation) clearly interlinking economic changes with principal changes in the society and especially humans-nature relationship. Not going deep into the history, at least one for modern world decision making conceptual change in relation to these epochal transitions has to be mentioned. This change is linked with the activities of the funders and members of the Club of Rome at the turn of sixties and seventies of the XX. Century, addressing dramatic scientific and technological changes happening in the world and in the same time lack of confidence that the problems faced by the world could be solved by development and technological progress. The background paper for their meeting in Rome in 1968 entitled "A Tentative Framework for Initiating System Wide Planning of World Scope" (Jantsch, 1968) prepared by the OEDC consultant Erich Jantsch catalysed discussion in OECD and formulation of "The Bellagio Declaration on Planning" (see: Schmelzer, 2017), emphasising the need to overcome global problems by prospective coordination and planning, as well as in 1972 published first report of the Club of Rome "The Limits to Growth" (Meadows, 2008). Increased public concern about the problem of natural resources limits turned the public attention from the social dimension to economic and environmental in the shadow of the 1973 oil crisis. But the understanding of necessary transformation as of the generalised meta-problem created by our past successes and failures determining the magnitude of gradually unfolded challenges problematized by Club of Rome during its formation sustained. The discussion on the social dimension of the transition processes as a meta-problem came alive in the 1980s, articulated by the US trade unions to protect workers affected by new water and air pollution regulations using the concept of "just transition". Recently, it gained traction with reference to the implementation of measures towards meeting the sustainable development goals (SDGs) as declared by the Agenda 2030, including greening the economy in a fair and inclusive process of transition, creating decent work opportunities, heading the net-zero economy goal, strengthening resilience and leaving no one behind. Recently, the justness, as the quality of transition processes, left the direct social connotations, creating the sound basis for fostering the dialogue on evidence-based and place-based approaches in developing a common vision for what a just transition recently means regarding different stakeholders, regions and places with their specific natural, cultural, and economic features. With the increasing recognition of the importance of such transition by the governments, the concept of the just transition starts to be operationalised in numerous short- and long-term strategies and plans (e.g. known as Nationally Determined Contributions, or NDCs) across different levels between global and local. Based on the areas in which the concept of just green transition can frame the adaptation processes to climate change and greening the economies defined by the UNDP in its document How Just Transition Can Help Deliver the Paris Agreement (Smith, 2022) the 5 main directions of future transition can be formulated as follows:

- bringing the public along, activating the public involvement into the co-creation and co-designing processes shaping the place-based and evidence-based approaches, strategies, plans and instruments,
- supporting a green jobs revolution in which the created alternative jobs have got guaranteed living wages, proper workplace safety protection and health benefits, and in which the technology-based solutions are alternated by nature close solutions sustainably using ecosystem services,
- laying the social groundwork for a resilient net-zero economy with transparent planning and active participation on solutions development and decision making addressing the intergenerational and inter-territorial justness in using the resources and limited capacity of our planet,
- driving local/regional solutions implemented by local/regional communities‘ actions using cooperation, sharing and work division for efficient addressing the challenges of just transition,
- *reinforcing the urgency for concerted efforts to combat climate change, combining the flexibility and addressability of short-term local actions with medium and long-term plans and strategies, with clear visions for the future at the regional, national, and global levels, in spite of growing uncertainties and development dynamics.*

In parallel with listing the directions framing the just green transition, the UNDP in its document Just Energy Transition: Governance needs and implications (UNDP, 2023) stresses broad agreement that governance approaches to manage a fair green transition are much less discussed and understood, listing several crucial questions, emerging in this context. The common denominator among them is the question of new quality of governance systems which does not only promote fair green transitions in terms of institutional arrangements and capabilities of public institutions but safeguards the achievements of transition goals in open decision- and policy-making and implementation processes, respecting justness of the interests and power dynamics between actors. It means a governance system able to facilitate and accelerate just transition through a participatory process of visioning, planning, experimenting and self-learning, efficient and effective implementing, and transferring solutions, using multiple reflective approaches to reflect multiple viewpoints in multi-dimensional, multi-level and polycentric, multi-actor, and multi-phase process of integrated multi-sectorial transformation

(Finka & Husar 2021; Foxon et al., 2008; Rotmans et al., 2001). The realistic view on the achievements reflecting our abilities to implement SDGs in real life (Pattberg & Bäckstrand, 2023) in the light of increased climate change dynamics, growing social, disparities, destroyed ecosystems across the world, serious economic crises, spread of poverty or large-scale migration in last decade shows that the transformation became our serious existential challenge urging us to think otherwise. This includes the role of the utopian impulse in enabling us to see the existing conditions not as how things are, but as how they are made to be, and how they might be unmade (Davoudi, 2023).

The Western Balkan states face the challenge of multiple transitions given by their historical development, state of art of their natural, social, economic, and political environments, current global changes and their European integration ambitions. Efficient mastering the transition can be supported by realistic view on running processes, their political and conceptual frameworks and available solutions based on experience from other countries, especially benefiting from similar situation in the Central European countries (Finka, 2011) at the turn of Millenia. This chapter tries to mediate their tacit knowledge by introducing realistic view on past and current transition concepts as a base for critical perception of the Green Deal as a political framework from the position of planning practice for multiple transition in Central Europe, in order to outline possible conceptual and practical solutions suitable for Western Balkan as inherent part of integrated Europe demonstrated on an example of Vision and Strategy of Development Slovakia 2030.

23.2 Green Deal—Green Transition Between Promises, Expectations, and Reality in the EU

Recent development in Europe is determined by multiple synergies between problems and challenges in environmental, social, economic, and territorial dimensions of the development including the unprecedented aggression and war in Ukraine with global effects on economy, security, environmental, and social issues.

Before the COVID-19 pandemic and Russian aggression in Ukraine, climate change was perceived by EU officials (Leyen, 2018) as the biggest challenge of our times and, at the same time, an opportunity to build a new economic model. When the European Commission published the European Green Deal, a new start towards meeting the goals of the Agenda 2030, commitments in the field of climate change and solving environmental challenges were announced, together with the transformation of the EU into a modern, resource-efficient and competitive economy. At the same time, the aim was declared to implement the Green Deal as fair and inclusive in the spirit of the universal principle of the Agenda 2030—let‘s not leave anyone and any place behind. Although the European Green Deal, addressing the climate change, became the first from 6 priorities defined by the European Commission for the period 2019/2024 and beyond, the positioning of the green transition in the complex societal transition on the way towards new development (not only economic) models for Europe has not been fully addressed. European Green Deal appeared on the list of seemingly complementary priorities as "An economy that works for people, A Europe fit for the digital age, Protecting our European way of life, A stronger Europe in the world, A new push for European democracy" (Leyen, 2018). Although being declared as a blueprint for rather complex transformational change affecting different sectors, followed subsequently by a set of sectorial documents (e.g. Industrial Strategy (COM, 2020a), Action Plan for the Circular Economy (COM, 2021), Long-term Climate Strategy of the European Union, or Farm-to-Fork Strategy (COM, 2020b), the Green Deal deals mostly with the environmental transition demands on economic sectors, some cross-sectorial effects and benefits (e.g. creating new opportunities for innovation, investment and green jobs, improving our health and well-being), but it is not addressing the synergies between green and other running transitional processes across different spheres of societal life (e.g. digitalisation, fuzzification, virtualisation, shift from government to governance, civil society development with elements of direct democracy, development of Manufacturing X, demographic change, etc.). This often led to the misperception of the green transition as a predominantly sectorial environmental strategy or "greening economy" strategy (Adamowicz, 2022) and less as a core of comprehensive development and transformation strategy with clear territorial embeddedness in respective scales considering transformative potential of integrated development management. The lack of attention to the synergies inside the complexity of the transformative processes and to the territorial specifics has provoked heavy criticism of the Green Deal‘s goals and aspects of implementation, up to its refusal as potentially being destructive to the European Union in its state (Gregoire, 2023; Oroschackoff & Hernandez-Morales, 2020). As the data published in the European Green Deal Barometer 2023 IEEP (2023) show, the European Green Deal implementation starves on multiple threats resulting from the above-mentioned insufficiencies, as well as insufficient commitment by the member states governments, top-down approach of Green Deal proponents not involving all stakeholders, different opinions on what sustainability means (economic, social, environmental) and absence of systematic approach to the governance of the EGD within the EU institutions (COM, 2023a). The missing embeddedness of the top-down composed Green Deal strategy opened the space for growing opposition to the EU Green Deal, especially in the context of appearance of the not properly balanced negative effects of its implementation and of European elections. Farmers, representatives of industry, the public, and even governments are questioning it and calling for a "regulatory pause" (Malingre, 2024), which is in contrast with recent concerns of the scientist about the growing speed of climate change reflecting the record temperatures in 2023 and at start of 2024 and more frequent weather events as an evidence of the acceleration of the climate crisis (Watts & Ahmedzade, 2024). The European Commission acknowledges that the stable long-term strategies across all levels of governance are crucial to help achieve the economic transformation needed and broader sustainable development goals (COM, 2024). Different levels of prospective view, definition of goals and priorities from local, regional via national, European, up to the global should reflect differences and needs for evidence and place-based approaches to the development of visions and strategies with green transition as the core element. The development of such prospective documents should be encouraged by the establishment of Just Transition Mechanism (JTM) as a key tool to ensure that the transition towards a climate-neutral economy happens, providing targeted support to help mobilise around €55 billion over the period 2021–2027 in the most affected regions, including the alleviation of the socio-economic impact of the transition. The update of the integrated national energy and climate plans elaborated by the EU member states and submitted in accordance with the EU's Governance regulation in 2019 should represent important step towards delivering the Green Deal. But in the assessment of the national energy and climate plans (NECP) by the European Commission (COM, 2023b) showed not only remaining ambition gaps to achieve the agreed increased targets and objectives for 2030 in climate and energy policies, but remaining needs to put in place sufficient policies and measures across different sectors. They must address elementary preconditions for Green Deal implementation, such as the increase of the renewable energy share in energy consumption, improvement of energy efficiency, comprehensive regulatory framework to reinforce electricity grids to integrate renewables and technological innovations, to encourage the uptake of flexibility assets, energy poverty and sustainable energy policies for vulnerable households, concentrate measures to attract private finance. Moreover, the assessment showed the gaps not only in the implementation phase, but in adequate policies, plans, and measures addressing just transition. The report shows that the EU Member States have only provided a partial assessment of the socio-economic impacts of the climate and energy transition on individuals, households and companies, and the related policies and measures often lack a strategic and forward-looking vision. Only a few Member States demonstrated concrete evidence of involvement of local and regional authorities, important partners for the implementation of energy and climate policies into the process of preparing the draft updated NECP, and even fewer are building on an established multi-level dialogue for this process. Such a process is the precondition not only for acceptance of the transformation policies framing the processes of societal change but, first of all, for active involvement of civil society in this multi-actor, multi-dimensional, multi-level, multi-sectorial, multi-territorial, and multi-phase transformational endeavour. To have the whole scale of actors on board, it needs to offer a realistic vision and plan transparently outlining the benefits and risks for each group in the society. This is, being confronted with the real effects of transformation gaps, lacking resilience and adaptation capacities, rather sensibilised and tending to follow the offered vision and participate in solutions' implementation as well as the hoaxes, misinterpretations, and disinformation. As the recent OECD Survey 2022 in 27 OECD countries (OECD, 2023) shows, climate change is seen as the fifth most chosen area for higher government prioritisation, with 59% of respondents calling for the government to prioritise the issue, 75% of respondents report being somewhat or very concerned about their household's finances and economic well-being when looking ahead through the next years. Looking ahead to the next decade, 75% of respondents, on average, worry about financial security in old age, while 71% worry about accessing good-quality healthcare in the future. Also, 67% and 65% of respondents, respectively, worry about accessing good-quality long-term care for themselves or for elderly family members in the next decade. Nearly half (48%) of respondents on average across countries believe that their government is not doing enough to tackle climate change when bearing in mind the costs and benefits of climate change policies. Across countries, 62% of respondents worry about job losses in carbon-intensive industries and 88% of respondents are concerned about price increases for energy, food and other goods. This is a clear expression of expectation and signals that a successful green transition has to be embedded into the overall societal transition processes, the governance of which must be integrative and integrated, involving all policy and transition areas as well as all societal groups. The social policy should not only play an important role in making sure that opportunities and challenges are shared across groups but also play a key role in getting support from those who may suffer losses from labour market friction and structural change caused by climate change mitigation and adaptation policies.

23.3 Visions and Strategies for the Development of Slovakia

The situation, as described above, represents a crucial challenge for policy-makers responsible not only for setting the goals of transition but also for transformation management, including the European level. As the problems perceived by the citizens and identified by the OECD survey have the character of existential questions, changing the way of thinking is needed. There is an urgent need to replace old schemes of sectorial policy-making implemented by top-down prescriptions, regulations and limitations only marginally reflecting the inter- and cross-sectorial relations and synergy effects and leaving minimum space for visioning, integrative, innovative, creative, place-based specific approaches, and experiments. As Davoudi (2023) wrote, planning systems, representing important tools of transformation management, have not been immune either. In many European countries, similarly to England, planning has been subject to not only debilitating cuts to its financial and human resources, but also demoralising attacks on its utility, validity, and legitimacy, losing its creative potential and connection to the public and becoming increasingly beholden to the developer rather than the people and public interest it is meant to serve. Davoudi introduced an alternative mode of planning as prefiguration seems to be one of the possible reflections of the above-identified challenges for the transformation process management. The concept of prefigurative planning as a revival of a "philosophy of the possible" and a call for the rejection of the given and the envisioning of alternatives represent the requested shift from pre-defining a postponed future, to performing the possibilities of the not-yet in the here and now. This mode of planning is about initiating, forging, enabling, and engaging in alternative ways of intervening in space–time from within the conditions of the present but without being limited to them. The combination of the prefigurative planning in the combination with proactive multilevel and multi-actors polycentric decision making (Finka & Kluvánková, 2015) can be the right alternative to the common approaches of reactive sequential planning of predefined events contrasting with high level of uncertainty we face recently.

The prefigurative planning is not only an academic construct. It provided a successful conceptual framework for the development of the document Vision and Strategy of the Development of Slovakia—Long Term Strategy for Sustainable Development (Slovakia 2030) as an implementation document of Agenda 2030 in Slovakia (MIRRI, 2020). It was approved by the Government of Slovak Republic in 2021 and the work continues with the development of new Vision Slovakia 2050. The participative process of the documents‘ development has faced high level of uncertainties in regard to framework conditions, multi-actors‘ decisions and political will in dynamically changing political environment. The use of the prefigurative planning concept created space for interplay between critique analyses of the present situation and clearly identified gaps in its sustainability and the engagement with the prospects of future transition processes whose exact forms are not known a priori.

At the heart of the document Slovakia 2030 is the outline of a vision and three integrated transition areas—fields of crucial changes as a precondition for achieving the vision. The vision of Slovakia as a country with a high standard of living, high-quality public services and a healthy sustainable environment for all residents relies on vital regions with educated and productive human capital, a functioning economy capitalising own country resources in a sustainable way and quality infrastructure. Economic development relying on innovation and industries with high added value is understood as the base for an increase in the standard of living of the inhabitants and a balanced ratio of work and family life (social) life.

It means:

- The country—a quality place to live, a state that effectively fulfils its functions and a public sector with quality and affordable services
- Residents—equal opportunities for every individual and equal access to everyone to an individual
- Communities—responsibility, cooperation, solidarity

- Regions—space for fulfilling human needs, successful, sustainably developing providing quality and a safe environment for a healthy and fulfilling life for all residents
- Development—improving the quality of life
- Sustainability—using resources in a way that does not limit future generations
- Environment—preservation and sustainable development of the natural environment and cultural heritage as a prerequisite for a healthy and fulfilling life
- Security—resilience and preparedness of the state for external and internal threats
- Full life—life lived in health, balance between work and other parts of life, sufficient space for education and cultural enjoyment.

This vision is not any pre-determined and fixed utopia of modernist planning; it is an outline of the future rooted in our present possibilities allowing the way towards in different trajectories of development processes in 3 transition areas as follows:

- protection and development of resources (natural, human, institutional, financial, others)
- use of resources (in economic and non-economic sectors),
- quality of life and development of communities (justness and affordability of the quality of life for all in all parts of the country)

They are understood as cross-cutting, integrating issues of different departments and different levels of state administration, regional, and local self-government.

Each transition programme, based on the results of extensive analytical work summarised in the introduction part of their description, is defined by:

- the goal of the integrated transition programme,
- the expected key changes that should be achieved through its implementation,
- key tasks, activities/actions that need to be implemented to achieve the desired change.

The descriptions of the 3 main areas of transition are in this context interlinking the rejection of given and envisioning the alternatives for the future representing the requested shift from pre-defining a postponed future, to performing the possibilities of the not-yet in the here being inspired by the concept of prefigurative planning.

The integrated transition area Protection and development of resources is defined by:

The goal: Growth and qualitative development of the population, with special emphasis on education, health, culture, civil development society and the rule of law and the protection and sustainable management and development of natural resources.

A key change: Enforcing the protection and development of human and natural resources as a superior public interest through strengthening the resilience of the state and society with an emphasis on the areas of:

- population growth and health, education and social development,
- rule of law and security,
- protection and improvement of natural and cultural resources 6 areas of necessary tasks, activities/actions to foster the change:
- stop the population decline and support the growth of the share of the economically active population
- improvement of the availability and quality of education and training of the population and harmonising them with the expected demand of the labour market
- improvement of the health status and length of active life of the population
- ensuring the full application of the principles of the rule and law and increase the security of the state and society
- ensuring effective and sustainable management of natural resources
- ensuring effective and sustainable management of cultural resources.

The integrated transition area Use of resources defined by

The goal: Transformation of the SR economy into a sustainable one, whose competitiveness is based on innovative and efficient exploitation of resources and which generates good wages and prosperity.

A key change: Transformation of the domestic economy towards:

- the use of innovations for the development and sustainable evaluation of the potential of regions,
- creation of high added value with the support of quality infrastructure,
- environmental and social sustainability.

Three areas of necessary tasks, activities/actions to foster the change include:

- support of the development of strong innovation-based regional economies
- completion of the infrastructure of an innovation-based economy

- improvement of sustainability and resilience of national and regional economies.

The integrated transition area Quality of life and development of communities defined by:

> The goal: Developing communities and increasing the quality of life for all social groups through rapprochement governance to citizens and social inclusion and accessible, high-quality and effective services;
>
> A key change: Increase support for the development of civil society and the participation of citizens, socio-economic partners and others interested actors in the creation, implementation and control of public policies;
>
> Strengthening of territorial principle as a basis for defining development policies, harmonizing and integrating sectoral policies and policies different hierarchical levels of public administration in the interest of citizens, communities and their sustainable development;
>
> Optimization of the system of state administration and self-government based on the distribution of powers, responsibilities and resources based on the principle of subsidiarity, partnership cooperation and synergy between individual levels and sectors public administration in order to effectively meet the needs of citizens;
>
> Availability, efficiency of operation and sustainability of infrastructure development for healthy and cohesive communities, creating conditions for the full-value life of the inhabitants of each region of Slovakia, regardless of their socio-economic background or disadvantage;

Three areas of necessary tasks, activities/actions to foster the change are defined as follows:

- improvement of accessibility, transparency, and efficiency of public administration
- strengthening the economic sustainability of communities, municipalities, cities, and regions
- ensuring the availability and sustainability of high-quality, modern and efficient infrastructure, services and housing for the healthy communities (social and health infrastructure, educational infrastructure, technical infrastructure, housing, public spaces, objects and facilities, services and retail, transport infrastructure, cultural infrastructure)

The use of the concept of prefigurative planning has been in harmony with the effort to look beyond Agenda 2030 towards the integration of sectorial policies based on a common denominator defined as Slovakia are people and places for their contented and full-fledged life. The aim is to strengthen its framework for implementing the Agenda 2030 while exploring the connection between well-being and sustainable development indicators as a base for the integration of transition processes. This shift in emphasis towards the interplay between sustainable development and well-being would enable a broader focus and the inclusion of different components, broadening the scope beyond the 17 Sustainable Development Goals (SDGs), and establishing longer-term goals specifically addressing the country's challenges.

23.4 Parallels, Inspirations, and Warnings

As we tried to show in this paper, despite all the specifics of European countries including the Western Balkan, the just green transformation is a challenging process we have to absolve jointly. It is a running process of learning by doing, bringing together multiple stakeholders with their specifics, different interests, many innovative solutions, and mistakes. It is a process of mutual learning in which the Western Balkan countries can bring their own tacit knowledge as Central Europe, including Slovakia, did. It can be not perceived as a West–East lesson learned; it is about the joint development of solutions built up on shared positive and negative experiences. The Balkan‘s sensitivity for justness and territoriality synthetising the place, evidence and historic specifics in the countries, regions and cities, in the combination with Central European tradition of multilevel polycentricity and prospective planning bringing the ratio into the territorial organisation and belonging, and with the Western European citizens‘ engagement experience can open new perspective for prefigurative planning as the framework for JGT planning. As the Slovakian example shows, the JGT planning referring to the inter-sectorial synergies and prospective acting of multi-actors, based not on ordinary planning‘s prescriptive pictures but on process oriented prefigurative planning concept, can help to overcome the perception of the appropriateness of sectorial architecture of governmental systems not only in Western Balkan countries. The appearance of parallel cross-sectorial planning and governance structures, e.g. focused on Agenda 2030 or Green Deal implementation, can represent the positive deviation points in the transformation of the sectorial governmental system towards territorial governance systems in Western Balkan and across Europe.

23.5 Conclusions

Recent development in Europe, with its multiple synergies between problems and challenges in environmental, social, economic and territorial dimensions of the development, requires new ways of thinking, opening the space for flexible reflection and implementation of innovative solutions, taking into account these synergies and mutual interdependences going beyond sectorial approaches and policies. The assessment of the achievements and gaps in the implementation of the Agenda 2030 shows how pressing is this change. This means that the green transition should not be perceived as environmental strategy, but as a core of comprehensive development strategy with clear territorial embeddedness in respective scale, and that the new approaches to the transition governance including planning are necessary. The contexts of JGT should create a core part of the integrated visions and comprehensive development strategies of 2050 reflecting the synergies between different transition processes and between different technology, social, environmental, economic, institutional, and other kinds of innovations. Mirroring critical reflection of assessment of the Agenda 2030 implementation process and of realistic ambitions for its extending to the year 2050 there is a principal shift in thinking about transformation management, especially planning and implementing needed including different levels of prospective view, definition of goals and priorities from regional via national, European up to the global underlying differences and needs for evidence and place-based approaches to the development of visions and strategies with green transition as the core element. The concept of prefigurative planning is offering an alternative innovative instrument of planning change as a reflection of identified problems and challenges. It can create a solid base for facing the growing uncertainties in multi-actors, multi-dimensional, multi-level, multi-sectorial, multi-territorial, and multi-phase transformational endeavour.

References

Adamowicz, M. (2022). Green deal, green growth and green economy as a means of support for attaining the sustainable development goals. *Sustainability 2022, 14, 5901.* https://doi.org/10.3390/su14105901. https://tepsa.eu/analysis/next-steps-for-the-european-green-deal-navigating-economic-and-security-challenges-keeping-a-focus-on-climate-and-justice/. Retrieved May 1, 2024.

COM. (2019). Communication from the Commission to the European Parliament, the European Council, The Council, The European Economic and Social Committee and the Committee of the Regions—The European Green Deal. COM/2019/640 final.

COM. (2020a). Communication from the Commission to the European Parliament, the European Council, the Council, The European Economic and Social Committee and the Committee of the Regions—A New Industrial Strategy for Europe, European Commission, Brussels.

COM. (2020b). Communication from the Commission to the European Parliament, the European Council, the Council, The European Economic and Social Committee and the Committee of the Regions—A Farm to Fork Strategy for a Fair, Healthy and Environmentally-friendly Food System, COM/2020/381 final, European Commission, Brussels.

COM. (2021). Communication from the Commission to the European Parliament, the European Council, the Council, The European Economic and Social Committee and the Committee of the Regions—A New Circular Economy Action Plan for a Cleaner and More Competitive Europe, COM/2020/98, European Commission, Brussels.

COM. (2023a). *796 final.* https://eurlex.europa.eu/legalcontent/EN/TXT/h?uri=COM%3A2023%3A796%3AFIN. Retrieved May 1, 2024.

COM. (2023b). Communication from the Commission to the European Parliament, the European Council, the Council, The European Economic and Social Committee and the Committee of the Regions—EU wide assessment of the draft updated National Energy and Climate Plans (NECP)—An important step towards the more ambitious 2030 energy and climate objectives under the European Green Deal and RePowerEU, Brussels, 18.12.2023.

COM. (2024). *National long-term strategies—EU countries' long-term strategies to meet their Paris Agreement commitments and the energy union objectives*, Overview, https://commission.europa.eu/energy-climate-change-environment/implementation-eu-countries/energy-and-climate-governance-and-reporting/national-long-term-strategies_en. Retrieved May 1, 2024.

Davoudi, S. (2023). Prefigurative planning: Performing concrete utopias in the here and now. *European Planning Studies, 31*(11), 2277–2290.

Filipov, D., & Dobritz, J. (2003). *Demographic consequences of Economic Transition in Countries of Central and Eastern Europe*. Council of Europe Publishing Strasbourg, p. 29

Finka, M. (2011). Evolving frameworks for regional development and spatial planning in the new regions of the EU. In *Territorial Development, Cohesion and Spatial Planning: Knowledge and policy development in an enlarged EU* (pp. 103–125). Routledge.

Finka, M., & Kluvánková, T. (2015). Managing complexity of urban systems: A polycentric approach. *In Land Use Policy, 42*(2015), 602–608.

Finka, M., & Husar, M. (2021). Multilevel polycentric governance in urban development policies—National urban policy structure in Slovakia. In *A modern Guide to National Urban Policies in Europe* (pp. 245–267).

Foxon, T., Hammond, G. P., & Pearson P. J. (2008). *Transition pathways for a low carbon energy system in the UK: Assessing the compatibility of large-scale and small-scale options*, 7th BIEE Academic Conference, St Johns College.

Gregoire, L. (2023). Opposition to EU Green deal grows as European elections approach. In *Euronews* 04/12/2023, https://www.euronews.com/my-europe/2023/12/04/opposition-to-eu-green-deal-grows-as-european-elections-approach Retrieved May 1, 2024.

IEEP. (2023). *European Green Deal Barometer 2023, Home Publication Institute for European Environmental Policy 18 June 2023*. https://ieep.eu/publications/european-green-deal-barometer-2023/. Retrieved May 1, 2024.

Jantsch, E. (1968). *A tentative framework for initiating system-wide planning on world scope* (unpublished draft), in AP/Box 150, 45 pp.

Lavigne, M. (1995). *The economies of transition: From socialist economy to market economy* (pp. 116, 117, 121). Macmillan.

Leyen, U. (2018). *My agenda for Europe—Political guidelines for the next European Commission 2019–2024*. https://commission.europa.eu/document/download/063d44e9-04ed-4033-acf9-639ecb187e87_en?filename=political-guidelines-next-commission_en.pdf. Retrieved May 1, 2024.

Malingre, V. (2024, January 29). Europe's green deal is attacked in all sides. In *Le Monde*. https://www.lemonde.fr/en/environment/article/2024/01/29/europe-s-green-deal-is-attacked-on-all-sides_6474382_114.html. Retrieved May 1, 2024.

Meadows, D. (2008). 30-Year of Limits to Growth finds global society in "Overshoot" *Club of Rome*. Archived from the original (PDF) on 2008–05–17. Retrieved June 30, 2016.

Meadows, D. H., Meadows, D. L., Randers, J., & Behrens, W. W. (1972). *The limits to growth—A report for the club of Rome's project on the predicament of mankind*. Universe Books.

MIRRI. (2020). *Vízia a stratégia rozvoja Slovenska do roku 2030—dlhodobá stratégia udržateľného rozvoja Slovenskej republiky—Slovensko 2030*. https://mirri.gov.sk/wp-content/uploads/2021/01/Slovensko-2030.pdf. Retrieved May 1, 2024.

OECD. (2023). *Main findings from the 2022 OECD Risks that matter survey*. http://oe.cd/rtm. Retrieved May 1, 2024.

Oroschackoff, K., & Hernandez-Morales, A. (2020, May 16). *EU climate law sparts political battles*. In Politico. https://www.politico.eu/article/eu-climate-law-sparks-political-battles/. Retrieved May 1, 2024.

Pattberg, P., & Bäckstrand, K. (2023). Enhancing the achievement of the SDGs: Lessons learned at the half-way point of the 2030 Agenda. *In Environ Agreements, 23*, 107–114 pp. https://doi.org/10.1007/s10784-023-09615-9. Retrieved May 1, 2024.

Raskin, P. (2016). *Journey to Earthland: A great transition to planetary civilization*. Tellus Institute Boston.

Rotmans, J., Kemp, R., & Asselt, M. (2001). More evolution than revolution: Transition management in public policy. *Foresight, 3*(1), 15–31.

Schmelzer, M. (2017). Born in the corridors of the OECD: The forgotten origins of the club of Rome, transnational networks and the 1970s in global history. *Journal of Global History, 12*(1), 26–48.

Smil, V. (2021). Epochal transitions. *Grand transitions: How the modern world was made* (online ed). Oxford Academic. https://doi.org/10.1093/oso/9780190060664.003.0001, Retrieved 1 May 2024.

Smith, K. J. (Ed.). (2022). *How just transition can help deliver the Paris Agreement*. UNDP.

UN. (2019). *Transforming our world: The 2030 agenda for sustainable development, A/RES/70/1*, https://sustainabledevelopment.un.org. Retrieved April 7, 2024.

UN. (2023). *Nationally Determined Contributions (NDCs)*. https://unfccc.int/process-and-meetings/the-paris-agreement/nationally-determined-contributions-ndcs. Retrieved May 1, 2024.

UNDP. (2023). *Just energy transition: Governance needs and implications* (UNDP Discussion paper for COP28). UNDP. https://www.undp.org/policy-centre/governance/publications/just-energy-transition-governance-needs-and-implications. Retrieved May 1, 2024.

Watts, J., & Ahmedzade, T. (2024, March 16). Scientists divided over whether record heat is acceleration of climate crisis. *The Guardian*. https://www.theguardian.com/science/2024/mar/16/scientists-divided-record-heat-acceleration-climate-crisis

CHAPTER 24

The Vulnerability of the Romanian Danube Valley to Heat-Related Phenomena in the Context of Just Green Transitions

Bianca Mitrică, Irena Roznovieţchi, Ines Grigorescu, Dana Micu, Paul-Răzvan Serban, and Nicoleta Damian

24.1 Introduction

Just Transition is not a new concept but it emerged in response to the fitful environmental transition that has been going on since the 1970s in the USA (Brecher, 2015, 2020) and the International Trade Union Confederation and the International Labour Organization's Workers' Office were

B. Mitrică (✉) · I. Roznovieţchi · P.-R. Serban
Human Geography and Regional Development Department, Institute of Geography, Romanian Academy, Bucharest, Romania
e-mail: biancadumitrescu78@yahoo.com

I. Roznovieţchi
e-mail: mocanitai@yahoo.com

P.-R. Serban
e-mail: paulrazvanserban@yahoo.co.uk

I. Grigorescu · N. Damian
Environmental Geography and GIS, Institute of Geography, Romanian Academy, Bucharest, Romania
e-mail: inesgrigorescu@yahoo.com

N. Damian
e-mail: icoleta_damian2002@yahoo.com

D. Micu
Department of Climatology, National Meteorological Administration (Meteo Romania), Bucharest, Romania
e-mail: dana.micu@meteoromania.ro

E. Berisha et al. (eds.), *The Palgrave Handbook of Just Green Transitions in the Western Balkans and Beyond*,
https://doi.org/10.1007/978-3-031-95075-9_24

those which made efforts to expand beyond USA the concept of just transition (Dimitris & Felli, 2016) and it "become a central element of global labour environmentalism, especially with respect to global climate policy" (Rosemberg, 2010, Felli, 2013 quoted by Dimitris & Felli, 2016, p. 35). Worldwide, a growing attention is paid to the idea of a just transition from fuel-era to the new post-fuel, the scholars devising various definitions, understandings, meanings of the term (Heffron et al., 2020; Jasanoff, 2018). Clarifying the different viewpoints, the scientific literature and the research agenda focused on the topic of just transition could be synthesized in five main themes: just transition as a labour-oriented concept, as an integrated framework for justice, as a theory of socio-technical transition, just transition as a governance strategy, and as public perception (Wang & Lo, 2021). The social dimension was initially included in the concept of sustainable development, but for almost two decades it has been neglected because it was subsumed within a discourse mainly focused on bio-physical environmental issues or it was covered by the great interests given to economic dimension (Kandachar, 2014; Valaance et al., 2011). Despite its long history, the Social Sustainability (SS) remains a chaos (Vallance et al., 2011) or a fuzzy concept (IPOL, Policy Department for Economic, Scientific and Quality of Life Policies, 2020). The dimensions of SS were topics approached in many recent research and studies (Åhman, 2013; Boström, 2012; Boyer et al., 2016; Najjar, 2022; Hajirasouli & Kumarasuriyar, 2016). Within the worldwide literature the needs in terms of the SS promotion were identified: encourage the community participation, promote improvements in social organizations and community well-being, use natural and recyclable materials in a sustainable way and transfer know-how to those who need it (Abdel-Mohsen et al., 2021). The links between just transitions, climate change, and deindustrialization (Doorey, 2016; Farrell, 2012) were approached through the lens of pair issues such as job transformation and unemployment, the green economy and its hinders caused by various challenges (the gray side of green growth, according to Demiral & Demiral, 2021). According to IPCC (2023), widespread and rapid changes have occurred in the atmosphere, ocean, cryosphere and biosphere and human-caused climate change is already affecting many weather and climate extremes in every region across the globe. Moreover, over the last decades almost every region has witnessed an increase in the occurrence and intensity of the effects of extreme climate phenomena (mostly heat-related) on society and economy (IPCC, 2023). Climate and weather-related hazards are affecting human settlements in different ways and with different intensities, depending on exposure, but particularly on the overall vulnerability. The already experienced and projected impacts underline the need for local assessments of climate risk and vulnerability, based on high-quality data (EEA, 2020). In Europe, the number of extreme precipitation events is expected to increase in the future, droughts will intensify, and heat waves will become more frequent and extended (Alcantara-Ayala et al., 2015; IPCC, 2012). Vulnerability has become a key concept in climate change research studies, particularly under the extension of climate change-related events and the intensification of their environmental and socio-economic consequences. As a result, vulnerability studies have become increasingly popular among scientists in their quest to better link the biophysical and the socio-economic systems, fostering the provision of effective solutions to decision-makers for better addressing disaster risk reduction and management. The vulnerability concept has been often applied to assess the environmental and/or socio-economic vulnerability to various or natural events, some climate change-related (i.e. floods, drought, heat/cold waves, and earthquakes). The Romanian Danube Valley (RDV) is an integral part of the Danube Region. The framework of the sustainable policy integration and coherent development of the territory is provided by the *EU Strategy for the Danube Region* (EUSRD), which sets out priority actions to make the Danube Region a competitive EU region for the twenty-first century. The Danube Region Strategy addresses a wide range of issues with 4 pillars and 12 priority areas, among which PA1A—Waterways/Mobility, P04—Water Quality, PA5—Environmental risks, PA06—Biodiversity, Landscapes and Air & Soil Quality, PA09—People & Skills and PA 10—Institutional capacity & Cooperation (http://www.danube-region.eu) are addressing various aspects related to this chapter. Aiming to analyze the socio-economic vulnerability to heat-related phenomena in the Romanian Danube valley, by computing the Heat Vulnerability Index (HVI), this paper uses and valorizes the methodology developed in studies and projects focused on vulnerability assessed at the national level (e.g. *National Risk Assessment-RO-RISK*, 2016–2018), in different regions of Romania (Dumitrașcu et al., 2018; Grigorescu et al., 2021; Stângă & Grozavu, 2012), and different sectors of the Romanian Danube Valley (Damian et al., 2023; Mocanu et al., 2021). The paper aims to develop a multi-criteria vulnerability assessment by identifying the levels of socio-economic vulnerability by computing a composite Heat Vulnerability Index (HVI). 20 indicators were selected and processed for the vulnerability assessment. The indicators were grouped into the three key components of vulnerability (Exposure, Sensitivity and Adaptive Capacity). Thus, the current research would provide useful scientific products for academics to improve the general understanding of the heat-related vulnerability in an area under the influence of one of Europe's most important rivers (to replicate and/or to improve the methodology for future studies), but also for local decision-makers and planners to include into the local climate change mitigation and adaptation strategies.

24.2 Study-Area

Romanian sector of the Danube Valley, located in the Southern and Eastern parts of the country, covers a large area where the life of resident communities is shaped by the Danube River (1,075 km long) (Fig. 24.1). The Romanian Danube Valley (RDV), delimited in term of geomorphic features by the eight terraces of Danube (Posea & Bugă, 2005), overlaps the area with a temperate continental climate, with several climatic nuances and topoclimatic features. The physical-geographic characteristics of the study area include a relatively homogeneous underlying surface, with a dominant Southern orientation, the presence of a permanent water body, the specific morpho-hydrography of the meadows, the predominant steppe vegetation, the silvosteppe on terraces and meadows, soils with a sandy and clay structure etc. (Dragotă, 2022). The Romanian sector of the Danube Valley includes part of four development regions (West, South-West Oltenia, South Muntenia, and South-East) and twelve counties (Caraş-Severin, Mehedinţi, Dolj, Olt, Teleorman, Giurgiu, Călăraşi, Ialomiţa, Constanţa, Brăila, Galaţi and Tulcea). The 266 Local Administrative Units (LAUs) in the RDV, grouped in 238 communes and 28 municipalities and cities, totalling a population of over 1.8 million inhabitants in 2022, of which 38.3% is rural and 61, 7% is urban (Damian et al., 2019; Mitrică & Nancu, 2022).

Starting with the 1990s, the Danube Valley, especially its urban areas experienced a deep-going restructuring process, urbanization acquiring new dimensions, and characteristics. The Danubian settlements lost one-fifth of their population. Besides the demographic decline, deindustrialisation (closing down of enterprises and labour migration) enhanced urban poverty. In the small towns, poverty had the highest record at the national level. The subsistence-based agriculture and the slow-going increase of services make a limited welfare. The farmers working on their own and the expansion of

Fig. 24.1 Geographical position of the study-area (*Source* authors' own elaboration)

poorly-paid work in the services sector made industry the only source of important income (Mitrică & Popescu, 2017).

24.3 Methodology and Database

Progress in socio-economic vulnerability assessment and the rising interest in the development of quantitative metrics have led to a variety of approaches for developing indices (Tate, 2012). Among all studies dedicated to vulnerability, most offer an indicator-based approach able to provide both a quantitative and a quantitative picture of vulnerability, very useful in building solutions to increase adaptation and resilience through implementation of adequate policies (Hinkel, 2011). Most of the indicator-based assessments use the IPCC (2007) vulnerability framework that analyzes the vulnerability of human settlements according to its three basic components (i.e. exposure, sensitivity and adaptive capacity) which offer a variety of spatial approaches, from macro-regional (EEA, 2020; Solecki et al., 2015) to national, regional and sub-regional (e.g. Damian et al., 2023; Dumitraşcu et al., 2018; Grigorescu et al., 2021; Mocanu et al., 2021). Theoretically, as agreed by the scientific community, the socio-economic vulnerability is influenced, among other factors, mainly by quantitative and structural demographic and socio-economic features of a territory and by those resulting from its characteristics in terms of land use/ cover, climate and biotic conditions, etc. (Cutter et al., 2003; Cutter et al., 2009; Grigorescu et al., 2021; Hinkel, 2011; Mitrică et al., 2016; Mocanu et al., 2020; Siagian et al., 2014). This study relies on the statistical data available at the lowest level of administrative-territorial units (LAU or NUTS5) provided by the National Institute of Statistics (Population Census 2011 and TEMPO-Online database) and on the climatic data provided by the National Meteorological Administration. Building the database is valorised through the computation of secondary and final indexes of socio-economic vulnerability taking into consideration three criteria (i.e. scientific & practical knowledge, availability, and consistency), each of them with its specific method and tools, and immediate results such as databases covering socio and economic statistical variables and indicators (Fig. 24.2).

Applying the criterion of scientific and practical knowledge, an initial extended database was built by searching and selecting within national database, such as TEMPO-Online statistics (National Institute of Statistics—NIS. Applying the criterion of data availability and considering that out of the 8 domains of the TEMPO-Online statistic national platform (containing almost 1,100 socio-economic and demographic indicators), 7 have a general necessary availability and only 3 domains present a local territorial level of availability (with 63 variables/indicators), the TEMPO-Online intermediate database was reduced to 13 indicators which will be completed by supplementary indicators from different official sources of data. This intermediate database is structured using the three key-components of the vulnerability′ filter: exposure, sensitivity and adaptive capacity (Cardona et al., 2012; IPCC,

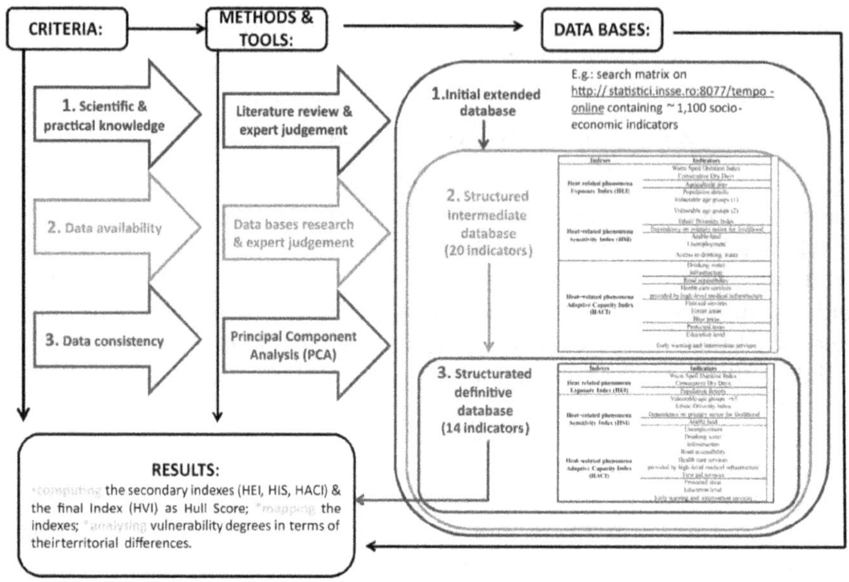

Fig. 24.2 The methodology applied in the development of socio-economic vulnerability to heat-related phenomena in the Romanian Danube Valley (*Source* authors' own elaboration)

2007; UNDP, 2004; UNISDR, 2004). Also, the intermediate database was subject to expert judgment, and it was completed by 2 climatic indicators (National Meteorological Administration—NMA), 3 indicators based on the Census 2011 (because the Census 2021 has incomplete final results and does not address at all the local territorial level) and by official data from Ministry of Development, Public Works and Administration. Finally, as Fig. 24.3 and Table 24.1 show, the structured intermediate database resulted have 20 indicators.

Applying the criterion of data consistency, the 20 variables were reduced to the components that most adequately explain the total variance, taking into account the correlations between the variables and eliminating the highly correlated ones. It emphasizes the need to select indicators that are as uncorrelated as possible to reduce redundancy (Kali, 2003). Thus, the structured definitive database contains the following 14 indicators for each of the three components of vulnerability:

- Heat Exposure Index (HEI): Warm Spell Duration Index (WSDI), Consecutive Dry Days (CDD) and population density (DENS);
- Heat Sensitivity Index (SI): Vulnerable age group (AGE65), Ethnic Diversity Index (EDI), Dependency on primary sector for livelihood (AGRIC-EMPL), Arable land (ARABLE), Unemployment (UNEMPL);

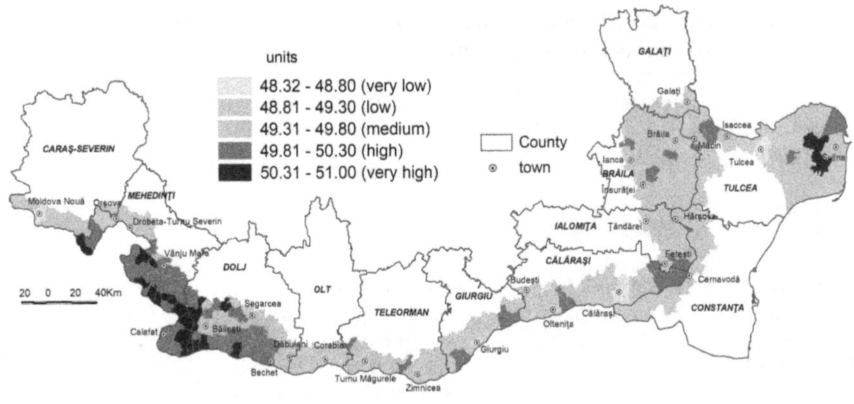

Fig. 24.3 The heat vulnerability index (*Source* authors' own elaboration)

- Heat Adaptive Capacity Index (ACI): Drinking water infrastructure (WATER-INFR), Road accessibility (ACCESS), First aid services (PHARM), Protected areas (PROTECT), Education level (EDUC), Early warning and intervention services (IES).

The variables of the statistical indicators are calculated using different measurement units, which brought about the necessity of a normalization procedure. In the case of indicators whose high values induce an increase in vulnerability, the following standardization formula was used: for the statistical indicator "X" for LAU "i": $X_{si} = (X_i - X_{min})/(X_{max} - X_{min})$, where X_i is the absolute value of the statistical indicator "X" for the LAU "i", X_{max} is the maximum value of X indicator in Romania and X_{min} is the minimum value of the X_i indicator at the level of entire country. If high values of the indicators cause a decrease in vulnerability, the following standardization formula was used: $X_{si} = (X_{max} - X_i)/(X_{max} - X_{min})$, where X_i, X_{max} and X_{min} have the same meanings as in the previous formula (Ianoş, 1997; Mitrică et al., 2017, 2020, 2021). Using the values registered at the national level for "min.-max." standardization technique offers the opportunity to compare the regional and local situations revealed by the indicators within study-area, namely the Romanian Danube Valley, with the national context, despite that when at the national level, some indices (e.g. those mirrored the employment features, Mocanu, 2016) are increasing, they may simultaneously decrease and/or increase at regional or local levels.

Consequently, the secondary indexes (Heat Exposure Index—HEI, Heat Sensitivity Index—HSI and Heat Adaptive Capacity Index—HACI) and the final Index (Heat Vulnerability Index—HVI) were computed as Hull Score with a mean of 50 and a standard deviation of 14, as follows:

- HEI = 50 + 14* (CDD + WSDI + DENS)/3

Table 24.1 Heat-related phenomena socio-economic Vulnerability Indexes and Indicators (*Source* authors' own elaboration)

Key components of vulnerability' filter/ secondary Indexes	*Indicators*	*Description (i), Data source (ii) Unit of measurement (iii)*	*Acronym (Influence)*
Heat-related phenomena **Exposure Index (HEI)**	1. Warm Spell Duration Index	(i) annual count of days contributing to "warm spells", when the maximum temperature (TX) remains above its climatological 90th percentile; (ii) NMA; (iii) days	**WSDI** (+)
	2. Consecutive Dry Days	(i) maximum number of consecutive dry days per time period with daily precipitation amount of less than 1 mm; (ii) NMA; (iii) days	**CDD** (+)
	3. Agricultural area	(i) all agricultural lands (e.g. arable, pastures, meadows, vineyards and vine nurseries, orchards and fruit tree nurseries) from the farm registers; (ii) NIS, TEMPO-Online time series; (iii) % of total area	**AGR** (+)
	4. Population density	(i) raport between total population and surface; (ii) NIS, TEMPO-Online time series; (iii) inh./km^2	**DENS** (+)
Heat-related phenomena **Sensitivity Index (HSI)**	5. Vulnerable age groups (1)	(i) persons aged 65 and above (% of the total population); (ii) NIS, TEMPO-Online time series; (iii) %	AGE65 (+)
	6. Vulnerable age groups (2)	(i) persons aged under 10 years old (% of the total population); (ii) NIS, TEMPO-Online time series; (iii) %	AGE10 (+)
	7. Ethnic Diversity Index	(i) the share of the minorities of the total population (ii) NIS, Population Census 2011; (iii) %	EDI (+)
	8. Dependency on primary sector for livelihood	(i) population employed in agriculture; (ii) NIS, Population Census 2011; (iii) %	AGRIC-EMPL (+)
	9. Arable land	(i) share arable land from agricultural land; (ii) NIS, TEMPO-Online time series; (iii) %	ARABLE (+)

(continued)

Table 24.1 (continued)

Key components of vulnerability' filter/ secondary Indexes	*Indicators*	*Description (i), Data source (ii) Unit of measurement (iii)*	*Acronym (Influence)*
	10. Unemployment	(i) unemployed population of the total labour resources aged 18–62 years (ii) NIS, TEMPO-Online time series; (iii) %	UNEMPL (+)
	11. Access to drinking water	(i) houses connected to the drinking water supply network (ii) NIS, TEMPO-Online time series; (iii) %	DRINK-WATER (−)
Heat-related phenomena **Adaptive Capacity Index (HACI)**	12. Drinking water infrastructure	(i) length of drinking water infrastructure/LAU (ii) NIS, TEMPO-Online time series; (iii) km/ 1,000 inh	WATER-INFR (−)
	13. Road accessibility	(i) The accessibility to the TEN-T core and comprehensiveness (gradient), combined with the accessibility to different non-TEN-T roads (motorway, European, National, County) (ii) Ministry of Development, Public Works and Administration; (iii) unit	ACCESS (−)
	14. Health care services provided by high-level medical infrastructure	(i) health care personnel (ii) NIS, TEMPO-Online time series; (iii) no. of physicians/1,000 inh	HEALTH (−)
	15. First aid services	(i) pharmacies considered as fist aid points; (ii) NIS, TEMPO-Online; (iii) no./ 1,000 inh	PHARM (−)
	16. Forest areas	(i) share of land covered by forest areas (ii) NIS, TEMPO-Online time series; (iii) %	FOREST (−)
	17. Blue areas	(i) Land covered with waters, ponds; (ii) NIS, TEMPO-Online time series; (iii) %	BLU (−)

(continued)

Table 24.1 (continued)

Key components of vulnerability' filter/ secondary Indexes	*Indicators*	*Description (i), Data source (ii) Unit of measurement (iii)*	*Acronym (Influence)*
	18. Protected areas	(i) share of land covered by protected areas (natural reserves, Natura 2000) (ii) NIS, TEMPO-Online time series; (iii) %	PROTECT (−)
	19. Education level	(i) share of illiterates; (ii) NIS, 2011 Population Census; (iii) % of illiterates/total over 10 year-old population	EDUC (+)
	20. Early warning and intervention services	(i) Inspectorates for Emergency Situations; (ii) Ministry of Internal Affairs; (iii) no./1000 inh	IES (−)

- HSI = 50 + 14* (AGE65 + EDI + AGRIC-EMPL + ARABLE + UNEMPL)/5
- HACI = 50 + 14* (WATER-INFR + ACCESS + PHARM + PROTECT – EDUC + IES)/6
- HVI = f (HEI + HSI- HACI) / 3

The Hull Score, with a mean of 50 and a standard deviation of 14, varies between 1 and 100 and is calculated as the sum of the direct or reverse relation of each statistical indicator or index in relation to the development process: "the indicators with a direct influence have been considered in the determination process as positive and those with a reverse influence were assumed to be negative" (Ianoş, 1997, p. 105, Regional disparities in Romania, 1990–1994, 1996). All indicators are weighted equally in the formula for computing the secondary and final indices, and the degrees of vulnerability are divided into five categories: very low, low, medium, high and very high. Uncertainties and limitations.

The approach of the socio-economic vulnerability to heat-related phenomena presents a series of uncertainties and limitations. The main uncertainties are the result of: the lack of a centralized inventory, which is now organized in various forms of storage and is accessed by many users or beneficiaries; the lack of existing data in digital format, which would allow for the creation of representative spatial database; the lack of a typological harmonization that would enable the achievement of representative inventories at local, regional or national level; the fragmented data collection system related to e.g.

population employed in tourism activity and tourism agencies. The main limitations are related to: the subjective selection of indicators, some vulnerability factors not being represented at all, as for instance issues of climate change strategies, GDP, and level of income from the agricultural activity; the vulnerability indicators that work well at the county and national scales are often too general or unspecific for local hazard mitigation practices that need more tailored approaches to individual circumstances being one obstacle to going down to a small scale (Berke & Godschalk, 2009).

24.4 Results

In terms of socio-economic vulnerability to heat-related phenomena, the Romanian Danube Valley is divided into two unequal parts. The western one is smaller and characterized by a territorial concentration of high and very high degrees of HVI in Mehedinţi and Dolj counties, because of the highest degrees of heat exposure and sensitivity indexes; the sector of Danube Gorges registers lower values of HIV due to the lowest exposure degrees (Table 4.2).

The eastern sector is more extensive, along most of the Romanian Plain, Dobrogea Plateau and continuing with Delta: there is an alternation between medium and high values with clusters in Brăila County (in its plain part) and Tulcea (partially, in the Danube Delta); the low values are registered especially in Brăila and Ialomiţa counties (in the Great Brăila Island and Ialomiţa Lake), Tulcea (partially, in the Danube Delta) and in Călăraşi and Constanţa counties (Fig. 24.3).

24.4.1 Exposure

The exposure to heat-related phenomena of the Romanian Danube Valley highlights two areas with very high and high values located: (i) in the southwestern part overlapping the Danube Gorge and the western part of Oltenia

Table 24.2 Synthesis on the Romanian Danube Valley local administrative units (LAU) in terms of degrees of socio-economic vulnerability to heat-related phenomena (HVI) (*Source* authors' own elaboration)

HIV values/ Degree of socio-economic vulnerability to heat-related phenomena	*Number of LAU*		*Total LAU/% of total LAU*
	Urban	*Rural*	
48.32–48.80/very low	4	2	6/2.3
48.81–49.30/low	12	84	96/36.1
49.31–49.80/medium	9	78	87/32.7
49.81–50.30/high	3	55	58/21.8
50.31–51.00/very high	–	19	19/7.1
Romanian Danube Valley	**28**	**238**	**266/100**

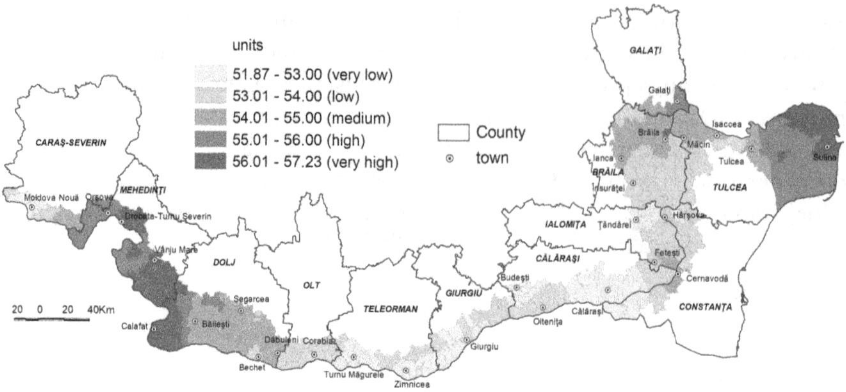

Fig. 24.4 The heat-related phenomena Exposure Index (*Source* authors' own elaboration)

Plain and (ii) eastern part overlapping the Danube Delta because of the high increase of heat waves duration and the consecutive dry days, the population density having relatively low values in these areas. In the class with very highest values (between 56.00 and 57.30 units) are included 35 LAUs (13.2% of the total RDV LAUs), the highest being registered in Calafat (Olt county), Obârșia de Câmp, Salcia, Vrața, Drobeta-Turnu Severin (Mehedinți county). The medium exposure is found mainly in the central and eastern part of Oltenia Plain and the northern part of Bărăgan Plain and Dobrogea Plateau, this class is grouping 85 LAUs representing 31.9% of the total RDV LAUs. The low exposure is found mainly in the Great Brăila Island Ialomița Lake and the eastern part of Oltenia Plain, in this class are grouping 64 LAUs representing 24.1% of the total RDV LAUs. 118 LAUs (44.4%) located along the central part of the RDV are featured by the very low exposure due to the low values of the three component elements, respectively duration of heat waves, consecutive dry days and population density (Fig. 24.4).

The most important increase of the duration of heat waves and of Consecutive Dry Days affected the south-western part, in the Oltenia area in Mehedinți and Dolj counties. The highest values of population density are registered by Brăila (4,477 inh./km^2), Drobeta Turnu-Severin (1,892 inh./km^2) and Galați (1,239 inh./km^2). The rural population density is influenced by the location and distance to urban centres.

24.4.2 *Sensitivity*

Heat Sensitivity Index (HSI) varies on a relatively large scale (Fig. 24.5) but territorially, the study-area doesn't register a diversity of local specificities.

On the contrary, the great majority of Danubian LAUs (171 LAUs representing 64.3% of the total LAUs) record high and very high sensitivity degrees

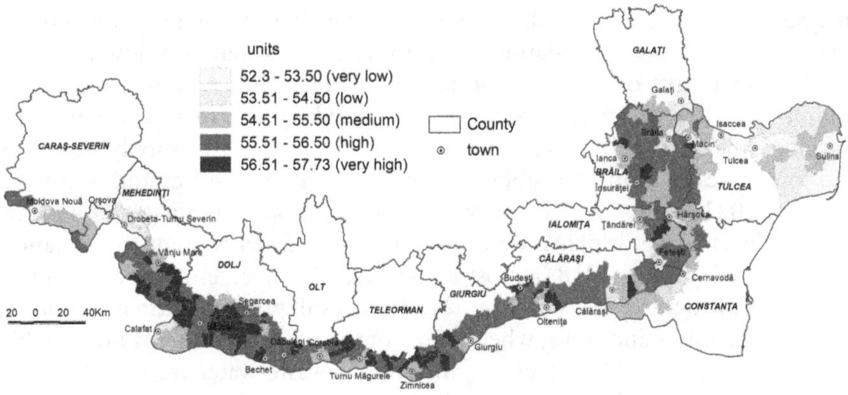

Fig. 24.5 The heat sensitivity index (*Source* authors' own elaboration)

to heat-related phenomena because the generalized high values of indicators reflecting the aged population, the dependency on occupancy in agriculture, the share of arable land of total land fund. Within the RDV, the dependency on agriculture for livelihood is high, as it is revealed by the preponderance of employment in agriculture. This high agriculture overemployment is linked with the predominance of the traditional household, implicitly, with a higher share of the contributing family worker and the self-employment category of the total employed population in agriculture. The medium values of this index are registered in 56 LAUs (21.0%) concentrated in Danube Gorges (Caraş-Severin and Mehedinţi counties) and in other scattered along the valley in the localities with the highest employment in agriculture, implicitly low unemployment. The study-area is experiencing alarming employment problems. The rural economy was and is dominated by subsistence occupancy, being nevertheless influenced by the urban economy, more strongly in the rural settlements located near the large cities. The state of the urban economy is the result of the cumulative negative effects of the restructuring of the communist urban industries and the economic-financial crisis from 2008 to 2014. To these causes are currently added structural changes in the industrial branches imposed at the EU level, by the European Green Deal—the new EU growth strategy, the provisions of which are transposed at the national level in various documents. The largest towns (e.g. Galaţi, Brăila, Drobeta Turnu-Severin) record low sensitivity degree, grace to their low dependency on agriculture, the concentration of tertiary activities, as a source of relatively secure and high incomes, low unemployment, high level of education.

24.4.3 *Adaptive Capacity*

Heat Adaptive Capacity Index (HSI) is driven by the individual or cumulated effect of drinking water infrastructure, road accessibility, pharmacies and drugstores, protected areas, illiteracy rate, and early warning and intervention services (Fig. 24.6). The adaptive capacity registers a generally low and very low level in almost over half of the RDV LAUs which has no uniform distribution. The low adaptive capacity is influenced by the inexistence of early warning and intervention services or public drinking water supply networks and low values of road accessibility, pharmacies and drugstores and protected areas. In 2021, almost 24% of RDV LAUs have a population without benefits from connection to drinking water infrastructure. Even though the situation has improved (e.g. in 2018, almost a 33% of LAUs were unconnected to this type of infrastructure, Mocanu et al. 2022), it is still problematic with drinking water from springs and wells, where, most of the time, the water is not potable (because of lack of sewage systems, insufficient waste water treatment, incorrect construction and location of wells, positioning and improper disposal of household waste/trash deposits).

The medium adaptive capacity is found in 92 LAUs (34.6% of total LAUs) spread mainly in western and eastern parts of the RDV having medium values of the indicators, except the early warning and intervention units found only 10 LAUs. This class also include urban areas where natural features (e.g. water bodies, forests) were gradually transformed by human intervention mainly for urban development processes. The high adaptive capacity features 31 LAUs (11.6% of total LAUs) are registered where the natural features are still preserved as environmental assets and especially as protected areas (natural reserves and Natura 2000 sites). This class is also characterized by a high level of drinking water infrastructure and pharmacies and drugstores. It is the case of LAUs with extended forest areas, wetlands and water bodies, mainly located in the Danube Gorges, Danube Delta which is a Biosphere Reserve and Ialomiţa Lake (Fig. 24.6). The two LAUs - Perişor (Dolj county) and Crişan (Tulcea county) with very high adaptive capacity has very high values

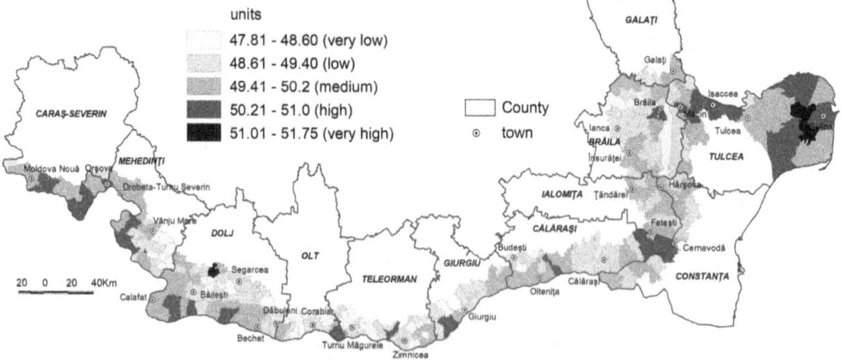

Fig. 24.6 The heat adaptive capacity index (*Source* authors' own elaboration)

of different adaptive capacity indicators (e.g. Perişor—pharmacies and drugstores, public drinking water supply network, road accessibility, Crişan—early warning and intervention units, protected areas).

24.5 Conclusions

The heat vulnerability index (HVI) shows a rather clear delineation of the very high and high level of vulnerability to heat-related phenomena registered by a total of 77 LAUs mainly concentrated in the south-western part of the study-area in Oltenia Plain. This part is obviously the most vulnerable to heat in terms of socio-economic dimension: very high sensitivity, one of the highest potential exposures, and a low to medium level of adaptive capacity. The high vulnerability is also found in the eastern part of the Danube Delta where the very high level of potential exposure is diminished by a low degree of sensitivity and high and very high adaptive capacity. The towns of RDV display generally low levels of vulnerability in relation to the low values of sensitivity indicators and the medium to high values of the adaptive capacity indicators. The highest vulnerability in the urban area is registered in Calafat town (Dolj County) due to the very high exposure, low sensitivity and average adaptive capacity. The two large cities, Galaţi and Brăila which overlap the Galaţi-Brăila Metropolitan Area are featured by low socio-economic vulnerability to heat-related phenomena influenced by the very low sensitivity, and medium to high adaptive capacity despite a high exposure. The central part of the RDV (Olt, Teleorman, Giurgiu, Călăraşi, and Ialomiţa counties) registers low and medium levels of HVI as a result of both low degrees of exposure and adaptive capacity even though the sensitivity is at the high levels.

Apart from the climate indicators use in the exposure analysis, the socio-economic indicators related to sensitivity and adaptive capacity are correlated to some general features of the study area: the socio-economic decline of a significant number of large urban centres and their ever lower role in decreasing the socio-economic vulnerability through a high adaptive capacity; the ever-greater regional demographic imbalances, age-structure degradation, and population ageing become alarming in the south-east regions of Romania increasing the socio-economic sensitivity; the loss of urban functions for many small and middle towns, mostly the one-industry towns affected by restructuring, frequently associated with great social problems which increase the socio-economic sensitivity; the distribution of historical development imbalances between the east and the west of the country, economic decoupling of some traditionally underdeveloped zones lying north, south-east of Romania and alongside the Danube; accessibility is a source of adaptive capacity, the Danube Axis rail transport is intersected by two railways at its ends plus a third—Giurgiu railway; no main rail connection throughout the central part of the Danube Valley from Calafat, Corabia, Turnu Măgurele, Zimnicea, Olteniţa, and Călăraşi to Tulcea; the road network is represented by roads of local and international importance—Giurgiu, Constanţa, Drobeta-Turnu Severin, Orşova and those running up to border points.

Indicator-based vulnerability assessments play a central role in the development of qualitative and quantitative studies, ideally with an important spatial and visual component (i.e. maps, graphs) to provide decision-makers with useful information in the development of mitigation and adaptation strategies and plans of action at different spatial scales. This category also includes the assessments of sectoral vulnerability to climate change in Romania from the latest national approach (RO-ADAPT) used as a scientific basis for the revision of the National Climate Change Adaptation Strategy and the Action Plan. This information will feed the European Climate Adaptation Platform Climate-ADAPT which supports Europe in adapting to climate change and responding to the EU Strategy on Adaptation to Climate Change. On the other hand, the Climate-ADAPT Platform offers the necessary information and support tools to the European regional and local authorities to prepare and plan for climate adaptation and resilience of communities through specific information. In this context, regional studies (such as the present one) that best reflect local contexts (place-based) have an essential role in substantiating and validating studies at macro levels (from national to European). For the West Balkans region, the interest aroused by the study of socio-economic vulnerability within a large fluvial valley, namely Danube valley, which represents a part of the northern limit of the Balkan Peninsula, could be used in the case of other important rivers valleys, such as the Sava River. Also, the Sava River represents the northern limit of the Balkan Peninsula. In this context of border areas, the cooperation between the national and regional entities involved in risk management within the Balkan Peninsula (the West Balkan region included) and the neighbourhood regions will have good opportunities to be developed. A study like the current one, based on scientific expertise capitalized as scientific basis for the revision of the National Climate Change Adaptation Strategy of Romania and its Action Plan, will could be a academic support for rising the connectivity potential with national, regional and local authorities within different other countries (those from West Balkans included) and with institutional entities managing the risks situations and more than that, the connectivity potential with stakeholders in the field of risk management and population. The socio-economic index developed in this paper has the potential to draw out meaning from large amounts of data to evaluate and project the vulnerability of a region (or national territory or local level) comprehensively so as to be useful for planning and management in any region and country. The social and economic vulnerability to heat-related phenomena would be able to identify key areas at regional and national levels, which have the prospective for vulnerable and areas less vulnerable to heat-related phenomena. By integrating an inclusive vulnerability assessment within spatial planning practices this study can be useful to provide in-depth knowledge and awareness on the issue at stake and reduce the uncertainties to develop a decision-support system for policymakers. Prioritization of the response actions according to their severity detailed information on vulnerability patterns to climate change impacts under both current and future climate change are needed (Kumar et al., 2016).

Acknowledgements The current study was carried out within the framework of the projects entitled "The Economic Restructuring of the Cities and Socio-economic Vulnerability" as part of the Research Plan of the Institute of Geography. The study partially valorized the methodological aspects from the projects entitled "Finalization of geomorphological and hydrological hazard maps within the perimeter of the Natural Hazards Research Center, Pătârlagele", as part of the Research Plan of the Institute of Geography. This scientific paper was published during the sustainability period of the project with the title: "Support Center for IEM research projects—competitive innovation in Horizon 2020", ID 107540. The project was co-financed from the European Regional Development Fund through the Competitiveness Operational Program 2014–2020.

References

Åhman, H. (2013). Social sustainability—society at the intersection of development and maintenance. *Local Environment, 18*(10), 1153–1166. https://doi.org/10.1080/13549839.2013.788480

Alcantara-Ayala, I., Altan, O., Baker, D., Briceno, S., Cutter, S., Gupta, H., Holloway, A., Ismail-Zadeh, A., Jimenez Diaz, V., Johnston, D., McBean, G, Ogawa, Y., Paton, D., Porio, E., & Silbereisen, R. (2015). *Disaster risks research and assessment to promote risk reduction and management* (p. 49). ICSU.

Berke, P., & Godschalk, D. (2009). Searching for the good plan: A meta-analysis of plan quality studies. *Journal of Planning Literature, 23*(3), 227–240.

Boström, M. (2012). A missing pillar? Challenges in theorizing and practicing social sustainability: Introduction to the special issue. *Sustainability: Science, Practice and Policy, 8*(1), 3–14. https://doi.org/10.1080/15487733.2012.11908080

Boyer, R. H. W., Peterson, N. D., Arora, P., & Caldwell, K. (2016). Five approaches to social sustainability and an integrated way forward. *Sustainability, 8*(9), 878. https://doi.org/10.3390/su8090878

Brecher, J. (2015). *A superfund for workers*. Dollars & Sense, November/December.

Brecher, J. (2020). No worker left behind: Protecting workers and communities in the green new deal. *New Labor Forum, 29*(2), 68–76.

Cardona, O. D., van Aalst, M. K., Birkmann, J., Fordham, M., McGregor, G., Perez, R., Pulwarty, R. S., Schipper, E. L. F., & Sinh, B. T. (2012). Determinants of risk: Exposure and vulnerability. In *Managing the risks of extreme events and disasters to advance climate change adaptation* (Field et al., Eds., pp. 65–108). Cambridge University Press.

Cutter, S. L., Boruff, B. J., & Shirley, W. L. (2003). Social vulnerability to environmental hazards. *Social Science Quarterly, 84*(2), 242–261.

Cutter, S. L., Emrich, C. T., Webb, J. J., Morath, D. (2009). *Social vulnerability to climate variability hazards: A review of the literature* (vol. 5). Final Report to Oxfam America.

Flanagan, B. E., Gregory, E. W., Hallisey, E. J., Heitgerd, J. L., Lewis, B. (2011). A social vulnerability index for disaster management. *Journal of Homeland Security and Emergency Management, 8*(1).

Damian, N., Mitrică, B., Mocanu, I., Grigorescu, I., & Dumitraşcu, M. (2023). An index-based approach to assess the vulnerability of socio-ecological systems to aridity and drought in the Danube Delta, Romania. *Environmental Development, 45*, Article 100799.

Damian, N., Mocanu, I., & Mitrică, B. (2019). Education and health care provision in the Romanian Danube Valley: Territorial disparities. *Area, 51*, 360–370.

Demiral, M., & Demiral, O. (2021). (2021), Where is the gray side of green growth? Theoretical insights, policy directions, and evidence from a multidimensional approach. *Environmental Science and Pollution Research, 28*, 63905–63930. https://doi.org/10.1007/s11356-021-13127-x

Dimitris S., & Felli. R. (2016). Green transitions, just transitions. *Kurswechsel, 3*, 35–45.

Doorey, D. (2016). A law of just transitions? Putting labor law to work on climate change. In Osgoode Hall Law School, Legal Studies Research Paper Series, Research paper No. 35, Vol. 12, Issue.

Dragotă, C. (2022). Diversitate climatică. In D. Bălteanu, & M. Dumitraşcu (Ed.), vol. Valea Dunării Româneşti. Perspective de dezvoltare durabilă (pp. 149–166). Editura Academiei Române, Bucureşti.

Dumitraşcu, M., Mocanu, I., Mitrică, B., Dragotă, C. S., Grigorescu, I., & Dumitrică, C. (2018). The assessment of socio-economic vulnerability to drought in Southern Romania (Oltenia Plain). *International Journal of Disaster Risk, 27*, 142–154. https://doi.org/10.1016/j.ijdrr.2017.09.049

EEA. (2020). *Urban adaptation in Europe: How cities and towns respond to climate change*. European Environment Agency, 186 pp.

EEA Report. (2020). *Urban adaptation in Europe: How cities and towns respond to climate change, 12*.

Farrell, C. (2012). Just Transition: Lessons learned from the environmental justice movement. *Duke Forum for Law & Social Change, 4*, 45–63.

Felli, R. (2013). An alternative socio-ecological Strategy? International trade unions' engagement with climate change. *Review of International Political Economy, 21*(2), 372–398.

Grigorescu, I., Mocanu, I., Mitrică, B., Dumitraşcu, M., Dumitrică, C., & Dragotă, C.-S. (2021). Socio-economic and environmental vulnerability to heat-related phenomena in Bucharest Metropolitan Area. *Environmental Research, 192*, Article 110268.

Hajirasouli, A., & Kumarasuriyar, A. (2016). The social dimention of sustainability: Towards some definitions and analysis. *Journal of Social Science for Policy Implications, 4*(2), 23–34. https://doi.org/10.15640/jsspi.v4n2a3

Heffron, R., Körner, M. F., Wagner, J., Weibelzahl, J., & Fridgen, G. (2020). Industrial demand-side flexibility: A key element of a just energy transition and industrial development. *Applied Energy, 269*, 115026.

Hinkel, J. (2011). Indicators of vulnerability and adaptive capacity: Towards a clarification of the science–policy interface. *Global Environmental Change, 21*(1), 198–208.

Ianoş, I. (1997). *Individualizarea şi analiza disparităţilor intraregionale. Aplicaţie la judeţul Alba*, Comunicări de geografie, I, University of Bucharest Publishing House, Bucharest.

IPCC. (2007). Climate Change 2007: Impacts, adaptation and vulnerability. Contribution of Working Group II to the Fourth Assessment Report of the Intergovernmental Panel on Climate Change (M. L. Parry, O. F. Canziani, J. P. Palutikof, P. J. van der Linden & C.E. Hanson, Eds., 976pp). Cambridge University Press.

IPCC. (2012). Managing the risks of extreme events and disasters to advance climate change adaptation. A special report of Working groups I and II of the Intergovernmental Panel on Climate Change (C. B. Field, Barros, V., Stocker, T. F., Dahe, Q., Dokken, D. J., Ebi, K. L., Mastrandrea, M. D., Mach, K. J., Plattner, G. K., Allen, S. K., et al., Eds., p. 582). Cambridge University Press.

IPCC. (2023). *AR6 Synthesis Report (SYR): Climate Change 2023.* https://www.ipcc.ch/report/sixth-assessment-report-cycle/

Jasanoff, S. (2018). Just transitions: A humble approach to global energy futures. *Energy Research & Social Science, 35*, 11–14. *44*, 31.

Kandachar, P. (2014). Materials and social sustainability. In *Materials Experience* (pp. 91–103). Butterworth-Heinemann.

Kali, R. (2003). Business groups, the financial market and modernization. *Economics of Transition, 11*(4), 671–696.

Kumar, P., Geneletti, D., & Nagendra, H. (2016). Spatial assessment of climate change vulnerability at city scale: A study in Bangalore, India. *Land Use Policy, 58*, 514–532. https://doi.org/10.1016/j.landusepol.2016.08.018

Mitrică, B., Damian, N., Mocanu, I., Şerban, P., Săgeată, R. (2016). Technical-urbanistic infrastructure in the Romanian Danube Valley. Urban vs. Rural territorial disparities. *Geographica Pannonica, 20*(4), 242–253.

Mitrică, B., Mocanu, I., Dumitraşcu, M., & Grigorescu, I. (2017). Socio-economic disparities in the development of the Romania's border areas. *Social Indicators Research, 134*(3), 899–916.

Mitrică, B., Mocanu, I., Grigorescu, I., Dumitraşcu, M., Pistol, A., Damian, N., & şerban P.-R. (2021b). Population vulnerability to SARS-COV-2 virus infection. A county-level geographical methodological approach. *GeoHealth, 5*(11), e2021GH000461. https://doi.org/10.1029/2021GH000461.

Mitrică, B., Mocanu, I., Grigorescu, I., Dumitraşcu, M., Pistol, A., Damian, N., & erban P.-R. (2021). Population vulnerability to SARS-COV- 2 virus infection. A county-level geographical methodological approach. *GeoHealth, 5*(11), e2021GH000461. https://doi.org/10.1029/2021GH000461

Mitrică, B., & Nancu, D. (2022). Potenţialul uman al localităţilor, în vol. Valea Dunării Româneşti. Perspective de dezvoltare durabilă (D. Bălteanu, & M. Dumitraşcu Eds., pp. 149–166). Editura Academiei Române, Bucureşti.

Mitrică, B., & Popescu, C. (2017). The economy of the Romanian Danube Valley—Evolution and trends. În vol. Lower Danube—Approaches to macroregional sustainability (pp. 150–165).

Mitrică, B., Şerban, P., Mocanu, I., Grigorescu, I., Damian, N., & Dumitraşcu, M. (2020). Social development and regional disparities in the rural areas of Romania: Focus on the social disadvantaged areas. *Social Indicators Research, 152*(1), 67–89. https://doi.org/10.1007/s11205-020-02415-7

Mocanu, I. (2016). Importance of the local territorial level in shaping the quality of employment patterns in Romania. *Romanian Journal of Geography, 60*(2), 119–131. Bucureşti.

Mocanu, I., Mitrică, B., Dumitraşcu, M., Damian, N., Persu, M, Dumitrică C. (5–7, October 2020). Socio-economic and environmental vulnerability to heat-related phenomena in Dobrogea. Romania, Proceedings of the GEOLINKS Conference Vision for new horizons. *International Conference on Environmental Sciences, Plovdiv, Bulgaria, 2*(2), Section Ecology and Environmental Studies, 143–154.

Mocanu, I., Grigorescu, I., Mitrică, B., Dumitraşcu, M., Dragotă, C. S., Mateeva, Z., & Dumitrică, C. (2021). Human health vulnerability to summer heat extremes in Romanian-Bulgarian cross-border area. *Natural Hazards Review, 22*(2), 05021003.

Najjar, R. (2022). Four dimensional spatial sustainability (4DSS): A revolutionary approach toward utopian sustainability. *Discover Sustainability, 3*(1), 21. https://doi.org/10.1007/s43621-022-00090-x

Policy Department for Economic, Scientific and Quality of Life Policies (IPOL). (2020). Study/Briefing: Dimensions and interpretations of social sustainability in EU policies. European Parliament. *Brussels.* PE 652.720

Posea, G., & Buga, D. (2005). Caracterizare generală a Văii Dunării. In G. Posea, O. Bogdan, I. Zăvoianu, M. Buza, D. Bălteanu, G. Niculescu (Coord.), In *Geografia României, Romanian Publishing House*(p. 488–537).

Rosemberg, A. (2010). Building a Just Transition: The linkages between climate change and employment. *International Journal of Labour Research, 2*(2), 125–162.

Siagian, T. H., Purhadi, P., Suhartono, S., & Ritonga, H. (2014). Social vulnerability to natural hazards in Indonesia: Driving factors and policy implications. *Natural Hazards, 70*(2), 1603–1617.

Solecki, W., Seto, K. C., Balk, D., Bigio, A., Boone, C. G., Creutzig, F., & Zwickel, T. (2015). A conceptual framework for an urban areas typology to integrate climate change mitigation and adaptation. *Urban Climate, 14*, 116–137.

Stângă, I. C., & Grozavu A. (2012). *Quantifying human vulnerability in rural areas: Case study of Tutova Hills (Eastern Romania). Natural Hazards and Earth System Sciences, 12*(6), 1987–2001. https://doi.org/10.5194/nhess-12-1987-2012

Tate, E. (2012). Social vulnerability indices: A comparative assessment using uncertainty and sensitivity analysis. *Natural Hazards, 63*(2), 325–347.

Vallance, S., Perkins, H. C., & Dixon, J. E. (2011). What is social sustainability? A clarification of concepts. *Geoforum, 42*(3), 342–348.

Wang, X., & Lo, K. (2021). Just transition: A conceptual review. *Energy Research & Social Science, 82*, 102291.

*** *Geography of Romania.* (2005). V, Institute of Geography, Bucharest.

*** *Regional Disparities in Romania 1990–1994.* (1996). Ramboll-Grupul de consultanţă, Bucharest.

*** UNISDR (United Nations International Strategy for Disaster Reduction). (2004). *Living with risk.* UNISDR. http://www.danube-region.eu

CHAPTER 25

Green Transition in the Adriatic Ionian Region By 2050: Lessons Learnt from the Futures

Maria Toptsidou and *Kai Böhme*

25.1 Introduction

Europe and the world are on the brink of a transformative era, which is characterised by profound changes that may redefine and reshape the future of societies, economies, environment, and international relations. The dawn of the last decade has brought numerous unprecedented events that have played their part in changing the framework conditions of the so-far *status quo*. Rapid technological advancements, shifting geopolitical landscapes, continuous threat by climate change and evolving social norms shape today's world. Although these topics are all but new, today more than ever, they demand innovative responses, immediate action, and new ways of thinking. Decisions and choices made today may determine the future paths of Europe, its places and its citizens.

Building on this momentum, climate change emerges as one of the most pressing challenges, demanding for immediate action. The European Union (EU) through its strong regulatory power manages to increase its influence to combat climate change, through numerous regulations and laws, forcing companies operating in the EU to adopt sustainability practices putting emphasis on green transition (KAIROS FUTURE, 2024). In addition, EU's Green Deal, its core overarching policy and ambitious package

M. Toptsidou (✉) · K. Böhme
Spatial Foresight, Heisdorf, Luxembourg
e-mail: maria.toptsidou@spatialforesight.eu

K. Böhme
e-mail: kai.boehme@spatialforesight.eu

E. Berisha et al. (eds.), *The Palgrave Handbook of Just Green Transitions in the Western Balkans and Beyond*,
https://doi.org/10.1007/978-3-031-95075-9_25

of measures towards a climate neutral EU by 2050 is the key policy beacon towards the green transition in the EU. It entails, among others, measures towards achieving net-zero greenhouse gas emissions, clean energy, sustainable mobility, energy efficient building and renovating, sustainable industry and biodiversity protection. Therefore, the green transition is often at the centre of political and policy discussions. How could the future of the EU and its territories look like if green transition is achieved? What could be a possible path towards green transition? Inspired by the results of the ESPON project 'TEVI2050—Territorial Scenarios for the Danube and the Adriatic Ionian Macro-regions' (ESPON, 2022), this chapter focuses on the Adriatic Ionian Region[1] and explores two different possible and plausible paths towards its green transition by 2050.

The aim of the chapter is to use foresight and scenarios to eventually to draw some lessons from the future. To do so, it first frames the core concept of green transition, gives the context of the Adriatic Ionian Region and explains how the two scenarios can offer glimpses into possible futures. It then goes through the two scenarios. First, it presents the hypotheses based on which the scenarios are developed and then look into two possible futures for green transition, operationalising the radical thinking used in scenario building. One scenario focuses on the green transition seen as part of people's wellbeing, while the other scenario focuses on the more radical approach of 'giving back to nature'. The two scenarios aim to offer 'food for thought' and provocative statements to inspire the reader, showing that there may be different options for the green transition, as well as different choices that can be made. The chapter closes with the lessons that can be drawn from the two scenarios with regards to policy making and future choices, as well as with some overall conclusions on the necessity of radical thinking and, at times, radical actions as a response to persistent challenges. For only then real change may be seen.

25.2 Towards Green Transition in the Adriatic Ionian Region by 2050

In the Anthropocene era, where humans are the only species responsible for the consequences that radically affect the environment, like overconsumption, overproduction, growth and overexploitation of resources, it has become evident that humanity has failed to engage in a sustainable way with nature (Dasgupta, 2021). Excessive temperature rise, altered precipitation patterns, sea level rise, droughts, and frequent and severe weather events are only some of the climate change consequences that impact substantially the ecosystems and biodiversity and, hence, human practices and lives. The Adriatic Ionian

[1] The Adriatic Ionian Region includes the following countries: Albania, Bosnia and Herzegovina, Croatia, Greece, Italy, Montenegro, North Macedonia, San Marino, Serbia and Slovenia.

Region is characterised by a unique natural environment with rich biodiversity, diverse flora and fauna, varied ecosystems, natural habitats and marine life. Nevertheless, this vibrant land and sea ecosystem remains highly vulnerable and increasingly at risk by the impacts of climate change. As a result, more sustainable approaches towards development and a more holistic approach towards green transition need to be considered (Moodie et al., 2025a, b). The way we address this existential threat will influence our collective future. Yet, often, the future is described as 'tempus nullius' or 'nobody's time' (Krznaric, 2020), where long-term thinking is challenging, especially as regards tackling longstanding challenges, as in fact is climate change (Randers, 2012). This presentism may result in short-sighted political decisions that fail to prioritise future generations, which stands for a 'tragedy of the time horizon' (Böhme, 2023). As an antidote, foresight and future thinking allow to explore different possible and plausible futures. Although predicting the future is impossible, preparing for it is a must today. Territorial scenarios are a useful tool for exploring the future and looking at how it may develop based on coherent and consistent assumptions. Scenarios do not make predictions but rather allow for looking into the future and how different trends or developments may interact or change. Their added value is that they stimulate thinking out-of-the-box and prompt social imagination, but above all they offer choices, as several futures can be developed which may help looking at different aspects and eventually taking more informed decisions.

Efforts towards green transition are a priority in the Adriatic Ionian Region, particularly in the framework of the EU Green Deal. However, as not all countries in the region are members of the EU or progress in a similar way (Berisha, 2025), the application of the Green Deal in practice varies (Berisha et al., 2025). At the same time, different economic, social, territorial disparities influence the transition towards greener solutions (Shaker, 2025). The Adriatic Ionian Region faces a number of challenges. Certainly, climate change and environmental risks pose threats in the region, with biodiversity loss, soil erosion, coastal and marine ecosystem degradation, but also high sea level rise putting additional pressure in the region. In addition, the high dependence on fossil fuels poses an additional risk towards the green transition and cleaner energy solutions. Although innovation and technological solutions could support leapfrogging towards the green transition, the Adriatic Ionian Region faces also a number of challenges in that respect. The digital divide and innovation gaps in the fields of digitalisation especially between urban and rural areas and the limited R&D investments hamper the region's potential for innovation. In addition to those, socio-economic disparities, connectivity issues and population decline are further challenges which put additional risks to territorial development and people's wellbeing. Last but not least, external factors, like external influence by non-EU players (Cotella & Berisha, 2019, 2021), through investments or other strategic projects in some of the countries of the Adriatic Ionian Region may create some complex geopolitical dynamics that may complicate any efforts towards the green transition or towards cohesion, in general (Berisha, 2023). Despite all these challenges, over the next years, it is expected that efforts towards green transition will be intensified, with its twin digital transition playing an equally important role to that direction. Against this background, the chapter uses foresight and scenarios to experiment with two different ideas towards a green transition in the Adriatic Ionian Region. The first scenario builds along the idea that green transition is part of people's wellbeing. This means that people have prioritised their quality of life against economic growth, and the latter goes beyond GDP figures. Instead, it is the ecological and social enablers that improve lives and adapting the economic model towards more sustainable solutions will be the foundation of wellbeing, combining social fairness, productivity, prosperity and environmental stability (European Commission, 2023). For that, citizen involvement has been necessary and enabled through the digital transition. The latter has also been largely utilised to achieve green transition. In this scenario, the EU, in its expanded role of strategic autonomy, has a strong regulatory presence and global presence regarding green transition. The second scenario offers a more radical approach, building on the extreme implementation of the green transition. In this scenario, citizens of the Adriatic Ionian Region have taken the decision to 'give back to nature', i.e. improve and restore ecosystems and rewild nature, by abandoning large territories and concentrating in the main urban areas and secondary cities. The new geography of work which allows working from anywhere has made it easier for people to move from left behind regions, but still staying in the region, reversing brain drain. At the same time, ecosystem services as tradable lead the way to the development of rural areas, while combined with community-based management (Walljasper, 2011) increases the involvement and participation of citizens. Despite the very distinct starting points and ideas, the two scenarios envision futures towards the green transition in the Adriatic Ionian Region with different territorial implications. The chapter is inspired by the ideas and scenarios developed by the ESPON (2022) project 'TEVI 2050 – Territorial scenarios for the Danube and the Adriatic Ionian Macro-regions'.

25.3 'Green' Is Part of Wellbeing in the Adriatic Ionian Region by 2050—A Scenario

Dramatic changes have challenged the EU and the Adriatic Ionian Region over the last years. Climate change effects have been dramatic over the last years. Europe experienced unprecedent droughts, floods, forest fires and heatwaves, which together with biodiversity loss and environmental pollution put EU's ecosystems at risk (European Commission, 2024). Climate change consequences have been affecting several economic sectors and policies, like agriculture, fisheries, forestry and bringing new economic challenges in sectors and territories highly exposed to climate change, like tourism in coastal areas. In addition, public health is also put at risk due to climate change implications, with particularly heat waves being a prominent threat. All these put particularly the most vulnerable groups at risk, further exacerbate regional disparities and affecting their wellbeing (Walljasper, 2011). Furthermore, the poly-crises of the last years have significantly affected the lives of people in the Adriatic Ionian Region. The restrictions and social distancing imposed during the COVID-19 pandemic have affected social life and people's mental health. In addition, the Russian war on Ukraine and the subsequent price increase in basic food products and energy supply has not only raised awareness about sustainable and autarkic agriculture, but also increased demand for energy safety and autonomy, making the mainstreaming of alternative energy sources a prerequisite. Citizens in the Adriatic Ionian Region have realised that it is not economic progress that brings joy, but happiness is to be found in good environmental conditions and social relations and interaction with people (Hanell, 2018). Thoughts about replacing GDP with alternative measurements for growth were prevailing. Putting quality of life at the forefront, citizens have prioritised improving environmental conditions and being close to nature as key enablers for better health and wellbeing. Hence, green transition had to be accelerated. With all of this in mind, the scenario focuses on the green transition being an important element of people's wellbeing. This transition towards a better quality of life by investing in environmental and social enablers has transformed many aspects of citizens' lives.

The path towards the green transition in the Adriatic Ionian Region hinges fundamentally on a pivotal shift in the mindset among its citizens. This mindset shift characterises changes in several aspects of people's lives. The principles of circular economy, along the lines of 'recycle-reuse-repair' has been part of the everyday common practices. Over time, such behavioural changes have resulted in less waste, improved environmental conditions, but also changed production and consumption patterns. Another element to that has been the increase of the sharing culture. This is largely due to the generation shift towards the GenZ (i.e. people born between 1997 and 2012), a generation which is not only deeply committed to environmental sustainability as such, but has also a strong sharing culture. This sharing culture has contributed to less materialisation, i.e. less crave for materials where people buy and consume less, but also reinforced community building. People feel more responsible for their community, their surrounding environment and the wellbeing of others. Nevertheless, the increase in circular and sharing cultures has not only been a characteristic of the newest generation, but has also been important for the older ones, as old professions revived to support the 'repair' culture which were rather more known by older generations. Places in the Adriatic Ionian Region with high manufacturing potential, as well as high young population were among the frontrunners.

This transformative mindset has also influenced the way people move, travel and have holidays. Mobility and tourism, two key economic sectors of the Adriatic Ionian Region. Over the decades the car usage has largely turned into the usage of electric cars been and in many cases even reduced significantly and replaced with car sharing activities. This, coupled with the increasing use of public transport, bikes and other forms of micromobility, especially in the urban areas of the Adriatic Ionian Region, have transformed transport and mobility in the area. In addition, green mobility through high speed rail networks has been encouraged. This has been supported by the development of good railway connections in the Adriatic Ionian Region over the years. Similarly, a shift in tourism practices towards more sustainable practices has also led to reduced environmental footprints in the region. Cruise tourism, which was a key tourism product for the region, has been radically reduced and replaced with domestic tourism opportunities, that do not require long flights or drives, Hence, less crowded rural areas, national parks and green areas have been preferred. Such shift in tourism is also particularly relevant to coastal areas in the region, to which tourism has been a key economic sector. In addition, sustainable tourism services, as reflected in accommodation and catering with eco-labels, less use of plastics, vegan meals, etc., in different tourism hotspots are highly appreciated. Such changes have not over time mitigated the environmental impact but also enhanced quality of life in the region.

Another critical aspect reflecting the mindset shift has been the transition towards small scale renewable energy sources. Here again, citizens have played an active role towards this transition, capitalising the Adriatic Ionian Region's high potential in hydropower and over time especially in solar and wind power. Although it has not been possible to cover all energy needs and additional sources were necessary, the adoption of renewable resources by citizens and households further empowered the community resilience and contributed to a more sustainable environment. This shift, however, has been more challenging for regions in the Adriatic Ionian Region depending more on coal and fossil fuels, while more opportunities are observed in areas with renewable energy potential, especially for wind and solar energy.

Besides people's fundamental mindset shift towards the green transition, the latter is also inextricably linked with it's the digital transition and even tripled with the demographic transition. The take up of digitalisation in the EU and by extension to the Adriatic Ionian Region has played a key role towards achieving the green transition. The digital transformation has allowed to use digital solutions for achieving more sustainable and environmentally friendly fixes. This has accelerated the green transition and also empowered regions that had little progress towards the two transitions to leapfrog and improve. Digitalisation has shaped several aspects of people's daily work, life and transformed industry. Starting with the workplace, the job roles and profiles as well as the industry demands have been revolutionised. More specifically, the digital transition has necessitated a high level of digital literacy. Although the majority of the younger population are digital natives, this has not been the case for the older population, whose rather more predictable or repetitive jobs have been replaced. Automation and digitalisation may have been the reasons for making a number of jobs obsolete however, they also created new types of jobs (Joint Research Centre, 2019). This, has demanded additional new skills, in particular non-cognitive skills such as emotional intelligence, ethical and critical thinking, advanced people skills (Joint Research Centre, 2019). Particularly, regions with high digital connectivity have been the frontrunners of this transition.

The use of digitalisation and automation has brought an evolution in the work environment. With many tasks being replaced by automation, employees would have more free time to focus on more rewarding aspects of their life and work. This does not only include working on more interesting tasks, but also prioritising their wellbeing. The hustle culture prevailing the earlier decades has now been replaced by less stress and a reshape of work-life balance.

The digital shift has also transformed the industrial and commercial landscapes. E-commerce and online shopping have advanced in the wake also of the COVID-19 pandemic, capitalising on more customised solutions to meet consumer needs. This has significantly affected smaller shops, putting the bricks-and-mortar retailers in need to adapt. This has made it faster and easier for consumers to shop and receive products directly at their doorstep. To counterbalance the sustainability challenges this shift may impose, additional efforts have been put in place to green the industry and the production lines. In particular, circular economy practices have been reinforced, while enhanced waste management has been put in place. However, greenwashing has been unavoidable in many cases.

None of the above would have been possible, if it weren't a strong policy framework and support to back the efforts. The EU through its strong regulatory power has been by 2050 a leading player in the green transition. In that framework, the EU has also become a stronger global player when it comes to combating climate change. Following the idea of EU strategic autonomy, the EU has shifted from only focusing on security and defence towards expanding to other policies, with environmental policy being a priority. This has played a key role in the Adriatic Ionian Region. The Green Deal, EU's overarching guiding policy towards green transition has been applied in all its territories with most of its goals being achieved. This has affected all countries of the Adriatic Ionian Region, as by 2050, all have become EU member states. The achievement of the Green Deal goals by 2050, has improved actions towards clean air, clean water, healthy soil, biodiversity, affordable food, better and more sustainable transport, energy efficient buildings, cleaner energy, and long-lasting products. Further EU policies, such as the EU circular economy action plan, as well as the RePOWER EU plan have borne fruit and ensured a strong regulated, a less dependent and a resilient green transition.

This scenario has given a glimpse over 2050, where green transition has been achieved in the name of the wellbeing of people. Certainly, green transition has neither affected territories and people equally, nor has it been a linear process. The scenario, nevertheless, shows that the mobilisation of different aspects is necessary to achieve such an important mission, as the green transition.

25.4 'Giving Back to Nature' in the Adriatic Ionian Region by 2050—A Scenario

The Adriatic Ionian Region is renowned for its rich land and marine biodiversity, home to an array of flora and fauna that contribute to the ecological balance of the region. The last years, however, have put this vital biodiversity at risk, with climate change being the key reason. Increasing threats from urbanisation, pollution, overfishing, practices like overproduction an overconsumption, but also increasing incidents of natural hazards, like floods, heatwaves, forest fires, and droughts have prompted a sense of urgency towards taking action to protect and preserve the natural heritage of the region. In the scenario, the earth temperature rise has risen more than 1,5 °C increasing further such phenomena, while human activities further intensify the pressure and are a key reason for this increase, as the unsustainable energy use, land use and land use, the lifestyles and consumption and production patters have been challenging the environment (IPCC, 2021, 2023). Action had to be taken underscoring the need for green transition, to on the one hand mitigate those negative consequences and on the other hand to ensure resilience and sustainability of the region's ecosystem. Protecting and restoring nature became a primary goal, as preserving the nature and biodiversity of the region would not only be critical for people's health but also for the economic sustainability, as well as for the production of sustainable food systems, resource management and the overall quality of life and prosperity of the current and future generations. As a result, efforts towards the green transition have been a one-way ticket for survival. First efforts were made through the investment in green and blue infrastructure. The region gradually developed a strategically planned network of natural areas to deliver ecosystem services (ESPON, 2019). Those ecological networks of natural areas, connected through green and blue corridors along the Adriatic Ionian Region have been the first step towards increasing the protected areas.

Nevertheless, citizens quickly realised that more radical decisions were necessary to reverse the negative effects of climate change and biodiversity degradation. The big leap to that direction has been human-driven and characterised by the revolutionary choice to restore and give back to mother nature. The Adriatic Ionian Region has for long suffered of areas in population decline, shrinking regions and places left behind (Rodríguez-Pose, 2018) or places that fell into development traps (European Commission, 2024) and could neither adjust to smart shrinking practices, revive economically. At the same time, rural areas that could not become more resilient over the years have also lacked economic diversification and progress. Hence, all these areas that were characterised by a lack of prospects have now been given back to nature and became fully reverted to natural areas. This restoration has not only enlarged the land and aquatic areas of the region, but also helped in rewilding its animal life. This has catalysed the revival of ecosystems, as well as of native species in the region, which gradually gained back its flourishing environment.

The ecological awakening was propelled primarily by the activism of the highly environmentally aware youth of the region. Realising that resources are not infinite, the youth of the region has been advocating that more radical solutions are necessary to restore nature. Amid also discussions on the need for new social contracts during the COVID-19 pandemic that put environmental awareness high in the agenda, as well as influenced by the uncertainties brought by the Russian war on Ukraine, the youth of the region has been dynamically involved in shaping the green transition movement in the region, shifting their mindset towards increased sustainability, supported de-growth and altered their consumption and production habits (Büchs & Koch, 2019), advocating for sustainable practices through their grassroots movements and participatory youth forums.

These more environmentally conscious lifestyle changes towards more sustainable living habits have played a key role in achieving the green transition and changing the consumption patterns. Young generations opt to shop less, own less and choose products with little environmental footprint. Driven by this behavioural change, businesses have also aligned their practices and responded to the new consumer preferences. To do so, they have increased their sustainability standards, offered eco-friendly and ethical products and went for greener alternatives in their production. As a results, a new understanding of prosperity and growth has been introduced which is not linked solely to economic growth and development, but rather to slow growth and de-growth, short supply chains and a shift to green economy.

Such tectonic shifts have further influenced the peoples' way of living. Giving back to nature meant that the majority of citizens have decided to move to larger and smaller urban areas, leaving the challenging areas that were now to be restored back to nature. This has contributed to urban sprawl and population pressures, allowed, however, the left behind areas to regenerate. Such urban migration was aided by the widespread technological advancements. The digital transition has contributed in its way towards rapidly achieving the green transition. In particular, the use of digitalisation in the 'work from anywhere' concept, which has advanced over the decades. The rise of digital nomads allowed people to work for large global corporations, while residing locally. With working from home being a norm, the need to migrate for employment opportunities has reduced, resulting in brain gain in the region. Being for decades challenged by brain drain, the population in the Adriatic Ionian Region capitalised their 'freedom to stay' (Letta, 2024) and contributed through their knowledge and talents to the local economies.

Further to that, the increased concentration of people on smaller urban areas has developed an important critical mass or innovation and manufacturing, with people gradually contributing to the re-industrialisation in the region, specialising on different sectors and making use of circular economy and approaches. In this case, digitalisation has also changed the way people innovate, as a more circular, ecological and social innovation has taken place in the region. In this case, urban areas and capital regions in the Adriatic Ionian Region have received most of the population, while smaller urban areas of the region have also been now affected by this migration flow.

Besides the urban areas, the rural areas and particularly those close to the ecosystems have played a key role in their preservation. More specifically, people have harnessed ecosystem services as tradeable commodities, i.e. they preserved their natural assets while at the same time generating some economic benefits. This approach contributed on the one hand to the environmental conservation, providing on the other hand a sustainable source of income for the local communities.

Further sectors, such as transport and tourism have also been affected by this radical change. In fact, the change in people's attitude towards transport. The use of transport has been reduced, as the population is rather concentrated to specific territories. As the road and rail connections have been responsible for putting nature networks at risks, they have been largely reduced. Instead, few road networks, connecting key core centres across the region have been maintained, linking them in particular with the ports of the region. Shipping, which constituted a key transport means in the area has rather focused on short sea shipping for the transport of goods across the region. In addition, shipping has over time become greener and more sustainable. This has been particularly relevant for the Adriatic Ionian Region's coastal areas and ports which have now gained a new momentum. Within the urban areas, and given the rise of digital connectivity, the physical transport has decreased substantially. Sharing practices, sustainable public transport, and micromobility solutions dominate the way people commute.

Mass tourism and cruise tourism, which has for decades been a reason for environmental pressure in the Adriatic Ionian Region has by 2050 been gradually replaced by more environmentally friendly choices. Infrastructure has been reduced and people choose nearby destinations, more conscious and affordable tourism products (European Commission, 2021), as well as prefer urban tourism, profiting from the good urban connections and urban attractions. Islands, coastal regions as well as touristic hotspots in the Adriatic Ionian Region as mostly affected by this change. When it comes to energy, by 2050 the Adriatic Ionian Region has capitalised to the extent possible on its renewable energy potential, particularly on solar and wind energy. In this paradigm shift, citizens take the role of prosumers, i.e. they are not only consumers of energy, but also producers. Gradually they have created small cross-border and neighbourhood energy communities, trading energy among the community.

Certainly, a great support to these efforts have been the strong EU policies in place, which have been the beacon for inspiring people and their actions. The Green Deal in particular, and its multifaceted priorities has been a guiding principle for people. In addition, EU Biodiversity Strategy has set the scene towards a stronger focus on the biodiversity preservation, while EU's New industrial policy offers the ground towards new approached for industry.

Visionary initiatives, such as the Rural Vision, the Circular Economy and Social Economy policies, as well as frameworks like the Territorial Agenda 2030 (TA2030, 2020) that has opted for a green and just Europe have contributed to this radical shift.

The scenario has given a glimpse to the Adriatic Ionian Region in 2050 where a radical change towards green transition has taken place and people have decided to give a substantial part of the region's territory back to nature in order to preserve the ecosystem and the natural heritage of the region. Such a shift has had ups and downs and influences the territories and people in substantial ways. The power and mobilisation of the youth has been crucial in defining this future, nevertheless a combination of different elements has been necessary for this transformation.

25.5 Lessons Learnt from the Futures

The two scenarios presented above show two different future paths towards green transition. In the scenario where green transition is seen as a means for wellbeing, people in the Adriatic Ionian Region have taken big steps towards improving their quality of life by improving the environmental condition of the region. Through their behavioural changes they have adopted more sustainable practices which had an impact on the economic and social spheres, too. In this scenario, digitalisation plays a crucial role by not only increasing citizen engagement and creating a stronger feeling of community, but also by creating the enabling conditions for the green transformation of businesses. Simultaneously, the EU's robust regulatory framework has allowed for monitoring the green transition and offering the base for it. People in this scenario prioritise leisure and mental health, alternative energy sources have contributed to cutting carbon emissions, while sectors like transport and tourism have adopted more sustainable practices. Prosperity gets a whole new meaning and is not counted through GDP numbers but by the overall improvement in quality of life. Nevertheless, the implications of this scenario are not the same for all territories in the Adriatic Ionian Region, as some may move faster to that direction than others. At the same time, bridging the digital gap across territories may also differ throughout the region, hence the progress in that respect also differs. Furthermore, this shift towards a slower living and a new approach to growth may result in a slower economic development which on the long run may have further economic and social implications.

In the scenario where the green transition is achieved through restoring the natural habitat, people have taken the radical decision to give large territories back to nature. Territories that did not adapt and smart shrinking approaches, fell into development traps or were scarcely populated have been given back to nature for restoring and rewilding purposes. Although this move has improved over time the environmental condition of the region, it has increased urbanisation to key urban centres, as people moved from the shrinking and rural regions to the urban ones. Coupled with the new geography of work, where employees can work from anywhere, and the advancements in digitalisation, a reverse brain drain was observed. Ecosystem services have become tradeable, enhancing their protection status and offering a social income to people who remained in the rural areas close to them. The migration of people to smaller urban areas has created a critical mass and advanced in eco-friendly, and social innovation. Short distance transport and short-term shipping have replaced most of other forms of transport, while mass tourism has gradually faded away. Certainly, the strong commitment of people, as well as the strong EU policy support have been fundamental for this shift. Such a revolutionary move has over time brought significant positive changes in the environmental condition of the region. However, it came at the cost of growth and development, as notions of de-growth prevailed by 2050.

The starting points and evolutions of the two scenarios are distinct, however, their red thread has been the achievement of the green transition. Such processes have not been linear or straight forward in either of the two scenarios. Nevertheless, some key lessons can be drawn from the two futures to walk the path towards a shift to a more sustainable future:

- Embracing change as the only constant. Today's world is characterised by an 'era of transformations', where constant changes are inevitable. Facing rapid environmental, geopolitical, technological and social shifts, these changes are fundamental and require equally dynamic responses. The urgency of the matters, like climate change and green transition, ask for proactive and bold solutions. Only by embracing change humanity may adapt to the new realities, be more resilient and adaptable to future challenges.
- Taking bold decisions as a means to act. In view of existential threats, as for instance climate change, radical decisions and mindset shifts are necessary. Dotting the i's and crossing the t's in policy documents may only be mediocre solutions, unable to bring real, substantial change. To address complex matters that require action, taking more extreme steps may be essential. Both scenarios show examples of such bold moves to the path for green transition, where the mindset shift of people has been the key step to change. Thinking in different, more radical ways can be an approach towards key solutions.
- Putting people at the centre of changes. For any important change and development to progress and take shape, the embarkment, active participation and involvement of people is necessary. Without a bottom-up support, any top-down decisions may backlash, creating unrest or discontent. The two scenarios have shown that the role of people in the path to green transition has been decisive. For a future for all places and people, strong place-and people sensitive decisions and policies need to be implemented. Hence, the engagement of communities, which are seen as drivers rather than followers is a prerequisite for more resilient and socially responsible futures.

- Having green and digital transitions in symbiosis. The pursuit of environmental goals cannot be isolated from the technological advancements. The green and digital transitions need to be seen in co-existence to maximise results. Digital solutions can play a crucial role for enhancing efficiency of green initiatives and accelerate progress. Certainly, the element of just transition needs to be added, as transitions need to fair for all societal groups.
- Strengthening the EU policy support. A holistic EU policy support system is essential for driving change and allow effective implementation of initiatives and practices across EU territories. The EU has the robust and strong regulatory framework to design, guide and monitoring actions at different governance levels, facilitate exchange, provide support and be the beacon for collective action.

The two scenarios have shown that a blend of different elements is pivotal for change and moving forward to a green transition. They highlight how different choices and decisions taken may evolve to different future pathways. Every choice may come at a consequence, yet unknown. However, there is no 'right or wrong', nor there is a 'one-size-fits-all' solution. It is about thinking collectively and positively towards the future, being prepared to take action and being willing to change.

Regardless of which scenario seems more desirable or possible, some challenges linked to demographic decline, climate change and digitalisation may persist. The scenarios have offered some glimpses to possible futures impacted by these challenges. In that respect, the Adriatic Ionian Region could consider preparing for these challenges, especially through cooperation frameworks, such as the EUSAIR macro-regional strategy. A few examples in that direction can be:

- Taking joint steps towards the twin transition. This can take the form of joint plans, or joint pilot actions that can be developed by the region and set the scene for a stepwise process towards the green and also the digital transition.
- Preparing for environmental risks. Climate change consequences, like natural disasters and biodiversity loss pose challenges in the territorial development of the region and the wellbeing of citizens. Joint actions and joint plans may be the first step towards a more prepared region for future environmental challenges.
- Think of impact on relevant economic sectors. On the way to the green transition, different economic sectors may be influenced in one way or another. Jointly thinking of the most impacted sectors, as for instance tourism, and developing concrete actions for reducing different negative effects or untapping potential may support a fairer transition.

The Western Balkans are no exception to all these challenges of the Adriatic Ionian Region. In this wider cooperation framework with other countries, the Western Balkans can also strengthen their regional cooperation and integration in the region, also in view of the future possible EU accessions, benefitting from joint actions and overarching policies. Furthermore, they can take joint actions towards issues of the twin transition, focusing on sustainable development and digital connectivity and skills. Investing in diversifying their economies with a focus on elements of the twin transition but also involving the youth in their future policy design would be steps towards more integration, cohesion and adaptability.

25.6 Conclusion

Future uncertainty is certain. Nobody knows how the future may unfold and what changes are to be expected. The poly-crises of the last years, from the COVID-19 pandemic to the Russian war on Ukraine, the economic upheavals, social unrest and rising inequalities have demonstrated that the future is unpredictable. They have also demonstrated that the consequences of the crises can be dramatic and even have long-lasting effects on people and places. This further highlights the importance and necessity to take action and make strategic choices, as the latter will define the future of the EU (ESPAS, 2024). Nothing is possible without action and a clear political commitment to move forward (Böhme & Toptsidou, 2022). This also highlights the need to be prepared for the future and develop informed decisions for all places and all people in the EU. Recognising this uncertainty is the first step towards building more resilient and prepared societies.

This uncertainty also holds for the green transition, which is in itself a challenging yet necessary endeavour. Nobody currently knows if and how the green transition will occur, as its realisation depends on several factors. The scenarios outlined in this paper describe two different and distinct paths towards green transition in the Adriatic Ionian Region by 2050, yet the actual path may lie somewhere in between or even in a totally unforeseen direction. The future has unlimited possibilities and this is what scenarios and foresight are for, to provide us with unlimited options. Taking the different future pathways into account, the Adriatic Ionian Region can design its own path to the future, making important decisions that are centred on the wishes of its people and the wider good of the territory.

Scenarios are not about making predictions and they should not be seen as that. Territorial scenarios are a plausible description of how the future might develop, based on coherent and internally consistent assumptions (scenario logic) about key relationships and driving forces (ESPON, 2022). They serve as eye openers offering different glimpses into possible and plausible futures, by taking into account how different trends and developments evolve and interact, how different factors may affect others or change the framework conditions. The aim of scenarios is to stimulate thinking out-of-the-box, prompt people to engage in creative and social imagination (Toptsidou, 2022b). They do so, by unfreezing any mental and personal barriers to allow thinking about the future (Toptsidou, 2022a). Scenarios, and in particular territorial scenarios, which add a focus on the territorial implications of future changes, are a useful tool to prepare for the future and eventually provide different options and alternatives.

The next step then is to transform these options and define new ideas about desirable futures, i.e. futures that are embraced both by policymakers and citizens. It is also about asking the right questions, namely which elements of the scenarios could be taken forward, which would we like to see in the future and which would we rather avoid? Engaging structurally in discussions about the future can ignite different ways of thinking and act as an antidote to presentism.

Policymakers often take decisions that may have implications in the long run. To avoid a policy myopia, scenarios can support and inform them through a better understanding of the present and a more structural and focused thinking about the future. Scenarios allow for this exploration and can hence support sound policies to avoid future dystopias and create desirable futures. In a continuously changing world, where every explicit or implicit choice matters, scenarios are a powerful tool to build back better and inform decision makers who ask about 'what-ifs' instead of later confronting the 'what-nows'.

References

Berisha E. (2023). The Adriatic Ionian Region: Challenges and opportunities in the time of geopolitical tensions and the rule of EUSAIR. *Economia Marche, Journal of Applied Economics, 42*(1), 59–68. https://doi.org/10.57638/1120-959 3AIREGION

Berisha, E. (2025). Green EU integration—The role of the green agenda in accelerating Just Green Transitions in the Western Balkans. In E. Berisha, J. Moodie, L. Allkja, & M. Jeftić (Eds.), *The Palgrave handbook of Just Green Transitions in the Western Balkans and beyond*. Palgrave Macmillan.

Berisha, E. Moodie, J., Allkja, L., & Jeftić, M. (2025). Just Green Transitions—An introduction. In E. Berisha, J. Moodie, L. Allkja, & M. Jeftić (Eds.), *The Palgrave handbook of Just Green Transitions in the Western Balkans and beyond*. Palgrave Macmillan.

Böhme, K. (2023). The tragedy of the time horizon: Navigating short-termism for long-term sustainability. *Transactions of the Association of European Schools of Planning, 7*.

Böhme, K., & Toptsidou, M. (2022). What if there had been a spatial vision for Europe? In D. Briesen & W. Strubelt (Eds.), *A new beginning? Spatial planning and research in Europe between 1945 and 1975* (pp. 467–485). Campus Verlag.

Büchs, M., & Koch, M. (2019). Challenges for the degrowth transition: The debate about wellbeing. *Futures, 105*, 155–165. https://doi.org/10.1016/j.futures.2018.09.002

Cotella, G., & Berisha, E. (2019). *From space in transition to space of transit. Risks and opportunities of EU and China investments in the Western Balkan Region*. Annual

Review of Territorial Governance in the Western Balkans. Polis University, Tirana, Albania. https://doi.org/10.32034/CP-TGWBAR-I01-02

Cotella, G., & Berisha, E. (2021). The impact of China's Belt and road initiative on the Western Balkan Region. A progressive erosion of EU conditionality? *European Spatial Research and Policy, 28*(2), 165–186. https://doi.org/10.18778/1231-1952.28.2.09

Dasgupta, P. (2021). *The economics of biodiversity: The Dasgupta review: Full report*. HM Treasury.

ESPAS. (2024). *Choosing Europe's Future. Global trends report 2040*. https://doi.org/10.2760/689540

ESPON. (2019). *GRETA—GReen infrastructure: Enhancing biodiversity and ecosysTem services for territoriAl development. Final Report*. ESPON EGTC.

ESPON. (2022). *Territorial scenarios for the Danube and Adriatic Ionian Macro-regions—TEVI 2050*. ESPON EGTC. Retrieved from https://www.espon.eu/TEVI2050

European Commission. (2021). *Regional impacts of the COVID-19 crisis on the tourist sector* (Final Report). https://doi.org/10.2776/179573

European Commission. (2023). *2023 Strategic Foresight Report. Sustainability and people's wellbeing at the heart of Europe's Open Strategic Autonomy*. https://doi.org/10.2792/0971

European Commission. (2024). *Ninth report on economic, social and territorial cohesion*. https://doi.org/10.2776/585966

Hanell, T. (2018). *Regional quality of life in the EU. Comprehending the European space beyond GDP through the capability approach*. Aalto University.

IPCC. (2021). *AR6 climate change 2021: The physical science basis*.

IPCC. (2023). *Climate change 2023: Synthesis report. Contribution of working groups I, II and III to the sixth assessment report of the intergovernmental panel on climate change*.

Joint Research Centre. (2019). *The changing nature of work and skills in the digital age*. European Commission. https://doi.org/10.2760/679150

KAIROS FUTURE. (2024, January). *The landscape of circular electronics towards 2034. Trend Report*.

Krznaric, R. (2020). *The Good Ancestor. How to think long term in a short-term world*. WH Allen, Penguin Random House. The Good Ancestor - Roman Krznaric, https://www.romankrznaric.com/good-ancestor

Letta, E. (2024). *Much more than a market. Speed, security, solidarity. Empowering the Single Market to deliver a sustainable future and prosperity for all EU citizens*. European Commission. Retrieved from https://www.consilium.europa.eu/media/ny3j24sm/much-more-than-a-market-report-by-enrico-letta.pdf

Moodie, J., Berisha, E., Allkja, L., & Jeftic, M. (2025a, in this volume). The multi-dimensionality of Just Green Transitions. In E. Berisha, J. Moodie, L. Allkja, & M. Jeftić (Eds.), *The Palgrave handbook of Just Green Transitions in the Western Balkans and beyond*. Palgrave Macmillan.

Moodie, J., Berisha, E., Allkja, L., & Jeftic, M. (2025b, in this volume.) The Just Green Transitions policy landscape. In E. Berisha, J. Moodie, L. Allkja, & M. Jeftić (Eds.), *The Palgrave handbook of Just Green Transitions in the Western Balkans and beyond*. Palgrave Macmillan.

Randers, J. (2012). *2052: A global forecast for the next forty years*. Chelsea Green Publishing.

Rodríguez-Pose, A. (2018). The revenge of the places that don't matter (and what to do about it). *Cambridge Journal of Regions, Economy and Society, 11*(1), 189–209. https://doi.org/10.1093/cjres/rsx024

Shaker, Y. (2025). The unknown knowns of the Just Green Transitions: A meta-governance approach for a possible enlargement of the European Union. In E. Berisha, J. Moodie, L. Allkja, & M. Jeftić (Eds.), *The Palgrave handbook of Just Green Transitions in the Western Balkans and beyond*. Palgrave Macmillan.

TA2030. (2020). *Territorial Agenda 2030: A future for all places*. Ministerial meeting of the EU under the German presidency of the Federal Ministry of the Interior, Building and Community. Retrieved from http://www.territorialagenda.eu

Toptsidou, M. (2022a, August). *Flourishing in green and social wellbeing: A scenario to just transition*. Retrieved from https://steadyhq.com/en/spatialforesight/posts/8431ef35-35cb-4c29-8cff-cf313ba5b89a

Toptsidou, M. (2022b, July). *Restoring nature for a revived ecosystem: A scenario for the Adriatic Ionian Region*. Retrieved from https://steadyhq.com/en/spatialforesight/posts/d5c3675d-0de4-4420-96d3-1b229a5da4f1

Walljasper, J. (2011, October). *Elinor Ostrom's 8 principles for managing a commons*. Retrieved from https://www.onthecommons.org/magazine/elinor-ostroms-8-principles-managing-commons

CHAPTER 26

Pathways to a Just Green Transitions in the Eastern Partnership: Guided by Western Balkan Experience

Ketevan Katcharava

26.1 Introduction

It is commonly understood that developed countries in Europe (EU) are leading the transition process to a low-carbon future. Nevertheless, the transition will affect the most developing regions due to economic fragility and dependency on carbon-intensive industries (Stefania Fiorentino, 2024). The EU, to alleviate the territorial discontents as part of the European Green Deal (EGD), introduced the Just Green Transition (JGT) mechanism. The JGT has become a global imperative (Romanello, 2021), and the EU, to bring about the initiative's ambitious goal, engages neighboring countries and partners to take a path contributing to their objectives. Among the external dimensions of the EU, the chapter focus is set on the EaP, a joint initiative introduced in 2008, to strengthen ties between the EU, its member states, and six Eastern partners: Armenia, Azerbaijan, Georgia, Moldova, Belarus, and Ukraine. The main objective of the partnership is to accelerate the "*Security, stability and prosperity, democracy and rule of law*" (European Union, 2008), with a given priority to common environmental challenges posed by climate change. The EaP stands at a juncture in its JGT path—a transformation for climate neutrality, social equity and economic resilience. This path is largerly influenced by dynamics of the EU integration; Ukraine and Moldova achieved candidate status and opened accession negotiations following their

K. Katcharava (✉)
Interuniversity Department of Regional and Urban Studies and Planning (DIST), Politecnico Di Torino, Turin, Italy
e-mail: keti.kacharava@gmail.com

E. Berisha et al. (eds.), *The Palgrave Handbook of Just Green Transitions in the Western Balkans and Beyond*,
https://doi.org/10.1007/978-3-031-95075-9_26

first intergovernmental conferences in 2024. Georgia, although granted candidate status, has stall accession process due to electoral shortcomings. Armenia and Azerbaijan maintain cooperative ties with the EU without formal accession status. Belarus remains excluded from the integration process due to its political links with Russia (COM, 2024). As the EaP moves closer to acquiring the EU candidate status, strong governance structures will become more relevant by the time. These frameworks must tackle climate challenges both domestically and within the broader context of the EU. The EaP need to prepare themselves to address climate issues not only internally but also as part of a wider partnership of shared benefits and responsibilities (EU4Climate, 2023). In that regard, the first section of the chapter observes the influence of climate policies on countries beyond the EU. While the second section specifically focuses on the WB context, which despite not being EU members, have pioneered their own Green Agenda. Political and governance challenges, as analyzed through the comparative framework in the following section, are rooted in shared socio-political histories and could be the foundation of the EaP's transition pathway. The chapter further explores the policy and governance frameworks needed to build multi-level partnership in the EaP and as the guiding tool, proposes the specifically tailored JGT roadmap to deal with regional specificities.

26.2 Influence of Climate Policies on Countries Beyond the EU

The influence of the EGD extends far beyond the EU's borders and impacts neighbouring countries and their approach to policies and sustainability practices (Walker, 2024). The EU uses international conventions and bilateral relationships as instruments to reinforce environmental profiles of neighbouring and partner countries. Enlargement process, also referred to as the preparation stage for EU accession, is accompanied by reforms in political-legal and socio-economic systems, and are essential for meeting the conditions of accession. The enlargement policy is today the Union's main instrument and normative power to export norms (including environmental ones), practices, and governance models beyond its borders (European Comission, n.d.). The policy outlines the criteria and conditions that candidates (Fig. 26.1), as well as potential candidate countries need to meet to be considered for membership.

The perspective of EU membership for the candidate/potential candidate countries requires compliance with the Environmental *acquis*, covering horizontal legislation, environmental impact assessment, strategic environmental assessment, access to information, waste management, climate change, water and air quality, nature protection, industrial pollution control and risk management, chemicals and genetically modified organisms, noise and forestry (European Union, 2023). On top of the environmental *acquis*, the EGD emphasizes reducing carbon emissions, and by the date of accession, candidates are expected to apply reforms (such as the European Climate

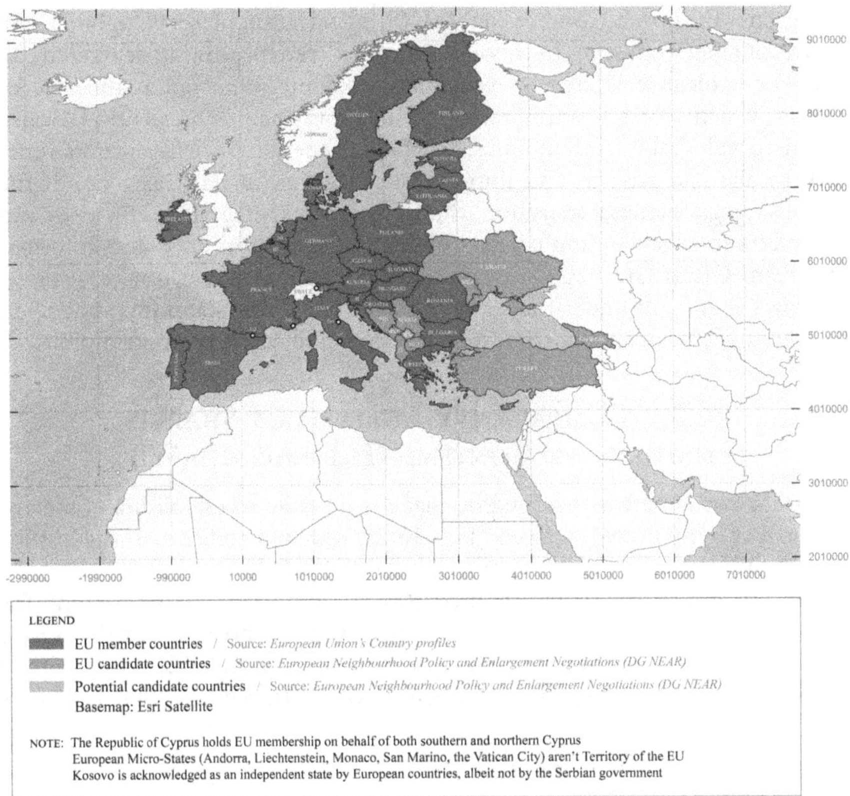

Fig. 26.1 The EU 27, candidate and potential candidate countries (*Source* author's own elaboration)

Law, Fit for 55 Package, TEN-E Regulation, etc.) that reflect its principles. This includes undergoing governance and institutional reforms, such as the policy integration and coherence (Do No Significant Harm Principle (EU Commission, 2021)), introducing the Just Transition Governance to address socio-economic disparities (Maria Yetano Roche, 2020), and emphasizing the Multi-Level Governance for improved coordination across different sectors. It means reinforcing governance structures, improving institutional capacity and policy implementation, reinforcing public participation and so on. Besides the environmental and institutional influence, the EGD has socio-economic impacts on candidate countries. The Just Transition Mechanism (JTM), introduced in 2020 by the EU Commission, is a key tool which targets the most affected regions and social groups. While its primary focus is on EU member states, the JTM also includes support for non-EU countries (European Comission, 2019). More specifically, the so-called "Sister initiative", Coal Regions in Transition, launched in December 2020, works for coal regions in the WB and Ukraine to support the just transition of coal-based energy production in

17 regions (European Commission, 2020b). Transition projects highlight the growth of green industries in those regions that were previously dependent on coal. The latter is facilitated by workforce reskilling, job creation and diversification, which works towards reducing unemployment rates. The JTM also extend to vulnerable communities, where the energy efficiency programmes work to alleviate poverty and improve social inclusion. For instance, in the case of Georgia's energy transition, the EU supports integrating efficiency and renewables in rehabilitating public buildings (Delegation Press and Information Team GEORGIA, 2023) to reduce electricity bills and total emissions. Georgia's alignment under the support of the EGD stimulates investments in the energy sector, which in turn helps to expand investment opportunities.

26.3 Comparative Analysis of Shared History and Context—The EaP and WB

Analyzing both contexts thoroughly made it possible to establish a comparative analysis between two regions (Fig. 26.2), which in addition to the lessons from the WB, could be a foundation for the EaP's JGT framework.

The initial years of independence was an important starting point for both the EaP and the WB. This period followed the break-up of the Soviet Republic for the EaP and the Yugoslav Republic for the WB, leading to a complex transition that sparked conflicts across both regions. The WB right

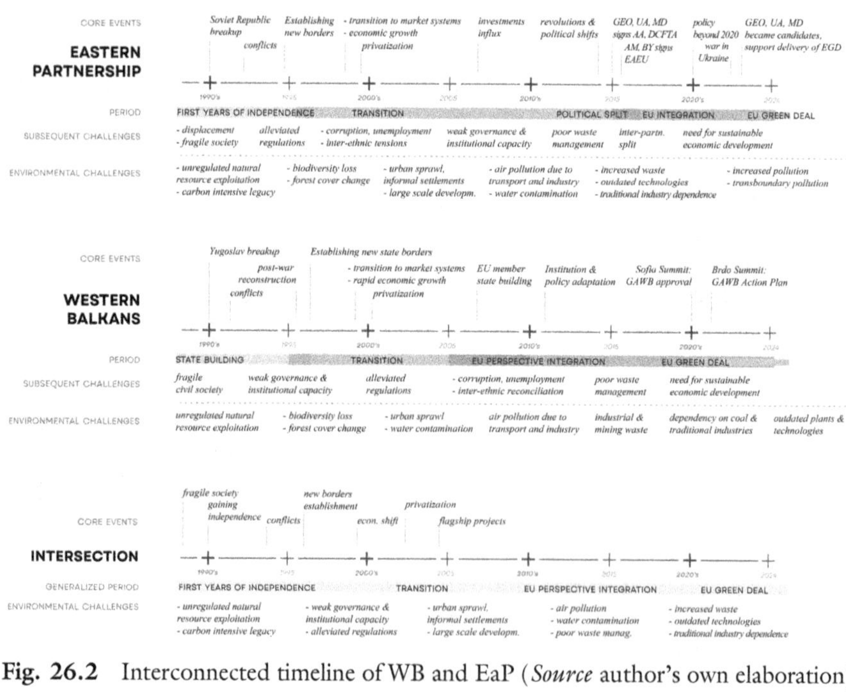

Fig. 26.2 Interconnected timeline of WB and EaP (*Source* author's own elaboration)

after gaining independence, focused on building new states, guided by international conventions and agreements that helped to set borders. In contrast, the EaP had longer-lasting conflicts, which delayed the rebuilding process. A common outcome of these conflicts in both cases was the displacement of populations and the feeling of instability, which shaped the fragile societies with limited public involvement in state-building and governance. The shift from socialist to democratic systems was a critical point for both regions. It required the creation of new institutions to ensure that the government was separate from party control and introduced market economies that differed from the centralized economic planning of the past. This period was challenging because of the need to implement fundamental reforms to move away from the institutional legacies of communism. Period was characterized by weak governance and limited institutional capacity. Countries struggled to simultaneously acquire the knowledge and strength for good governance and at the same time deal with the impacts of recent conflicts. Challenges were mainly in the political and social areas. The next phase, known as the transition period, saw both regions moving towards market economies, characterized by a rush to privatize state-owned assets and an increase in investments. Land privatization and economic reforms in the mid-1990s aimed to stabilize the economy and attract foreign investment, which eventually led to the privatization of services, manufacturing, and industry (David Gogishvili, 2018a). This process exposed the lack of readiness of institutional bodies and the absence of solid planning frameworks, which was further intensified by corruption and informal practices. The impact of the privatization wave was the most visible in urban areas. Major cities faced gentrification, displacement, and changes to traditional urban spaces to accommodate the growing real estate market. The period between 2005 and 2010 was a decisive time for both regions when their paths began to diverge after 15 years of facing similar political, social, and governance challenges. The EaP countries of Belarus and Armenia leaned towards Russian politics, while Ukraine worked to free itself from Russian influence. On the other hand, WB countries began to secure spots on candidate lists for EU membership. The WB initiated the EU integration much earlier than the EaP. The process for EaP countries was slower and affected by conflicts in Ukraine and agreements. Those developments eventually granted the candidate status to Moldova, Ukraine, and Georgia. Thus, an important difference between the EaP and WB emerges in their approach to EU integration, whereas the EaP also faces division among its members regarding their orientation (towards the EU or Russia).

26.3.1 Shared Political and Governance Constraints and Challenges in the EaP and WB

Despite providing support, the implementation effectiveness of initiatives varies across the EaP countries (Zofia Gródek-Szostak, 2025). Certain challenges hinder the transmission and acceptance of these measures. Economic constraints are the primary reason (OECD, 2024a). Examining the Income Per Capita data for 2010 reveals great disparities within the regions. Internal inequality is particularly visible in Ukraine (Fig. 26.3), where coastal and capital city areas starkly contrast. Most economically struggling countries have the lowest income per capita indicator and the highest dependency on fossil fuels due to their affordability (Energy Institute Statistical Review of World Energy, 2021).

Political instability is another factor that challenges both—the policy coherence in the region and the implementation of long-term sustainable strategies. In recent years, several conflicts have been unfolding in the EU and its neighbours, including the interstate border dispute between Armenia and Azerbaijan over Nagorno-Karabakh (Waal, 2023), and Russian invasion of Ukraine (since 2014). Additionally, Georgia still faces territorial disputes with

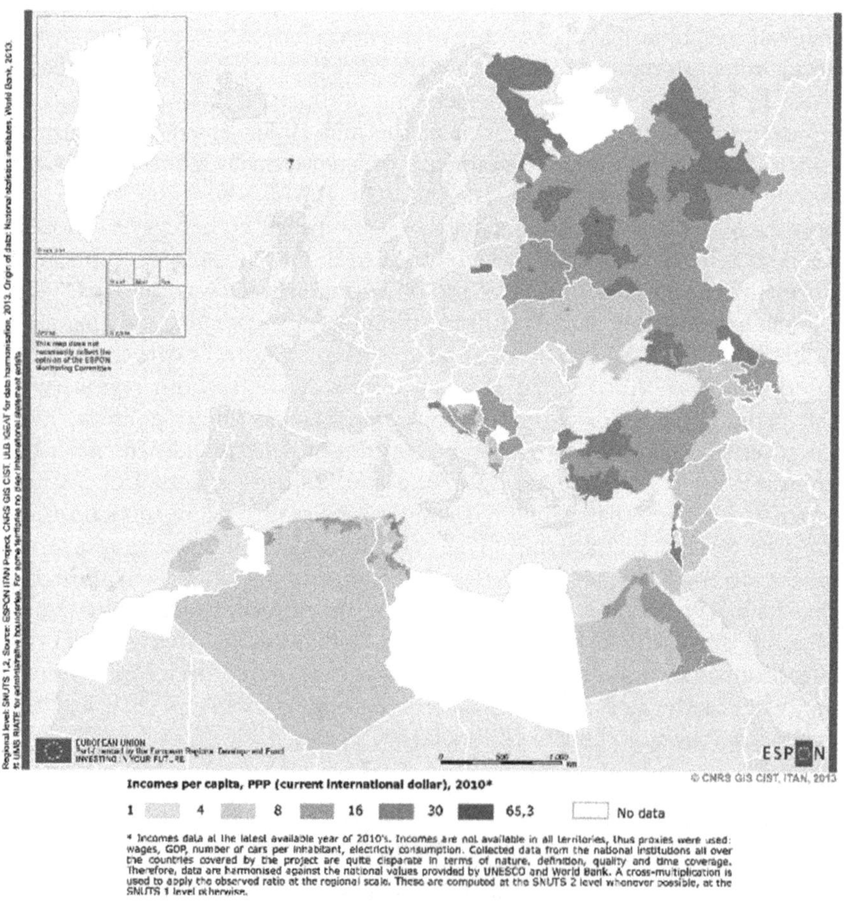

Fig. 26.3 Incomes Per Capita (*Source* ESPON, 2010)

Russia for Abkhazia and South Ossetia. The war in Ukraine, according to the Ministry of Environmental Protection and Natural Resources (2022), has caused severe damage to the environment, both immediate and longer-term consequences. Most of them have transboundary impacts on neighbours and beyond. Along with environmental damage, conflicts pose a risk on long term strategies, as they "weaken, constrain or destroy social stabilization mechanisms" (Olga Litvishko, 2023). Corruption level in most EaP countries presents another challenge to implementing measures. Challenges also arise in institutional arrangements, where state institutions specifically designated for climate change are missing. For instance, in Georgia and Azerbaijan, there's an absence of a structural unit (or working group) that solely focuses on climate change (The Ministry of Environmental Protection and Agriculture of Georgia, 2021). The responsibility for related topics is assigned to a single member, who also shares responsibilities for other duties. This weak institutional arrangement leads to another challenge associated with a lack of specialized structure and, therefore, training specialists on climate change. This issue is particularly problematic in Belarus and Moldova, where inadequate staff members lack the training and skills to tackle the green challenges (Anna Maria Augustyn, 2022). Moreover, significant challenge arises for a gap between the climate change issues and their inclusion in the regulatory documents and sectoral programmes. There's an absence of a common regulatory framework that covers sectoral activities. Low-carbon measures exist sectorally (air, energy, water waste), although coordination among sectoral public agencies is weak (OECD, 2024b). For instance, in Armenia, the lack of a cohesive strategic planning framework for climate change and green transition hinders a full understanding and evaluation of national efforts of climate-related policies that have been put into place (Ministry of Environment of the Republic of Armenia, 2020). The non-coherence of the frameworks and poor coordination spread over the region and lead to the lack of a shared vision in the partnership and, therefore, weak regional cooperation. Another important challenge following the latter is issues in the identification of financial needs. Due to the absence of measures at the national and regional levels, authorities are facing problems with an exploration of financial sources for solving climate actions. The technological constraints are mainly related to the Soviet legacy of carbon-intensive industries, unskilled workers, energy dependency, and insufficient monitoring systems (OECD, 2024b). For instance, coal still accounts for a great share of Ukraine's energy source, making up 70% of primary energy consumption (IRENA, 2022). Energy dependency is one of the biggest technological constraints for the EaP. Most countries don't meet the national energy demand,[1] which makes them dependent on imported sources (IEA,

[1] According to the data, Armenia—covers 27% of energy demand with domestic energy production (2020); Azerbaijan—has one of the world's highest energy self-sufficiency ratios, with production exceeding demand almost four times; *Belarus*—only 15% of the country's energy demand was met by domestic production (2018); *Georgia*—energy

2021). In renewable energy sources, challenges are posed by outdated technologies. Meanwhile, in the energy refurbishment sector, issues arise from the old housing stock, which was constructed during the Soviet era (before the 1990s), and lacks energy-efficient building standards and certifications for construction materials. The refurbishment constraints stem from the absence of thermal points and outdated heating systems that do not allow for the regulation of heat consumption, coupled with urgent renovation needs. The municipal management companies lack the financial and technical resources to provide adequate maintenance for these systems (Republic of Moldova, 2023).

Additionally, insufficient monitoring systems pose an additional obstacle (European Commission, 2023). The effective implementation of measures and the accurate assessment of their impact necessitate regularly updated data (OECD, 2024b). For instance, in Armenia, forest issues are on hold due to inadequate data on the effects of climate change. Although some data covers 2–5 years, this time span isn't sufficient to provide a reliable analysis (Republic of Armenia Ministry of Environment, 2020). The Measurement, Reporting, and Verification system has been proposed by the Policy Beyond 2020 (EU4Climate, 2020) and countries are encouraged to implement it. Nevertheless, the implementation of such systems is challenged by the economic and financial gaps, due to their high investment cost and long payback periods.

26.3.2 *Common Environmental Challenges*

Despite geographical differences, both regions experienced similar developmental paths for the first 25 years post-independence. During the formative years of independence regulations worked for smoother transition and investments attraction (David Gogishvili, 2018b), which lead to unregulated natural resource exploitation across the EaP and WB. Challenges were more intense in the transition period, due to smoother regulations and low institutional capacity. This period saw forest and land cover changes, biodiversity loss, urban sprawl, poor waste management, air and water pollution, and the persistence of carbon-intensive industries (Secretary-General of the OECD, 2021). Urbanization and the shift to capitalism notably impacted the ecology (David Gogishvili, 2018a, 2018b). Market reforms and widespread privatization put the primary pressure on public green spaces. The lack of comprehensive urban planning and inadequate monitoring raised the illegal constructions, while existing housing stock deteriorated due to insufficient funding for renovations. Air quality and contamination in both regions were mainly due to transportation and industrial activities. Mining and manufacturing plants, mainly

production covers about one-fifth of its energy demand (21.1% in 2020); *Moldova*—around 20% of its energy demand is covered by domestic production; *Ukraine*—nearly 65% of total energy demand is covered by domestic production.

utilizing low-quality coal, contributed to increased emissions and air contamination (Maja Colovic Daul, 2019). Concerning the transportation system, the issue is linked to the importation of second-hand, aged, and highly polluting vehicles, mainly sourced from the EU and chosen for their affordability. An increased motorisation rate, along with a low quality of fuel, became the major sources of air contamination and health problems in the regions (Fabrègue, 2019). Despite the 10-year lag in the start of the EU integration process between the regions, they continue to face similar environmental and governance issues. These include air and water pollution, increased waste, reliance on coal, and outdated technologies. A dominant challenge, more pronounced in the WB, is the dependence on coal (Frey, 2024). Coal is the primary fossil fuel in the WB, supporting traditional industries, especially thermal power plants (TPPs) (Bechev, 2023). Despite some attempts at updates, they remain less efficient and more outdated than their EU counterparts (Emerging Europe, 2021). The inadequacy of air filtration and wastewater treatment systems in these plants presents challenges to the environment and leads to cross-contamination of water and soil. Neverthless, these traditional industries are the major employment providers in the regions, which increases the social vulnerability for the shift towards greener alternatives (EU Joint Research Centre, 2024). Additionally, both regions are encountering growing industrial and mining waste challenges. Landfilling is the dominant waste management strategy, and illegal dumping is the problem across practically all EaP (EU4Environment, 2021) and WB countries (EEA, 2023). Municipal waste collection is insufficient, especially in rural areas, while outdated waste facilities further intensify the problem. Water quality and sanitation present additional challenges. Decaying infrastructure intensifies the situation and leads to the cross-contamination of water between sewage and freshwater systems (WAREG, 2021). Many large cities in both regions lack adequate wastewater treatment facilities, while in rural areas, service systems are often completely absent (EEA, 2020). Despite allocated funds and progress in the development of wastewater systems, the steps remain insufficient in most countries. Lastly, the trend in CO2 emissions presents a shared challenge, albeit more acutely in EaP countries (Fig. 26.4). While medium and smaller-sized countries in these regions contribute less to global climate issues compared to more developed nations, their CO2 output, when considered in relation to their economic output, positions EaP and WB countries (Martin, 2023) as disproportionately high contributors to GHG emissions within the EU 27 (World Bank, 2022).

26.3.3 Policy Framework and Priorities for the JGT

In terms of policy framework, the EaP and the WB have different approaches to govern the JGT. Both regions have been influenced by the EU's environmental directives mainly due to their accession requirements. The WB's newly emerged Agenda aligns with the EGD's objectives and is specifically tailored for the region's environmental, economic and political context. On another

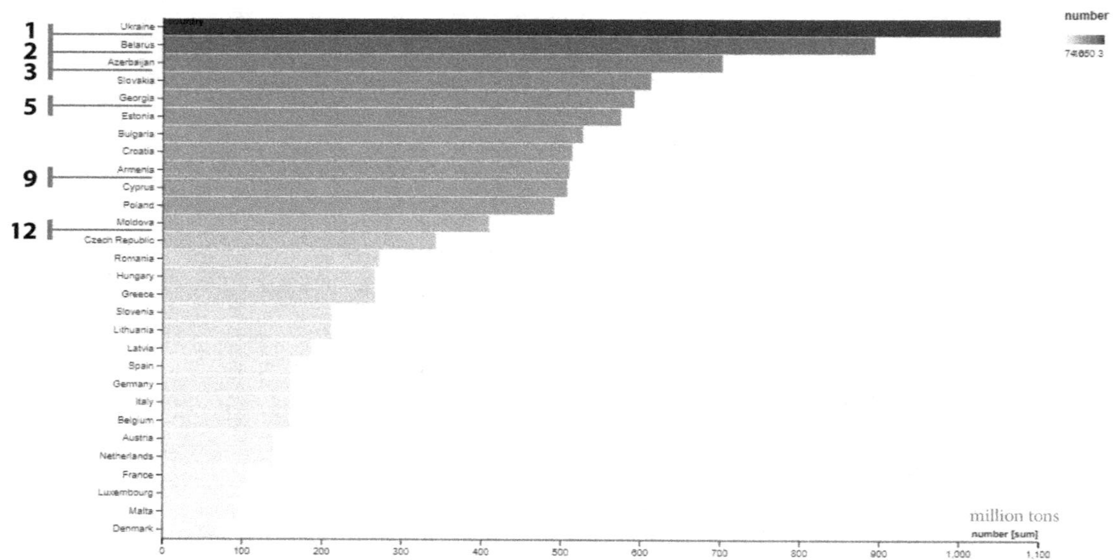

Fig. 26.4 GHG emissions per million dollars of GDP for EU-27 & EaP (*Source* World Bank)

hand, although such a specifically tailored agenda is absent for EaP, it participates in different EU funded initiatives ('EU4Climate', 'EU4Environment'). Under the umbrella of the EU's EaP, have developed national and regional strategies that reflect their commitment to climate action, albeit with varying degrees of integration. Those initiatives also align with the EGD by sharing common goals of achieving lower emissions and a greener future. However, currently the policy beyond 2020 is the closest stand initiative that directly supports the EGD's delivery (European Commission, 2020a).

26.3.4 Challenges and Barriers for Developing and Implementing Green Transition Policies

In terms of challenges and barriers for implementing policies, several overlapping factors arise. Both regions face institutional weaknesses that impede effectiveness of the JGT measures. In the WB, a rapid transition failed to establish powerful institutions, which created a capacity void of governance, as detailed in the "Report on Western Balkans Just Green Transition Conceptualisation". Similarly, the EaP struggles with fragmented institutional frameworks.

Financial limitations are the factor that significantly impedes the process across both regions. In the WB, experts have highlighted a preference for grants over loans, which points to concerns about economic uncertainty. The EaP faces similar constraints, where limited funding impacts essential services like waste management and water treatment, as noted in national reports from Armenia and Moldova. Both regions struggle to attract investments to overcome basic needs and attract potential investors. Societal preparedness and stakeholder engagement are critically low in both regions. The WB's historical centralisation has weakened public participation. The EaP also shows weak public engagement, which, as evidenced by the EU4Climate reports, is increased by inadequate information sharing on JGT and undermines the acceptance of the initiatives (Kampel, 2023). Moreover, both regions are burdened by outdated and carbon-intensive industries. The WB's reliance on coal and the EaP's struggles with illegal logging are indicative of environmental management issues. Technological constraints are also prevalent. Political instability also affects both regions but manifests differently. In the WB, frequent shifts in political and administrative leadership hinder the effectiveness of strategies. In the EaP, varying political orientations and political fragmentation, particularly between countries aligned more closely with the EU versus those less committed, create uneven progress in JGT. This division is sharply illustrated by different levels of engagement and implementation in EU-aligned initiatives, as detailed in EU policy review documents. Furthermore, while both regions share a commitment to engage a wide range of stakeholders, their methodologies are different. The WB's governance system has a highly structured approach that emphasizes regional ownership and engagement across various policy domains. They have a more structured cross-sectoral collaboration that supports the alignment with the latest developments and needs. Contrastingly, the EaP's governance lacks centralized coordination mechanisms; their strategy is managed through high-level summits, official meetings, and the EaP Civil Society Forum. This approach appears broader and less targeted compared to the WB's. Another important point is the regulatory framework governing climate action. The EaP has a fragmented approach due to the lack of a unified regulatory framework, which leads to less coherent national and transnational strategy. Conversely, the WB, with its tailored Green Agenda, has a more unified strategy for implementing JGT measures. In terms of the financial coordination, the EaP's challenges are due to absence of clear measures at the national and regional levels, which makes it harder to explore sources for climate action. This contrasts with the WB's implementation of a Donor Coordination platform, which organizes annual meetings to streamline financial support and facilitate dialogue on financial needs.

26.4 Successful JGT Practices from the WB and Recommendations for the EaP

Despite shared challenges, certain practices within the WB have been successful for the JGT, such as the Donor Coordination Platform, Regional Cooperation Council, Multi-Stakeholder Platforms, and so on. The proposals rely on interviews with regional experts, the Reports on Western Balkans Just Green Transition Conceptualization, and the GAWB Action Plan Implementation Report (Regional Cooperation Council, 2023). Regarding financial strategies, the WB experience proves that dependency on external funding is necessary for a financial planning redesign. For instance, the progress of initiatives under the GAWB were uneven due to the difficulties in getting sufficient funds. Yet, the Donor Coordination Platform under the GAWB governance model, could be considered as a good practice. It links the funding sources with strategic needs and makes sure that the financial resources are used efficiently and transparently. This platform works as the central hub for gathering and distributing funds from different international donors, thus avoiding the duplication of funding on similar projects and making sure that every initiative gets the required financial support. Besides, the platform increases the transparency of the funding process by providing a clear overview of where and how funds are spent. Essentially, it centralizes and rationalizes funds management and ensures that funding is both adequate and aligned with environmental goals. Additionally, for designing a resilient system, the interviews prioritized diversifying funding sources, such as public–private partnerships, EU funds, and innovative financing mechanisms (green bonds, environmental taxes). They can work together to fill the gap and maintain the continuity of the strategies. More specifically, innovative financing mechanisms of green bonds can attain a wider cycle of investors and mobilize capital flows. Environmental taxes could also be applied to generate steady revenue streams that can be reinvested in further initiatives to create a self-sustaining cycle of investment in sustainability. To encourage Stakeholder Engagement, the WB utilized workshops and public events as the main tools to involve various groups of government bodies, academia, and NGOs. The GAWB Action Plan presents multi-stakeholder forums where the progress discussion is held and the future actions are being coordinated. These forums tend to improve participatory decision-making processes by integrating stakeholder feedback directly into policy design and ensuring that strategies are reflective of the community's collective needs. The Action Plan also highlights capacity-building initiatives to bridge knowledge gaps by providing local stakeholders with the skills needed to manage and implement green technologies. Altogether, ensures that policies are supported by the communities they affect. Engagement activities should be aligned with local cultural norms. This could be done by designing public awareness and educational campaigns that are culturally sensitive and resonate with local values. As well as encouraging community-led environmental initiatives that align with regional customs and encourage more participation and co-ownership. They can build trust in fragmented societies, and eventually ensure better support for JGT initiatives. Moreover, the EaP context is stressing the need for collaborating diverse political alignments and integration Levels. In that regard, establishment of forums that motivate dialogue and consensus-building among EaP countries is suggested. They can be used to share best practices, discuss common challenges, and develop a sense of regional mutual trust for sustainability efforts. Besides, regional projects that require cooperation, such as cross-border environmental monitoring and management initiatives, should also be encouraged as they could serve as neutral ground to promote cooperation without necessitating full policy alignment with the EU. Incentive mechanisms, such as regional awards or grants for the collaborative environmental projects, could be adapted to motivate countries to work together in achieving the common environmental goals and shared benefits. Lastly, for Improving Regional Cooperation, the success story from the WB is the Regional Cooperation Council (RCC), which includes monitoring, donor coordination, partnerships, and communication to achieve socio-economic development and regional coordination across diverse sectors (Regional Cooperation Council, 2024). Its approach works to incorporate diverse perspectives into regional initiatives and strengthens their relevance. This framework could serve as a blueprint for the EaP by providing a structured approach to effectively share best practices, data, and resources and ultimately foster a cohesive regional response to emerging challenges.

26.5 Green Transition Roadmap Development and Applicability to EaP Context

Having analysed the good practices and recommendations from the WB experience, it's now possible to delineate the JGT roadmap, designed explicitly for the EaP's context. For that it's fundamental to reconsider regional peculiarities. Among these factors, the highest risks are associated with frequent governmental shifts, political instability, and tensions with neighbouring states, which might lead to an unsustained commitment to JGT. The proposed framework breaks down the road map into sequential preparatory, planning, implementation, and monitoring stages and further explains specific points within (Fig. 26.5).

Preparation Phase

Initiating from the preparation phase, the JGT framework, first and foremost, should be adjusted to diverse political climates, economic conditions, and unique cultural features. The Framework, therefore, needs to be flexible and adaptable to accommodate the varying degrees of EU integration and the geopolitical sensitivities within the region. The most adjustments are expected in the preparation phase. Initially, the framework should incorporate mechanisms that allow for varied levels of engagement and commitment. It should have a flexible nature that will reflect the distinct political aspirations of the member countries. For example, Belarus, Armenia and Azerbaijan may require a more tailored approach that considers their stance towards EU integration and emphasizes regional benefits and independence. Due to the same reasons, as well as the fragmented governance and coordination, it's essential to have a body or group designated for regional cooperation. The structure can benefit from WB experience of Regional Framework for Cooperation, which in case of EaP context, will serve for dialogue and consensus-building among the members; It will be responsible for an exchanging of best practices, addressing common challenges, fostering a sense of mutual trust, and regional cooperation. After establishing the coordination group, the focus shifts to national levels, where due to the absence of special institutions working on climate change, setting up working structures on the JGT topic becomes essential. The following step is the contextualization of the JGT. At this stage, local knowledge and expertise are invaluable, and thus, it is essential to involve stakeholders who possess local expertise and can incorporate various perspectives. The further step is the definition of baseline situation, which includes the evaluation of current environmental and socio-economic conditions in specific context. These assessments' outcome will help establish a factual basis for setting priorities, formulating policies, and monitoring progress. Integrating these data will give the possibility to address the unique challenges and opportunities of the region, ensuring that subsequent steps in the GT process are grounded in reality and strategically focused. In parallel to the baseline scenario definition, EaP countries could benefit from gathering information and data from sectoral initiatives or projects. For example,

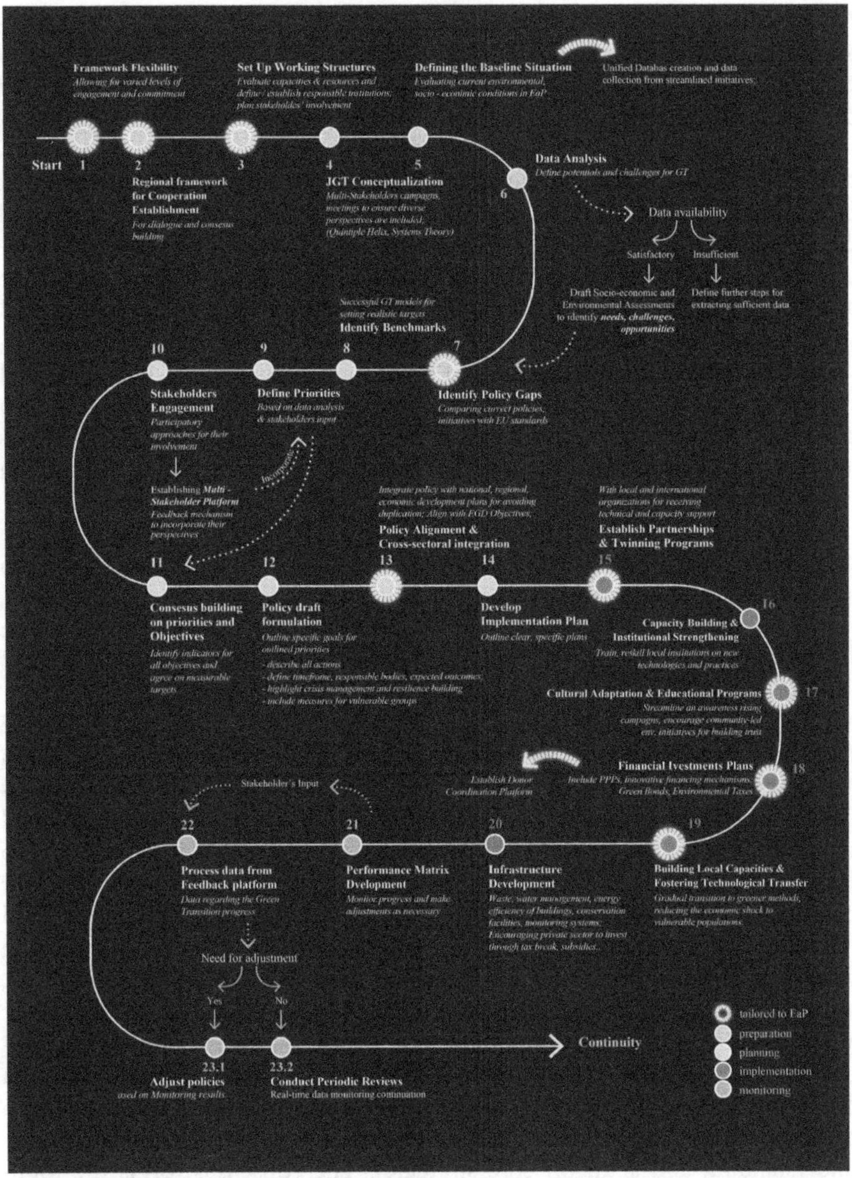

Fig. 26.5 Just Green Transition Roadmap for the EaP (*Source* author's own elaboration)

EU funded initiatives could provide sector-specific data which can be consolidated in unified databases. This approach would improve data quality and consistency and facilitate more accurate analysis. The consolidated data could then be used to identify trends, evaluate existing policies, and develop new strategies tailored to the region's needs. Managing information from scattered initiatives, the EaP can build a data-driven foundation for the upcoming processes. Subsequent steps encompass the data analysis, where potentials and challenges need to be evaluated. This involves an assessment of the data's availability and volume, which runs parallel to the main process. Depending on the adequacy of the data, this phase may dictate the need for gathering additional information or, if sufficient, proceeding with socio-economic and environmental assessments. This will set the stage for identifying specific needs and opportunities that will shape the planning process. The last tailored step in the preparation phase is identification of policy gaps, where EaP can assess and compare current policies with EU standards. This involves a comparison of regulatory frameworks, enforcement mechanisms, and sector-specific policies in energy, transportation, waste management, and etc. Identifying these gaps can highlight areas needing improvement and help understand barriers to aligning with EU directives. This bridges the preparation and planning phases, lays the groundwork for developing targeted strategies and better formulate JGT goals.

Planning Phase

The planning stage opens with identification of benchmarks and successful JGT models in order to set realistic targets. This could help to establish measurable standards that guide progress evaluation, while successful models inform better strategies. This dual approach ensures that targets are adapted to local contexts and enhances their relevance and feasibility. The next step involves defining priority areas tailored to specific contexts, informed by prior data analysis and stakeholder input. This process is inherently participatory and reinforces stakeholder engagement in parallel with priority setting. By utilizing bottom-up approaches, this stage integrates the perspectives of communities, local citizens, vulnerable groups, employers, and governments, ensuring that all voices are contributing. The circular nature of this process allows for continuous integration of local experience and knowledge and moves away from a traditional top-down approach. After stakeholders' visions are integrated into the definition of priority areas, consensus building becomes essential to unify understanding and align on core strategies. The consensus forms the foundation for drafting policies where specific goals are aligned with established priorities. Policy formulation should detail the responsible entities, set clear timeframes, outline expected outcomes, and define strategies for supporting vulnerable groups. Following these specifications, it's essential to integrate JGT policies with national, regional, and economic development plans for avoiding duplication, as in EaP context sectoral policies stand as siloes. This involves a review and alignment of existing policies to identify overlaps, gaps, and potential areas for collaboration. It paves the way for the final step in the planning phase: developing the implementation plan—which should specify tasks, assign responsibilities, establish deadlines, and allocate resources needed to achieve the transition objectives.

26.6 Implementation Phase

Once the policy draft is finalized and the implementation plan is in place, the project moves to the implementation stage. At this point, the focus shifts to the establishing partnerships and twinning programmes with local and international organizations for receiving technical and capacity support. This will ensure that those countries have the necessary tools and resources to successfully implement their initiatives. Along with this, the redevelopment model should be launched l where capacity-building programmes are essential tools to reskill local institutions in new technologies and practices. Parallel to the reinforcement of institutions, there is a need to introduce financial investment plans such as Public Private Partnerships and innovative funding mechanisms. The GAWB governance model can introduce a structure similar to the Donor Coordination Platform, which could connect funding with needs and ensure that financial resources are allocated efficiently. The latter could be an essential for the EaP context due to increased corruption levels. This platform can serve as a hub for the collection of funds from international donors—by this means preventing the duplication across similar projects and ensuring that each initiative receives the necessary financial support. After the financial investments are gathered and redirected, it's essential to build local capacities and foster technological transfer. The framework should address disparities in industrial and infrastructural development and focus on building local capacities, aligned with specific needs and capabilities of each country. This approach will reduce the economic shock to vulnerable populations. Referring to the cultural acceptance, due to the past traumatic transitions and cultural barriers, it's essential that the JGT framework should promote policies that are developed through inclusive and participatory processes. The framework should promote co-ownership initiative for gaining public trust and fostering a collective sense of ownership over environmental initiatives. Parallel to that, educational programmes and awareness campaigns need to communicate the benefits of the JGT in a way that is accessible and engaging for all stakeholders, particularly in areas with low environmental awareness. Despite the financial resources under the EGD, additional funding is necessary to support further development projects. Thus, in the implementation phase, mobilizing financial instruments becomes a priority step to ensure smooth progression and prevent disruptions in ongoing projects. Once a stable financial flow is established, they can be redirected towards essential infrastructure development projects.

Monitoring Phase

In the monitoring phase, a top priority is establishing a performance matrix to track progress and launch necessary adjustments. This phase mirrors the circular approach of the planning stage, where stakeholder feedback and matrix data mutually inform ongoing assessments. Continuous feedback loop allows for the timely adjustment of policies or prepares them for future periodic reviews based on the inputs. This systematic approach ensures that policies stay relevant and effective across varying contexts and communities, and dynamically adapt to emerging needs and outcomes.

26.7 Conclusions

The chapter has examined the pathways towards the JGT in the EaP, using the experiences of the WB as a comparative lens. The study focused on analyzing the influence of EU green policies, identifying key drivers and barriers towards these practices, and proposing a roadmap tailored to the EaP's unique regional dynamics. Central to the research is the recognition that while the EaP, as well as the WB, countries are diverse in their socio-political contexts, they share common challenges, including reliance on carbon-intensive industries, fragmented governance, weak institutional capacities and socio-economic vulnerabilities. Analyzing these interplay, based on the WB experience, the study reveals potential pathways for the EaP context. This includes the importance of multi-level governance, transparent financial coordination, and participatory stakeholder engagement to effectively drive the transition. The tailored JGT roadmap is another outcome of the study, which summarizes the findings and once again emphasizes the need for flexible, context-sensitive strategies that can adapt to varying levels of EU integration and geopolitical contexts. However, the study faced certain limitations including the data availability. Interviews were conducted solely for the WB context, while conclusions for the EaP were primarily taken from secondary data (policy documents and reports), which often do not fully capture the dynamic processes. Furthermore, differences in political orientation among EaP countries further complicated the generalization of the findings—which represent a snapshot in time, and may not fully capture the future development. For these reasons, future research should prioritize longitudinal studies to evaluate the long-term impacts of JGT policies and explore strategies for policy coherence and cross-sectoral coordination. Additionally, it should focus on establishing mechanisms for regional cooperation, such as the WB's successful model of the Regional Cooperation Council and investigate culturally sensitive public awareness initiatives to accelerate societal acceptance of transitional policies.

References

Augustyn, D. L. J. L. (2022). *Localising the green agenda in the Eastern Partnership*. s.n.

Bechev, D. (2023). *The Green Transition and the Western Balkans*. s.n.

COM. (2024). *2024 Communication on EU enlargement policy*. s.n.

David Gogishvili, A. C. (2018a). Cities of the South Caucasus: A view from Georgia. *QU3—iQuaderni di U3, 15* (ISSN: 2531-7091), 5–73.
David Gogishvili, A. C. (2018b). Cities of the South Caucasus: A view from Georgia. *QU3—iQuaderni di U3,* 105.
Delegation Press and Information Team GEORGIA. (2023). *Representation of the European Union in Georgia.* [Online]. https://www.eeas.europa.eu/georgia/evrokavshiri-da-sakartvelo_ka?s=221
EEA. (2020). *Water availability, surface water quality and water use in the Eastern Partnership countries: An indicator-based assessment.* s.n.
EEA. (2023). *Municipal waste management in the Western Balkan countries.* s.n.
Emerging Europe. (2021). *Western Balkans coal plants pollute twice as much as those in the EU.* s.n.
Energy Institute Statistical Review of World Energy. (2021). *Fossil fuel consumption.* s.n.
EU Commission. (2021). Establishing the recovery and resilience facility. *Official Journal of the European Union,* 17–57.
EU Joint Research Centre. (2024). *Green transition in the Western Balkans requires stronger focus on sustainable innovation.* s.n.
EU4Climate. (2020). *Roadmap for the development of a GHG Inventory System: How this is addressed in the EU and EaP countries.* s.n.
EU4Climate. (2023). *Progress and ways forward in the Eastern Partnership countries.* s.n.
EU4Environment. (2021). *EU4Environment fights for waste reduction in the EaP countries.* s.n.
European Comission. (2019). *Initiative for coal regions in transition.* [Online]. https://energy.ec.europa.eu/topics/oil-gas-and-coal/eu-coal-regions/initiative-coal-regions-transition_en
European Commission. (2020a). *Eastern Partnership policy beyond 2020: Reinforcing Resilience—An Eastern Partnership that delivers for all.* s.n.
European Commission. (2020b). *SISTERS project: Systemic innovations for a sustainable reduction of European food wastage.* s.n.
European Commission. (2023). *European Union for Environment in Eastern Partnership countries—Water resources and environmental data.* s.n.
European Union. (2008). *Extraordinary European council.* s.n.
European Union. (2023). *WECOOP—Central Asia water, environment and climate change cooperation.* [Online]. https://wecoop.eu/glossary/eu/
Fabrègue, B. (2019). *Pollution in the Western Balkans (Part 1): An overview.* s.n.
Frey, B. (2024). *The energy transition in the Western Balkans: The status quo, major challenges and how to overcome them.* s.n.
IEA. (2021). *Energy security.* [Online]. https://www.iea.org/. [Accessed 2025].
IRENA. (2022). *Ukraine energy profile.* s.n.
Kampel, E. (2023). *Climate mainstreaming activities in EaP countries.* s.n.
Maja Colovic Daul, M. K. O. K. (2019). *Air pollution and human health: The case of the Western Balkans.* s.n.
Maria Yetano Roche, W. I. (2020). *Governance of transitions toolkit.* s.n.
Martin, M. (2023). *A long way to go: the Western Balkans and the green transition.* s.n.
Ministry of Environment of the Republic of Armenia. (2020). *Fourth national communication on climate change.* s.n.

OECD. (2024a). *Ensuring compliance with environmental regulations in the Eastern Partnership countries.* s.n.

OECD. (2024b). *Environment at a glance in the EU Eastern partnership countries.* s.n.

Olga Litvishko, E. P. (2023). Conflict communication in disrupting sustainable development of regions: Discourse analysis framework. *Journal of Geography.*

Regional Cooperation Council. (2023). *Green agenda for the Western Balkans.* [Online].https://www.rcc.int/greenagenda

Regional Cooperation Council. (2024). *Annual report of the secretary general.* Regional Cooperation Council.

Republic of Armenia Ministry of Environment. (2020). *Fourth national communication on climate change.* UNDP.

Republic of Moldova. (2023). *Fifth national communication of the Republic of Moldova.* s.n.

Romanello M., K. v. D. D. M. (2021). Tracking progress on health and climate change in Europe. *Lancet Public Health.*

Secretary-General of the OECD. (2021). *Sustainable infrastructure for low-carbon development in the EU Eastern Partnership: Hotspot analysis and needs assessment.* OECD Publishing.

Stefania Fiorentino, A. K. G. L. L. R. M. P. T. (2024). 'Left behind places': What can be done about them? *Cambridge Journal of Regions, Economy and Society, 17*, 259–274.

The Ministry of Environmental Protection and Agriculture of Georgia. (2021). *Fourth national communication of Georgia.* s.n.

Waal, T. D. (2023). *The end of Nagorno-Karabakh.*

Walker, C. (2024). Global impacts of the European Green Deal. *Nature Plants,* Issue ISSN 2055-0278.

WAREG. (2021). Water sector challenges in the Western Balkans—Editorial. *WAREG Newsletter.*

World Bank. (2022). *Carbon intensity: CO_2 emissions per dollar of GDP.* s.n.

Zofia Gródek-Szostak, J. A. K. K.-W. A. N. K. W. O. A. S.-S. (2025). *The European green deal and the Eastern Partnership: Towards resilient, sustainable and integrated economies.* Taylor & Francis ed. s.n.

CHAPTER 27

Pathways to Just Green Transitions: Reflections and Future Directions

Erblin Berisha, John Moodie, Ledio Allkja, and Marija Jeftić

27.1 Introduction

The chapters presented in this volume collectively explore the multifaceted challenges and opportunities associated with Just Green Transitions (JGT). Even though not clearly defined by the literature (Shaker, 2025; Shaker & Berisha, 2024, 2025), the concept of JGT addresses the need for an equitable transition to sustainability, balancing socio-economic resilience, environmental stewardship, and territorial cohesion (Berisha et al., 2025b). According to the European Commission (2019), the transition to a sustainable and equitable future is one of the most pressing challenges of our time. Through the

E. Berisha (✉)
Interuniversity Department of Regional and Urban Studies and Planning (DIST), Politecnico Di Torino, Turin, Italy
e-mail: erblin.berisha@polito.it

J. Moodie
Nordregio Research Institute for Regional Development and Planning, Nordregio, Stockholm, Sweden
e-mail: john.moodie@nordregio.org

L. Allkja
Co-Plan Institute for Habitat Development and Polis University, Tirana, Albania
e-mail: ledio_allkja@universitetipolis.edu.al

M. Jeftić
Faculty of Geography, University of Belgrade, Belgrade, Serbia
e-mail: marija.jeftic@gef.bg.ac.rs

E. Berisha et al. (eds.), *The Palgrave Handbook of Just Green Transitions in the Western Balkans and Beyond*,
https://doi.org/10.1007/978-3-031-95075-9_27

diverse lenses of case studies and theoretical frameworks, this volume underscores the complexity of navigating a global paradigm shift towards greener economies and societies while ensuring inclusivity and fairness. The chapters have explored the complexities of JGT, offering a rich tapestry of perspectives that highlight the socio-economic, environmental, and governance dimensions of this global shift. By examining diverse contexts across the Western Balkan (WB), Europe and beyond, these contributions provide invaluable insights into the opportunities and challenges of aligning ambitious climate goals with social justice and territorial cohesion. The collection of chapters has indeed allowed us to reflect on JGT's different conceptual dimensions, bridging the gap between theory, institutional discourse, and practices. This concluding chapter summarises the main findings and warnings, giving special attention to the conditions of the WB. It is structured into three main sections. Besides this brief introduction, the first section focuses on defining the features of effective JGTs, such as multiscalar and flexibility, inclusive governance and participation, and multisectoral and future-orientedness. The second section addresses the risks of unjust green transitions, presenting the main warnings on unfair transition processes. Section three instead presents the main lesson for the WB based on the various experiences the handbook has illustrated. The chapter concludes with reflections on future research directions, emphasising the need for interdisciplinary and cross-sectoral approaches to refine and implement JTG further.

27.2 Features of Effective Just Green Transitions

27.2.1 Multiscalarity and Flexibility

What emerged from this collection of chapters is that each context reflects JGT's measures according to its path-dependent logic, whether contingent or contextual. Reading the variety of contributions from the WB or European Union (EU) Member States, it clearly emerges the necessity to adopt a flexible approach when understanding the topic and its policy and political implications (Berisha et al., 2025a). Even though marked by a similar and interconnected supranational umbrella, the EU Green Deal and Just Transition Mechanisms for the EU Member States and the Green Agenda for the Western Balkans, how the topic has been implemented largely differs across the various experiences that have been highlighted in this handbook. While in some of the EU countries, JGT have been addressed through the Territorial Just Transition Plans, in the WB, the countries have focused on specific and punctual solutions without a strategic and comprehensive vision. Indeed, a central theme running through this volume is the need for tailored, region-specific approaches to JGT. Whether addressing the high-carbon industries of Italy and Poland (Berisha et al., 2025b) or the climate vulnerabilities of the Romanian Danube Valley (Mitrică et al., 2025), the diversity of contexts underscores the inadequacy of one-size-fits-all solutions. Instead, flexible policies that account for local socio-economic and environmental conditions are critical. For instance, Italy and Poland prioritise economic resilience and workforce retraining to mitigate the socio-economic impacts of decarbonisation, while in Bosnia and Herzegovina, North Macedonia, and Serbia, the priority is given to the impact of closing coal mines and providing energy. Similarly, the Slovak Vision 2030 (Finka, 2025) demonstrates how past experiences with economic transitions can inform contemporary strategies, offering a model of iterative learning and adaptation that balances ecological, social, and economic priorities. In this respect, in their chapter, Toska et al. (2025) open a circular economy perspective by turning waste to energy in Montenegro, while Temali Kraja (2025) discusses the use of nature-based solutions in coastal areas for adapting to climate change. Bridging the gap between universal challenges and regional responses clearly shows the need for a heterogeneous.

27.2.2 Inclusive Governance and Public Participation

Governance emerges as a pivotal factor in the success of JGT efforts. JGTs are complex multi-actor processes conducted across multiple levels of governance (Geels, 2002; Geels et al., 2016). How to effectively govern these transitions is potentially problematic in a WB context where countries face various geopolitical, sociopolitical, economic, and institutional capacity challenges (Berisha, 2023), which can negatively impact JGT spatial planning and territorial governance processes. Shaker (Chapter 3) argues that traditional EU Multi-level Governance models do not possess the necessary socio-territorial, institutional policy mixes, and (macro) regional capacities to govern energy transitions. The author points to the need for a meta-level governance framework to homogenise a strategic approach to JGTs across WB countries to aid EU accession processes by enhancing alignment and cohesion in the implementation of the European Green Deal, Green Agenda for the Western Balkans, and New EU Growth Plan for the Western Balkans. A meta-level governance approach advocates a holistic, adaptive, flexible, and inclusive intergovernmental macro-regionalisation framework aimed at strengthening vertical and horizontal coordination across scales, sectors, actors, and levels of decision-making (Gänzle & Kern, 2016; Jessop, 2011; Tahirović, 2024). Further research is needed to explore how the proposed meta-level governance policy approach links to and interacts with Geels' (2002) multi-level perspective approach developed in the field of science and technology studies. Furthermore, intergovernmental macroregional approaches have faced criticism for fostering top-down policymaking with a limited role for subnational-level actors (McMaster & van der Zwet, 2016; Moodie et al., 2021a). National governments in the WBs already have a prominent position in policymaking, so more research is required to examine how subnational-level actors can be more engaged and integrated into macroregional processes to ensure that JGT policies reflect and address relevant regional and local challenges and opportunities. Indeed, the role of sub-national level institutions and stakeholders, including regional and municipal level authorities, has been curtailed in JGT processes dominated by national-level institutions and industrial sector representatives (Moodie et al., 2021b). The engagement of sub-national-level actors is especially pertinent in a WB context where power is usually centralised at the national and rarely devolved to regional and local levels. Existing power asymmetries can create conditions whereby national and sectoral priorities dominate the JGT policy agenda. Shaneh et al. (2025) argue that this dynamic has been seen in the development of EU Territorial Just Transition Plans, where national and sectoral actors have the technical and economic dimensions over the social dimension. Consequently, some of these plans have been criticised for not adequately addressing regional and local socio-economic challenges and opportunities or the needs of the citizens. As Tapia et al. (2025) note the territorial-specific nature of JGTs means that targeted policy measures are needed that address the specific characteristics of the region, industry, or social groups affected by the transition.

A just green transition under the European Green Deal is sustainable considering, as Brunetta et al. (2025) state, that distributive justice is overemphasised, while procedural justice is neglected. Namely, although equal distribution of resources is important, the insufficient involvement of all social groups in the decision-making process limits the fairness of the transition. Continuing on the previous, authors believe that introducing participatory planning and active involvement of marginalised groups through education and consultation is essential. Spatial justice implies an even geographic distribution of resources and opportunities. However, the current approach does not pay enough attention to it, and it is necessary to carry out localised analyses of the impact of green policies while ensuring that all communities have equal access to new infrastructures and services. Instead of seeing justice as a secondary element, it needs to become the primary goal, with a focus on the structural injustices that cause environmental and social problems. Also, it is necessary to ensure a fair allocation of financial resources based on local needs and degree of vulnerability to avoid problems in distributing public and private funds. Introducing mechanisms for equal distribution of funds can significantly contribute to a fairer transition.

To do that, JGT policymaking processes are expected to be open, equal, and inclusive, ensuring broad participation of different actors who bring a range of knowledge types to policy discussions, including technical expertise, practical, and local knowledge (Cedergren et al., 2022). Valkenburg (2025) supports this perspective arguing that JGTs pose complex challenges that need to be approached using interdisciplinary policymaking models which can integrate a broad range of types of knowledge (Moodie et al., 2025a, 2025b). He continues that epistemic justice needs to become one of several key conceptual ideas that guide JGT policymaking processes to ensure that different sorts of knowledge are equally and fairly included (Valkenburg, 2025). The open and inclusive participatory planning approach to JGT policymaking suggested by Jaššo and Husár (2025) is built on epistemic justice's logic. They note that traditional hierarchical spatial planning methods, which rely on regulations and legislation, are not sufficient to address the challenges presented by JGTs; more bottom-up, flexible and adaptable JGT planning strategies and tools are required that engage local communities and empower them to shape their own lives through participation, education, information, and technology (Jaššo & Husár, 2025). Promoting participatory policymaking and planning models is especially relevant within centralised governance systems across the WBs, where JGT policymaking is dominated by national governments that collaborate with large industrial sectors. Participatory planning and policymaking processes advocating a more democratic approach to JGTs are important, as citizen engagement in JGT processes has, up to this point, been limited (Georgieva & Ferry, 2023, 2024). Tapia et al. (2025) highlight an interesting paradox in the evolution of JGT policies in relation to citizen involvement in the policymaking process. They point out that early preventative and mitigative policy measures dealt more directly with the socio-economic impacts of JGTs on citizens and workers while not engaging them directly in the policy process. In contrast, more recent adaptive and transformational policies advocate for wide-ranging citizen engagement in JGT policy processes while focusing more on the technical and economic dimensions of JGTs, limiting the potential for citizen engagement.

27.2.3 Multisectoral and Future-Oriented

As humanity confronts the dual crises of climate change and socio-economic inequality, the insights from this volume offer a roadmap for achieving a just and sustainable future. JGTs are not merely technical or economic endeavours; they are deeply political and social processes that require collaboration, innovation, and a commitment to fairness. Embracing the principles of flexibility, inclusivity, and long-term thinking, regions across Europe and beyond can chart pathways that balance environmental goals with the needs of their communities. As the chapters in this volume demonstrate, the journey towards sustainability is complex and challenging—but it is also an opportunity to build a more equitable and resilient world for future generations. The need to balance environmental and social requirements is manageable through a just energy transition (RCC, 2021) but not only. The potential of renewable energy sources is, for instance, recognised as a key basis for decarbonising Bosnia and Hercegovina's energy system, reducing dependence on fossil fuels, thoroughly elaborated by Novikau and Krupalija (2025). As the authors emphasise, managing the coal mine closure process represents one of the biggest challenges in the country, requiring support for local communities through retraining programs and economic diversification. Due to the interconnected challenges, the authors suggested adopting a multidisciplinary approach to achieve energy sustainability and social justice through sustainable policies, efficient management and transparent processes. Introducing mandatory and transparent environmental impact assessments and greater involvement of local communities in decision-making is crucial for small hydropower plants. Harmonisation with the legal framework of the European Union is essential for access to funds and support for the transition, as well as intensifying efforts in the implementation of EU directives and regulations, while simultaneously strengthening capacities for monitoring and reporting. The process of coal mine closure must be carefully managed to avoid social and economic shocks. In this sense, the creation of regional transition plans that include alternative industries, infrastructure projects and programs to support workers is significant as well as the creation of regional transition plans that include alternative industries, infrastructure projects and programs to support workers. Besides, cross-border public services (related to environmental protection, disaster risk management, sustainable agriculture, and the circular economy), for instance, play a key role in accelerating a just green transition. They require a coordinated approach and the active role of national and regional actors, specifically considering negative migration trends, climate change and insufficient infrastructure. However, cross-border areas also represent, as Nikolov states, "laboratories of European integration" where new concepts and models of cooperation are tested to achieve common green goals. Regional initiatives, such as the IPA and INTERREG programs, provide significant opportunities for financing and implementing activities that support the Western Balkans Green Agenda. The involvement of regional bodies, such as the RCC (Regional Cooperation Council), and the formation of new working groups can also improve coordination and efficiency (Nikolov, 2025). Similarly to Chapters 9 (Novikau & Krupalija, 2025) and 10 (Nikolov, 2025), Chapters 13 (Garvanlieva Andonova et al., 2025), 16 (Oruc et al., 2025), 17 (Toska et al., 2025) and 18 (Ndreka & Imami, 2025), discuss the interrelationship between JGT, energy transition and waste management in contexes such as Bosnia & Hercegovina, Montenegro, North Macedonia, Kosovo*[1] and Albania. Undoubtedly, in the WB, the link between energy transition and JGT is strong, as most countries emphasise a strong focus on energy policies. Meanwhile, these come with high social costs, as demonstrated in Chapters 13 and 16 and should be carefully supported by cost–benefit analysis. Energy transition should not be seen as a sectorial policy objective but in a comprehensive and integrated manner with socioeconomic policies. Both chapters highlight the importance of strengthening policy and institutional coordination, development of re-skilling and alternative employment, subnational-level fiscal resilience as well as addressing social and spatial inequalities. Additionally, Garvanlieva Andonova et al. (2025) suggest that although initially these efforts might seem as costly, when taking in consideration medium to long-term impacts, they are the most cost-effective type of

[1] This designation is without prejudice to positions on status, and is in line with UNSCR 1244 and the ICJ Opinion on the Kosovo declaration of independence.

action. Meanwhile, Ndreka and Imami (2025) and Toska et al. (2025) discuss the interrelations between energy transition and waste management. Using waste as a source of energy production in some of the high-consuming sectors is seen as a prospect and cost-efficient alternative for supporting the Just Green Transition. This approach helps solve problems of one of the most vulnerable sectors in WB, such as waste management, while at the same time offering a plausible alternative for energy production. Also, from a social perspective, the waste-to-energy perspective creates new jobs or alternative (green) employment and businesses where other sectors easily transfer skills.

A future-oriented approach is paramount, as presented in Maričić and Pantić (2025), Finka (2025), and Toptsidou and Böhme (2025). All chapters reflect on the role of strategic planning in addressing JGT and climate change. While Maričić and Pantić (2025) provide the starting points for developing a future comprehensive JGT Strategy in Serbia, Finka (2025) discusses the case of Slovakia. In the case of Serbia, the authors pointed out the importance of enabling energy independence, economic sustainability and social justice in mining regions (Maričić & Pantić, 2025), among which discussing the cases of Kolubara and Kostolac that are considered the most important ones from the economic point of view. The exploitation of natural resources illustrates the complexity of balancing economic, ecological and social priorities in green transition. As stated in the case of Kopaonik (Protić et al., 2024), due to the depletion of ore reserves, the irrationality and lack of competitiveness of mining production, political problems and restrictions imposed by the protection regime of the National Park, mining on Kopaonik is not a significant economic activity in the region. The potential activation and access to exploit the deposits would raise a number of new issues ranging from economic viability to environmental justification and sustainability (Protić et al., 2024). Similarly, the chapter of Toptsidou and Böhme (2025) on the Adriatic Ionian Region grapples with socioeconomic disparities and demographic challenges as it envisions pathways to a sustainable future. Scenarios that prioritise well-being and ecological restoration highlight the potential for innovative, community-led solutions to bridge gaps between environmental goals and human needs.

Temali Kraja (2025) and Mitrică et al. (2025) paid particular attention to the interplay between climate change and socioeconomic vulnerabilities. The Romanian Danube Valley (Mitrică et al., 2025) exemplifies how climate risks, such as heat-related phenomena, intersect with economic fragility, amplifying inequalities and threatening community resilience. Developing a Heat Vulnerability Index in this region provides a data-driven approach to identifying priority areas for intervention, underscoring the need for localised, evidence-based strategies. Chapter 19 (Temali Kraja, 2025) focuses on the fragility of coastal areas and the impact of climate change. The adaptation through the use of green technologies is seen as a solution, and various case studies are taken into consideration. The author identifies three main strategies of intervention for protecting coastal areas: defend, retreat, and co-exist. The latter is a more sustainable solution through the application of green technologies. This chapter discusses several case studies from different regions such as Northwest Europe (Rotterdam), Southern Europe (Marche Region), and initiatives in the WB such as the Toolbox for the resilience of the Sava River. The chapter advocates that while climate change adaptation in coastal areas has major cost implications, the use of green technologies (nature-based solutions) such as planting of native trees and building of flexible spaces, in essence enhancing eco-system services to adapt, is a viable and cost-effective solution.

27.3 The Risks of Unjust Green Transitions

Several studies have underscored the risks of an unjust green transition (Schuster et al., 2023). Decarbonisation and environmental sustainability, if not carefully managed, may disproportionately impact the most vulnerable territories, certain socioeconomic groups, and even ecosystems and landscapes (Buza & Allkja, 2025), effectively shaping the distribution of "transition costs" (Rodríguez-Pose & Bartalucci, 2023). These costs are not merely economic but also social and political, often exacerbating pre-existing inequalities and territorial disparities. In particular, the way in which the JGT has been implemented in the WB raises critical concerns. Various chapters in this volume (including Chapters 7, 14, etc.) delve into how specific policies and strategies—while aimed at fostering sustainability—may have inadvertently reinforced socio-spatial imbalances, marginalised already disadvantaged communities, and generated resistance to green reforms. The discussion highlights the need for a more inclusive and context-sensitive approach, ensuring that environmental goals do not come at the expense of social justice and territorial cohesion.

Economic Displacement and Energy Poverty

The Western Balkans heavily rely on coal for electricity generation, with countries like Serbia, Bosnia and Herzegovina, and North Macedonia heavily dependent on lignite power plants (Berisha, 2025a, b). A rapid shift from coal without proper planning and social protections could lead to mass job losses, particularly in mining communities with scarce alternative employment opportunities as shown by Oruc et al. (2025) in this volume. Given some areas' high unemployment rates, shutting down fossil fuel industries without sufficient retraining programs could worsen poverty and social instability. At the same time, energy poverty is a pressing issue (see Garvanlieva Andonova et al., 2025). Many households in the WB struggle to afford energy, particularly during harsh winters, and a poorly designed green transition could increase energy costs (Berisha, 2025a). If renewable energy projects are not made accessible and affordable, vulnerable populations could face higher electricity bills, while wealthier segments benefit from incentives for solar panels, heat pumps, and electric vehicles.

Unequal Access to Green Technologies and Investments

There is a risk that the green transition will deepen inequalities between urban and rural areas, as well as between socioeconomic classes. Novikau and Krupalija (2025), in their Chapter 7, affirm that to soften the potentially unfair impacts of JGT, it is imperative to ensure a transition process that will not threaten social stability, which is possible by establishing mechanisms for social protection and retraining of workers from the coal sector and economic diversification in regions dependent on coal. It is recommended that investments in solar and wind energy be increased while stricter criteria for small hydropower projects are introduced to reduce their negative environmental impacts. Indeed, access to funding and technology for renewable energy projects, sustainable transport, and energy efficiency measures is often concentrated in larger cities, while rural areas—where much of the population still relies on outdated energy infrastructure—struggle to attract investment. Additionally, while EU funds and international financial institutions support green projects, bureaucratic barriers, corruption risks, and weak institutional capacities in the WB can hinder the equitable distribution of resources. Large companies or well-connected businesses might benefit, while smaller enterprises, cooperatives, and local initiatives are left behind.

Rhetoric JGT and Political Manipulation

In the absence of strong governance and transparent policies, the green transition could be exploited for political and economic gain rather than for real sustainability (Živanović et al., 2025) Some governments may promote "green" projects that do not deliver meaningful environmental benefits but serve as opportunities for corruption and greenwashing. Chapter 14 by Imami et al. (2025) argues that there is also the risk of "purposive" transitions in the context of WB. Taking Albania as a case study, JGT is seen as a "critical juncture" for the country in terms of leapfrogging to a new stage of development. According to the authors, this transition should not be seen as a mere technocratic process but rather a radical socio-political and cultural change (Berisha, 2025a, b). While it holds up a series of opportunities, it can also be undermined by rhetorical policy discussions that do not touch on the roots of the problems. Rhetoric JGT implementation can be observed when it comes to introducing some hydropower projects in the WB that have been promoted as green solutions. However, they have caused environmental degradation and displacement of local communities, particularly in ecologically sensitive areas such as the Vjosa River in Albania and the Tara River in Montenegro. Poorly planned renewable projects that disregard environmental and social impacts could create resistance to the green transition rather than support for it.

Geopolitical and Energy Security Challenges

The green transition in the WB is closely linked to geopolitical influences. The region has historically depended on Russian gas and other external energy sources, making it vulnerable to energy price shocks and supply chain disruptions (Berisha, 2025a). If the transition away from fossil fuels is not well managed, there is a risk of increased energy dependency on external actors like China, which has been investing in coal and hydropower projects in the region. At the same time, political instability in the WB could delay or distort the green transition process. Governments might prioritise short-term economic or political gains over long-term sustainability, leading to fragmented or inconsistent policies that hinder regional cooperation on climate action.

Social Backlash and Resistance

JGT is sometimes perceived as unjust or elite-driven, which could trigger public resistance and social unrest. If communities feel that environmental policies are imposed without consultation or disproportionately burden lower-income groups, opposition to green reforms may grow. This could slow down the transition, as seen in other parts of Europe where "yellow vest" protests and farmer uprisings emerged in response to unfair climate policies. In the WB, where trust in institutions is already low, a lack of public engagement in green policymaking could fuel scepticism and populist narratives against the EU's Green Deal and climate commitments. Politicians may capitalise on public fears, using anti-EU rhetoric or nationalist sentiments to delay or resist sustainability measures.

27.4 Lessons for the Western Balkans

The JGTs' experiences presented in this volume provide a valuable foundation for policymakers and decision-makers in the Western Balkans. Regional actors can leverage the insights and discussions outlined in this handbook to refine and enhance the ongoing political and technical discourse surrounding the JGT in the region. Drawing from these shared experiences, stakeholders can foster more informed decision-making, encourage cross-border collaboration, and contribute to developing more effective and sustainable governance frameworks.

The Importance of Multi-Level Governance and the Role of Sub-National Actors

Shaneh et al. (2025) point out, WB policymakers can enhance the role of sub-national level actors and citizens by approaching JGT policymaking through a territorial governance and place-based policymaking lens (Barca, 2009; Barca et al., 2012; Schmitt & van Well, 2016), based on the principle of active subsidiarity (Aleksieva & Orejarena, 2024; Moodie et al., 2021c). In this respect, Jeftić et al. (2019) argue that effective governance based on flexibility, capability and capacity for participatory planning and decision-making enables the creation of dynamic and sustainable urban environments. Effective governance can improve the quality of neighbourhood planning, the environment and economic productivity, thereby supporting other segments of development (Jeftić et al., 2019). However, the implementation of this concept, as Brunetta et al. (2025) state, is often limited by the structural features of the governance system, such as rigid planning laws and institutional inefficiencies, which makes it all difficult to implement fair policies in particular in the region (Berisha et al., 2021). Organising training sessions for citizens to become familiar with their procedural rights and active participation in decision-making is crucial in terms of capacity building and informing citizens. It is recommended that institutional coordination be strengthened and bureaucratic obstacles removed for the more efficient implementation of spatial justice policies (Brunetta, Caldarice & Mohabat Doost, 2025).

Enhance Cooperation

Cross-border cooperation is also of key importance for an effective green transition in the WB through establishing joint projects with neighbouring countries and the exchange of best practices, with an emphasis on innovation and sustainable development (Novikau & Krupalija, 2025). Maričić and Pantić (2025) point to the same direction, stating that for Serbia, for example, the cross-border cooperation and exchange of experiences with countries that have successfully implemented the green transition, such as Germany and Poland, represent a significant opportunity. Joint projects with neighbouring countries and learning from their examples can accelerate the implementation of effective measures and encourage the development of innovative solutions (Maričić & Pantić, 2025). Nikolov (2025) supports this perspective, adding that cross-border cooperation is indispensable in the process of a just green transition in the Western Balkans, providing the basis for a sustainable future and inclusive development. Investigating the cross-border dimension of the just green transition in the Western Balkans, she focuses on the cross-border areas of the Republic of Serbia. As Gatarić et al. (2022) state, the geostrategic importance of border areas, such as, for example, eastern part of Serbia, along the border with Romania, is the basis for establishing cross-border partnerships that can enable more efficient use of resources, improvement of environmental protection and support for local communities in the process of green transition. In that respect, Nikolov identified key themes as essential to implementing this transition. She argues that the cross-border dimension of the green transition relies on political and strategic documents that promote territorial cohesion, the reduction of regional disproportions and sustainable development in accordance with the Green Agenda of the WB.

Enhance Public Engagement in Decision-Making

The participatory planning models presented by Jaššo and Husár (2025) provide WB policymakers with practical examples of community-driven JGT policies, which have successfully raised public awareness and knowledge of the benefits of renewable energy and the environmental importance of reducing coal dependence. Proactive citizen engagement in JGT processes is essential within a WB setting where the term "transition" has become loaded with negative connotations (Toto et al., 2023). Policymakers across the WBs must proactively engage and communicate with citizens throughout JGT processes to ensure acceptance and support for JGT policies. Moodie et al. (2022d) note that there is strong potential for enhancing citizen involvement in EU cohesion policymaking (which provides the framework for the EU Just Transition Mechanism and Funds) using a range of new digital and traditional citizen engagement tools. WB policymakers can also learn from research projects applying innovative citizen engagement tools in JGT policymaking; for example, the ongoing Horizon Europe DUST project is experimenting with Regional Future Literacy Labs as a tool for promoting citizen involvement in sustainability transition policymaking (Balz et al., 2023). Similarly, the recently published "Nordic Tool Box for Just Green Transition" also provides policymakers outside the Nordics with practical tips and tools for developing more inclusive and fairer green transition policies (Sánchez Gassen et al., 2024). Local participation and public dialogue must be at the centre of the transition process. Involving local communities in planning and decision-making will enable a better understanding and acceptance of the green transition. At the same time, informing and educating the population about the benefits of this process can alleviate resistance to change (Maričić & Pantić, 2025).

Importantly, preservation of public interest in the process of spatial planning is the basis for the successful implementation of the green transition, especially in the countries of the WB region facing numerous challenges, including institutional inefficiency, lack of transparency, and the dominance of narrow economic interests over the long-term public good. In some countries, public interest in planning is particularly questionable, often leading to insufficient involvement of local communities and insufficient consideration of environmental aspects in development policies. Starting from the dilemma of the extent to which the public interest is preserved in planning (Jeftić et al., 2019), it is important to establish a clear connection between planning, institutions, laws, planners and citizens. Planners, as actors of public interest, have a key role in cooperation with institutions in order to increase the resilience of urban and regional areas in the context of the green transition (Jeftić et al., 2019).

Education and Workers Rescaling

JGTs require developing new skills and qualifications that will enable workers to be competitive in renewable energy sources, energy efficiency and sustainable production. In this context, Maričić and Pantić (2025) state that introducing new educational programs and courses focused on renewable energy technologies, sustainable development and energy management can significantly improve workers' qualifications and ensure their competitiveness in the labour market. The education system also requires transformation in order to respond to the new needs of the labour market. Depopulating areas, which are increasing not only in Serbia (Gatarić et al., 2022) but also in other countries of the Western Balkans, lead to a weakening of the local economy and demographic aging. This shows that new economic strategies are needed, including reorientating the workforce towards sectors that are maintained and connected to the green economy. The authors also emphasise that technological innovation and the introduction of sustainable alternatives represent an important aspect of the green transition. The development of solar and wind power plants on reclaimed mining areas, as well as the use of geothermal energy, can contribute to the stability of the energy sector and the reduction of harmful gas emissions.

27.5 Conclusion: Future Research Directions

The transition to a low-carbon and sustainable future is an ongoing and dynamic process that requires continuous adaptation and innovation. This volume has comprehensively examined JGTs in the WB and beyond, highlighting the challenges and opportunities inherent in this transformation. However, several areas require further research and policy attention. First, there is a need for more profound interdisciplinary research on the interactions between the environmental, economic, and social dimensions of JGTs. Future studies should integrate environmental science, political economy, and social justice theory insights to develop holistic transition frameworks. Second, additional research is needed to assess the long-term impacts of JGT policies, particularly in coal-dependent regions, to understand how transitions unfold over time and what adjustments may be necessary. Finally, further work is required to understand the geopolitical implications of JGTs, particularly in the context of EU enlargement and regional integration. The WB's path towards European integration is closely linked to its ability to implement just and green transition policies (Berisha, 2025a, 2025b). Future research should examine how geopolitical dynamics shape the feasibility and effectiveness of these transitions, ensuring that justice remains at the forefront of sustainability agendas. To sum up, JGTs are not merely about technological shifts but fundamentally about reconfiguring socio-economic and governance systems to be more inclusive, democratic, and sustainable. The pathways outlined in this handbook serve as a roadmap for future research, policy interventions, and collaborative actions to realise the vision of a just and sustainable future for all.

References

Aleksieva, R., & Orejarena, G. (2024). *Opportunities for and barriers to active subsidiarity in just sustainability transition policies* (Horizon Europe DUST Project Policy Brief 1).

Balz, V., Verbeek, T., Qu. L., Baugh, R., Dąbrowski, M., Jungsberg, L., Ferry, M., Georgieva, N., Serdült, U., Trifonova, M., Amin, S., & Clouet, Y. (2023). *DUST theoretical and conceptual framework* (Deliverable 1.1. Horizon Europe DUST project).

Barca, F. (2009). An agenda for a reformed Cohesion Policy. A place-based approach to meeting European Union challenges and expectations. Independent Report Prepared at the Request of Danuta Hubner, Commissioner for Regional Policy, Directorate General for Regional Policy, European Commission.

Barca, F., McCann, P., & Rodríguez-Pose, A. (2012). The case for regional development intervention: Place-based versus placeneutral approaches. *Journal of Regional Science, 52*(1), 134–152. https://doi.org/10.1111/j.1467-9787.2011.00756.x

Berisha, E. (2023). The Adriatic Ionian region: Challenges and opportunities in the time of geopolitical tensions and the rule of EUSAIR. *Economia Marche-Journal of Applied Economics, 42*(1). https://doi.org/10.57638/1120-9593AIREGION

Berisha, E. (2025a). *Moving towards an inclusive Green Agenda in the Western Balkans*. Foundation for European Progressive Studies (FEPS). https://doi.org/10.13140/RG.2.2.12092.76162

Berisha, E. (2025b, in this volume). Green EU integration—The role of the green agenda in accelerating Just Green Transitions in the Western Balkans. In E. Berisha, J. Moodie, L. Allkja, & M. Jeftić (Eds.), *The Palgrave handbook of Just Green Transitions in the Western Balkans and beyond*. Palgrave Macmillan.

Berisha, E., Cotella, G., & Solly, A. (2021). Governing territorial development in the Western Balkans: Conclusive remarks and future research perspectives. In E. Berisha, G. Cotella, & A. Solly (Eds.), *Governing territorial development in the Western Balkans* (pp. 357–381). Springer. https://doi.org/10.1007/978-3-030-72124-4_17

Berisha, E., Cotella, G., Puntillo, E. & Shaker, Y. (2025b, in this volume). Just Green Transitions in the EU: Comparing the cases of Italy, Sweden, Poland, and The Netherlands. In E. Berisha, J. Moodie, L. Allkja, & M. Jeftić (Eds.), The *Palgrave handbook of Just Green Transitions in the Western Balkans and beyond*. Palgrave Macmillan.

Berisha, E., Moodie J., Allkja, L., & Jeftic, M. (2025a, in this volume). Just Green Transitios—An introduction. In E. Berisha, J. Moodie, L. Allkja, & M. Jeftić (Eds.), *The Palgrave handbook of Just Green Transitions in the Western Balkans and beyond*. Palgrave Macmillan.

Brunetta, G., Caldarice, O., & Mohabat Doost, D. (2025, in this volume). Between light and shadow: exploring spatial and multidimensional justice in the EU's green transition. In E. Berisha, J. Moodie, L. Allkja, & M. Jeftić (Eds.), *The Palgrave handbook of Just Green Transitions in the Western Balkans and beyond*. Palgrave Macmillan.

Buza, D., & Allkja, L. (2025, in this volume). Landscape design as a tool to foster spatial transformations in light of just green transition in the Western Balkan. . In E. Berisha, J. Moodie, L. Allkja, & M. Jeftić (Eds.), *The Palgrave handbook of Just Green Transitions in the Western Balkans and beyond*. Palgrave Macmillan.

Cedergren, E., Tapia, C., Sánchez Gassen, N., & Lundgren, A. (2022). *Just Green Transition–key concepts and implications in the Nordic Region* (Nordregio Working Paper 2002:2).

European Commission. (2019). *Communication from the Commission—The European Green deal, COM, 2019, 640 Final.*

Finka M. (2025, in this volume). Synergy effects of Just Green Transition—from overall challenges to innovative responses in the case of Slovakia 2030. In E. Berisha, J. Moodie, L. Allkja, & M. Jeftić (Eds.), *The Palgrave handbook of Just Green Transitions in the Western Balkans and beyond*. Palgrave Macmillan.

Gänzle, S., & Kern, K. (2016). Macro-regions, 'Macro-regionalization' and Macro-regional Strategies in the European Union: Towards a New Form of European Governance?. In S. Gänzle, & K. Kern (Eds.), *A 'Macro-regional' Europe in the Making. Palgrave Studies in European Union Politics*. Palgrave Macmillan. https://doi.org/10.1007/978-1-137-50972-7_1

Garvanlieva Andonova, V., Gotovska, E., & Nikolov, M. (2025, in this volume). Just Green Transitions in North Macedonia: A place-based case study from South-west Planning region. In E. Berisha, J. Moodie, L. Allkja, & M. Jeftić (Eds.), *The Palgrave handbook of Just Green Transitions in the Western Balkans and beyond*. Palgrave Macmillan.

Gatarić, D., Đerčan, B., Živković, M. B., Ostojić, M., Manojlović, S., Sibinović, M., Lukić, T., Jeftić, M., Lutovac, M., & Lutovac, M. (2022). Can depopulation stop deforestation? The impact of demographic movement on forest cover changes in the settlements of the South Banat District (Serbia). *Frontiers in Environmental Science, 10*, 897201. https://doi.org/10.3389/fenvs.2022.897201

Geels, F. W., Kern, F., Fuchs, G., Hinderer, N., Kungl, G., Mylan, J., Neukirch, M., & Wassermann, S. (2016). The enactment of socio-technical transition pathways: A reformulated typology and a comparative multi-level analysis of the German and UK low-carbon electricity transitions (1990–2014). *Research Policy, 45*(4), 896–913. https://doi.org/10.1016/j.respol.2016.01.015

Geels, F. W. (2002). Technological transitions as evolutionary reconfiguration processes: A multi-level perspective and a case-study. *Research Policy, 31*, 1257–1274. https://doi.org/10.1016/S0048-7333(02)00062-8

Georgieva, N., & Ferry, M. (2023). *Identification and assessment of participatory processes in sustainability transition measures in case study regions*, Deliverable 3.1. Horizon Europe DUST Project.

Georgieva, N., & Ferry, M. (2024). *Factors impacting the scale and quality of participation of least engaged communities in deliberative governance of transition policies*, Deliverable 3.1. Horizon Europe DUST Project.

Imami, F., Aliaj, B., & Bejko (Gjoka), A. (2025, in this volume). The risks of purposive transitions in Albania—What to expect by the Just Green Transitions. In E. Berisha, J. Moodie, L. Allkja, & M. Jeftić (Eds.), *The Palgrave handbook of Just Green Transitions in the Western Balkans and beyond*. Palgrave Macmillan.

Jaššo, M., & Husár, M. (2025, in this volume). Participatory planning culture and its role and potential for fostering just green transition, In E. Berisha, J. Moodie, L. Allkja, & M. Jeftić (Eds.), *The Palgrave handbook of Just Green Transitions in the Western Balkans and beyond*. Palgrave Macmillan.

Jeftić, M., Tošić, D., & Nikolić, T. (2019). Resilient functional urban regions: Spatial planning in the light of climate change. In W. Leal Filho et al. (Eds.), *International perspectives on climate change, climate change management* (pp. 113–126). https://doi.org/10.1007/978-3-030-03383-5_8. Springer International Publishing Switzerland. Print ISBN 978-3-030-03382-8, Online ISBN 978-3-030-03383-5

Jessop, B. (2011). Metagovernance. In *The SAGE Handbook of Governance* (pp. 106–123). SAGE Publications Ltd. https://doi.org/10.4135/9781446200964.n8

Maričić T., & Pantić, M. (2025, in this volume). Just Green Transitions of coal-mining cities. In E. Berisha, J. Moodie, L. Allkja, & M. Jeftić (Eds.), *The Palgrave handbook of Just Green Transitions in the Western Balkans and beyond*. Palgrave Macmillan.

McMaster, I., & van der Zwet, A. (2016). Macro-regions and the European Union: The role of Cohesion Policy. In S, Gänzle & K. Kern (Eds.), *A 'macro-regional' Europe in the making*. Palgrave Macmillan. https://doi.org/10.1007/978-1-137-50972-7_3

Mitrică, B., Roznovieţchi, I., Grigorescu, I., Micu, D., & Damian, N. (2025, in this volume). The vulnerability of the Romanian Danube Valley to heat-related phenomena. A socio-economic index-based approach. In E. Berisha, J. Moodie, L. Allkja, & M. Jeftić (Eds.), *The Palgrave handbook of Just Green Transitions in the Western Balkans and beyond*. Palgrave Macmillan.

Moodie, J., Berisha, E., Allkja, L., & Jeftic, M. (2025a, in this volume). The multidimensionality of Just Green Transitions. In E. Berisha, J. Moodie, L. Allkja, & M. Jeftić (Eds.), *The Palgrave handbook of Just Green Transitions in the Western Balkans and beyond*. Palgrave Macmillan.

Moodie, J., Berisha, E., Allkja, L., & Jeftic, M. (2025b, in this volume). The Just Green Transitions policy landscape. In E. Berisha, J. Moodie, L. Allkja, & M. Jeftić (Eds.), *The Palgrave handbook of Just Green Transitions in the Western Balkans and beyond*. Palgrave Macmillan.

Moodie, J. R., Salenius, V., & Wøien Meijer, M. (2021a). Why territory matters for implementing active subsidiarity in EU regional policy. *Regional Studies, 56*(5), 866–878. https://doi.org/10.1080/00343404.2021.1966404

Moodie, J. R., Wøien Meijer, M., Salenius, V., & Kull, M. (2021b). Territorial governance and smart specialisation: Empowering the sub-national level in EU regional policy. *Territory, Politics, Governance, 11*(7), 1392–1412. https://doi.org/10.1080/21622671.2021.1905542

Moodie, J., Tapia, C., Löfving, L., Gassen, N. S., & Cedergren, E. (2021c). Towards a territorially just climate transition—Assessing the swedish EU territorial just transition plan development process. *Sustainability, 13*(13), 7505. https://doi.org/10.3390/su13137505

Ndreka, E., & Imami, F. (2025, in this volume). Implementation of the energy efficiency plan and Waste Management in Kosovo. In E. Berisha, J. Moodie, L. Allkja, & M. Jeftić (Eds.), *The Palgrave handbook of Just Green Transitions in the Western Balkans and beyond*. Palgrave Macmillan.

Nikolov, A. (2025, in this volume) Cross-border dimension of Just Green Transition in the Western Balkans—Case study of the Republic of Serbia. In E. Berisha, J. Moodie, L. Allkja, & M. Jeftić (Eds.), *The Palgrave handbook of Just Green Transitions in the Western Balkans and beyond*. Palgrave Macmillan.

Novikau, A., & Krupalija, M. (2025, in this volume). Green energy transition in the Western Balkans: Challenges and opportunities for Bosnia and Herzegovina. In E. Berisha, J. Moodie, L. Allkja, & M. Jeftić (Eds.), *The Palgrave handbook of Just Green Transitions in the Western Balkans and beyond*. Palgrave Macmillan.

Oruc, N., Imami, F., Bejko, A., Kurta A., & Kurtovic, I. (2025, in this volume). Local labour market and fiscal impacts of decarbonisation in Bosnia-Herzegovina: The case of Zenica and Banovici. In E. Berisha, J. Moodie, L. Allkja, & M. Jeftić (Eds.), *The Palgrave handbook of Just Green Transitions in the Western Balkans and beyond*. Palgrave Macmillan.

Protić, B., Lukić, B., Popović, V., Ristić, D., Šećerov, V., Jeftić, M. R., Đordević, D. S., & Vračarević, B. (2024). Balancing Act: Assessing the impacts of Winter Tourism on Natural Heritage in Kopaonik National Park and charting a sustainable path forward. *Sustainability, 16*(4), 1509. https://doi.org/10.3390/su16041509

Regional Cooperation Council—RCC. (2021). *Action plan for the implementation of the Sofia declaration on the Green Agenda for the Western Balkans*. https://www.rcc.int/docs/596/action-plan-for-the-implementation-of-the-sofia-declaration-on-the-green-agenda-for-the-western-balkans-2021-2030. Accessed 16 January 2025.

Rodríguez-Pose, A., & Bartalucci, F. (2023). The green transition and its potential territorial discontents. *Cambridge Journal of Regions, Economy and Society, 17*(2), 339–358. https://doi.org/10.1093/cjres/rsad039

Sánchez Gassen, N., Rohrer, L., Berlina, A., Tapia, C., Cavicchia, R., Lundgren, A. & Ögland, L. (2024). Nordic Toolbox for a Just Green Transition: Navigate challenges, listen to those affected, and get inspired through this interactive, research-based toolbox for policymakers, *Nordregio Report 2024*, 21.

Schmitt, P., & van Well, L. (2016). Territorial governance across Europe. *Routledge*. https://doi.org/10.4324/9781315716220

Schuster, A., Zoll, M., Otto, I. M., & Stölzel, F. (2023). The unjust just transition? Exploring different dimensions of justice in the lignite regions of Lusatia, Eastern Greater Poland, and Gorj. *Energy Research & Social Science, 104*, Article 103227. https://doi.org/10.1016/j.erss.2023.103227

Shaker, Y. (2025, in this volume). The unknown knows of the Just Green Transitions: A Meta-governance approach for a possible enlargement of the European Union. In E. Berisha, J. Moodie, L. Allkja, & M. Jeftić (Eds.), *The Palgrave handbook of Just Green Transitions in the Western Balkans and beyond*. Palgrave Macmillan.

Shaker, Y., & Berisha, E. (2024). Transizioni verdi giuste: Tra imprecisioni semantiche e necessità di nuovi modelli di governance. *Società Italiana degli Urbanisti: PLANUM*, 193–203. https://hdl.handle.net/11583/2989395

Shaker, Y., & Berisha, E. (2025). Just Green Transitions: Between terminological inexactitudes, conceptual fragmentation, and the exigency for a theoretical framework. European Spatial Research and Policy, 32(1), 121–138. https://doi.org/10.18778/1231-1952.32.1.05

Shaneh, A., Moodie, J., Tapia, C., Giacometti, A., & Cedergen, E. (2025, in this volume). Understanding territorial governance dynamics in the Swedish EU territorial just transition planning processes—Lessons for the Western Balkans. In E. Berisha, J. Moodie, L. Allkja, & M. Jeftić (Eds.), *The Palgrave Handbook of Just Green Transitions in the Western Balkans and beyond*. Palgrave Macmillan.

Tahirović, M. (2024). The main security challenges and threats in the Western Balkans and possible model for solving them in the process of European Integration. Studia Europejskie. *Centre for Europe, University of Warsaw, 28*(3). ISSN 1428-149X.

Tapia, C. Giacometti, G. Moodie, J.R., Cedergen, E., & Shaneh, A. (2025, in this volume). From mitigation to transformation: An evolutionary perspective on the Just Green Transition. In E. Berisha, J. Moodie, L. Allkja, & M. Jeftić (Eds.), *The*

Palgrave Handbook of Just Green Transitions in the Western Balkans and beyond. Palgrave Macmillan.

Temali Kraja, I. (2025, in this volume). Climate change adaptation green Technologies for coastal risk Prevention—Case studies summary from Coastal countries. In E. Berisha, J. Moodie, L. Allkja, & M. Jeftić (Eds.), *The Palgrave handbook of Just Green Transitions in the Western Balkans and beyond.* Palgrave Macmillan.

Toptsidou, M., & Böhme, K. (2025, in this volume). Green transition in the Adriatic-Ionian Region: A scenario. In E. Berisha, J. Moodie, L. Allkja, & M. Jeftić (Eds.), *The Palgrave handbook of Just Green Transitions in the Western Balkans and beyond.* Palgrave Macmillan.

Toska, M., Gjoka, R., Gjiknuri, O., & Vuković, A. (2025, in this volume). The Montenegrin waste management sector transformation and its impacts towards Just Green Transitions. In E. Berisha, J. Moodie, L. Allkja, & M. Jeftić (Eds.), *The Palgrave handbook of Just Green Transitions in the Western Balkans and beyond.* Palgrave Macmillan.

Toto, R., Imami, F., Poro, E., et al. (2023). *Just Green Transition conceptualisation in the Western Balkans.* GreenFORCE Horizon project, Deliverable 4.1.

Valkenburg, G. (2025, in this volume). Epistemic work and epistemic Justice in Green Transitions. In E. Berisha, J. Moodie, L. Allkja, & M. Jeftić (Eds.), *The Palgrave handbook of Just Green Transitions in the Western Balkans and beyond.* Palgrave Macmillan.

Živanović, Z., Popović, V., Protić, B., Djordjević, A., Šećerov, V., & Jeftić, M. (2025, in this volume). Signs by the roadside" on the road to Just Green transitions in Serbians-border dimension of Just Green Transitions in the Western Balkans. In E. Berisha, J. Moodie, L. Allkja, & M. Jeftić (Eds.), *The Palgrave handbook of Just Green Transitions in the Western Balkans and beyond.* Palgrave Macmillan.

www.ingramcontent.com/pod-product-compliance
Lightning Source LLC
Chambersburg PA
CBHW070727261125
35997CB00004B/118
9783031950742